A Step-by-Step Approach to Using the SAS® System for Factor Analysis and Structural Equation Modeling

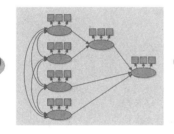

Larry Hatcher, Ph.D.

Comments or Questions?

The author assumes complete responsibility for the technical accuracy of the content of this book. If you have any questions about the material in this book, please write to the author at this address:

SAS Institute Inc.
Books by Users
Attn: Larry Hatcher
SAS Campus Drive
Cary, NC 27513

If you prefer, you can send e-mail to sasbbu@sas.com with "comments for Larry Hatcher" as the subject line, or you can fax the Books by Users program at (919) 677-4444.

The correct bibliographic citation for this manual is as follows: Hatcher, Larry, *A Step-by-Step Approach to Using the SAS® System for Factor Analysis and Structural Equation Modeling*, Cary, NC: SAS Institute Inc., 1994. 588 pp.

A Step-by-Step Approach to Using the SAS® System for Factor Analysis and Structural Equation Modeling

Copyright © 1994 by SAS Institute Inc., Cary, NC, USA.

ISBN 1-55544-643-4

1st printing, September 1994
2nd printing, September 1996
3rd printing, January 1998

Note that text corrections may have been made at each printing.

Contents

ACKNOWLEDGMENTS

I learned about structural equation modeling while on a sabbatical at Bowling Green State University during the 1990-1991 academic year. My thanks to Joe Cranny, who was chair of the Psychology Department at BGSU at the time, and who helped make the sabbatical possible.

My department chair, Mel Goldstein, encouraged me to complete this book and made many accommodations in my teaching schedule so that I would have time to do so. My friend and department colleague, Heidar Modaresi, encouraged me to begin this project and offered useful comments on how to proceed. My secretary, Cathy Carter, eased my workload by performing many helpful tasks. My friend, Nancy Stepanski, edited an early draft of Chapter 4, and provided many constructive comments that helped shape the final book. My thanks to all. Special thanks to my wife Ellen, who, as usual, offered encouragement and support every step of the way.

Many people at SAS Institute were very helpful in reviewing and editing chapters, and in answering hundreds of questions. These include David Baggett, Jennifer Ginn, Jeff Lopes, Blanche Phillips, Jim Ashton, Cathy Maahs-Fladung, and David Teal. All of these were consistently positive, patient, and constructive, and I appreciate their contributions.

DEDICATION

I dedicate this book to my parents, who worked hard and sacrificed so that I would have the opportunities that they never had.

Using This Book

Purpose

A step-by-step approach to using the SAS system for factor analysis and structural equation modeling is designed to provide an easy-to-understand introduction to some of the more advanced statistical procedures used in social science research. Part I illustrates the use of exploratory factor analysis and its close relative, principal component analysis. Part II covers path analysis with manifest variables, along with procedures that are used to test LISREL-type models (such as confirmatory factor analysis and path analysis with latent variables).

This text assumes that you have no prior knowledge of these procedures. Therefore, the text covers the procedures at an introductory level. Nonetheless, after completing a chapter on a given topic, you will understand the basic issues related to the analysis. You will also be able to write SAS programs to perform the analysis, to interpret the results, and to prepare tables and text that summarize the results according to the guidelines of the *Publication manual of the American Psychological Association* (the most widely used publication format in the social science literature).

Audience

This text is designed for students who are learning about these procedures for the first time as well as for researchers who need to perform the analyses in applied research. Although this material requires little prior knowledge concerning other statistical procedures, the text will probably be more easily understood by those who have completed an elementary statistics course. This text does not require the use of matrix algebra or complex mathematical formulas. The few mathematical computations required of the reader typically involve simple operations such as addition or division.

You do not need to be familiar with the SAS System to use this text. Appendices A.1 – A.5 in the book describe virtually everything you need to know to write SAS programs that

- input data (either as raw data or as correlation or covariance matrices)
- transform variables
- create new variables
- create data subsets
- perform simple descriptive statistics
- create scattergrams
- perform simple correlations

With these appendices, this volume is a *self-contained* introduction to the SAS System that shows you how to create data sets and perform advanced statistical procedures.

Organization

The book begins with six chapters covering factor analysis, path analysis, and latent-variable causal models. If you are already familiar with the SAS System, you can start with this section. If you are new to the SAS System, you may want to begin instead with the appendices that introduce SAS System basics.

Material covered in each chapter is summarized as follows:

Part I: Exploratory Factor Analysis and Related Procedures

Chapter 1: Principal Component Analysis

provides an introduction to principal component analysis: a variable-reduction procedure similar to factor analysis. This chapter offers guidelines regarding the necessary sample size and the recommended number of items per component. The chapter also shows how to determine the number of components to retain, how to interpret the rotated solution, how to create factor scores, and how to summarize the results. Fictitious data from two studies are analyzed to illustrate these procedures. This chapter deals only with the creation of **orthogonal** (uncorrelated) components; **oblique** (correlated) solutions are covered in the following chapter on exploratory factor analysis.

Chapter 2: Exploratory Factor Analysis

shows how you can use exploratory factor analysis to investigate the factor structure underlying a set of observed variables. This chapter discusses important concepts related to the common factor model and explains the differences between factor analysis and principal component analysis. Chapter 2 provides guidelines regarding the necessary sample size as well as the number of variables that should load on each factor. The chapter shows how to determine the number of factors to retain, how to interpret the rotated solution, how to create factor scores, and how to summarize the results. The examples provided here show how to perform and interpret a promax rotation that results in oblique (correlated) factors.

Chapter 3: Assessing Scale Reliability with Coefficient Alpha

shows how PROC CORR can be used to compute the coefficient alpha reliability index for a multiple-item scale. (This chapter follows the chapters on principal component analysis and factor analysis because those procedures are often used to *create* multiple-item scales). Chapter 3 reviews basic issues regarding the assessment of reliability and describes the circumstances under which a measure of internal consistency reliability is likely to be high. Fictitious questionnaire data are analyzed to demonstrate how the results of PROC CORR can be used to perform an item analysis, thereby improving the reliability of the scale.

Part II: Structural Equation Modeling

Chapter 4: Path Analysis with Manifest Variables

shows how to use the CALIS procedure to test path models with **manifest** (observed) variables. This chapter introduces important terms and concepts in path analysis and describes the necessary conditions for performing the analysis. Chapter 4 demonstrates how to prepare a program figure that specifies the parameters to be estimated and how to turn this figure into a SAS program that tests the model of interest. This chapter also shows you how to prepare the PROC CALIS statement and related statements, how to interpret the results of the analysis, how to modify a model to achieve a better fit, and how to summarize the results for a paper. Chapter 4 deals only with the testing of recursive, or unidirectional, models.

Chapter 5: Developing Measurement Models with Confirmatory Factor Analysis

shows how the CALIS procedure can be used to develop, test, and improve the fit of a measurement model. A **measurement model** is essentially a factor model in which a number of manifest variables serve as indicators for a smaller number of underlying constructs. Developing an acceptable measurement model is the first step in performing a path analysis with latent variables (which is covered in the following chapter); therefore, Chapters 5 and 6 combined provide a comprehensive introduction to the testing of causal models with latent variables. Chapter 5 shows how to use confirmatory factor analysis to determine whether your measurement model provides a good fit to the data, and it discusses the use of modification indices that can be used to improve the model's fit. This chapter also shows how to assess the reliability, convergent validity, and discriminant validity of the model's indicators and constructs, and it describes the characteristics of an ideal fit for a measurement model.

Chapter 6: Path Analysis with Latent Variables

shows how the CALIS procedure can be used to perform path analysis for causal models that include latent, or unobserved, variables (some texts refer to these models as LISREL models). This chapter builds on concepts introduced in Chapter 5 by showing how to test **theoretical models**: modified measurement models in which causal relationships are hypothesized to exist between some of the underlying constructs. The chapter shows how to determine whether the theoretical model, as a whole, provides an acceptable fit to the data, whether specific causal paths are statistically significant, and whether specific model modifications are likely to result in an improved fit. Chapter 6 illustrates a variety of goodness of fit indices and parsimony indices, and it shows you how to analyze **standard models** (in which all of the variables in the "structural" portion of the model are latent factors), as well as **nonstandard models** (in which some of these variables are manifest variables). The chapter describes characteristics of an ideal fit and provides guidelines for summarizing the results of the analysis in figures, tables, and text.

Appendices

Appendix A: Basics for Using the SAS System

Appendix A.1: Introduction to SAS Programs, SAS Logs, and SAS Output
describes three types of files that you will work with when using the SAS System: the SAS program, the SAS log, and the SAS output file. Appendix A.1 presents a simple SAS program, along with the log and output files produced by the program. This appendix also provides the big picture regarding the steps that you will follow when submitting SAS programs.

Appendix A.2: Data Input
shows how to key data and write the statements that create a SAS data set. This chapter provides instructions on inputting data with either the CARDS statement or the INFILE statement, and it shows how information can be read either as raw data or in the form of a correlation or covariance matrix. By the conclusion of the appendix, you should be able to input virtually any type of data that is commonly encountered in social science research.

Appendix A.3: Working with Variables and Observations in SAS Data Sets
shows how to modify a data set so that existing variables are transformed or recoded or so that new variables are created. Appendix A.3 shows you how to write statements that eliminate unwanted observations from a data set so that analyses are performed on only a specified subgroup or on subjects that have no missing data. This chapter demonstrates the correct use of arithmetic operators, IF-THEN control statements, and comparison operators. Procedures for concatenating and merging data sets are also reviewed.

Appendix A.4: Exploring Data with PROC MEANS, PROC FREQ, PROC PRINT, and PROC UNIVARIATE
discusses the use of four procedures:

- PROC MEANS, which can be used to calculate means, standard deviations, and other descriptive statistics for numeric variables
- PROC FREQ, which can be used to construct frequency distributions
- PROC PRINT, which enables you to create a printout of the raw data set
- PROC UNIVARIATE, which can be used to test for normality and to prepare stem-and-leaf plots.

Once data are keyed, these procedures can be used to screen data for errors as well as to obtain simple descriptive statistics to be reported in the research article.

Appendix A.5: Preparing Scattergrams and Computing Correlations
describes the circumstances under which it is appropriate to compute Pearson correlations, and it shows how to prepare bivariate scattergrams with PROC PLOT to verify that the relationship between two variables is linear. Appendix A.5 also shows how to use the CORR procedure to compute correlations between variables and discusses some of the options available with PROC CORR.

Appendix B: Data Sets
    provides data sets used in Chapters 1–3.

Appendix C: Critical Values of the Chi-Square Distribution
    provides a table of critical values of the chi-square distribution.

Reference

American Psychological Association (1987). *Publication manual of the American Psychological Association, third edition*. Washington, DC: American Psychological Association.

Chapter 1

PRINCIPAL COMPONENT ANALYSIS

Overview. This chapter provides an introduction to principal component analysis: a variable-reduction procedure similar to factor analysis. It provides guidelines regarding the necessary sample size and number of items per component. It shows how to determine the number of components to retain, interpret the rotated solution, create factor scores, and summarize the results. Fictitious data from two studies are analyzed to illustrate these procedures. The present chapter deals only with the creation of orthogonal (uncorrelated) components; oblique (correlated) solutions are covered in Chapter 2, "Exploratory Factor Analysis".

Introduction: The Basics of Principal Component Analysis

Principal component analysis is appropriate when you have obtained measures on a number of observed variables and wish to develop a smaller number of artificial variables (called principal components) that will account for most of the variance in the observed variables. The principal components may then be used as predictor or criterion variables in subsequent analyses.

A Variable Reduction Procedure

Principal component analysis is a variable reduction procedure. It is useful when you have obtained data on a number of variables (possibly a large number of variables), and believe that there is some redundancy in those variables. In this case, redundancy means that some of the variables are correlated with one another, possibly because they are measuring the same construct. Because of this redundancy, you believe that it should be possible to reduce the observed variables into a smaller number of principal components (artificial variables) that will account for most of the variance in the observed variables.

Because it is a variable reduction procedure, principal component analysis is similar in many respects to exploratory factor analysis. In fact, the steps followed when conducting a principal component analysis are virtually identical to those followed when conducting an exploratory factor analysis. However, there are significant conceptual differences between the two procedures, and it is important that you do not mistakenly claim that you are performing factor analysis when you are actually performing principal component analysis. The differences between these two procedures are described in greater detail in a later section titled "Principal Component Analysis is *Not* Factor Analysis."

An Illustration of Variable Redundancy

A specific (but fictitious) example of research will now be presented to illustrate the concept of variable redundancy introduced earlier. Imagine that you have developed a 7-item measure of job satisfaction. The instrument is reproduced here:

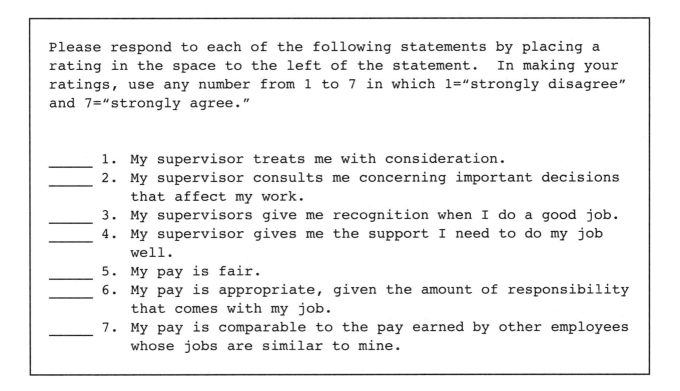

```
Please respond to each of the following statements by placing a
rating in the space to the left of the statement.  In making your
ratings, use any number from 1 to 7 in which 1="strongly disagree"
and 7="strongly agree."


_____   1.  My supervisor treats me with consideration.
_____   2.  My supervisor consults me concerning important decisions
             that affect my work.
_____   3.  My supervisors give me recognition when I do a good job.
_____   4.  My supervisor gives me the support I need to do my job
             well.
_____   5.  My pay is fair.
_____   6.  My pay is appropriate, given the amount of responsibility
             that comes with my job.
_____   7.  My pay is comparable to the pay earned by other employees
             whose jobs are similar to mine.
```

Perhaps you began your investigation with the intention of administering this questionnaire to 200 or so employees, and using their responses to the seven items as seven separate variables in subsequent analyses (for example, perhaps you intended to use the seven items as seven separate predictor variables in a multiple regression equation in which the criterion variable was "intention to quit the organization").

There are a number of problems with conducting the study in this fashion, however. One of the more important problems involves the concept of redundancy that was mentioned earlier. Take a close look at the content of the seven items in the questionnaire. Notice that items 1-4 all deal with the same topic: the employees' satisfaction with their supervisors. In this way, items 1-4 are somewhat redundant to one another. Similarly, notice that items 5-7 also all seem to deal with the same topic: the employees' satisfaction with their pay.

Empirical findings may further support the notion that there is redundancy in the seven items. Assume that you administer the questionnaire to 200 employees and compute all possible correlations between responses to the 7 items. The resulting fictitious correlations are reproduced in Table 1.1:

Table 1.1

Correlations among Seven Job Satisfaction Items

			Correlations				
Variable	1	2	3	4	5	6	7
1	1.00						
2	.75	1.00					
3	.83	.82	1.00				
4	.68	.92	.88	1.00			
5	.03	.01	.04	.01	1.00		
6	.05	.02	.05	.07	.89	1.00	
7	.02	.06	.00	.03	.91	.76	1.00

Note: $\underline{N}$ = 200.

When correlations among several variables are computed, they are typically summarized in the form of a **correlation matrix**, such as the one reproduced in Table 1.1. This is an appropriate opportunity to review just how a correlation matrix is interpreted. The rows and columns of

Table 1.1 correspond to the seven variables included in the analysis: Row 1 (and column 1) represents variable 1, row 2 (and column 2) represents variable 2, and so forth. Where a given row and column intersect, you will find the correlation between the two corresponding variables. For example, where the row for variable 2 intersects with the column for variable 1, you find a correlation of .75; this means that the correlation between variables 1 and 2 is .75.

The correlations of Table 1.1 show that the seven items seem to hang together in two distinct groups. First, notice that items 1-4 show relatively strong correlations with one another. This could be because items 1-4 are measuring the same construct. In the same way, items 5-7 correlate strongly with one another (a possible indication that they all measure the same construct as well). Even more interesting, notice that items 1-4 demonstrate very weak correlations with items 5-7. This is what you would expect to see if items 1-4 and items 5-7 were measuring two different constructs.

Given this apparent redundancy, it is likely that the seven items of the questionnaire are not really measuring seven different constructs; more likely, items 1-4 are measuring a single construct that could reasonably be labelled "satisfaction with supervision," while items 5-7 are measuring a different construct that could be labelled "satisfaction with pay."

If responses to the seven items actually displayed the redundancy suggested by the pattern of correlations in Table 1.1, it would be advantageous to somehow reduce the number of variables in this data set, so that (in a sense) items 1-4 are collapsed into a single new variable that reflects the employees' satisfaction with supervision, and items 5-7 are collapsed into a single new variable that reflects satisfaction with pay. You could then use these two new artificial variables (rather than the seven original variables) as predictor variables in multiple regression, or in any other type of analysis.

In essence, this is what is accomplished by principal component analysis: it allows you to reduce a set of observed variables into a smaller set of artificial variables called principal components. The resulting principal components may then be used in subsequent analyses.

What is a Principal Component?

How principal components are computed. Technically, a **principal component** can be defined as a linear combination of optimally-weighted observed variables. In order to understand the meaning of this definition, it is necessary to first describe how subject scores on a principal component are computed.

In the course of performing a principal component analysis, it is possible to calculate a score for each subject on a given principal component. For example, in the preceding study, each subject would have scores on two components: one score on the satisfaction with supervision component, and one score on the satisfaction with pay component. The subject's actual scores on the seven questionnaire items would be optimally weighted and then summed to compute their scores on a given component.

Below is the general form for the formula to compute scores on the first component extracted (created) in a principal component analysis:

$$C_1 = b_{11}(X_1) + b_{12}(X_2) + \ldots b_{1p}(X_p)$$

where

C_1 = the subject's score on principal component 1 (the first component extracted)

b_{1p} = the regression coefficient (or weight) for observed variable p, as used in creating principal component 1

X_p = the subject's score on observed variable p.

For example, assume that component 1 in the present study was the "satisfaction with supervision" component. You could determine each subject's score on principal component 1 by using the following fictitious formula:

$$
\begin{aligned}
C_1 = \quad & .44\ (X_1) + .40\ (X_2) + .47\ (X_3) + .32\ (X_4) \\
+ \quad & .02\ (X_5) + .01\ (X_6) + .03\ (X_7)
\end{aligned}
$$

In the present case, the observed variables (the "X" variables) were subject responses to the seven job satisfaction questions; X_1 represents question 1, X_2 represents question 2, and so forth. Notice that different regression coefficients were assigned to the different questions in computing subject scores on component 1: Questions 1–4 were assigned relatively large regression weights that range from .32 to 44, while questions 5–7 were assigned very small weights ranging from .01 to .03. This makes sense, because component 1 is the satisfaction with supervision component, and satisfaction with supervision was assessed by questions 1–4. It is therefore appropriate that items 1–4 would be given a good deal of weight in computing subject scores on this component, while items 5–7 would be given little weight.

Obviously, a different equation, with different regression weights, would be used to compute subject scores on component 2 (the satisfaction with pay component). Below is a fictitious illustration of this formula:

$$
\begin{aligned}
C_2 = \quad & .01\ (X_1) + .04\ (X_2) + .02\ (X_3) + .02\ (X_4) \\
+ \quad & .48\ (X_5) + .31\ (X_6) + .39\ (X_7)
\end{aligned}
$$

The preceding shows that, in creating scores on the second component, much weight would be given to items 5–7, and little would be given to items 1–4. As a result, component 2 should account for much of the variability in the three satisfaction with pay items; that is, it should be strongly correlated with those three items.

At this point, it is reasonable to wonder how the regression weights from the preceding equations are determined. The SAS System's PROC FACTOR solves for these weights by using a special type of equation called an **eigenequation**. The weights produced by these eigenequations are optimal weights in the sense that, for a given set of data, no other set of weights could produce a set of components that are more successful in accounting for variance in the observed variables. The weights are created so as to satisfy a principle of least squares similar (but not identical) to the principle of least squares used in multiple regression. Later, this chapter will show how PROC FACTOR can be used to extract (create) principal components.

It is now possible to better understand the definition that was offered at the beginning of this section. There, a principal component was defined as a linear combination of optimally weighted observed variables. The words "linear combination" refer to the fact that scores on a component are created by adding together scores on the observed variables being analyzed. "Optimally weighted" refers to the fact that the observed variables are weighted in such a way that the resulting components account for a maximal amount of variance in the data set.

Number of components extracted. The preceding section may have created the impression that, if a principal component analysis were performed on data from the 7-item job satisfaction questionnaire, only two components would be created. However, such an impression would not be entirely correct.

In reality, the number of components extracted in a principal component analysis is equal to the number of observed variables being analyzed. This means that an analysis of your 7-item questionnaire would actually result in seven components, not two.

However, in most analyses, only the first few components account for meaningful amounts of variance, so only these first few components are retained, interpreted, and used in subsequent analyses (such as in multiple regression analyses). For example, in your analysis of the 7-item job satisfaction questionnaire, it is likely that only the first two components would account for a meaningful amount of variance; therefore only these would be retained for interpretation. You would assume that the remaining five components accounted for only trivial amounts of variance. These latter components would therefore not be retained, interpreted, or further analyzed.

Characteristics of principal components. The first component extracted in a principal component analysis accounts for a maximal amount of total variance in the observed variables. Under typical conditions, this means that the first component will be correlated with at least some of the observed variables. It may be correlated with many.

The second component extracted will have two important characteristics. First, this component will account for a maximal amount of variance in the data set that was not accounted for by the first component. Again under typical conditions, this means that the second component will be correlated with some of the observed variables that did not display strong correlations with component 1.

The second characteristic of the second component is that it will be *uncorrelated* with the first component. Literally, if you were to compute the correlation between components 1 and 2, that correlation would be zero.

The remaining components that are extracted in the analysis display the same two characteristics: each component accounts for a maximal amount of variance in the observed variables that was not accounted for by the preceding components, and is uncorrelated with all of the preceding components. A principal component analysis proceeds in this fashion, with each new component accounting for progressively smaller and smaller amounts of variance (this is why only the first few components are usually retained and interpreted). When the analysis is complete, the resulting components will display varying degrees of correlation with the observed variables, but are completely uncorrelated with one another.

What is meant by "total variance" in the data set? To understand the meaning of "total variance" as it is used in a principal component analysis, remember that the observed variables are standardized in the course of the analysis. This means that each variable is transformed so that it has a mean of zero and a variance of one. The "total variance" in the data set is simply the sum of the variances of these observed variables. Because they have been standardized to have a variance of one, each observed variable contributes one unit of variance to the "total variance" in the data set. Because of this, the total variance in a principal component analysis will always be equal to the number of observed variables being analyzed. For example, if seven variables are being analyzed, the total variance will equal seven. The components that are extracted in the analysis will partition this variance: perhaps the first component will account for 3.2 units of total variance; perhaps the second component will account for 2.1 units. The analysis continues in this way until all of the variance in the data set has been accounted for.

Orthogonal versus Oblique Solutions

This chapter will discuss only principal component analyses that result in orthogonal solutions. An **orthogonal solution** is one in which the components remain uncorrelated (orthogonal means "uncorrelated").

It is possible to perform a principal component analysis that results in correlated components. Such a solution is called an **oblique solution**. In some situations, oblique solutions are superior to orthogonal solutions because they produce cleaner, more easily-interpreted results.

However, oblique solutions are also somewhat more complicated to interpret, compared to orthogonal solutions. For this reason, the present chapter will focus only on the interpretation of orthogonal solutions. To learn about oblique solutions, see Chapter 2. The concepts discussed in this chapter will provide a good foundation for the somewhat more complex concepts discussed in that chapter.

Principal Component Analysis is *Not* Factor Analysis

Principal component analysis is sometimes confused with factor analysis, and this is understandable, because there are many important similarities between the two procedures: both are variable reduction methods that can be used to identify groups of observed variables that tend to hang together empirically. Both procedures can be performed with the SAS System's FACTOR procedure, and they sometimes even provide very similar results.

Nonetheless, there are some important conceptual differences between principal component analysis and factor analysis that should be understood at the outset. Perhaps the most important deals with the **assumption of an underlying causal structure**: factor analysis assumes that the covariation in the observed variables is due to the presence of one or more latent variables (factors) that exert causal influence on these observed variables. An example of such a causal structure is presented in Figure 1.1:

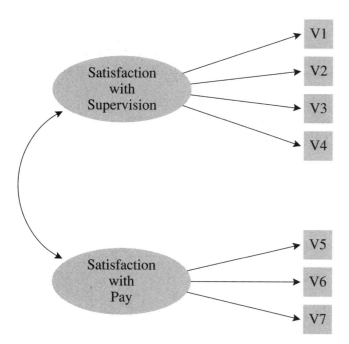

Figure 1.1: Example of the Underlying Causal Structure that is Assumed in Factor Analysis

The ovals in Figure 1.1 represent the latent (unmeasured) factors of "satisfaction with supervision" and "satisfaction with pay." These factors are latent in the sense that they are assumed to actually exist in the employee's belief systems, but cannot be measured directly. However, they do exert an influence on the employee's responses to the seven items that constitute the job satisfaction questionnaire described earlier (these seven items are represented as the squares labelled V1-V7 in the figure). It can be seen that the "supervision" factor exerts influence on items V1-V4 (the supervision questions), while the "pay" factor exerts influence on items V5-V7 (the pay items).

Researchers use factor analysis when they believe that certain latent factors exist that exert causal influence on the observed variables they are studying. Exploratory factor analysis helps the researcher identify the number and nature of these latent factors.

In contrast, principal component analysis makes no assumption about an underlying causal model. Principal component analysis is simply a variable reduction procedure that (typically) results in a relatively small number of components that account for most of the variance in a set of observed variables.

In summary, both factor analysis and principal component analysis have important roles to play in social science research, but their conceptual foundations are quite distinct.

Example: Analysis of the Prosocial Orientation Inventory

Assume that you have developed an instrument called the Prosocial Orientation Inventory (POI) that assesses the extent to which a person has engaged in helping behaviors over the preceding six-month period. The instrument contains six items, and is reproduced here.

```
Instructions:   Below are a number of activities that people
sometimes engage in.  For each item, please indicate how
frequently you have engaged in this activity over the preceding
six months.  Make your rating by circling the appropriate number
to the left of the item, and use the following response format:

    7 = Very Frequently
    6 = Frequently
    5 = Somewhat Frequently
    4 = Occasionally
    3 = Seldom
    2 = Almost Never
    1 = Never


 1 2 3 4 5 6 7    1.    Went out of my way to do a favor for a
                        coworker.

 1 2 3 4 5 6 7    2.    Went out of my way to do a favor for a
                        relative.
```

```
1 2 3 4 5 6 7     3.    Went out of my way to do a favor for a
                        friend.

1 2 3 4 5 6 7     4.    Gave money to a religious charity.

1 2 3 4 5 6 7     5.    Gave money to a charity not associated with
                        a religion.

1 2 3 4 5 6 7     6.    Gave money to a panhandler.
```

When you developed the instrument, you originally intended to administer it to a sample of subjects and use their responses to the six items as six separate predictor variables in a multiple regression equation. However, you have recently learned that this would be a questionable practice (for the reasons discussed earlier), and have now decided to instead perform a principal component analysis on responses to the six items to see if a smaller number of components can successfully account for most of the variance in the data set. If this is the case, you will use the resulting components as the predictor variables in your multiple regression analyses.

At this point, it may be instructive to review the content of the six items that constitute the POI to make an informed guess as to what you are likely to learn from the principal component analysis. Imagine that, when you first constructed the instrument, you assumed that the six items were assessing six different types of prosocial behavior. However, inspection of items 1-3 shows that these three items share something in common: they all deal with the activity of "going out of one's way to do a favor for an acquaintance." It would not be surprising to learn that these three items will hang together empirically in the principal component analysis to be performed. In the same way, a review of items 4-6 shows that all of these items involve the activity of "giving money to the needy." Again, it is possible that these three items will also group together in the course of the analysis.

In summary, the nature of the items suggests that it may be possible to account for the variance in the POI with just two components: An "acquaintance helping" component, and a "financial giving" component. At this point, we are only speculating, of course; only a formal analysis can tell us about the number and nature of the components measured by the POI.

(Remember that the preceding fictitious instrument is used for purposes of illustration only, and should not be regarded as an example of a good measure of prosocial orientation; among other problems, this questionnaire obviously deals with very few forms of helping behavior).

Preparing a Multiple-Item Instrument

The preceding section illustrates an important point about how *not* to prepare a multiple-item measure of a construct: Generally speaking, it is poor practice to throw together a questionnaire, administer it to a sample, and then perform a principal component analysis (or factor analysis) to see what the questionnaire is measuring.

Better results are much more likely when you make a priori decisions about what you want the questionnaire to measure, and then take steps to ensure that it does. For example, you would have been more likely to obtain desirable results if you:

- had begun with a thorough review of theory and research on prosocial behavior

- used that review to determine how many types of prosocial behavior probably exist

- wrote multiple questionnaire items to assess each type of prosocial behavior.

Using this approach, you could have made statements such as "There are three types of prosocial behavior: acquaintance helping, stranger helping, and financial giving." You could have then prepared a number of items to assess each of these three types, administered the questionnaire to a large sample, and performed a principal component analysis to see if the three components did, in fact, emerge.

Number of Items per Component

When a variable (such as a questionnaire item) is given a great deal of weight in constructing a principal component, we say that the variable **loads** on that component. For example, if the item "Went out of my way to do a favor for a coworker" is given a lot of weight in creating the acquaintance helping component, we say that this item loads on the acquaintance helping component.

It is highly desirable to have at least three (and preferably more) variables loading on each retained component when the principal component analysis is complete. Because some of the items may be dropped during the course of the analysis (for reasons to be discussed later), it is generally good practice to write at least five items for each construct that you wish to measure; in this way, you increase the chances that at least three items per component will survive the analysis. Note that we have unfortunately violated this recommendation by apparently writing only three items for each of the two a priori components constituting the POI.

One additional note on scale length: the recommendation of three items per scale offered here should be viewed as an absolute minimum, and certainly not as an optimal number of items per scale. In practice, test and attitude scale developers normally desire that their scales contain many more than just three items to measure a given construct. It is not unusual to see individual scales that include 10, 20, or even more items to assess a single construct. Other things held constant, the more items in the scale, the more reliable it will be. The recommendation of three items per scale should therefore be viewed as a rock-bottom lower bound, appropriate only if practical concerns (such as total questionnaire length) prevent you from including more items. For more information on scale construction, see Spector (1992).

Minimally Adequate Sample Size

Principal component analysis is a large-sample procedure. To obtain reliable results, the minimal number of subjects providing usable data for the analysis should be the larger of 100 subjects or five times the number of variables being analyzed.

To illustrate, assume that you wish to perform an analysis on responses to a 50-item questionnaire (remember that, when responses to a questionnaire are analyzed, the number of variables is equal to the number of items on the questionnaire). Five times the number of items on the questionnaire equals 250. Therefore, your final sample should provide usable (complete) data from at least 250 subjects. It should be remembered, however, that any subject who fails to answer just one item will not provide usable data for the principal component analysis, and will therefore be dropped from the final sample. A certain number of subjects can always be expected to leave at least one question blank (despite the most strongly worded instructions to the contrary!). To ensure that the final sample includes at least 250 usable responses, you would be wise to administer the questionnaire to perhaps 300-350 subjects.

These rules regarding the number of subjects per variable again constitute a lower bound, and some have argued that they should apply only under two optimal conditions for principal component analysis: when many variables are expected to load on each component, and when variable communalities are high. Under less optimal conditions, even larger samples may be required.

What is a communality? A **communality** refers to the percent of variance in an observed variable that is accounted for by the retained components (or factors). A given variable will display a large communality if it loads heavily on at least one of the study's retained components. Although communalities are computed in both procedures, the *concept* of variable communality is more relevant in a factor analysis than in principal component analysis.

SAS Program and Output

You may perform a principal component analysis using either the PRINCOMP or FACTOR procedures. This chapter will show how to perform the analysis using PROC FACTOR since this is a somewhat more flexible SAS System procedure (it is also possible to perform an exploratory factor analysis with PROC FACTOR). Because the analysis is to be performed using the FACTOR procedure, the output will at times make references to factors rather than to principal components (i.e., component 1 will be referred to as FACTOR1 in the output, component 2 as FACTOR2, and so forth). However, it is important to remember that you are nonetheless performing a principal component analysis.

This section will provide instructions on writing the SAS program, along with an overview of the SAS output. A subsequent section will provide a more detailed treatment of the steps followed in the analysis, and the decisions to be made at each step.

Writing the SAS Program

The DATA step. To perform a principal component analysis, data may be input in the form of raw data, a correlation matrix, a covariance matrix, as well as other some other types of data (for details, see Chapter 21 on "The FACTOR Procedure" in the *SAS/STAT users guide, version 6, fourth edition, volume 1* [1989]). In this chapter's first example, raw data will be analyzed.

Assume that you administered the POI to 50 subjects, and keyed their responses according to the following keying guide:

Line	Column	Variable Name	Explanation
1	1-6	V1-V6	Subjects' responses to survey questions 1 through 6. Responses were made using a 7-point "frequency" scale.

Here are the statements that will input these responses as raw data. The first three and the last three observations are reproduced here; for the entire data set, see Appendix B.

```
1     DATA D1;
2        INPUT    #1    @1    (V1-V6)    (1.)  ;
3     CARDS;
4     556754
5     567343
6     777222
7     .
8     .
9     .
10    767151
11    455323
12    455544
13    ;
```

The data set in Appendix B includes only 50 cases so that it will be relatively easy for interested readers to key the data and replicate the analyses presented here. However, it should be remembered that 50 observations will normally constitute an unacceptably small sample for a principal component analysis. Earlier it was said that a sample should provide usable data from the larger of either 100 cases or 5 times the number of observed variables. A small sample is being analyzed here for illustrative purposes only.

The PROC FACTOR statement. The general form for the SAS program to perform a principal component analysis is presented here:

```
PROC FACTOR    DATA=data-set-name
               SIMPLE
               METHOD=PRIN
               PRIORS=ONE
               MINEIGEN=p
               SCREE
               ROTATE=VARIMAX
               ROUND
               FLAG=desired-size-of-"significant"-factor-loadings ;
   VAR  variables-to-be-analyzed ;
   RUN;
```

Options used with PROC FACTOR. The PROC FACTOR statement begins the FACTOR procedure, and a number of options may be requested in this statement before it ends with a semicolon. Some options that may be especially useful in social science research are:

FLAG=desired-size-of-"significant"-factor-loadings
> causes the printer to flag (with an asterisk) any factor loading whose absolute value is greater than some specified size. For example, if you specify

> FLAG=.35

> an asterisk will appear next to any loading whose absolute value exceeds .35. This option can make it much easier to interpret a factor pattern. Negative values are not allowed in the FLAG option, and the FLAG option should be used in conjunction with the ROUND option.

METHOD=factor-extraction-method
> specifies the method to be used in extracting the factors or components. The current program specifies METHOD=PRIN to request that the principal axis (principal factors) method be used for the initial extraction. This is the appropriate method for a principal component analysis.

MINEIGEN=p

>   specifies the critical eigenvalue a component must display if that component is to be retained (here, p = the critical eigenvalue). For example, the current program specifies

>   MINEIGEN=1

>   This statement will cause PROC FACTOR to retain and rotate any component whose eigenvalue is 1.00 or larger. Negative values are not allowed.

NFACT=*n*

>   allows you to specify the number of components to be retained and rotated, where *n* = the number of components.

OUT=name-of-new-data-set

>   creates a new data set that includes all of the variables of the existing data set, along with factor scores for the components retained in the present analysis. Component 1 is given the varible name FACTOR1, component 2 is given the name FACTOR2, and so forth. It must be used in conjunction with the NFACT option, and the analysis must be based on raw data.

PRIORS=prior-communality-estimates

>   specifies prior communality estimates. Users should always specify PRIORS=ONE to perform a principal component analysis.

ROTATE=rotation-method

>   specifies the rotation method to be used. The preceding program requests a varimax rotation, which results in orthogonal (uncorrelated) components. Oblique rotations may also be requested; oblique rotations are discussed in Chapter 2.

ROUND

>   causes all coefficients to be limited to two decimal places, rounded to the nearest integer, and multiplied by 100 (thus eliminating the decimal point). This generally makes it easier to read the coefficients because factor loadings and correlation coefficients in the matrices printed by PROC FACTOR are normally carried out to several decimal places.

SCREE

>   creates a plot that graphically displays the size of the eigenvalue associated with each component. This can be used to perform a scree test to determine how many components should be retained.

SIMPLE

>   requests simple descriptive statistics: the number of usable cases on which the analysis was performed, and the means and standard deviations of the observed variables.

The VAR statement. The variables to be analyzed are listed in the VAR statement, with each variable separated by at least one space. Remember that the VAR statement is a *separate* statement, not an option within the FACTOR statement, so don't forget to end the FACTOR statement with a semicolon before beginning the VAR statement.

Example of an actual program. The following is an actual program, including the DATA step, that could be used to analyze some fictitious data from your study. Only a few sample lines of data appear here; the entire data set may be found in Appendix B.

```
1      DATA D1;
2         INPUT      #1      @1     (V1-V6)       (1.)   ;
3      CARDS;
4      556754
5      567343
6      777222
7      .
8      .
9      .
10     767151
11     455323
12     455544
13     ;
14     PROC FACTOR     DATA=D1
15                     SIMPLE
16                     METHOD=PRIN
17                     PRIORS=ONE
18                     MINEIGEN=1
19                     SCREE
20                     ROTATE=VARIMAX
21                     ROUND
22                     FLAG=.40    ;
23        VAR V1 V2 V3 V4 V5 V6;
24        RUN;
```

Results from the Output

If printer options are set so that LINESIZE=80 and PAGESIZE=60, the preceding program would produce four pages of output. Here is a list of some of the most important information provided by the output, and the page on which it appears:

- Page 1 includes simple statistics.

- Page 2 includes the eigenvalue table.

- Page 3 includes the scree plot of eigenvalues.

- Page 4 includes the unrotated factor pattern and final communality estimates.

- Page 5 includes the rotated factor pattern.

The output created by the preceding program is reproduced here as Output 1.1:

```
                              The SAS System                              1

                Means and Standard Deviations from 50 observations

                    V1         V2         V3         V4         V5         V6
Mean              5.18        5.4       5.52       3.64       4.22        3.1
Std Dev     1.39518121 1.10656667 1.21621695 1.79295674 1.66953495 1.55511008
```

```
                                                                          2


Initial Factor Method: Principal Components

                 Prior Communality Estimates: ONE

        Eigenvalues of the Correlation Matrix:  Total = 6  Average = 1

                                1          2          3
              Eigenvalue    2.2664     1.9746     0.7973
              Difference    0.2918     1.1773     0.3581
              Proportion    0.3777     0.3291     0.1329
              Cumulative    0.3777     0.7068     0.8397

                                4          5          6
              Eigenvalue    0.4392     0.2913     0.2312
              Difference    0.1479     0.0601
              Proportion    0.0732     0.0485     0.0385
              Cumulative    0.9129     0.9615     1.0000

        2 factors will be retained by the MINEIGEN criterion.
```

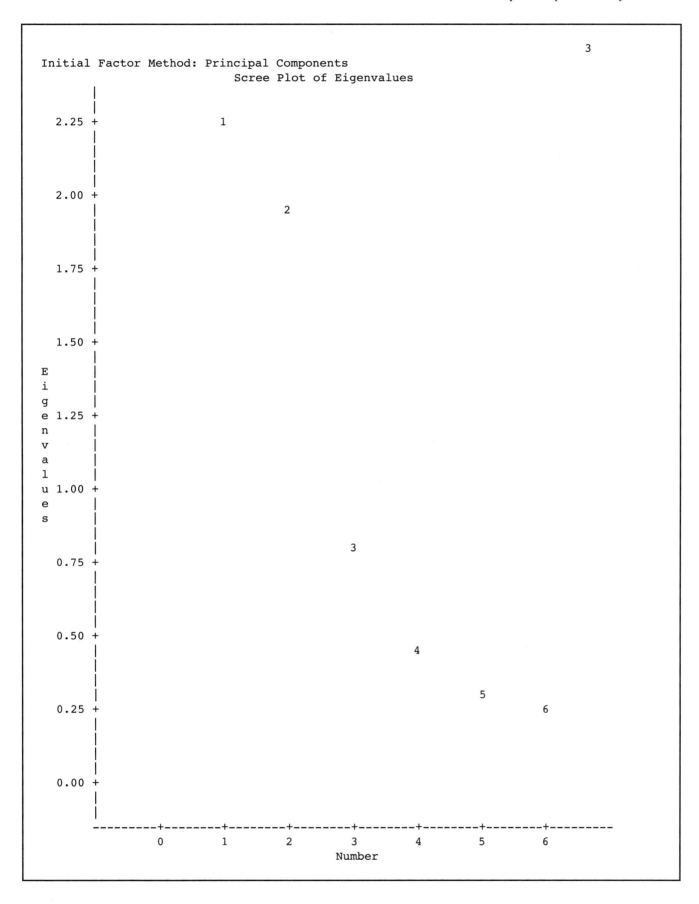

4

Initial Factor Method: Principal Components

Factor Pattern

	FACTOR1	FACTOR2
V1	58 *	70 *
V2	48 *	53 *
V3	60 *	62 *
V4	64 *	-64 *
V5	68 *	-45 *
V6	68 *	-46 *

NOTE: Printed values are multiplied by 100 and rounded to the nearest integer.
    Values greater than 0.4 have been flagged by an '*'.

Variance explained by each factor

FACTOR1	FACTOR2
2.266436	1.974615

Final Communality Estimates: Total = 4.241050

V1	V2	V3	V4	V5	V6
0.823418	0.508529	0.743990	0.822574	0.665963	0.676575

The SAS System   5

Rotation Method: Varimax

Orthogonal Transformation Matrix

	1	2
1	0.76914	0.63908
2	-0.63908	0.76914

Rotated Factor Pattern

	FACTOR1	FACTOR2
V1	0	91 *
V2	3	71 *
V3	7	86 *
V4	90 *	-9
V5	81 *	9
V6	82 *	8

```
NOTE: Printed values are multiplied by 100 and rounded to the nearest integer.
      Values greater than 0.4 have been flagged by an '*'.

                       Variance explained by each factor

                          FACTOR1    FACTOR2
                          2.147248   2.093803


          Final Communality Estimates: Total = 4.241050

              V1         V2         V3         V4         V5         V6
           0.823418   0.508529   0.743990   0.822574   0.665963   0.676575
```

Output 1.1: Results of the Initial Principal Component Analysis of the Prosocial Orientation Inventory (POI) Data

Page 1 from Output 1.1 provides simple statistics for the observed variables included in the analysis. Once the SAS log has been checked to verify that no errors were made in the analysis, these simple statistics should be reviewed to determine how many usable observations were included in the analysis and to verify that the means and standard deviations are in the expected range. The top line of Output 1.1, page 1, says "Means and Standard Deviations from 50 Observations", meaning that data from 50 subjects were included in the analysis.

Steps in Conducting Principal Component Analysis

Principal component analysis is normally conducted in a sequence of steps, with somewhat subjective decisions being made at many of these steps. Because this is an introductory treatment of the topic, it will not provide a comprehensive discussion of all of the options available to you at each step. Instead, specific recommendations will be made, consistent with practices often followed in applied research. For a more detailed treatment of principal component analysis and its close relative, factor analysis, see Kim and Mueller (1978a; 1978b), Rummel (1970), or Stevens (1986).

Step 1: Initial Extraction of the Components

In principal component analysis, the number of components extracted is equal to the number of variables being analyzed. Because six variables are analyzed in the present study, six components will be extracted. The first component can be expected to account for a fairly large amount of the total variance. Each succeeding component will account for progressively smaller amounts of variance. Although a large number of components may be extracted in this way, only the first few components will be important enough to be retained for interpretation.

Page 2 from Output 1.1 provides the eigenvalue table from the analysis (this table appears just below the heading "Eigenvalues of the Correlation Matrix: Total = 6 Average = 1"). An **eigenvalue** represents the amount of variance that is accounted for by a given component. In the row headed "Eignenvalue" (running from left to right), the eigenvalue for each component is presented. Each column in the matrix (running up and down) presents information about one of the six components: The column headed "1" provides information about the first component extracted, the column headed "2" provides information about the second component extracted, and so forth.

Where the row headed EIGENVALUE intersects with the columns headed "1" and "2," it can be seen that the eigenvalue for component 1 is 2.27, while the eigenvalue for component 2 is 1.97. This pattern is consistent with our earlier statement that the first components extracted tend to account for relatively large amounts of variance, while the later components account for relatively smaller amounts.

Step 2: Determining the Number of "Meaningful" Components to Retain

Earlier it was stated that the number of components extracted is equal to the number of variables being analyzed, necessitating that you decide just how many of these components are truly meaningful and worthy of being retained for rotation and interpretation. In general, you expect that only the first few components will account for meaningful amounts of variance, and that the later components will tend to account for only trivial variance. The next step of the analysis, therefore, is to determine how many meaningful components should be retained for interpretation. This section will describe four criteria that may be used in making this decision: the eigenvalue-one criterion, the scree test, the proportion of variance accounted for, and the interpretability criterion.

A. The eigenvalue-one criterion. In principal component analysis, one of the most commonly used criteria for solving the number-of-components problem is the eigenvalue-one criterion, also known as the Kaiser criterion (Kaiser, 1960). With this approach, you retain and interpret any component with an eigenvalue greater than 1.00.

The rationale for this criterion is straightforward. Each observed variable contributes one unit of variance to the total variance in the data set. Any component that displays an eigenvalue greater than 1.00 is accounting for a greater amount of variance than had been contributed by one variable. Such a component is therefore accounting for a meaningful amount of variance, and is worthy of being retained.

On the other hand, a component with an eigenvalue less than 1.00 is accounting for less variance than had been contributed by one variable. The purpose of principal component analysis is to reduce a number of observed variables into a relatively smaller number of components; this cannot be effectively achieved if you retain components that account for less variance than had been contributed by individual variables. For this reason, components with eigenvalues less than 1.00 are viewed as trivial, and are not retained.

The eigenvalue-one criterion has a number of positive features that have contributed to its popularity. Perhaps the most important reason for its widespread use is its simplicity: You do not make any subjective decisions, but merely retain components with eigenvalues greater than one.

On the positive side, it has been shown that this criterion very often results in retaining the correct number of components, particularly when a small to moderate number of variables are being analyzed and the variable communalities are high. Stevens (1986) reviews studies that have investigated the accuracy of the eigenvalue-one criterion, and recommends its use when less than 30 variables are being analyzed and communalities are greater than .70, or when the analysis is based on over 250 observations and the mean communality is greater than or equal to .60.

There are a number of problems associated with the eigenvalue-one criterion, however. As was suggested in the preceding paragraph, it can lead to retaining the wrong number of components under circumstances that are often encountered in research (e.g., when many variables are analyzed, when communalities are small). Also, the mindless application of this criterion can lead to retaining a certain number of components when the actual difference in the eigenvalues of successive components is only trivial. For example, if component 2 displays an eigenvalue of 1.001 and component 3 displays an eigenvalue of 0.999, then component 2 will be retained but component 3 will not; this may mislead you into believing that the third component was meaningless when, in fact, it accounted for almost exactly the same amount of variance as the second component. In short, the eigenvalue-one criterion can be helpful when used judiciously, but the thoughtless application of this approach can lead to serious errors of interpretation.

With the SAS System, the eigenvalue-one criterion can be implemented by including the MINEIGEN=1 option in the PROC FACTOR statement, and not including the NFACT option. The use of MINEIGEN=1 will cause PROC FACTOR to retain any component with an eigenvalue greater than 1.00.

The eigenvalue table from the current analysis appears on page 2 of Output 1.1. The eigenvalues for components 1, 2, and 3 were 2.27, 1.97, and 0.80, respectively. Only components 1 and 2 demonstrated eigenvalues greater than 1.00, so the eigenvalue-one criterion would lead you to retain and interpret only these two components.

Fortunately, the application of the criterion is fairly unambiguous in this case: The last component retained (2) displays an eigenvalue of 1.97, which is substantially greater than 1.00, and the next component (3) displays an eigenvalue of 0.80, which is clearly lower than 1.00. In this analysis, you are not faced with the difficult decision of whether to retain a component that demonstrates an eigenvalue that is close to 1.00, but not quite there (e.g., an eigenvalue of .98). In situations such as this, the eigenvalue-one criterion may be used with greater confidence.

B. The scree test. With the scree test (Cattell, 1966), you plot the eigenvalues associated with each component and look for a "break" between the components with relatively large eigenvalues and those with small eigenvalues. The components that appear *before* the break are assumed to be meaningful and are retained for rotation; those apppearing *after* the break are assumed to be unimportant and are not retained.

Sometimes a scree plot will display several large breaks. When this is the case, you should look for the *last* big break before the eigenvalues begin to level off. Only the components that appear before this last large break should be retained.

Specifying the SCREE option in the PROC FACTOR statement causes the SAS System to print an eigenvalue plot as part of the output. This appears as page 3 of Output 1.1.

You can see that the component numbers are listed on the horizontal axis, while eigenvalues are listed on the vertical axis. With this plot, notice that there is a relatively small break between component 1 and 2, and a relatively large break following component 2. The breaks between components 3, 4, 5, and 6 are all relatively small.

Because the large break in this plot appears between components 2 and 3, the scree test would lead you to retain only components 1 and 2. The components appearing after the break (3-6) would be regarded as trivial.

The scree test can be expected to provide reasonably accurate results, provided the sample is large (over 200) and most of the variable communalities are large (Stevens, 1986). However, this criterion has its own weaknesses as well, most notably the ambiguity that is often displayed by scree plots under typical research conditions: Very often, it is difficult to determine exactly where in the scree plot a break exists, or even if a break exists at all.

The break in the scree plot on page 3 of Output 1.1 was unusually obvious. In contrast, consider the plot that appears in Figure 1.2.

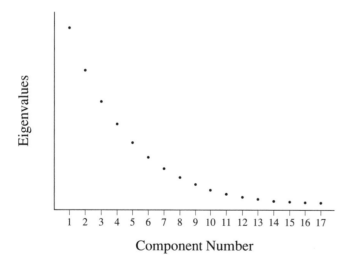

Figure 1.2: A Scree Plot with No Obvious Break

Figure 1.2 presents a fictitious scree plot from a principal component analysis of 17 variables. Notice that there is no obvious break in the plot that separates the meaningful components from the trivial components. Most researchers would agree that components 1 and 2 are probably meaningful, and that components 13–17 are probably trivial, but it is difficult to decide exactly where you should draw the line.

Scree plots such as the one presented in Figure 1.2 are common in social science research. When encountered, the use of the scree test must be supplemented with additional criteria, such as the variance accounted for criterion and the interpretability criterion, to be described later.

Why do they call it a "scree" test? The word "scree" refers to the loose rubble that lies at the base of a cliff. When performing a scree test, you normally hope that the scree plot will take the form of a cliff: At the top will be the eigenvalues for the few meaningful components, followed by a break (the edge of the cliff). At the bottom of the cliff will lie the scree: eigenvalues for the trivial components.

In some cases, a computer printer may not be able to prepare an eigenvalue plot with the degree of precision that is necessary to perform a sensitive scree test. In such cases, it may be best to prepare the plot by hand. This may be done simply by referring to the eigenvalue table on output page 2. Using the eigenvalues from this table, you can prepare an eigenvalue plot following the same format used by the SAS System (component numbers on the horizontal axis, eigenvalues on the vertical). Such a hand-drawn plot may make it easier to identify the break in the eigenvalues, if one exists.

C. Proportion of variance accounted for. A third criterion in solving the number of factors problem involves retaining a component if it accounts for a specified proportion (or percentage) of variance in the data set. For example, you may decide to retain any component that accounts for at least 5% or 10% of the total variance. This proportion can be calculated with a simple formula:

$$\text{Proportion} = \frac{\texttt{Eigenvalue for the component of interest}}{\texttt{Total eigenvalues of the correlation matrix}}$$

In principal component analysis, the "total eigenvalues of the correlation matrix" is equal to the total number of variables being analyzed (because each variable contributes one unit of variance to the analysis).

Fortunately, it is not necessary to actually compute these percentages by hand, since they are provided in the results of PROC FACTOR. The proportion of variance accounted for by each component is printed in the eigenvalue table from output page 2, and appears to the right of the "Proportion" heading.

The eigenvalue table for the current analysis appears on page 2 of Output 1.1. From the "Proportion" line in this eigenvalue table, you can see that the first component alone accounts for 38% of the total variance, the second component alone accounts for 33%, the third component accounts for 13%, and the fourth component accounts for 7%. Assume that you have decided to retain any component that accounts for at least 10% of the total variance in the data set. For the present results, using this criterion would cause you to retain components 1, 2, and 3 (notice that use of this criterion would result in retaining more components than would be retained with the two preceding criteria).

An alternative criterion is to retain enough components so that the *cumulative* percent of variance accounted for is equal to some minimal value. For example, remember that components 1, 2, 3, and 4 accounted for approximately 38%, 33%, 13%, and 7% of the total variance, respectively. Adding these percentages together results in a sum of 91%. This means that the *cumulative* percent of variance accounted for by components 1, 2, 3, and 4 is 91%. When researchers use the "cumulative percent of variance accounted for" as the criterion for solving the number-of-components problem, they usually retain enough components so that the cumulative percent of variance accounted for at least 70% (and sometimes 80%).

With respect to the results of PROC FACTOR, the "cumulative percent of variance accounted for" is presented in the eigenvalue table (from page 2), to the right of the "Cumulative" heading. For the present analysis, this information appears in the eigenvalue table on page 2 of Output 1.1. Notice the values that appear to the right of the heading "Cumulative": Each value in this line indicates the percent of variance accounted for by the present component, as well as all preceding components. For example, the value for component 2 is .7068 (this appears at the intersection of the row headed "Cumulative" and the column headed "2"). This value of .7068 indicates that approximately 71% of the total variance is accounted for by components 1 and 2 combined. The corresponding entry for component 3 is .8397, meaning that approximately 84% of the variance is accounted for by components 1, 2, and 3 combined. If you were to use 70% as the "critical value" for determining the number of components to retain, you would retain components 1 and 2 in the present analysis.

The proportion of variance criterion has a number of positive features. For example, in most cases, you would not want to retain a group of components that, combined, account for only a minority of the variance in the data set (say, 30%). Nonetheless, the critical values discussed earlier (10% for individual components and 70%-80% for the combined components) are obviously arbitrary. Because of these and related problems, this approach has sometimes been criticized for its subjectivity (Kim & Mueller, 1978b).

D. The interpretability criteria. Perhaps the most important criterion for solving the "number-of-components" problem is the **interpretability criterion**: interpreting the substantive meaning of the retained components and verifying that this interpretation makes sense in terms of what is known about the constructs under investigation. The following list provides four rules to follow in doing this. A later section (titled "Step 4: Interpreting the Rotated Solution") shows how to actually interpret the results of a principal component analysis; the following rules will be more meaningful after you have completed that section.

1. **Are there at least three variables (items) with significant loadings on each retained component?** A solution is less satisfactory if a given component is measured by less than three variables.

2. **Do the variables that load on a given component share the same conceptual meaning?** For example, if three questions on a survey all load on component 1, do all three of these questions seem to be measuring the same construct?

3. **Do the variables that load on different components seem to be measuring different constructs?** For example, if three questions load on component 1, and three other questions load on component 2, do the first three questions seem to be measuring a construct that is conceptually different from the construct measured by the last three questions?

4. **Does the rotated factor pattern demonstrate "simple structure?"** Simple structure means that the pattern possesses two characteristics: (a) Most of the variables have relatively high factor loadings on only one component, and near zero loadings on the other components, and (b) most components have relatively high factor loadings for some variables, and near-zero loadings for the remaining variables. This concept of simple structure will be explained in more detail in a later section titled "Step 4: Interpreting the Rotated Solution."

Recommendations. Given the preceding options, what procedure should you actually follow in solving the number-of-components problem? We recommend combining all four in a structured sequence. First, use the MINEIGEN=1 options to implement the eigenvalue-one criterion. Review this solution for interpretability, and use caution if the break between the components with eigenvalues above 1.00 and those below 1.00 is not clear-cut (i.e., if component 2 has an eigenvalue of 1.001, and component 2 has an eigenvalue of 0.998).

Next, perform a scree test and look for obvious breaks in the eigenvalues. Because there will often be more than one break in the scree plot, it may be necessary to examine two or more possible solutions.

Next, review the amount of common variance accounted for by each individual component. You probably should not rigidly use some specific but arbitrary cutoff point such as 5% or 10%. Still, if you are retaining components that account for as little as 2% or 4% of the variance, it may be wise to take a second look at the solution and verify that these latter components are of truly substantive importance. In the same way, it is best if the combined components account for at least 70% of the cumulative variance; if less than 70% is accounted for, it may be wise to consider alternative solutions that include a larger number of components.

Finally, apply the interpretability criteria to each solution that is examined. If more than one solution can be justified on the basis of the preceding criteria, which of these solutions is the most interpretable? By seeking a solution that is both interpretable and also satisfies one (or more) of the other three criteria, you maximize chances of retaining the correct number of components.

Step 3: Rotation to a Final Solution

Factor patterns and factor loadings. After extracting the initial components, PROC FACTOR will create an unrotated **factor pattern matrix**. The rows of this matrix represent the variables being analyzed, and the columns represent the retained components (these components are referred to as FACTOR1, FACTOR2 and so forth in the output).

The entries in the matrix are factor loadings. A **factor loading** is a general term for a coefficient that appears in a factor pattern matrix or a factor structure matrix. In an analysis that results in oblique (correlated) components, the definition for a factor loading is different depending on whether it is in a factor *pattern* matrix or in a factor *structure* matrix. However, the situation is simpler in an analysis that results in orthogonal components (as in the present chapter): In an orthogonal analysis, factor loadings are equivalent to bivariate correlations between the observed variables and the components.

For example, the factor pattern matrix from the current analysis appears on page 4 of Output 1.1. Where the rows for observed variables intersect with the column for FACTOR1, you can see that the correlation between V1 and the first component is .58; the correlation between V2 and the first component is .48, and so forth.

Rotations. Ideally, you would like to review the correlations between the variables and the components and use this information to *interpret* the components; that is, to determine what construct seems to be measured by component 1, what construct seems to be measured by component 2, and so forth.   Unfortunately, when more than one component has been retained in an analysis, the interpretation of an unrotated factor pattern is usually quite difficult. To make interpretation easier, you will normally perform an operation called a rotation. A **rotation** is a linear transformation that is performed on the factor solution for the purpose of making the solution easier to interpret.

PROC FACTOR allows you to request several different types of rotations. The preceding program that analyzed data from the POI study included the statement

```
ROTATE=VARIMAX
```

which requests a **varimax rotation**. A varimax rotation is an orthogonal rotation, meaning that it results in uncorrelated components. Compared to some other types of rotations, a varimax rotation tends to maximize the variance of a column of the factor pattern matrix (as opposed to a row of the matrix). This rotation is probably the most commonly used orthogonal rotation in the social sciences. The results of the varimax rotation for the current analysis appear on page 5 of Output 1.1.

Step 4: Interpreting the Rotated Solution

Interpreting a rotated solution means determining just what is measured by each of the retained components. Briefly, this involves identifying the variables that demonstrate high loadings for a given component, and determining what these variables have in common. Usually, a brief name is assigned to each retained component that describes its content.

The first decision to be made at this stage is to decide how large a factor loading must be to be considered "large." Stevens (1986) discusses some of the issues relevant to this decision, and even provides guidelines for testing the statistical significance of factor loadings. Given that this is an introductory treatment of principal component analysis, however, simply consider a loading to be "large" if its absolute value exceeds .40.

The rotated factor pattern for the POI study appears on page 5 of Output 1.1. The following text provides a structured approach for interpreting this factor pattern.

A. Read across the row for the first variable. All "meaningful loadings" (i.e., loadings greater than .40) have been flagged with an asterisk ("*"). This was accomplished by including the FLAG=.40 option in the preceding program. If a given variable has a meaningful loading on more than one component, scratch that variable out and ignore it in your interpretation. In many situations, researchers want to drop variables that load on more than one component, because the variables are not pure measures of any one construct. In the present case, this means looking at the row headed "V1", and reading to the right to see if it loads on more than one component. In this case it does not, so you may retain this variable.

B. Repeat this process for the remaining variables, scratching out any variable that loads on more than one component. In this analysis, none of the variables have high loadings on more than one component, so none will have to be dropped.

C. Review all of the surviving variables with high loadings on component 1 to determine the nature of this component. From the rotated factor pattern, you can see that only items 4, 5, and 6 load on component 1 (note the asterisks). It is now necessary to turn to the questionnaire itself and review the content of the questions in order to decide what a given component should be named. What do questions 4, 5, and 6 have in common? What common construct do they seem to be measuring? For illustration, the questions being analyzed in the present case are reproduced here. Remember that question 4 was represented as V4 in the SAS program, question 5 was V5, and so forth. Read questions 4, 5, and 6 to see what they have in common.

```
1 2 3 4 5 6 7      1.  Went out of my way to do a favor for a
                       coworker.

1 2 3 4 5 6 7      2.  Went out of my way to do a favor for a
                       relative.

1 2 3 4 5 6 7      3.  Went out of my way to do a favor for a
                       friend.
```

```
1 2 3 4 5 6 7     4.  Gave money to a religious charity.

1 2 3 4 5 6 7     5.  Gave money to a charity not associated with a
                      religion.

1 2 3 4 5 6 7     6.  Gave money to a panhandler.
```

Questions 4, 5, and 6 all seem to deal with "giving money to the needy." It is therefore reasonable to label component 1 the "financial giving" component.

D. Repeat this process to name the remaining retained components. In the present case, there is only one remaining component to name: component 2. This component has high loadings for questions 1, 2, and 3. In reviewing these items, it becomes clear that each seems to deal with helping friends, relatives, or other acquaintances. It is therefore appropriate to name this the "acquaintance helping" component.

E. Determine whether this final solution satisfies the interpretability criteria. An earlier section indicated that the overall results of a principal component analysis are satisfactory only if they meet a number of interpretability criteria. In the following list, the adequacy of the rotated factor pattern presented on page 5 of Output 1.1 is assessed in terms of these criteria.

1. **Are there at least three variables (items) with significant loadings on each retained component?** In the present example, three variables loaded on component 1, and three also loaded on component 2, so this criterion was met.

2. **Do the variables that load on a given component share some conceptual meaning?** All three variables loading on component 1 are clearly measuring giving to the needy, while all three loading on component 2 are clearly measuring prosocial acts performed for acquaintances. Therefore, this criterion is met.

3. **Do the variables that load on different components seem to be measuring different constructs?** The items loading on component 1 clearly are measuring the respondents' financial contributions, while the items loading on component 2 are clearly measuring helpfulness toward acquaintances. Because these seem to be conceptually very different constructs, this criterion seems to be met as well.

4. **Does the rotated factor pattern demonstrate "simple structure?"** Earlier, it was said that a rotated factor pattern demonstrates simple structure when it has two characteristics. First, most of the variables should have high loadings on one component, and near-zero loadings on the other components. It can be seen that the pattern obtained here meets that requirement: items 1-3 have high loadings on

component 2, and near-zero loadings on component 1. Similarly, items 4-6 have high loadings on component 1, and near-zero loadings on component 2. The second characteristic of simple structure is that each component should have high loadings for some variables, and near-zero loadings for the others. Again, the pattern obtained here also meets this requirement: component 1 has high loadings for items 4-6 and near-zero loadings for the other items, while component 2 has high loadings for items 1-3, and near-zero loadings on the remaining items. In short, the rotated component pattern obtained in this analysis does seem to demonstrate simple structure.

Step 5: Creating Factor Scores or Factor-Based Scores

Once the analysis is complete, it is often desirable to assign scores to each subject to indicate where that subject stands on the retained components. For example, the two components retained in the present study were interpreted as a financial giving component and an acquaintance helping component. You may want to now assign one score to each subject to indicate that subject's standing on the financial giving component, and a different score to indicate that subject's standing on the acquaintance helping component. With this done, these component scores could be used either as predictor variables or as criterion variables in subsequent analyses.

Before discussing the options for assigning these scores, it is important to first draw a distinction between factor scores versus factor-based scores. In principal component analysis, a **factor score** (or **component score**) is a linear composite of the optimally-weighted observed variables. If requested, PROC FACTOR will compute each subject's factor scores for the two components by

- determining the optimal regression weights

- multiplying subject responses to the questionnaire items by these weights

- summing the products.

The resulting sum will be a given subject's score on the component of interest. Remember that a separate equation, with different weights, is developed for each retained component.

A **factor-based score**, on the other hand, is merely a linear composite of the variables that demonstrated meaningful loadings for the component in question. For example, in the preceding analysis, items 4, 5, and 6 demonstrated meaningful loadings for the financial giving component. Therefore, you could calculate the factor-based score on this component for a given subject by simply adding together his or her responses to items 4, 5, and 6. Notice that, with a factor-based score, the observed variables are not multiplied by optimal weights before they are summed.

Computing factor scores. Factor scores are requested by including the NFACT= and OUT= options in the PROC FACTOR statement. Here is the general form for a SAS program that uses the NFACT= and OUT= option to compute factor scores:

```
PROC FACTOR     DATA=data-set-name
                SIMPLE
                METHOD=PRIN
                PRIORS=ONE
                NFACT=number-of-components-to-retain
                ROTATE=VARIMAX
                ROUND
                FLAG=desired-size-of-"significant"-factor-loadings
                OUT=name-of-new-SAS-data-set    ;
   VAR  variables-to-be-analyzed ;
   RUN;
```

Here are the actual program statements (minus the DATA step) that could be used to perform a principal component analysis and compute factor scores for the POI study.

```
1       PROC FACTOR     DATA=D1
2                       SIMPLE
3                       METHOD=PRIN
4                       PRIORS=ONE
5                       NFACT=2
6                       ROTATE=VARIMAX
7                       ROUND
8                       FLAG=.40
9                       OUT=D2    ;
10      VAR V1 V2 V3 V4 V5 V6;
11      RUN;
```

Notice how this program differs from the original program presented earlier in the chapter (in the section titled "SAS Program and Output"): the MINEIGEN=1 option has been dropped, and has been replaced with the NFACT=2 option; and the OUT=D2 option has been added.

Line 9 of the preceding programs asks that an output data set be created and given the name D2. This name was arbitrary; any name consistent with SAS System requirements would have been acceptable. The new data set named D2 will contain all of the variables contained in the previous data set (D1), as well as new variables named FACTOR1 and FACTOR2. FACTOR1 will contain factor scores for the first retained component, and FACTOR2 will contain scores for the second component. The number of new "FACTOR" variables created will be equal to the number of components retained by the NFACT statement.

The OUT= option may be used to create component scores only if the analysis has been performed on a raw data set (as opposed to a correlation or covariance matrix). The use of the NFACT= option is also required.

Having created the new variables named FACTOR1 and FACTOR2, you may be interested in seeing how they relate to the study's original observed variables. This can be done by appending PROC CORR statements to the SAS program, following the last of the PROC FACTOR statements. The full program (minus the DATA step) is now reproduced:

```
1      PROC FACTOR    DATA=D1
2                     SIMPLE
3                     METHOD=PRIN
4                     PRIORS=ONE
5                     NFACT=2
6                     ROTATE=VARIMAX
7                     ROUND
8                     FLAG=.40
9                     OUT=D2    ;
10        VAR V1 V2 V3 V4 V5 V6;
11        RUN;
12
13     PROC CORR    DATA=D2;
14        VAR FACTOR1 FACTOR2;
15        WITH V1 V2 V3 V4 V5 V6 FACTOR1 FACTOR2;
16        RUN;
```

Notice that the PROC CORR statement on line 13 specifies DATA=D2. This data set (D2) is the name of the output data set created on line 9 in the PROC FACTOR statement. The PROC CORR statements request that the factor score variables (FACTOR 1 and FACTOR2) be correlated with the subjects' responses to questionnaire items 1-6 (V1-V6), as well as with themselves (FACTOR1 and FACTOR2).

With printer options of LINESIZE=80 and PAGESIZE=60, the preceding program would again produce four pages of output. Pages 1-2 provide simple statistics, the eigenvalue table, and the unrotated factor pattern, identical to those produced with the first program. Page 3 provides the rotated factor pattern and final communalities (same as before), along with the standardized scoring coefficients used in creating the factor scores. Finally, page 4 provides the correlations requested by the CORR procedure. Pages 3 and 4 of the output created by the preceding program are reproduced here as Output 1.2.

```
                        The SAS System                                3
Rotation Method: Varimax

                 Orthogonal Transformation Matrix

                              1          2

                   1       0.76914    0.63908
                   2      -0.63908    0.76914

                    Rotated Factor Pattern

                          FACTOR1     FACTOR2

            V1              0          91 *
            V2              3          71 *
            V3              7          86 *
            V4             90 *        -9
            V5             81 *         9
            V6             82 *         8
```

NOTE: Printed values are multiplied by 100 and rounded to the nearest integer.
      Values greater than 0.4 have been flagged by an '*'.

```
                 Variance explained by each factor

                       FACTOR1    FACTOR2
                       2.147248   2.093803


          Final Communality Estimates: Total = 4.241050

       V1         V2         V3         V4         V5         V6
    0.823418   0.508529   0.743990   0.822574   0.665963   0.676575


            Scoring Coefficients Estimated by Regression
       Squared Multiple Correlations of the Variables with each Factor

                       FACTOR1    FACTOR2
                       1.000000   1.000000

                 Standardized Scoring Coefficients

                          FACTOR1     FACTOR2

            V1           -0.03109     0.43551
            V2           -0.00726     0.34071
            V3            0.00388     0.41044
            V4            0.42515    -0.07087
            V5            0.37618     0.01947
            V6            0.38020     0.01361
```

```
                                                          4
                    Correlation Analysis

 8 'WITH' Variables:  V1       V2       V3      V4       V5       V6
                      FACTOR1  FACTOR2
 2 'VAR'  Variables:  FACTOR1  FACTOR2


                    Simple Statistics

Variable       N       Mean      Std Dev         Sum      Minimum      Maximum

V1            50     5.18000     1.39518     259.00000     1.00000      7.00000
V2            50     5.40000     1.10657     270.00000     3.00000      7.00000
V3            50     5.52000     1.21622     276.00000     2.00000      7.00000
V4            50     3.64000     1.79296     182.00000     1.00000      7.00000
V5            50     4.22000     1.66953     211.00000     1.00000      7.00000
V6            50     3.10000     1.55511     155.00000     1.00000      7.00000
FACTOR1       50           0     1.00000           0     -1.87908      2.35913
FACTOR2       50           0     1.00000           0     -2.95892      1.58951


  Pearson Correlation Coefficients / Prob > |R| under Ho: Rho=0 / N = 50

                              FACTOR1            FACTOR2

             V1             -0.00429            0.90741
                             0.9764             0.0001

             V2              0.03328            0.71234
                             0.8185             0.0001

             V3              0.06720            0.85993
                             0.6429             0.0001

             V4              0.90274           -0.08740

                             0.0001             0.5462

             V5              0.81055            0.09474
                             0.0001             0.5128

             V6              0.81834            0.08303
                             0.0001             0.5665

             FACTOR1         1.00000            0.00000
                             0.0                1.0000

             FACTOR2         0.00000            1.00000
                             1.0000             0.0
```

because orthogonal?

Output 1.2: Output Pages 3 and 4 from the Analysis of POI Data in Which Factor Scores Were Created

The simple statistics for the CORR procedure appear at the top of page 4 in Output 1.2. Notice that the simple statistics for the observed variables (V1-V6) are identical to those that appeared at the beginning of the FACTOR output discussed earlier (at the top of Output 1.1, page 1). In contrast, note the simple statistics for FACTOR1 and FACTOR2 (the factor score variables for components 1 and 2, respectively): both have means of 0 and standard deviations of 1. Obviously, these variables were constructed in such a way as to be standardized variables.

The correlations between FACTOR1 and FACTOR2 and the original observed variables appear on the bottom half of page 4. You can see that the correlations between FACTOR1 and V1–V6 on page 4 of Output 1.2 are identical to the factor loadings of V1–V6 on FACTOR1 on page 5 of Output 1.1, under "Rotated Factor Pattern". This makes sense, as the elements of a factor pattern (in an orthogonal solution) are simply correlations between the observed variables and the components themselves. Similarly, you can see that the correlations between FACTOR2 and V1–V6 from page 4 of Output 1.2 are also identical to the corresponding factor loadings from page 5 of Output 1.1.

Of special interest is the correlation between FACTOR1 and FACTOR2, as computed by PROC CORR. This appears on page 4 of Output 1.2, where the row for FACTOR2 intersects with the column for FACTOR1. Notice the observed correlation between these two components is zero. This is as expected: the rotation method used in the principal component analysis was the varimax method, which produces orthogonal, or uncorrelated, components.

Computing factor-based scores. A second (and less sophisticated) approach to scoring involves the creation of new variables that contain factor-based scores rather than true factor scores. A variable that contains factor-based scores is sometimes referred to as a **factor-based scale**.

Although factor-based scores can be created in a number of ways, the following method has the advantage of being relatively straightforward and is commonly used:

1. To calculate factor-based scores for component 1, first determine which questionnaire items had high loadings on that component.

2. For a given subject, add together that subject's responses to these items. The result is that subject's score on the factor-based scale for component 1.

3. Repeat these steps to calculate each subject's score on the remaining retained components.

Although this may sound like a cumbersome task, it is actually made quite simple through the use of data manipulation statements contained in a SAS program. For example, assume that you have performed the principal component analysis on your survey responses, and have obtained the findings reported in this chapter. Specifically, you found that survey items 4, 5, and 6 loaded on component 1 (the financial giving component), while items 1, 2, and 3 loaded on component 2 (the acquaintance helping component).

You would now like to create two new SAS variables. The first variable, called FINANCE, will include each subject's factor-based score for financial giving. The second variable, called ACQUAINT, will include each subject's factor-based score for acquaintance helping. Once these variables are created, you can use them as criterion variables or predictor variables in subsequent analyses. To keep things simple in the present example, assume that you are simply interested in determining whether there is a significant correlation between FINANCE and ACQUAINT.

At this time, it may be useful to review Appendix A.3, "Working with Variables and Observations in SAS Data Sets," particularly the section on creating new variables from existing variables. Such a review should make it easier to understand the data manipulation statements used here.

Assume that earlier statements in the SAS program have already input subject responses to the six questionnaire items. These variables are included in a data set called D1. The following are the subsequent lines that would go on to create a new data set called D2. The new data set will include all of the variables in D1, as well as the newly created factor-based scales called FINANCE and ACQUAINT.

```
14
15      DATA D2;
16         SET D1;
17
18      FINANCE    = (V4 + V5 + V6);
19      ACQUAINT   = (V1 + V2 + V3);
20
21      PROC CORR    DATA=D2;
22         VAR FINANCE ACQUAINT;
23         RUN;
```

Lines 15 and 16 request that a new data set called D2 be created, and that it be set up as a duplicate of existing data set D1. In line 18, the new variable called FINANCE is created. For each subject, his or her responses to items 4, 5, and 6 are added together. The result is the subjects' score on the factor-based scale for the first component. These scores are stored in a variable called FINANCE. The component-based scale for the acquaintance helping component is created on line 19, and these scores are stored in the variable called ACQUAINT. Line 21–23 request the correlations between FINANCE and ACQUAINT be determined. FINANCE and ACQUAINT may now be used as predictor or criterion variables in subsequent analyses.

To save space, the results of this program will not be reproduced here. However, note that this output would probably display a significant correlation between FINANCE and ACQUAINT. This may come as a surprise, because earlier it was shown that the factor scores contained in FACTOR1 and FACTOR2 (counterparts to FINANCE and ACQUAINT) were completely uncorrelated.

[handwritten margin note: It doesn't show a significant correlation?]

The reason for this apparent contradiction is simple: FACTOR1 and FACTOR2 are true principal components, and true principal components (created in an orthogonal solution) are always created with optimally weighted equations so that they will be mutually uncorrelated.

In contrast, FINANCE and ACQUAINT are not true principal components that consist of true factor scores; they are merely artificial varibles that were *based* on the results of a principal component analysis. Optimal weights (that would ensure orthogonality) were not used in the creation of FINANCE and ACQUAINT. This is why factor-based scales will often demonstrate nonzero correlations with one another, while true principal components (from an orthogonal solution) will not.

Recoding reversed items prior to analysis. It is generally best to recode any reversed items before conducting any of the analyses described here. In particular, it is essential that reversed items be recoded prior to the program statements that produce factor-based scales. For example, the three questionnaire items that assess financial giving appear again here:

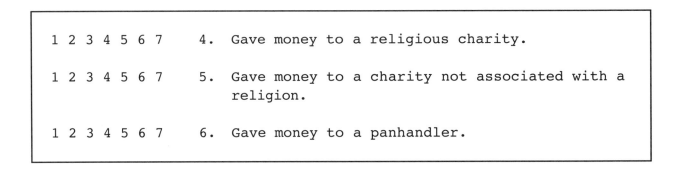

```
1 2 3 4 5 6 7     4.  Gave money to a religious charity.

1 2 3 4 5 6 7     5.  Gave money to a charity not associated with a
                      religion.

1 2 3 4 5 6 7     6.  Gave money to a panhandler.
```

None of these items are reversed; with each item, a response of "7" indicates a high level of financial giving. In the following, however, item 4 is a reversed item: with item 4, a response of "7" indicates a *low* level of giving:

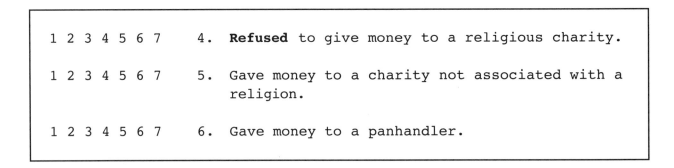

```
1 2 3 4 5 6 7     4.  Refused to give money to a religious charity.

1 2 3 4 5 6 7     5.  Gave money to a charity not associated with a
                      religion.

1 2 3 4 5 6 7     6.  Gave money to a panhandler.
```

If you were to perform a principal component analysis on responses to these items, the factor loading for item 4 would most likely have a sign that is the opposite of the sign of the loadings for items 5 and 6 (e.g., if items 5 and 6 had positive loadings, item 4 would have a negative loading). This would complicate the creation of a component-based scale: with items 5 and 6, higher scores indicate greater giving; with item 4, lower scores indicate greater giving. Clearly, you would not want to sum these three items as they are presently coded. First, it will be necessary to reverse item 4. Notice how this is done in the following program (assume that the data have already been input in a SAS data set named D1):

```
15      DATA D2;
16          SET D1;
17
18      V4 = 8 - V4;
19
20      FINANCE    = (V4 + V5 + V6);
21      ACQUAINT   = (V1 + V2 + V3);
22
23      PROC CORR    DATA=D2;
24          VAR FINANCE ACQUAINT;
25          RUN;
```

Line 18 of the preceding program created a new, recoded version of variable V4. Values on this new version of V4 will be equal to the quantity 8 minus the value of the old version of V4. Therefore, for subjects whose score on the old version of V4 was 1, their value on the new version of V4 will be 7 (because $8 - 1 = 7$); for subjects whose score on the old version of V4 was 7, their value on the new version of V4 will be 1 (because $8 - 7 = 1$); and so forth.

The general form of the formula used when recoding reversed items is

```
Variable-name  =  constant  -  variable-name ;
```

In this formula, the "constant" is the following quantity:

The number of points on the response scale used with the questionnaire item plus 1

Therefore if you are using the 4-point response format, the constant is 5; if using a 9-point scale, the constant is 10.

If you have prior knowledge about which items are going to appear as reversed items (with reversed component loadings) in your results, it is best to place these recoding statements early in your SAS program, before the PROC FACTOR statements. This will make interpretation of the components a bit more straightforward because it will eliminate significant loadings with opposite signs from appearing on the same component. In any case, it is essential that the statements that recode reversed items appear before the statements that create any factor-based scales.

Step 6: Summarizing the Results in a Table

For published articles that summarize the results of your analysis, it is generally desirable to prepare a table that presents the rotated factor pattern. When the variables being analyzed contain responses to questionnaire items, it can be helpful to actually reproduce the questionnaire items themselves within this table. This is done in Table 1.2:

Table 1.2

Rotated Factor Pattern and Final Communality Estimates from
Principal Component Analysis of Prosocial Orientation Inventory

Component				
1	2	h²		Items
.00	.91	.82	1.	Went out of my way to do a favor for a coworker.
.03	.71	.51	2.	Went out of my way to do a favor for a relative.
.07	.86	.74	3.	Went out of my way to do a favor for a friend.
.90	-.09	.82	4.	Gave money to a religious charity.
.81	.09	.67	5.	Gave money to a charity not associated with a religion.
.82	.08	.68	6.	Gave money to a panhandler.

Note: N = 50. Communality estimates appear in column headed h².

The final communality estimates from the analysis are presented under the heading "h²" in the table. These estimates appear in the SAS output following the "Rotated Factor Pattern" and "Variance explained by each factor" (page 3 of Output 1.2).

Very often, the items that constitute the questionnaire are so lengthy, or the number of retained components is so large, that it is not possible to present both the factor pattern, the communalities, and the items themselves in the same table. In such situations, it may be preferable to present the factor pattern and communalities in one table, and the items in a second (or in the text of the paper). Shared item numbers may then be used to associate each item with its corresponding factor loadings and communality.

Step 7: Preparing a Formal Description of the Results for a Paper

The preceding analysis could be summarized in the following way for a published paper:

> Responses to the 6-item questionnaire were subjected to a principal component analysis using ones as prior communality estimates. The principal axis method was used to extract the components, and this was followed by a varimax (orthogonal) rotation.
>
> Only the first two components displayed eigenvalues greater than 1, and the results of a scree test also suggested that only the first two components were meaningful. Therefore, only the first two components were retained for rotation. Combined, components 1 and 2 accounted for 71% of the total variance.
>
> Questionnaire items and corresponding factor loadings are presented in Table 1.2. In interpreting the rotated factor pattern, an item was said to load on a given component if the factor loading was .40 or greater for that component, and was less than .40 for the other. Using these criteria, three items were found to load on the first component, which was subsequently labelled the financial giving component. Three items also loaded on the second component, which was labelled the acquaintance helping component.

purpose ?

An Example with Three Retained Components

The Questionnaire

The next example involves a piece of fictitious research that investigates the investment model (Rusbult, 1980). The investment model identifies variables that are believed to affect a person's commitment to a romantic relationship. In this context, **commitment** refers to the person's intention to maintain the relationship and stay with the current romantic partner.

One version of the investment model predicts that commitment will be affected by three antecedent variables: satisfaction, investment size, and alternative value. **Satisfaction** refers to the subject's affective response to the relationship; among other things, subjects report high levels of satisfaction when their current relationship comes close to their ideal relationship. **Investment size** refers to the amount of time, energy, and personal resources that an individual has put into the relationship. For example, subjects report high investments when they have spent a lot of time with their current partner and have developed a lot of mutual friends that may be lost if the relationship were to end. Finally, **alternative value** refers to the attractiveness of one's alternatives to the current partner; a subject would score high on alternative value if, for example, it would be attractive to date someone else or perhaps to not be dating at all.

Assume that you wish to conduct research on the investment model, and are in the process of preparing a 12-item questionnaire that will assess levels of satisfaction, investment size, and alternative value in a group of subjects involved in romantic associations. Part of the instrument used to assess these constructs is reproduced here:

```
Indicate the extent to which you agree or disagree with each of the
following statements by writing the appropriate response number in the
space to the left of the statement.  Please use the following response
format in making these ratings:

     7 = Strongly Agree
     6 = Agree
     5 = Slightly Agree
     4 = Neither Agree Nor Disagree
     3 = Slightly Disagree
     2 = Disagree
     1 = Strongly Disagree


_____   1.   I am satisfied with my current relationship.


_____   2.   My current relationship comes close to my ideal relationship.


_____   3.   I am more satisfied with my relationship than is the average
               person.


_____   4.   I feel good about my current relationship.


_____   5.   I have invested a great deal of time in my current
               relationship.


_____   6.   I have invested a great deal of energy in my current
               relationship.


_____   7.   I have invested a lot of my personal resources (e.g., money)
               in developing my current relationship.


_____   8.   My partner and I have developed a lot of mutual friends which
               I might lose if we were to break up.


_____   9.   There are plenty of other attractive people around for me to
               date if I were to break up with my current partner.
```

```
_____  10.  It would be attractive for me to break up with my current
            partner and date someone else.

_____  11.  It would be attractive for me to break up with my partner and
            just be alone for a while.

_____  12.  It would be attractive for me to break up with my partner and
            "play the field" for a while.
```

In the preceding questionnaire, items 1–4 were designed to assess satisfaction, items 5-8 were designed to assess investment size, and items 9–12 were designed to assess alternative value. Assume that you administer this questionnaire to 300 subjects, and now want to perform a principal component analysis on their responses.

Writing the Program

Earlier, it was mentioned that it is possible to perform a principal component analysis on a correlation matrix as well as on raw data; this section shows how this is done. The following program inputs the correlation matrix that provides all possible correlations between responses to the 12 items on the questionnaire, and performs a principal component analysis on these responses (these correlations are based on fictitious data):

```
1    DATA D1(TYPE=CORR) ;
2        INPUT    _TYPE_    $
3                 _NAME_    $
4                 V1-V12    ;
5    CARDS;
6    N    .   300  300  300  300  300  300  300  300  300  300  300  300
7    STD  .   2.48 2.39 2.58 3.12 2.80 3.14 2.92 2.50 2.10 2.14 1.83 2.26
8    CORR V1  1.00   .    .    .    .    .    .    .    .    .    .    .
9    CORR V2   .69 1.00   .    .    .    .    .    .    .    .    .    .
10   CORR V3   .60  .79 1.00   .    .    .    .    .    .    .    .    .
11   CORR V4   .62  .47  .48 1.00   .    .    .    .    .    .    .    .
12   CORR V5   .03  .04  .16  .09 1.00   .    .    .    .    .    .    .
13   CORR V6   .05 -.04  .08  .05  .91 1.00   .    .    .    .    .    .
14   CORR V7   .14  .05  .06  .12  .82  .89 1.00   .    .    .    .    .
15   CORR V8   .23  .13  .16  .21  .70  .72  .82 1.00   .    .    .    .
16   CORR V9  -.17 -.07 -.04 -.05 -.33 -.26 -.38 -.45 1.00   .    .    .
17   CORR V10 -.10 -.08  .07  .15 -.16 -.20 -.27 -.34  .45 1.00   .    .
18   CORR V11 -.24 -.19 -.26 -.28 -.43 -.37 -.53 -.57  .60  .22 1.00   .
19   CORR V12 -.11 -.07  .07  .08 -.10 -.13 -.23 -.31  .44  .60  .26 1.00
20   ;
21   PROC FACTOR   DATA=D1
22                 METHOD=PRIN
23                 PRIORS=ONE
```

```
24                    MINEIGEN=1
25                    SCREE
26                    ROTATE=VARIMAX
27                    ROUND
28                    FLAG=.40     ;
29        VAR   V1-V12;
30        RUN;
```

The PROC FACTOR statement in the preceding program follows the general form recommended for the initial analysis of a data set. Notice that the MINEIGEN=1 statement requests that all components with eigenvalues greater than one be retained, and the SCREE option requests a scree plot of the eigenvalues. These options are particularly helpful for the initial analysis of the data, as they can help determine the correct number of components to retain. If the scree test (or the other criteria) suggests retaining some number of components other than what would be retained using the MINEIGEN=1 option, that option may be dropped and replaced with the NFACT= option.

Results of the Initial Analysis

The preceding program produced four pages of output, with the following information appearing on each page:

- Page 1 includes the eigenvalue table.

- Page 2 includes the scree plot of eigenvalues.

- Page 3 includes the unrotated factor pattern and final communality estimates.

- Page 4 includes the rotated factor pattern.

The eigenvalue table from this analysis appears on page 1 of Output 1.3. The eigenvalues themselves appear in the row to the right of the "Eigenvalue" heading. From the values appearing in this row, you can see that components 1, 2, and 3 demonstrated eigenvalues of 4.47, 2.73, and 1.70, respectively. Further, you can see that only these first three components demonstrated eigenvalues greater than one. This means that three components will be retained by the MINEIGEN criterion. Notice that the first nonretained component (component 4) displays an eigenvalue of approximately 0.85 which, of course, is well below 1.00. This is encouraging, as you can have more confidence in the eigenvalue-one criterion when the solution does not contain components with "near-miss" eigenvalues of , say, .98 or .99.

The SAS System   1

Initial Factor Method: Principal Components

                Prior Communality Estimates: ONE

        Eigenvalues of the Correlation Matrix: Total = 12   Average = 1

	1	2	3	4
Eigenvalue	4.4706	2.7306	1.7017	0.8463
Difference	1.7400	1.0289	0.8555	0.2256
Proportion	0.3725	0.2276	0.1418	0.0705
Cumulative	0.3725	0.6001	0.7419	0.8124

	5	6	7	8
Eigenvalue	0.6206	0.4110	0.3450	0.3029
Difference	0.2096	0.0660	0.0421	0.0701
Proportion	0.0517	0.0343	0.0288	0.0252
Cumulative	0.8642	0.8984	0.9272	0.9524

	9	10	11	12
Eigenvalue	0.2328	0.1869	0.1062	0.0453
Difference	0.0460	0.0806	0.0609	
Proportion	0.0194	0.0156	0.0089	0.0038
Cumulative	0.9718	0.9874	0.9962	1.0000

        3 factors will be retained by the MINEIGEN criterion.

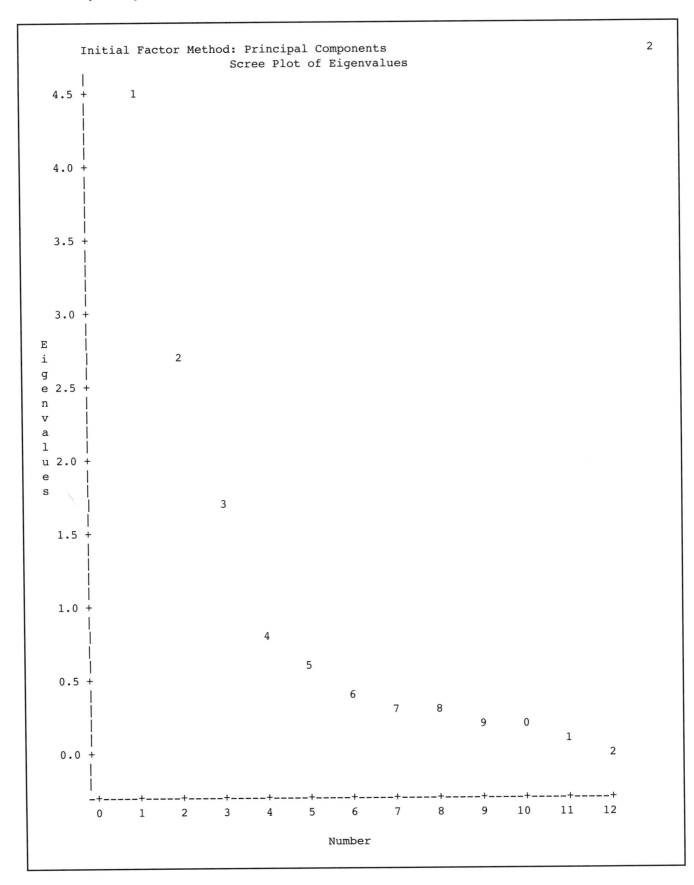

Initial Factor Method: Principal Components
Scree Plot of Eigenvalues

```
                        The SAS System                                    3

Initial Factor Method: Principal Components

                        Factor Pattern

                    FACTOR1    FACTOR2    FACTOR3

          V1           39        76  *     -14
          V2           31        82  *     -12
          V3           34        79  *       9
          V4           31        69  *      15
          V5           80  *    -26         41  *
          V6           79  *    -32         41  *
          V7           87  *    -27         26
          V8           88  *    -14          9
          V9          -61  *     14         47  *
          V10         -43  *     23         68  *
          V11         -72  *     -6         12
          V12         -40        19         72  *

NOTE: Printed values are multiplied by 100 and rounded to the nearest integer.
      Values greater than 0.4 have been flagged by an '*'.

                   Variance explained by each factor

                    FACTOR1     FACTOR2    FACTOR3
                    4.470581    2.730623   1.701734


             Final Communality Estimates: Total = 8.902938

          V1          V2         V3          V4          V5          V6
      0.755221    0.782123   0.747982    0.598878    0.871668    0.899804


          V7          V8         V9         V10         V11         V12
      0.899918    0.796680   0.611250    0.694877    0.532084    0.712453
```

```
                        The SAS System                                    4

Rotation Method: Varimax

              Orthogonal Transformation Matrix

                        1          2          3

              1      0.83139    0.34426   -0.43620
              2     -0.29475    0.93866    0.17902
              3      0.47107   -0.02026    0.88186
```

```
                     Rotated Factor Pattern

                   FACTOR1     FACTOR2     FACTOR3

        V1              3        85 *       -16
        V2             -4        88 *       -10
        V3              9        86 *         8
        V4             13        75 *        12
        V5             93 *        2         -3
        V6             95 *       -4         -4
        V7             93 *        4        -19
        V8             81 *       17        -33
        V9            -32        -9         71 *
        V10           -11         6         82 *
        V11           -52 *      -30         41 *
        V12            -5         3         84 *
```

NOTE: Printed values are multiplied by 100 and rounded to the nearest integer.
      Values greater than 0.4 have been flagged by an '*'.

```
                 Variance explained by each factor

                   FACTOR1     FACTOR2     FACTOR3
                   3.704983    2.936412    2.261543
```

```
          Final Communality Estimates: Total = 8.902938

            V1         V2         V3         V4         V5         V6
        0.755221   0.782123   0.747982   0.598878   0.871668   0.899804

            V7         V8         V9        V10        V11        V12
        0.899918   0.796680   0.611250   0.694877   0.532084   0.712453
```

Output 1.3: Results of the Initial Principal Component Analysis of the Investment Model Data

The eigenvalue table in Output 1.3 also shows that the first three components combined account for approximately 74% of the total variance (this variance value can be observed at the intersection of the row headed "Cumulative" and column headed "3"). According to the "percentage of variance accounted for" criterion, this once again suggests that it may be appropriate to retain three components.

The scree plot from this solution appears on page 2 of Output 1.3. This scree plot shows that there are several large breaks in the data following components 1, 2, and 3, and then the line begins to flatten out beginning with component 4. The last large break appears after component 3, suggesting that only components 1–3 account for meaningful variance. This indicates that only these first three components should be retained and interpreted. Notice how it is almost possible to draw a straight line through components 4–12? The components that lie along a semi-straight line such as this are typically assumed to be measuring only trivial variance (components 4–12 constitute the "scree" of your scree plot!).

So far, the results from the eigenvalue-one criterion, the variance accounted for criterion, and the scree plot have converged in suggesting that a three-component solution may be appropriate. It is now time to review the rotated factor pattern to see if such a solution is interpretable. This matrix is presented on page 4 of Output 1.3.

Following the guidelines provided earlier, you begin your interpretation by looking for factorially complex items; that is, items with meaningful loadings for more than one component. A review shows that item 11 (variable V11) is a complex item, loading on both components 1 and 3. Item 11 is therefore scratched out. Except for this item, the solution is otherwise fairly clean.

To interpret component 1, you read down the column for FACTOR1 and see that items 5–8 display significant loadings for this component (remember that item 11 has been scatched out). These items are:

_____ 5. I have invested a great deal of time in my current relationship.

_____ 6. I have invested a great deal of energy in my current relationship.

_____ 7. I have invested a lot of my personal resources (e.g., money) in developing my current relationship.

_____ 8. My partner and I have developed a lot of mutual friends which I might lose if we were to break up.

All of these items deal with the investments that subjects have made in their relationships, so it makes sense to label this the "investment size" component.

The rotated factor pattern shows that items 1–4 displayed meaningful loadings for component 2. These items are:

_____ 1. I am satisfied with my current relationship.

_____ 2. My current relationship comes close to my ideal relationship.

_____ 3. I am more satisfied with my relationship than is the average person.

_____ 4. I feel good about my current relationship.

Given the content of the preceding items, it seems reasonable to label component 2 the "satisfaction" component.

Finally, component 3 displayed large loadings for items 9, 10, and 12 (again, remember that item 11 has been scratched out). These items are:

```
_____    9.   There are plenty of other attractive people around for me to date
                if I were to break up with my current partner.

_____   10.   It would be attractive for me to break up with my current partner
                and date someone else.

_____   12.   It would be attractive for me to break up with my partner and
                "play the field" for a while.
```

These items all seem to deal with the attractiveness of one's alternatives to the current relationship, so it makes sense to label this the "alternative value" component.

You may now step back and determine whether this solution satisfies the interpretability criteria presented earlier:

1. Are there at least three variables with meaningful loadings on each retained component?

2. Do the variables that load on a given component share the same conceptual meaning?

3. Do the variables that load on different components seem to be measuring different constructs?

4. Does the rotated factor pattern demonstrate "simple structure"?

In general, the answer to each of the preceding questions is "yes," indicating that the current solution is in most respects satisfactory. There was, however, a problem with item 11, which loaded on both components 1 and 3. This problem prevented the current solution from demonstrating a perfectly "simple structure" (criterion 4 from above). To eliminate this problem, it may be desirable to repeat the analysis, this time analyzing all of the items *except* for item 11. This will be done in the second analysis of the investment model data, to be described below.

Results of the Second Analysis

To repeat the current analysis with item 11 deleted, it is necessary only to modify the VAR statement of the preceding program. This may be done by changing the VAR statement so that it appears as follows:

```
VAR V1-V10 V12;
```

All other aspects of the program would remain as they were previously. The eigenvalue table, scree plot, the unrotated factor pattern, the rotated factor pattern, and final communality estimates obtained from this revised program appear in Output 1.4:

```
                              The SAS System                                  1

Initial Factor Method: Principal Components

                    Prior Communality Estimates: ONE

          Eigenvalues of the Correlation Matrix:   Total = 11   Average = 1

                              1           2           3           4
          Eigenvalue       4.0241      2.7270      1.6898      0.6838
          Difference       1.2970      1.0372      1.0060      0.1274
          Proportion       0.3658      0.2479      0.1536      0.0622
          Cumulative       0.3658      0.6137      0.7674      0.8295

                              5           6           7           8
          Eigenvalue       0.5564      0.3963      0.3074      0.2668
          Difference       0.1601      0.0889      0.0406      0.0798
          Proportion       0.0506      0.0360      0.0279      0.0243
          Cumulative       0.8801      0.9161      0.9441      0.9683

                              9          10          11
          Eigenvalue       0.1869      0.1131      0.0486
          Difference       0.0739      0.0645
          Proportion       0.0170      0.0103      0.0044
          Cumulative       0.9853      0.9956      1.0000

             3 factors will be retained by the MINEIGEN criterion.
```

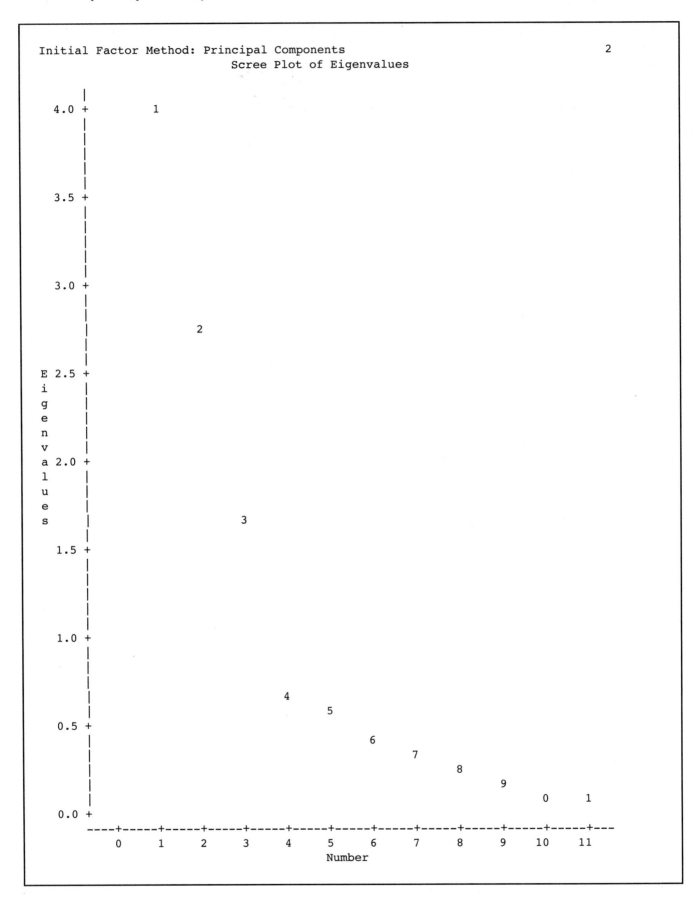

```
                         The SAS System                              3

Initial Factor Method: Principal Components

                         Factor Pattern

                    FACTOR1     FACTOR2     FACTOR3

         V1            38         77 *       -17
         V2            30         83 *       -15
         V3            32         80 *         8
         V4            29         70 *        15
         V5            83 *      -23          38
         V6            83 *      -30          38
         V7            89 *      -24          24
         V8            88 *      -12           7
         V9           -56 *       13          47 *
         V10          -44 *       22          70 *
         V12          -40         18          74 *

NOTE: Printed values are multiplied by 100 and rounded to the nearest integer.
      Values greater than 0.4 have been flagged by an '*'.

                   Variance explained by each factor

                 FACTOR1     FACTOR2     FACTOR3
                 4.024086    2.727039    1.689791


          Final Communality Estimates: Total = 8.440916

         V1          V2          V3          V4          V5          V6
      0.772386    0.798289    0.748233    0.591921    0.882544    0.921349


         V7          V8          V9          V10         V12
      0.904096    0.796623    0.553800    0.736193    0.735482
```

```
                         The SAS System                              4

Rotation Method: Varimax

                  Orthogonal Transformation Matrix

                         1          2          3

               1      0.84709    0.32928   -0.41715
               2     -0.27787    0.94351    0.18051
               3      0.45303   -0.03699    0.89073
```

```
                    Rotated Factor Pattern

                  FACTOR1    FACTOR2    FACTOR3

        V1           3         86 *      -17
        V2          -4         89 *      -11
        V3           8         86 *        8
        V4          12         75 *       14
        V5          94 *        4         -4
        V6          96 *       -2         -6
        V7          93 *        5        -20
        V8          81 *       18        -33
        V9         -30         -8         68 *
        V10        -12          4         85 *
        V12         -5          1         86 *
```

NOTE: Printed values are multiplied by 100 and rounded to the nearest integer.
      Values greater than 0.4 have been flagged by an '*'.

```
                Variance explained by each factor

                  FACTOR1    FACTOR2    FACTOR3
                  3.444866   2.866255   2.129795


         Final Communality Estimates: Total = 8.440916

        V1         V2         V3         V4         V5         V6
     0.772386   0.798289   0.748233   0.591921   0.882544   0.921349

        V7         V8         V9         V10        V12
     0.904096   0.796623   0.553800   0.736193   0.735482
```

Output 1.4: Results of the Second Analysis of the Investment Model Data

The results obtained when item 11 was dropped from the analysis are very similar to those obtained when it was included. The eigenvalue table of Output 1.4 shows that the eigenvalue-one criterion would again result in retaining three components. The first three components account for 77% of the total variance, which means that three components would also be retained if you used the variance-accounted-for criterion. Finally, the scree plot from page 2 of Output 1.4 is just a bit cleaner than had been observed with the initial analysis: The break between components 3 and 4 is now slightly more distinct, and the eigenvalues again level off after this break. This means that three components would also likely be retained if the scree test were used to solve the number-of-components problem.

The biggest change can be seen in the rotated factor pattern, which appears on page 4 of Output 1.4. The solution is now cleaner, in the sense that no item now loads on more than one component. In this sense, the current results demonstrate a somewhat simpler structure than had been demonstrated by the initial analysis of the investment model data.

Conclusion

Principal component analysis is a powerful tool for reducing a number of observed variables into a smaller number of artificial variables that account for most of the variance in the data set. It is particularly useful when you need a data reduction procedure that makes no assumptions concerning an underlying causal structure that is responsible for covariation in the data. When it is possible to postulate the existence of such an underlying causal structure, it may be more appropriate to analyze the data using exploratory factor analysis.

Both principal component analysis and factor analysis are often used to construct multiple-item scales from the items that constitute questionnaires. Regardless of which method is used, once these scales have been developed it is often desirable to assess their reliability by computing **coefficient alpha**: an index of internal consistency reliability. Chapter 3 shows how this can be done using the SAS System's PROC CORR.

Appendix: Assumptions Underlying Principal Component Analysis

Because a principal component analysis is performed on a matrix of Pearson correlation coefficients, the data should satisfy the assumptions for this statistic. These assumptions are described in detail in Chapter 6, "Measures of Bivariate Association," of Hatcher & Stepanski (1994) and are briefly reviewed here:

- **Interval-level measurement**. All analyzed variables should be assessed on an interval or ratio level of measurement.

- **Random sampling**. Each subject will contribute one score on each observed variable. These sets of scores should represent a random sample drawn from the population of interest.

- **Linearity**. The relationship between all observed variables should be linear.

- **Normal distributions**. Each observed variable should be normally distributed. Variables that demonstrate marked skewness or kurtosis may be transformed to better approximate normality (see Rummel, 1970).

- **Bivariate normal distribution**. Each pair of observed variables should display a bivariate normal distribution; e.g., they should form an elliptical scattergram when plotted. However, the Pearson correlation coefficient is robust against violations of this assumption when the sample size is greater than 25.

References

Cattell, R. B. (1966). The scree test for the number of factors. *Multivariate Behavioral Research, 1,* 245-276.

Hatcher, L. & Stepanski, E. (1994). *A step-by-step approach to using the SAS system for univariate and multivariate statistics*. Cary, NC: SAS Institute Inc.

Kaiser, H. F. (1960). The application of electronic computers to factor analysis. *Educational and Psychological Measurement, 20,* 141-151.

Kim, J. O. & Mueller, C. W. (1978a). *Introduction to factor analysis: What it is and how to do it.* Beverly Hills, CA: Sage.

Kim, J. O. & Mueller, C. W. (1978b). *Factor analysis: Statistical methods and practical issues.* Beverly Hills, CA: Sage.

Rummel, R. J. (1970). *Applied factor analysis.* Evanston, IL: Northwestern University Press.

Rusbult, C.E. (1980). Commitment and satisfaction in romantic associations: A test of the investment model. *Journal of Experimental Social Psychology, 16,* 172-186.

SAS Institute Inc. (1989). *SAS/STAT users guide, version 6, fourth edition, volume 1.* Cary, NC: SAS Institute Inc.

Spector, P.E. (1992). *Summated rating scale construction: An introduction.* Newbury Park, CA: Sage.

Stevens, J. (1986). *Applied multivariate statistics for the social sciences.* Hillsdale, NJ: Lawrence Erlbaum Associates.

Chapter 2

EXPLORATORY FACTOR ANALYSIS

Overview. This chapter shows how to use exploratory factor analysis to investigate the factor structure underlying a set of observed variables. It discusses important concepts related to the common factor model, and explains the differences between factor analysis and principal component analysis. It provides guidelines regarding the necessary sample size as well as the number of variables that should load on each factor. The chapter shows how to determine the number of factors to retain, interpret the rotated solution, create factor scores, and summarize the results. The examples provided here show how to perform and interpret a promax rotation that results in oblique (correlated) factors.

Introduction: When Is Exploratory Factor Analysis Appropriate?

Exploratory factor analysis may be appropriate when you have obtained measures on a number of variables, and want to identify the number and nature of the underlying factors that are responsible for covariation in the data. In other words, exploratory factor analysis is appropriate when you want to identify the factor structure underlying a set of data.

For example, imagine that you are a political scientist who has developed a 50-item questionnaire to assess political attitudes. You administer the questionnaire to 500 people, and perform a factor analysis on their responses. The results of the analysis suggest that, although the questionnaire contained 50 items, it was really just measuring two underlying factors, or constructs. You decided to label the first construct the **social conservatism** factor. Individuals who scored high on this construct tended to agree with statements such as "People should be married before living together" and "Children should respect their elders." You chose to label the second construct the **economic conservatism** factor. Individuals who scored high on this factor tended to agree with statments such as "The size of the federal government should be reduced" and "Our taxes should be lowered."

In short, by performing a factor analysis on responses to this questionnaire, you were able to determine the number of constructs measured by this questionnaire (two) as well as the nature of those constructs. The results of the analysis showed you which questionnaire items were measuring the social conservatism factor, and which were measuring the economic conservatism factor.

The use of factor analysis assumes that each of the observed variables being analyzed are measured on an interval or ratio scale. Some additional assumptions underlying the use of factor analysis are listed in an appendix at the end of this chapter.

Note: You will see a lot of similarity between the issues discussed in this chapter and those discussed in the preceding chapter on principal component analysis. This is because although there are conceptual differences between principal component analysis and exploratory factor analysis, there are also many similarities in terms of how the procedures are conducted. Some of these differences and similarities are discussed in a later section, "Exploratory Factor Analysis versus Principal Component Analysis."

It is likely that some users will read this chapter without first reviewing the chapter on principal component analysis; this made it necessary to present in this chapter much of the material that was already covered in the principal component chapter. Readers who have already covered the principal component chapter should be able to skim over this material fairly quickly.

Introduction to the Common Factor Model

Example: Investment Model Questionnaire

Exploratory factor analysis will be demonstrated by performing a factor analysis on fictitious data from a questionnaire designed to measure investment model constructs (Rusbult, 1980). The investment model was introduced in the preceding chapter; remember that this model describes certain constructs that affect an individual's **commitment** to a romantic relationship, that is, his or her intention to maintain the relationship. Two of the constructs that are believed to influence commitment are alternative value and investment size. **Alternative value** refers to the attractiveness of a person's alternatives to his or her current romantic partner. For example, a woman would score high on alternative value if it would be appealing for her to leave her current partner for a different partner, or simply leave her current partner and be unattached to anyone. **Investment size** refers to the time or personal resources that a person has put into his or her relationship with the current partner. For example, a woman would score high on investment size if she has invested a lot of time and effort in developing her current relationship, or if she and her partner have many mutual friendships that would be lost if the relationship were to end.

Imagine that you have developed a short questionnaire to assess alternative value and investment size. The questionnaire is to be completed by persons who are currently involved in romantic associations. With this questionnaire, items 1-3 were designed to assess investment size, and items 4-6 were designed to assess alternative value. Part of the questionnaire is reproduced here:

Please rate each of the following items to indicate the extent to which you agree or disagree with each statement. Use a response scale in which 1=Strongly Disagree and 7=Strongly Agree.

_____ 1. I have invested a lot of time and effort in developing my relationship with my current partner.

_____ 2. My current partner and I have developed interest in a lot of fun activities that I would lose if our relationship were to end.

_____ 3. My current partner and I have developed a lot of mutual friendships that I would lose if our relationship were to end.

_____ 4. It would be more attractive for me to be involved in a relationship with someone else rather than continue in a relationship with my current partner.

_____ 5. It would be more attractive for me to be by myself than to continue my relationship with my current partner.

_____ 6. In general, my alternatives to remaining in this relationship are quite attractive.

Assume that this questionnaire was administered to 200 subjects, and their responses were keyed so that responses to question 1 were coded as variable V1, responses to question 2 were coded as variable V2, and so forth. The correlations between the six variables are presented in Table 2.1.

Table 2.1.

Correlations between Questions Assessing Investment Size and Alternative Value

			Intercorrelations			
Question	V1	V2	V3	V4	V5	V6
V1	1.00					
V2	.81	1.00				
V3	.79	.92	1.00			
V4	-.03	-.07	-.01	1.00		
V5	-.06	-.01	-.11	.78	1.00	
V6	-.10	-.08	-.04	.79	.85	1.00

Note. N = 200.

The preceding matrix of correlations consists of six rows (running horizontally) and six columns (running vertically). Where the row for one variable intersects with the column for a second variable, you can find the correlation for that pair of variables. For example, where the row for V2 intersects with the column for V1, you can see that the correlation between these variables is .81.

Notice the pattern of intercorrelations in Table 2.1. Questions 1, 2, and 3 are strongly correlated with one another, but these variables are essentially uncorrelated with questions 4, 5, and 6. Similarly, questions 4, 5, and 6 are strongly correlated with one another, but are essentially uncorrelated with questions 1, 2, and 3. Reviewing the complete matrix reveals that there are two sets of variables that seem to hang together: Variables 1, 2, and 3 form one group, and variables 4, 5, and 6 form the second. But why are the variables grouping together in this way?

The Common Factor Model: Basic Concepts

One possible explanation for this pattern of intercorrelations may be found in the path model of Figure 2.1. In this figure, responses to questions 1 through 6 are represented as the six squares labeled V1 through V6. This path model suggests that variables V1, V2, and V3 are correlated with one another because they are all influenced by the same underlying factor. A **factor** is an unobserved variable (or latent variable). Being latent means that you cannot measure a factor directly, as you would measure an observed variable such as height or weight. A factor is a hypothetical construct: you believe that it exists, and you believe that it influences certain manifest variables (or observed variables) that you can measure directly. In the present study, the manifest variables, or observed variables, are subject responses to items 1 through 6.

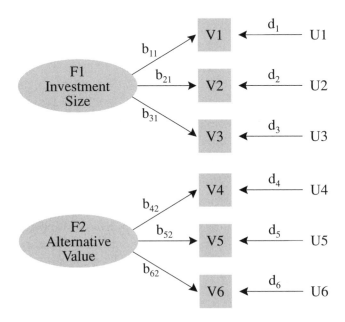

Figure 2.1: Path Model for a 6-Variable, 2-Factor Model, Orthogonal Factors, Factorial Complexity = 1

When representing factor models in figures, it is conventional to represent observed variables as squares or rectangles, and to represent latent factors as circles or ovals. You can see how the two factors appear in Figure 2.1. The first is labeled "F1: Investment Size," and the second is labeled "F2: Alternative Value."

Now return to the original question: Why do variables V1, V2, and V3 correlate so strongly with one another? According to the path model presented in Figure 2.1, these variables are intercorrelated because they are all caused by the same latent factor: the subjects' standing on the underlying investment size construct. This model proposes that, within each subject's belief system, there is a construct that you might call "investment size." Furthermore, this construct has a powerful influence on the way that subjects respond to questions 1, 2, and 3 (notice the arrows going from the oval factor to the squares). Therefore, even though you cannot directly measure someone's standing on the factor (it is a hypothetical construct, after all), you can infer that it really does exist by

- noting that questions 1, 2, and 3 correlate highly with one another

- reviewing the content of questions 1, 2, and 3 (that is, noting what these questions actually say)

- noting that all three questions seem to be measuring the same basic construct, a construct that could reasonably be named "investment size."

Note: The preceding is not a description of how one actually performs a factor analysis; it is just an attempt to help convey the conceptual meaning of the path model presented in the figure.

Common factors. The investment size factor (F1) presented in Figure 2.1 is known as a common factor. A **common factor** is a factor that influences more than one observed variable. In this case, you can see that variables V1, V2, and V3 are all influenced by the investment size factor. It is called a *common* factor because more than one variable has it in common. Because of this terminology, the type of analysis discussed in this chapter is often referred to as **common factor analysis**.

In the lower half of Figure 2.1, you can see that there is a second common factor (F2) representing the alternative value hypothetical construct. This factor affects responses to items 4, 5, and 6 (notice the directional arrows). In short, variables V4, V5, and V6 are intercorrelated because they have this alternative value factor in common. Variables V4, V5, and V6 are not influenced by the investment size factor (notice that there are no causal arrows going from F1 to these variables), and, similarly, V1, V2, and V3 are not influenced by the alternative value factor, F2. This should help clarify why variables V1, V2, and V3 tended to be uncorrelated with variables V4, V5, and V6.

Orthogonal versus oblique models. A few more points must be made in order to more completely understand the factor model presented in Figure 2.1. Notice that there is no arrow connecting F1 and F2. If it were hypothesized that the factors were correlated with one another, there would be a curved, double-headed arrow (a bidirectional arrow) connecting the two ovals. A double-headed arrow indicates that the researcher believes that the two constructs are correlated, but the arrow is not specifying any cause-and-effect relationship. The lack of such an arrow in Figure 2.1 means that the researcher expects these factors to be uncorrelated, or **orthogonal**. If a double-headed arrow did connect them, you would say that the factors are correlated, or **oblique**. Oblique factor models are discussed later in this chapter.

In some factor models a single-headed arrow connects two latent factors, indicating that one factor is expected to have a causal effect on the other. However, such models are normally not investigated with exploratory factor analysis, and will not be discussed in this chapter. For information on models that predict causal relationships between latent factors, see Chapter 5, "Developing Measurement Models with Confirmatory Factor Analysis," and Chapter 6, "Path Analysis with Latent Variables."

Unique factors. Notice that the two common factors are not the only factors that influence the observed variables. For example, you can see that there are actually two factors that influence variable V1: the common factor, F1, and a second factor labeled "U1." Here, U1 is a **unique factor**: a factor that influences only one observed variable. A unique factor represents all of the independent factors that are unique to that single variable (including the error component that is unique to that variable). In the figure, the unique factor U1 affects only V1, U2 affects only V2, and so forth.

Factor loadings. In Figure 2.1, each of the arrows going from a common factor to a variable is identified with a specific coefficient such as b_{11}, b_{21}, or b_{42}. The convention used in labeling these coefficients is quite simple: the first number in the subscript represents the number of the variable that the arrow points toward, and the second number in the subscript represents the number of the factor where the arrow originates. In this way, the coefficient b_{21} represents the arrow that goes to variable 2 from factor 1, the coefficient b_{52} represents the arrow that goes to variable 5 from factor 2, and so forth.

These coefficients represent **factor loadings**. But what exactly is a factor loading? Technically, it is a coefficient that appears in either a factor pattern matrix or a factor structure matrix (these matrices are included in the output of an oblique factor analysis). When you conduct an oblique factor analysis, the loadings in the pattern matrix will have a definition that is different from the definition given to loadings in the structure matrix (these definitions are discussed later in the chapter). To keep things simple, however, skip the oblique analysis for the moment, and focus on what the loadings represent when you perform an analysis in which the factors are orthogonal (uncorrelated). Factor loadings have a more simple interpretation in an orthogonal solution.

When investigating orthogonal factors, the b coefficients may be thought of in a number of different ways. For example, they may be viewed as:

- **Standardized regression coefficients**. The factor loadings obtained in an analysis with orthogonal factors may be thought of as standardized regression weights. If all of the variables (including the factors) are standardized to have unit variance (variance = 1.00), the b coefficients are analogous to the standardized regression coefficients (or regression weights) obtained in regression analysis. In other words, the b weights may be thought of as optimal linear weights by which the F factors are multiplied in calculating subject scores on the V variables (i.e., the weights used in predicting the variables from the factors).

- **Correlation coefficients**. Factor loadings also represent the product-moment correlation between an observed variable and an underlying factor. For example, if $b_{52} = .85$, this would indicate that the correlation between V5 and F2 is .85. This may surprise you if you are familiar with multiple regression, because most textbooks on multiple regression point out that standardized multiple regression coefficients and correlation coefficients are two different things. However, standardized regression coefficients are equivalent to correlation coefficients when the predictor variables are completely uncorrelated with each other. And that is the case in factor analysis with orthogonal factors: the factors serve as *predictor* variables in predicting the observed variables. And because the factors are uncorrelated with each other, the factor loadings may be interpreted as both standardized regression weights and as correlation coefficients.

- **Path coefficients**. Finally, the b coefficients are also analogous to the path coefficients obtained in path analysis. That is, they may be seen as standardized linear weights that represent the size of the effect that an underlying factor has in causing variability in the observed variable (path analysis is covered in Chapter 4, "Path Analysis with Manifest Variables").

Factor loadings are important because they help you interpret the factors that are responsible for the covariation in the data. This means that, after the factors are rotated, you can review the nature of the variables that have significant loadings for a given factor (i.e., the variables that are most strongly related to the factor). The nature of these variables will help you understand the nature of that factor.

Factorial complexity. **Factorial complexity** is a characteristic of an observed variable. The factorial complexity of a variable refers to the number of common factors that have a significant loading for that variable. For example, in Figure 2.1, you can see that the factorial complexity of V1 is one: V1 displays a significant loading for F1, but not for F2. The factorial complexity of V4 is also one: it displays a significant loading for F2, but not for F1.

Although the factor model of Figure 2.1 is fairly simple, a more complex model is presented in Figure 2.2. As with the previous model, two common factors are again responsible for the covariation in the data set. However, you can see that each of the common factors in Figure 2.2 has significant loadings on all six observed variables. In the same way, you can see that each variable is influenced by both common factors. Because each variable in the figure has significant loadings for two common factors, it may be said that each variable has a factorial complexity of two.

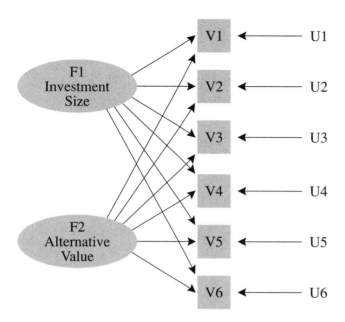

Figure 2.2: Path Model for a 6-Variable, 2-Factor Model, Orthogonal Factors, Factorial Complexity = 2

Observed variables as linear combinations of underlying factors. It is possible to think of a given observed variable, such as V1, as being a weighted sum of the underlying factors included in the factor model. For example, notice that in Figure 2.2, there are three factors that affect V1: two common factors (F1 and F2), and one unique factor (U1). By multiplying these factors by the appropriate weights, it is possible to calculate any subject's score on V1. In algebraic form, this would be done with the following equation:

$$V1 = b_{11}(F1) + b_{12}(F2) + d_1(U1)$$

In this equation, b_{11} is the regression weight for F1 (the amount of weight given to F1 in the prediction of V1), b_{12} is the regression weight for F2, and d_1 is the regression weight for the unique factor associated with V1. You can see that a given person's score on V1 is determined by multiplying the underlying factors by the appropriate regression weights, and summing the resulting products. This is why, in factor analysis, the observed variables are viewed as linear combinations of the underlying factors.

The preceding equation is therefore similar to the multiple regression equation as it is described in most statistics texts; in factor analysis, the observed variable (i.e., V1) serves as counterpart to the criterion variable (Y) in multiple regression, and the latent factors (i.e., F1, F2, and U1) serve as counterparts to the predictor variables (the X variables) in multiple regression. You would generally expect to obtain a different set of factor weights, and thus a different predictive equation, for each observed variable in a factor analysis.

Where do you find the regression weights for the common factors in factor analysis? In the **factor pattern matrix**. An example of a pattern matrix is presented here:

Factor Pattern

Variable	Factor 1	Factor 2
V1	.87	.26
V2	.80	.48
V3	.77	.34
V4	−.56	.49
V5	−.58	.52
V6	−.50	.59

You can see that the rows (running horizontally) in the factor pattern represent the different observed variables such as V1 and V2. The columns in the factor pattern represent the different factors such as F1 and F2. Where a row and column intersect, you will find a factor loading (or standardized regression coefficient). For example, in determining values of variable V1, F1 is given a weight of .87, and F2 is given a weight of .26; in determining values of V2, F1 is given a weight of .80, and F2 is given a weight of .48.

Communality versus the unique component. A **communality** is a characteristic of an observed variable. It refers to the variance in an observed variable that is accounted for by the common factors. If a variable demonstrates a large communality, it means that this variable is strongly influenced by at least one of the common factors. The symbol for communality is h^2. The communality for a given variable is computed by squaring that variable's factor loadings for all retained common factors, and summing these squares. For example, using the factor loadings from the previous factor pattern, you may compute the communality for V1 in the following way:

$$h_1^2 = b_{11}^2 + b_{12}^2$$

$$= (.87)^2 + (.26)^2$$

$$= .756 + .068$$

$$= .82$$

So the communality for V1 is approximately 82. This means that 82% of the variance in V1 is accounted for by the two common factors. You can now compute the communality for each variable, and add these values to the table that contains the pattern matrix:

Factor Pattern

Variable	Factor 1	Factor 2	h^2
V1	.87	.26	.82
V2	.80	.48	.87
V3	.77	.34	.71
V4	−.56	.49	.55
V5	−.58	.52	.61
V6	−.50	.59	.60

In contrast to the communality, the **unique component** refers to the proportion of variance in a given observed variable that is not accounted for by the common factors. Once communalities are computed, it is easy to calculate the unique component: simply subtract the communality from one. The unique component for V1 can be calculated in this way:

$$d_1{}^2 = 1 - h_1{}^2$$

$$= 1 - .82$$

$$= .18$$

This shows that 18% of the variance in V1 is not accounted for by the common factors; alternatively, you could say that 18% of the variance in V1 is accounted for by the unique factor, U1.

If you then proceed to take the square root of the unique component, you can compute the coefficient, d. This should look familiar, because earlier d was defined as the weight given to a unique factor in determining values on the observed variable. For variable V1, the unique component was calculated as .18. The square root of .18 is approximately .42. Therefore, the unique factor U1 would be given a weight of .42 in determining values of V1 (that is, $d_1 = .42$).

Exploratory Factor Analysis
versus Principal Component Analysis

Some readers are likely to notice the many similarities between exploratory factor analysis and principal component analysis. In fact, these similarities have even led some researchers to incorrectly report that they have conducted factor analysis when, in fact, they have conducted principal component analysis. Because of this common misunderstanding, this section reviews some of the similarities and differences between the two procedures.

How Factor Analysis Differs from Principal Component Analysis

Purpose. Only factor analysis may be used to identify the factor structure underlying a set of variables. In other words, if you want to identify the number and nature of the latent factors that are responsible for covariation in the data set, then factor analysis, and not principal component analysis, should be used.

Principal components versus common factors. A principal component is an artificial variable; it is a linear combination of the (optimally weighted) observed variables. It is possible to calculate where a given subject stands on a principal component by simply summing that subject's (optimally weighted) scores on the observed variable being analyzed. For example, you could determine each subject's score on principal component 1 by using the following formula:

$$C_1 = b_{11}(X_1) + b_{12}(X_2) + \ldots b_{1p}(X_p)$$

where

C_1 = the subject's score on principal component 1 (the first component extracted)

b_{1p} = the regression coefficient (or weight) for observed variable p, as used in creating principal component 1

X_p = the subject's score on observed variable p.

In contrast, a common factor is a hypothetical latent variable that is assumed to be responsible for the covariation between two or more observed variables. Because factors are unmeasured latent variables, you may never know exactly where a given subject stands on an underlying factor (although it is possible to arrive at estimates, as you will see later).

In common factor analysis, the factors are not assumed to be linear combinations of the observed variables (as is the case in principal component analysis). Factor analysis assumes just the opposite, that the observed variables are linear combinations of the underlying factors. This is illustrated in the following equation:

$$X_1 = b_1(F_1) + b_2(F_2) + \ldots b_q(F_q) + d_1(U_1)$$

where

X_1 = the subject's score on observed variable 1

b_q = the regression coefficient (or weight) for underlying common factor q, as used in determining the subject's score on X_1

F_q = the subject's score on underlying factor q

d_1 = the regression weight for the unique factor associated with X_1

U_1 = the unique factor associated with X_1.

Because similar steps are followed in extracting principal components and common factors, it is easy to incorrectly assume that they are conceptually identical. However, the preceding equations show that they differ in an important way: in principal components analysis, the principal components are linear combinations of the observed variables, whereas the factors of factor analysis are not viewed in this way. In factor analysis the observed variables are viewed as linear combinations of the underlying factors.

You might be confused by this point if you know that it is possible to compute factor scores in exploratory factor analysis, and that these factor scores are essentially linear composites of observed variables. However, in reality these factor scores are merely *estimates* of where the subjects stand on the underlying factors. These so-called factor scores generally do not correlate perfectly with scores on the actual underlying factor (for this reason, they will be referred to as **estimated factor scores** in this text).

On the other hand, the principal component scores obtained in principal component analysis are not estimates; they are perfect representations of the extracted components. Remember that a principal component is simply a mathematical transformation (a linear combination) of the observed variables. So a given subject's component score represents with perfect accuracy where that subject stands on the principal component. It is therefore proper to discuss *actual* component scores rather than estimated component scores.

Variance accounted for. Factor analysis and principal component analysis also differ with respect to the type of variance accounted for. The factors of factor analysis account for common variance in a data set, while the components of principal component analysis account for total variance in the data set. This difference may be understood with reference to Figure 2.3.

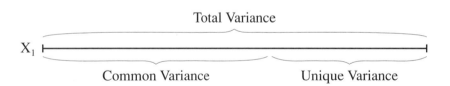

Figure 2.3: Total Variation in Variable X_1, as Divided into Common and Unique Components

Assume that the length of the line in Figure 2.3 represents the total variance in observed variable X_1, and that variables X_1 through X_6 are subjected to factor analysis. The figure shows that the total variance in X_1 may be divided into two components: the common variance and the unique variance. The **common variance** corresponds to the communality of X_1: the proportion of total variance in the variable that is accounted for by the common factors. The remaining variance is the unique component: that variance (whether systematic or random) which is specific to variable X_1.

In factor analysis, factors are extracted to account for only the common variance, and the remaining unique variance remains unanalyzed. This is accomplished by analyzing an **adjusted correlation matrix**: a correlation matrix with communality estimates on the diagonal. You cannot know a variable's actual communality prior to the factor analysis, and so it must be estimated using one of a number of alternative procedures. This chapter recommends that squared multiple correlations be used as prior communality estimates. A variable's squared multiple correlation is obtained by using multiple regression to regress it on the remaining observed variables (later, you will find that these values can be obtained easily by using the PRIORS= option with PROC FACTOR). The adjusted correlation matrix that is analyzed in factor analysis has correlations between the observed variables off the diagonal, and communality estimates on the diagonal.

In principal component analysis, however, components are extracted to account for the **total variance** in the data set, not just the common variance. This is accomplished by analyzing an **unadjusted correlation matrix**: a correlation matrix with ones (1.00) on the diagonal. Why ones? Since all variables are standardized in the analysis, each has a variance of one. Because the correlation matrix contains ones (rather than communalities) on the diagonal, 100% of each variable's variance will be accounted for by the combined components, not just the variance that the variable shares in common with other variables.

It is this difference that explains why only factor analysis, and not principal component analysis, can be used to identify the number and nature of the factors that are responsible for covariation in the data set. Because principal component analysis makes no attempt to separate the common component from the unique component of each variable's variance, this procedure can provide a misleading picture of the factor structure underlying the data. Either procedure may be used to reduce a number of variables to a more manageable number; however, if you want to identify the factor structure of a data set (such as that portrayed in the path model of Figure 2.1), only factor analysis will do.

How Factor Analysis Is Similar to Principal Component Analysis

Purpose (in some cases). Both factor analysis and principal component analysis may be used as **variable reduction procedures**; that is, both may be used to reduce a number of variables to a smaller, more manageable number. This is why both procedures are so widely used in analyzing data from multiple-item questionnaires in the social sciences. Both procedures can be used to reduce a large number of survey questions into a smaller number of scales.

Extraction methods (in some cases). In this chapter, you first learn how to use the principal axis method for extracting factors. This is the same mathematical procedure used to extract principal components in Chapter 1 (a later example will also show how to use the maximum likelihood method: an extraction method that is typically used only with factor analysis).

Results (in some cases). Principal component analysis and factor analysis often lead to similar conclusions regarding the appropriate number of factors (or components) to retain, as well as similar conclusions regarding how the factors (or components) should be interpreted. This is especially true when the variable communalities are high (near 1.00). The reason for this should be obvious: when the principal axis extraction method is used, the only real difference between the two procedures involves the values that appear on the diagonal of the correlation matrix. If the communalities are very high (near 1.00), there is little difference between the matrix that is analyzed in principal component analysis and the matrix that is analyzed in factor analysis; hence, the similar solutions.

Preparing and Administering the Investment Model Questionnaire

As was discussed at the beginning of this chapter, assume that you are interested in measuring two constructs that constitute important components of the investment model (Rusbult, 1980). One construct is investment size: the amount of time or personal resources that the person has put into his or her relationship with the current partner. The other construct is alternative value: the attractiveness of a person's alternatives to his or her current romantic partner.

Writing the Questionnaire Items

The questionnaire used earlier in this chapter is again reproduced here. Note that items 1–3 were designed to assess investment size, and items 4–6 were designed to assess alternative value.

Please rate each of the following items to indicate the extent to which you agree or disagree with each statement. Use a response scale in which 1=Strongly Disagree and 7=Strongly Agree.

_____ 1. I have invested a lot of time and effort in developing my relationship with my current partner.

_____ 2. My current partner and I have developed interest in a lot of fun activities that I would lose if our relationship were to end.

_____ 3. My current partner and I have developed a lot of mutual friendships that I would lose if our relationship were to end.

_____ 4. It would be more attractive for me to be involved in a relationship with someone else rather than continue in a relationship with my current partner.

_____ 5. It would be more attractive for me to be by myself than to continue my relationship with my current partner.

_____ 6. In general, my alternatives to remaining in this relationship are quite attractive.

Number of Items per Factor

As was mentioned in Chapter 1, it is highly desirable to have at least three (and preferably more) variables loading on each factor when the analysis is complete. Because some of the items may be dropped during the course of the analysis, it is generally good practice to write at least five items for each construct that you wish to measure; in this way, you increase the chances that at least three items per factor will survive the analysis (you can see that the preceding questionnaire violates this recommendation by including only three items for each factor at the outset).

Note: Remember that the recommendation of three items per scale actually constitutes a *lower bound*. In practice, test and attitude scale developers normally desire that their scales contain many more than just three items to measure a given construct. It is not unusual to see individual scales that include 10, 20, or even more items to assess a single construct. Other things held constant, the more items in the scale, the more reliable it will be. The recommendation of three items per scale should therefore be viewed as a rock-bottom lower bound, appropriate only if practical concerns (such as total questionnaire length) prevent you from including more items. For more information on scale construction, see Spector (1992).

Minimally Adequate Sample Size

Factor analysis is a large-sample procedure, so it is important to use guidelines to choose the sample size which will be minimally adequate for an analysis. *The minimal number of subjects in the sample should be the larger of 100 subjects, or 5 times the number of variables being analyzed.* If responses to a questionnaire are being analyzed, then the number of variables is equal to the number of items on the questionnaire. To illustrate, assume that you want to perform an analysis on responses to a 50-item questionnaire. Five times the number of items on the questionnaire equals 250. Therefore, it would be best if your final sample provides usable (complete) data from at least 250 subjects. It should be remembered, however, that any subject who fails to answer just one item will not provide usable data for the factor analysis, and will therefore be dropped from the final sample. A certain number of subjects can always be expected to leave at least one question blank; therefore, to ensure that the final sample includes at least 250 usable responses, you should administer the questionnaire to perhaps 300 subjects.

These rules regarding the number of subjects per variable again constitute a lower bound, and some have argued that they should apply only under two optimal conditions for factor analysis: (a) when many variables are expected to load on each factor, and (b) when variable communalities are high. Under less optimal conditions, larger samples may be required.

SAS Program and Analysis Results

This section provides instructions on writing the SAS program, along with an overview of the SAS output. A subsequent section will provide a more detailed treatment of the steps followed in the analysis, and the decisions to be made at each step.

Writing the SAS Program

The DATA step. To perform an exploratory factor analysis, data may be input in the form of raw data, a correlation matrix, a covariance matrix, as well as other types of data sets (for details, see Chapter 21 of the *SAS/STAT users guide, volume 1* [1989]). In this chapter's first example, raw data will be analyzed.

Assume that you administered your questionnaire to 50 subjects, then keyed their responses to each question. The SAS names given to these variables, and the format used in keying the data, are presented here:

Line	Column	Variable Name	Explanation
1	1-6	V1-V6	Subject's responses to survey questions 1 through 6. Responses were made using a 7-point scale, where higher scores indicate stronger agreement with the statement.
	8-9	COMMIT	Subject scores on the commitment variable. Scores may range from 4 to 28, and higher scores indicate higher levels of commitment to maintain the relationship.

At this point, you are interested only in variables V1-V6 (subject responses to the six questionnaire items). Scores on the commitment variable (COMMIT) are also included in the data set because you will later compute correlations between estimated factor scores and COMMIT.

Here are the statements that will input these responses as raw data. The first three and the last three observations are reproduced here; for the entire (fictitious) data set, see Appendix B, "Data Sets."

```
1      DATA D1;
2         INPUT    #1    @1    (V1-V6)    (1.)
3                        @8    (COMMIT)   (2.) ;
4      CARDS;
5      776122 24
6      776111 28
7      111425  4
8      .
9      .
10     .
```

```
11        433344 15
12        557332 20
13        655222 13
14        ;
```

The data set in Appendix B includes only 50 cases so that it will be relatively easy for interested readers to key the data and replicate these analyses. However, it should be remembered that 50 observations will normally constitute an unacceptably small sample for an exploratory factor analysis. Earlier it was said that a sample should provide usable data from the larger of either 100 cases or 5 times the number of observed variables. A small sample is being analyzed here for illustrative purposes only.

The PROC FACTOR statement. The general form for the SAS program to perform an exploratory factor analysis with oblique rotation is:

```
PROC FACTOR    DATA=data-set-name
               SIMPLE
               METHOD=factor-extraction-method
               PRIORS=prior-communality-estimates
               NFACT=n
               SCREE
               ROTATE=PROMAX
               ROUND
               FLAG=desired-size-of-"significant"-factor-loadings ;
     VAR variables-to-be-analyzed ;
     RUN;
```

Here is an actual program, including the DATA step, that could be used to analyze some fictitious data from the investment model study:

```
1      DATA D1;
2         INPUT    #1   @1   (V1-V6)    (1.)
3                       @8   (COMMIT)   (2.) ;
4      CARDS;
5      776122 24
6      776111 28
7      111425  4
8      .
9      .
10     .
11     433344 15
12     557332 20
13     655222 13
14     ;
15     PROC FACTOR    DATA=D1
16                    SIMPLE
17                    METHOD=PRIN
18                    PRIORS=SMC
19                    NFACT=2
20                    SCREE
21                    ROTATE=PROMAX
22                    ROUND
23                    FLAG=.40    ;
24        VAR V1 V2 V3 V4 V5 V6;
25        RUN;
```

Options used with PROC FACTOR. The PROC FACTOR statement begins the FACTOR procedure, and a number of options may be requested in this statement before it ends with a semicolon. Some options that may be especially useful in social science research are:

FLAG=desired-size-of-"significant"-factor loadings
> causes the printer to flag (with an asterisk) any factor loading whose absolute value is greater than some specified size. For example, if you specify
>
> FLAG=.35
>
> then an asterisk will appear next to any loading whose absolute value exceeds .35. This option can make it much easier to interpret a factor pattern. Negative values are not allowed in the FLAG option, and the FLAG option should be used in conjunction with the ROUND option.

```
METHOD=factor-extraction-method
```
specifies the method to be used in extracting the factors. The current program specifies

```
            METHOD=PRIN
```

to request that the principal axis (principal factors) method be used for the initial extraction. Although the principal axis method is probably the most popular extraction method, some researchers prefer the maximum likelihood method because it provides a significance test for solving the "number of factors" problem, and it is also believed to provide better parameter estimates. The maximum likelihood method may be requested with the option

```
            METHOD=ML
```

```
MINEIGEN=p
```
specifies the critical eigenvalue that a factor must display if that factor is to be retained (here, p = the critical eigenvalue). Negative values are not allowed.

```
NFACT=n
```
allows you to specify the number of factors to be retained and rotated, where n = the number of factors.

```
OUT=name-of-new-data-set
```
creates a new data set that includes all of the variables of the existing data set, along with estimated factor scores for the retained factors. Factor 1 is given the variable name FACTOR1, factor 2 is given the name FACTOR2, and so forth. OUT= must be used in conjunction with the NFACT option, and the analysis must be based on raw data.

```
PRIORS=prior-communality-estimates
```
specifies prior communality estimates. The preceding specifies SMC to request that the squared multiple correlations between a given variable and the other observed variables be used as that variable's prior communality estimate.

`ROTATE=rotation-method`

specifies the rotation method to be used. The preceding program requests a promax rotation that results in oblique (correlated) factors. This option is requested by specifying

`ROTATE=PROMAX`

Orthogonal rotations may also be requested; Chapter 1 showed how to request an (orthogonal) rotation by specifying

`ROTATE=VARIMAX`

`ROUND`

causes all coefficients to be limited to two decimal places, rounded to the nearest integer, and multiplied by 100 (thus eliminating the decimal point). This generally makes it easier to read the coefficients. Factor loadings and correlation coefficients in the matrices printed by PROC FACTOR are normally carried out to several decimal places.

`SCREE`

creates a plot that graphically displays the size of the eigenvalue associated with each factor. This can be used to perform a scree test to determine how many factors should be retained.

`SIMPLE`

requests simple descriptive statistics: the number of usable cases on which the analysis was performed, and the means and standard deviations of the observed variables.

The VAR statement. The variables to be analyzed are listed in the VAR statement, with each variable separated by at least one space. Remember that the VAR statement is a different statement, not an option within the FACTOR statement, so don't forget to end the FACTOR statement with a semicolon before beginning the VAR statement.

Results from the Output

When specifying LINESIZE=80 and PAGESIZE=60 in the OPTIONS statement, the preceding program would produce an output 7 pages long. The following table lists some of the information included in this output, and the page on which it appears:

- Page 1 includes simple statistics.

- Page 2 includes prior communality estimates and the eigenvalue table.

- Page 3 includes the scree plot of eigenvalues.

- Page 4 includes the unrotated factor pattern and final communality estimates.

- Page 5 includes the results of the orthogonal prerotation method (varimax rotation) such as the rotated factor pattern for the varimax solution, along with final communality estimates.

- Page 6 includes some results from the oblique rotation method (promax rotation) such as the inter-factor correlations and the rotated factor pattern (standardized regression coefficients).

- Page 7 includes more results of the promax rotation such as the reference structure (semipartial correlations) and the factor structure (correlations).

The following section reviews the steps in which exploratory factor analysis is conducted. Integrated into this discussion will be excerpts from the preceding output, along with guidelines for interpreting this output.

Steps in Conducting Exploratory Factor Analysis

Factor analysis is normally conducted in a sequence of steps, with somewhat subjective decisions being made at many of these steps. Because this is an introductory treatment of the topic, it will not provide a comprehensive discussion of all the options available to you at each step. Instead, specific recommendations will be made, consistent with practices often followed in applied research. For a more detailed treatment of exploratory factor analysis, see Kim and Mueller (1978a; 1978b), Loehlin (1987), and Rummel (1970).

Step 1: Initial Extraction of the Factors

The first step of the analysis involves the initial extraction of the factors. The preceding program specified the option

```
METHOD=PRIN
```

which calls for the principal factors, or principal axis, method. This is the same method used to extract the components of principal component analysis.

As with component analysis, the number of factors extracted will be equal to the number of variables being analyzed. Because six variables are being analyzed in the present study, six factors will be extracted. The first factor can be expected to account for a fairly large amount of the common variance. Each succeeding factor will account for progressively smaller amounts of variance. Although a large number of factors may be extracted in this way, only the first few factors will be important enough to be retained for interpretation.

As with principal components, the extracted factors will have two important properties: (a) Each factor will account for a maximum amount of the variance that has not already been accounted for by other, previously extracted, factors, and (b) each factor will be uncorrelated with all of the previously extracted factors. This second characteristic may come as a surprise, because earlier it was said that you were going to obtain an oblique solution (by specifying ROTATE=PROMAX) in which the factors would be correlated. However, in this analysis the factors are in fact orthogonal (uncorrelated) at the time they are extracted. It is only later in the analysis that their orthogonality is relaxed, and they are allowed to become oblique. This will be discussed in more detail in a subsequent section on factor rotation.

These concepts will now be related to some of the results that appeared in the output created by the preceding program. Pages 1-2 of the output provided simple statistics, the eigenvalue table, and some additional information regarding the initial extraction of the factors. Those pages are reproduced here as Output 2.1.

```
                        The SAS System                              1

            Means and Standard Deviations from 50 observations

                V1          V2          V3          V4          V5          V6
Mean          4.62        4.38        4.36        2.76        2.36        2.56
Std Dev  1.53715879   1.5103723  1.63831671  1.25454277  1.10213152  1.37261853
```

```
Initial Factor Method: Principal Factors                            2

                 Prior Communality Estimates: SMC

        V1          V2          V3          V4          V5          V6
    0.782395    0.817056    0.676621    0.479189    0.523803    0.498715

            Eigenvalues of the Reduced Correlation Matrix:
              Total = 3.77777847   Average = 0.62962975

                              1           2           3
            Eigenvalue      2.8753      1.2766     -0.0124
            Difference      1.5987      1.2890      0.0748
            Proportion      0.7611      0.3379     -0.0033
            Cumulative      0.7611      1.0990      1.0957

                              4           5           6
            Eigenvalue     -0.0873     -0.1241     -0.1502
            Difference      0.0369      0.0261
            Proportion     -0.0231     -0.0329     -0.0398
            Cumulative      1.0726      1.0398      1.0000

        2 factors will be retained by the NFACTOR criterion.
```

Output 2.1: Simple Statistics, Prior Communalities, and Eigenvalue Table from Analysis of Investment Model Questionnaire

On page 1 of Output 2.1, the simple statistics section shows that the analysis was based on 50 observations. Means and standard deviations are also provided.

The first line of page 2 says "Initial Factor Method: Principal Factors." This indicates that the principal factors method was used for the initial extraction of the factors.

Next, the prior communality estimates are printed. Because the program included the option PRIORS=SMC, the prior communality estimates are squared multiple correlations.

Below that, the eigenvalue table is printed. An **eigenvalue** represents the amount of variance that is accounted for by a given factor. In the row headed "Eigenvalue" (running horizontally), the eigenvalue for each factor is presented. Each column in the matrix (running vertically) presents information about one of the six factors: the column headed "1" provides information about the first factor extracted, the column headed "2" provides information about the second factor extracted, and so forth.

Where the row headed "Eigenvalue" intersects with the columns headed "1" and "2", you can see that the eigenvalue for factor 1 is approximately 2.88, while the eigenvalue for factor 2 is approximately 1.28. This pattern is consistent with the earlier statement that the first factors extracted tend to account for relatively large amounts of variance, while the later factors account for relatively smaller amounts.

Step 2: Determining the Number of "Meaningful" Factors to Retain

As with principal component analysis, the number of factors extracted is equal to the number of variables being analyzed, necessitating that you decide just how many of these factors are truly meaningful and worthy of being retained for rotation and interpretation. In general, we expect that only the first few factors will account for meaningful amounts of variance, and that the later factors will tend to account for relatively small amounts of variance (largely error variance). The next step of the analysis, therefore, is to determine how many meaningful factors should be retained for interpretation.

(The preceding program specified NFACT=2 so that two factors would be retained; because this was the initial analysis, you had no empirical reason to expect two meaningful factors, and specified NFACT=2 on a hunch. If the empirical results suggest a different number of meaningful factors, the NFACT= option may be changed on subsequent analyses.)

Chapter 1 discussed four options that can be used to help make the "number of factors" decision, and the first of these was the **eigenvalue-one** criterion, or Kaiser criterion. In using this criterion, you retain any principal component with an eigenvalue greater than 1.00.

The eigenvalue-one criterion made sense in principal component analysis because each variable contributed one unit of variance to the analysis. This criterion ensured that you would not retain any component that accounted for less variance than had been contributed by one variable.

For the same reason, however, you can see that the eigenvalue-one criterion is less appropriate in common factor analysis. Remember that each variable does not contribute one unit of variance to this analysis, but instead contributes its prior communality estimate. This estimate will be less than 1.00, and so it makes little sense to use the value of 1.00 as a cutting point for retaining factors. Without the eigenvalue-one criterion, you are left with the following three options:

A. The scree test. With the scree test (Cattell, 1966), you plot the eigenvalues associated with each factor, and look for a break between the factors with relatively large eigenvalues and those with smaller eigenvalues. The factors that appear before the break are assumed to be meaningful and are retained for rotation; those appearing after the break are assumed to be unimportant and are not retained.

Specifying the SCREE option in the PROC FACTOR statement causes the SAS System to print an eigenvalue plot as part of the output. This scree plot is reproduced as Output 2.2.

You can see that the factor numbers are listed on the horizontal axis, while eigenvalues are listed on the vertical axis. With this plot, notice that there is a relatively large break between factors 1 and 2, another large break between factors 2 and 3, but that there is not a large break between factors 3 and 4, 4 and 5, or 5 and 6. Because factors 3 through 6 have relatively small eigenvalues, and the data points for factors 3 through 6 could almost be fitted with a straight line, they can be assumed to be relatively unimportant factors. Because there is a relatively large break between factors 2 and 3, factor 2 can be viewed as a relatively important factor. Given this plot, a scree test would suggest that only factors 1 and 2 be retained because only these factors appear before the last big break. Factors 3 through 6 appear after the break, and thus will not be retained.

In some cases, a computer printer may not be able to prepare an eigenvalue plot with the degree of precision that is necessary to perform a sensitive scree test. In such cases, it may be best to prepare the plot by hand. Fortunately, the necessary eigenvalues are printed in the eigenvalue table that appears at the beginning of the output, and you can simply use these values to prepare the scree plot by hand. Such a hand-drawn plot may make it easier to identify the break in the eigenvalues, if one exists.

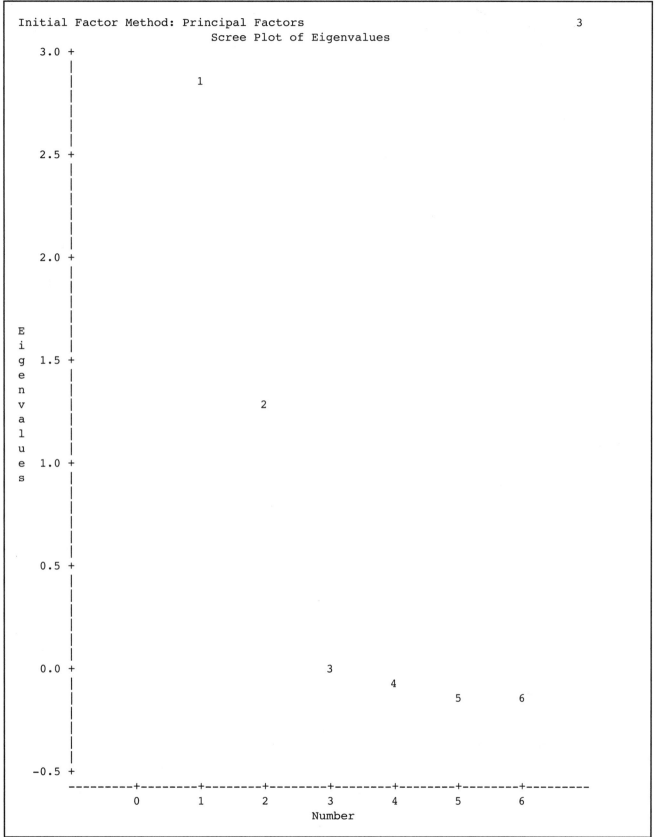

Output 2.2: Scree Plot of Eigenvalues from Analysis of Investment Model Questionnaire

B. Proportion of variance accounted for. A second criterion in making the number of factors decision involves retaining a factor if it accounts for a certain **proportion (or percentage) of the variance in the data set**. For example, you may decide to retain any factor that accounts for at least 5% or 10% of the common variance. This proportion can be calculated with a simple formula:

$$\text{Proportion} = \frac{\text{Eigenvalue for the factor of interest}}{\text{Total eigenvalues of the correlation matrix}}$$

In principal component analysis, the "total eigenvalues of the correlation matrix" was equal to the total number of variables being analyzed (because each variable contributed one unit of variance to the data set). In common factor analysis, however, the total eigenvalues will be equal to the sum of the communalities that appear on the main diagonal of the matrix being analyzed.

The proportion of common variance accounted for by each factor is printed in the eigenvalue table from output page 2, and appears to the right of the word "Proportion". The eigenvalue table for the preceding analysis is reproduced again as Output 2.3.

```
         Eigenvalues of the Reduced Correlation Matrix:
             Total = 3.77777847   Average = 0.62962975


                              1          2          3
         Eigenvalue       2.8753     1.2766    -0.0124
         Difference       1.5987     1.2890     0.0748
         Proportion       0.7611     0.3379    -0.0033
         Cumulative       0.7611     1.0990     1.0957


                              4          5          6
         Eigenvalue      -0.0873    -0.1241    -0.1502
         Difference       0.0369     0.0261
         Proportion      -0.0231    -0.0329    -0.0398
         Cumulative       1.0726     1.0398     1.0000

      2 factors will be retained by the NFACTOR criterion.
```

Output 2.3: Eigenvalue Table from Analysis of Investment Model Questionnaire

From the "Proportion" row of the preceding eigenvalue table, you can see that the first factor alone accounts for 76% of the common variance, the second factor alone accounts for 34%, and the third factor accounts for less than 1% (in fact, factor 3 actually has a negative percentage; see the following box for an explanation). If you were using, say, 10% as the criterion for deciding whether a factor should be retained, only factors 1 and 2 would be retained in the present analysis. Despite the apparent ease of use of this criterion, however, remember that this approach has been criticized for its subjectivity (Kim & Mueller, 1978b).

How can you account for over 100% of the common variance? The bottom row of the eigenvalue table (labelled "Cumulative") provides the cumulative percent of common variance accounted for by the factors. Output 2.3 shows that factor 1 accounts for 76% of the common variance (the value in the table is 0.7611), and factors 1 and 2 combined account for 110% (the actual value in the table is 1.099). But how can two factors account for over 100% of the common variance?

In brief, this is because the prior communality estimates were not perfectly accurate. Consider this: if your prior communality estimates were perfectly accurate estimates of the variables' actual communalities, and if the common factor model was correctly estimated, then the factors that you retained would have to account for exactly 100% of the common variance, and the remaining factors would have to account for 0%. The fact that this did not happen in the present analysis is probably because your prior communality estimates (squared multiple correlations) were not perfectly accurate.

You may also be wondering why some of the factors seem to be accounting for a negative percent of the common variance (i.e., why they have negative eigenvalues). This is because the analysis is constrained so that the "Cumulative" proportion must equal 1.00 after the last factor is extracted. Since this cumulative value exceeds 1.00 at some points in the analysis, it was necessary mathematically that some factors have negative eigenvalues.

C. Interpretability criteria. Perhaps the most important criteria to use when solving the number of factors problem is the **interpretability criteria:** interpreting the substantive meaning of the retained factors and verifying that this interpretation makes sense in terms of what is known about the constructs under investigation. Here are four rules to follow in doing this (a later section of this chapter will provide a step-by-step illustration of how to interpret a factor solution; the following rules will be more meaningful at that time):

1. **Are there at least three variables (items) with significant loadings on each retained factor?** A solution is less satisfactory if a given factor is measured by less than three variables.

2. **Do the variables that load on a given factor share some conceptual meaning?** For example, if three questions on a survey all load on factor 1, do all three of these questions seem to be measuring the same underlying construct?

3. **Do the variables that load on different factors seem to be measuring different constructs?** For example, if three questions load on factor 1, and three other questions load on factor 2, do the first three questions seem to be measuring a construct that is conceptually different from the construct measured by the last three questions?

4. **Does the rotated factor pattern demonstrate "simple structure?"** Simple structure means that the pattern possesses two characteristics: (a) most of the variables have relatively high factor loadings on only one factor, and near-zero loadings for the other factors, and (b) most factors have relatively high factor loadings for some variables, and near-zero loadings for the remaining variables. This concept of simple structure will be explained in more detail in a later section.

Recommendations. Given the preceding options, what procedure should you actually follow in solving the number of factors problem? This text recommends combining all three in a structured sequence. First, perform a scree test and look for obvious breaks in the data. Because there will often be more than one break in the eigenvalue plot, it may be necessary to examine two or more possible solutions. Next, review the amount of common variance accounted for by each factor. If you are retaining factors that account for as little as 2% or 3% of the variance, it may be wise to take a second look at the solution and verify that these latter factors are of truly substantive importance. Finally, apply the interpretability criterion. If more than one solution can be justified on the basis of a scree test or the "variance accounted for" criteria, which of these solutions are the most interpretable? By seeking a solution that satisfies all three of these criteria, you maximize chances of correctly identifying the factor structure of the data set.

Step 3: Rotation to a Final Solution

After extracting the initial factors, the computer will print an unrotated factor pattern matrix. The rows of this matrix represent the variables being analyzed, and the columns represent the retained factors. The entries in the matrix are factor loadings. In a factor *pattern* matrix, the observed variables are assumed to be linear combinations of the common factors, and the factor loadings are standardized regression coefficients for predicting the variables from the factors. (Later, you will see that the loadings have a different interpretation in a factor *structure* matrix.) With PROC FACTOR, the unrotated factor pattern is printed under the heading "Factor Pattern", and appears on output page 4. The factor pattern for the present analysis is reproduced as Output 2.4.

```
Initial Factor Method: Principal Factors                              4

                          Factor Pattern

                          FACTOR1     FACTOR2

                  V1         87 *        26
                  V2         80 *        48 *
                  V3         77 *        34
                  V4        -56 *        49 *
                  V5        -58 *        52 *
                  V6        -50 *        59 *
```

```
NOTE: Printed values are multiplied by 100 and rounded to the nearest integer.
      Values greater than 0.4 have been flagged by an '*'.

                       Variance explained by each factor

                              FACTOR1    FACTOR2
                              2.875329   1.276585


            Final Communality Estimates: Total = 4.151914

              V1         V2         V3         V4         V5         V6
          0.816776   0.874178   0.704434   0.558828   0.607056   0.590642
```

Output 2.4: Unrotated Factor Pattern from Analysis of Investment Model Questionnaire

When more than one factor has been retained, an unrotated factor pattern is usually difficult to interpret. Factor patterns are easiest to interpret when some of the variables in the analysis have very high loadings on a given factor, and the remaining variables have near-zero loadings on that factor. Unrotated factor patterns often fail to display this type of pattern. For example, consider the loadings in the column headed "FACTOR1" in Output 2.4. Notice that variables V1, V2, and V3 do display fairly high loadings for this factor, which is good. Unfortunately, however, variables V4, V5, and V6 do not display near-zero loadings for this factor; the loadings for these three variables range from –.50 to –.58, which is to say that they are of moderate size. For reasons that will be made clear shortly, this would make it difficult to interpret factor 1.

To make interpretation easier, you will normally perform a linear transformation on the factor solution called a **rotation**. Chapter 1 demonstrated the use of an orthogonal rotation. It was explained that orthogonal rotations result in components (or factors) that are uncorrelated with one another.

In contrast, this chapter illustrates the use of the promax rotation, which is a specific type of oblique rotation. Oblique rotations generally result in factors (or components) that are correlated with one another.

A promax rotation is actually conducted in two steps. The first step involves an orthogonal varimax prerotation. At this point in the analysis, the extracted factors are still uncorrelated. During the second step (the promax rotation), the orthogonality of the factors is relaxed, and they are allowed to become correlated. You will see that the interpretation of an oblique solution is more complicated than the interpretation of an orthogonal solution, although oblique rotations often provide better results (at least in those situations in which the actual, underlying factors truly are correlated).

Step 4: Interpreting the Rotated Solution

Orthogonal solutions. During the prerotation step, the SAS System produces a rotated factor pattern similar to that which would be produced if you had specified ROTATE=VARIMAX. This matrix appears on output page 5 of the current output, and is reproduced as Output 2.5.

```
Prerotation Method: Varimax                                          5

                    Orthogonal Transformation Matrix

                             1          2

                    1     0.82009   -0.57223
                    2     0.57223    0.82009

                    Rotated Factor Pattern

                          FACTOR1    FACTOR2

             V1              86 *       -28
             V2              93 *        -7
             V3              82 *       -16
             V4             -18          73 *
             V5             -17          76 *
             V6              -7          77 *

NOTE: Printed values are multiplied by 100 and rounded to the nearest integer.
      Values greater than 0.4 have been flagged by an '*'.

                    Variance explained by each factor

                       FACTOR1    FACTOR2
                      2.351818   1.800096


           Final Communality Estimates: Total = 4.151914

          V1         V2         V3         V4         V5         V6
      0.816776   0.874178   0.704434   0.558828   0.607056   0.590642
```

Output 2.5: Varimax (Orthogonal) Rotated Factor Pattern from Analysis of Investment Model Questionnaire

If you were interested in an orthogonal solution, it would be perfectly acceptable to interpret this rotated factor pattern in the manner described in Chapter 1. Interested readers may turn to that chapter for a detailed discussion of how this is done. Because this chapter deals with oblique rotations, however, it will instead focus on how to interpret the results of the promax procedure.

Oblique solutions. Before interpreting the meaning of the retained factors, you should first check the inter-factor correlations that appear on output page 6. The results for the current analysis are reproduced here as Output 2.6.

```
                        Inter-factor Correlations                        6

                             FACTOR1    FACTOR2

                    FACTOR1    100 *     -34
                    FACTOR2    -34       100 *

NOTE: Printed values are multiplied by 100 and rounded to the nearest integer.
      Values greater than 0.4 have been flagged by an '*'.
```

Output 2.6: Inter-Factor Correlations from Analysis of Investment Model Questionnaire

In Output 2.6, look in the section headed "Inter-factor Correlations". Where the row headed "FACTOR1" intersects with the column headed "FACTOR2" you will find a correlation coefficient of –.34. This means that there is a correlation of –.34 between the two factors. At this point in the analysis, you do not know exactly what this correlation means because you have not yet interpreted the meaning of the factors themselves. You will therefore return to this correlation after the interpretation of the factors has been completed.

In a sense, interpreting the nature of a given factor is relatively straightforward. You begin by looking for variables (survey items) that have high loadings on that factor. A high loading means that the variable is "measuring" that factor. You must review all of the variables with high loadings on that factor, and attempt to determine what the variables have in common. What underlying construct do all of the items seem to be measuring? In naming this construct, you name the factor.

As always, however, somewhat subjective decisions often must be made. For example, how large must a factor loading be before you will conclude it is a "high" loading? As with the preceding chapter, you can consider loadings equal to or greater than .40 as meaningful loadings, and you can ignore loadings under .40. As you gain expertise in performing factor analyses, you should explore the more sophisticated procedures for identifying significant loadings, such as those discussed by Stevens (1986).

With an orthogonal rotation, factor interpretation was fairly straightforward: you simply reviewed one matrix, the factor pattern matrix, to identify the variables with significant loadings on a given factor. With oblique rotations, however, the situation is somewhat more complex because you must interpret two (and in some cases three) different matrices in order to fully understand the results. In all cases, the rotated factor pattern and factor structure matrices should be reviewed; in some cases, it may also be necessary to review the reference structure matrix.

First, you should review the **rotated factor pattern matrix**. This matrix appeared at the bottom of page 6 of the output for the current analysis. It is reproduced here as Output 2.7.

```
                   Rotated Factor Pattern (Std Reg Coefs)

                              FACTOR1     FACTOR2

                    V1         85 *        -14
                    V2         97 *         10
                    V3         84 *         -1
                    V4         -5           73 *
                    V5         -4           76 *
                    V6          7           79 *

NOTE: Printed values are multiplied by 100 and rounded to the nearest integer.
      Values greater than 0.4 have been flagged by an '*'.
```

Output 2.7: Promax (Oblique) Rotated Factor Pattern from Analysis of Investment Model Questionnaire

Notice that the abbreviation "Std Reg Coefs" appears in parentheses as part of the heading for this matrix. This stands for "standardized regression coefficients," and should help you remember that the loadings that appear in this factor pattern are regression coefficients of the variables on the factors. In common factor analysis, the observed variables are viewed as linear combinations of the factors, and the elements of the factor pattern are regression weights associated with each factor in the prediction of these variables. The loadings in this matrix are also called **pattern loadings**, and may be said to represent the unique contribution that each factor makes to the variance of the observed variables (Rummel, 1970).

You should rely most heavily on this rotated factor pattern matrix to interpret the meaning of each factor. The rotated factor pattern is more likely to display simple structure than the structure matrix (discussed later) and will be more useful in determining what names should be assigned to the factors.

Chapter 1 provided a structured procedure to follow in interpreting a rotated factor pattern. These guidelines are reproduced again here:

A. **Read across the row for the first variable**. All meaningful loadings (i.e., loadings greater than .40) have been flagged with an asterisk ("*"). This was accomplished by including the FLAG=.40 option in the preceding program. If a given variable has a meaningful loading on more than one factor, scratch out that variable and ignore it in your interpretation. In many situations, researchers drop variables that load on more than one factor because the variables are not "pure" measures of any one construct. In the present case, this means reviewing the row headed V1, and reading to the right to see if it loads on more than one factor. In this case it does not, so you may retain this variable.

B. Repeat this process for the remaining variables, scratching out any variable that loads on more than one factor. In this analysis, none of the variables have high loadings for more than one factor, so none will have to be dropped.

C. Review all of the surviving variables with high loadings on factor 1 to determine the nature of this factor. From the rotated factor pattern, you can see that only items 1, 2, and 3 load on factor 1 (note the asterisks). It is now necessary to turn to the questionnaire itself and review the content of the questions in order to decide what a given factor should be named. What do questions 1, 2, and 3 have in common? What common construct do they seem to be measuring? For illustration, the questions being analyzed in the present case are reproduced again here. Remember that question 1 was represented as V1 in the SAS program, question 2 was V2, and so forth. To interpret factor 1, you must read questions 1, 2, and 3 to see what they have in common.

Please rate each of the following items to indicate the extent to which you agree or disagree with each statement. Use a response scale in which 1=Strongly Disagree and 7=Strongly Agree.

_____ 1. I have invested a lot of time and effort in developing my relationship with my current partner.

_____ 2. My current partner and I have developed interest in a lot of fun activities that I would lose if our relationship were to end.

_____ 3. My current partner and I have developed a lot of mutual friendships that I would lose if our relationship were to end.

_____ 4. It would be more attractive for me to be involved in a relationship with someone else rather than continue in a relationship with my current partner.

_____ 5. It would be more attractive for me to be by myself than to continue my relationship with my current partner.

_____ 6. In general, my alternatives to remaining in this relationship are quite attractive.

Questions 1, 2, and 3 all seem to be dealing with the size of the investment that the respondent has put into the relationship. It is therefore reasonable to label factor 1 the "investment size" factor.

D. Repeat this process to name the remaining retained factors. In the present case, there is only one remaining factor to name: factor 2. This factor has high loadings for questions 4, 5, and 6. In reviewing these items, it becomes clear that each seems to deal with the attractiveness of one's alternatives to the current relationship. It is therefore reasonable to label this the "alternative value" factor.

E. Determine whether this solution satisfies the interpretability criteria. The overall results of a principal factor analysis are satisfactory only if they meet the following interpretability criteria:

1. **Are there at least three variables (items) with significant loadings on each retained factor?** In the present example, three variables loaded on factor 1, and three also loaded on factor 2, so this criterion was met.

2. **Do the variables that load on a given factor share some conceptual meaning?** All three variables loading on factor 1 are clearly measuring investment size, while all four loading on factor 2 are clearly measuring alternative value. Therefore, this criterion is met.

3. **Do the variables that load on different factors seem to be measuring different constructs?** Because the items loading on the investment size factor seem to be conceptually very different from the items loading on the alternative value factor, this criterion seems to be met as well.

4. **Does the rotated factor pattern demonstrate simple structure?** A rotated factor pattern demonstrates simple structure when it has two characteristics. First, most of the variables should have high loadings on one factor, and near-zero loadings on the other factors. You can see that the pattern obtained here meets that requirement: items 1–3 have high loadings on factor 1, and near-zero loadings on factor 2. Similarly, items 4–6 have high loadings on factor 2, and near-zero loadings on factor 1. The second characteristic of simple structure is that each factor should have high loadings for some variables, and near-zero loadings for the others. Again, the pattern obtained here also meets this requirement: factor 1 has high loadings for items 1–3 and near-zero loadings for the other items, while factor 2 has high loadings for items 4–6, and near-zero loadings on the remaining items. In short, the rotated factor pattern obtained in this analysis does seem to demonstrate simple structure.

As was stated earlier, the rotated factor pattern should be the first matrix reviewed by you in naming the factors. However, it does have one limitation: the pattern loadings of this matrix are not constrained to range between +1.00 and –1.00. In some rare cases in which the factors are strongly correlated, some loadings may be as large as 10.00 or even larger. In such cases the interpretation of the pattern matrix may be difficult.

When faced with such a situation, it is generally easier to review the **reference structure matrix** instead. This appears under the heading "Reference Structure (Semipartial Correlations)" on output page 7. The reference structure for the current analysis of the investment model questionnaire is reproduced here as Output 2.8.

```
              Reference Structure (Semipartial Correlations)                7

                              FACTOR1    FACTOR2

                    V1         80 *       -13
                    V2         91 *        10
                    V3         78 *        -1
                    V4         -5          68 *
                    V5         -4          72 *
                    V6          6          74 *

NOTE: Printed values are multiplied by 100 and rounded to the nearest integer.
      Values greater than 0.4 have been flagged by an '*'.
```

Output 2.8: Reference Structure (Semipartial Correlations) from Analysis of Investment Model Questionnaire

The heading for the reference structure includes the words "Semipartial Correlations" in parentheses. This is because the coefficients in this matrix represent the semipartial correlations "between variables and common factors, removing from each common factor the effects of other common factors" (*SAS/STAT users guide, volume 1*, 1989, p. 800).

The steps followed in interpreting the reference structure are identical to those followed in reviewing the factor pattern. Notice that the size of the loadings in the current reference structure are very similar to those in the rotated factor pattern. It is clear that interpreting the reference structure in this study would have led to exactly the same interpretation of factors as was obtained using the rotated factor pattern.

In addition to interpreting the rotated factor pattern (and reference structure, if necessary), you should also review the **factor structure matrix**. The structure matrix for the present study also appears on page 7, and is presented here as Output 2.9.

```
              Factor Structure (Correlations)

                        FACTOR1    FACTOR2

          V1              89 *      -43 *
          V2              93 *      -23
          V3              84 *      -30
          V4             -30         75 *
          V5             -30         78 *
          V6             -20         77 *


NOTE: Printed values are multiplied by 100 and rounded to the nearest integer.
      Values greater than 0.4 have been flagged by an '*'.
```

Output 2.9: Factor Structure (Correlations) from Analysis of Investment Model Questionnaire

The word "Correlations" appears in parentheses in the heading for this matrix, because the structure loadings that it contains represent the product-moment correlations between the variables and common factors. For example, where the row for V1 intersects with the column for FACTOR1, a structure loading of .89 appears. This reveals that the correlation between item 1 and factor 1 is +.89.

The structure matrix is generally less useful for interpreting the meaning of the factors (compared to the rotated pattern matrix) because it often fails to demonstrate simple structure. For example, notice that the "low" loadings in this structure matrix are not really that low: the loading of V1 on factor 2 is −.43; the corresponding loading from the rotated pattern matrix was considerably lower at −.14. Comparing the rotated pattern matrix to the structure matrix reveals the superiority of the former in achieving simple structure.

If this is the case, then why review the structure matrix at all? Because the pattern matrix and the structure matrix provide different information about the relationships between the observed variables and the underlying factors. The factor pattern reveals the unique contribution that each factor makes to the variance of the variable. The pattern loadings in this matrix are essentially standardized regression coefficients comparable to those obtained in multiple regression.

The factor structure, on the other hand, reveals the correlation between a given factor and variable. It helps you understand the big picture of how the variables are really related to the factors. For example, consider the rotated factor pattern matrix which appeared on page 6 of the current output. It is reproduced again here as Output 2.10.

```
           Rotated Factor Pattern (Std Reg Coefs)

                        FACTOR1     FACTOR2

           V1             85 *        -14
           V2             97 *         10
           V3             84 *         -1
           V4             -5           73 *
           V5             -4           76 *
           V6              7           79 *

NOTE: Printed values are multiplied by 100 and rounded to the nearest integer.
      Values greater than 0.4 have been flagged by an '*'.
```

Output 2.10: Promax (Oblique) Rotated Factor Pattern from Analysis of Investment Model Questionnaire

Notice that the pattern loading for V1 on factor 2 is only −.14. Do not allow this very weak pattern loading to mislead you into believing that V1 and factor 2 are completely unrelated. Because this is a *pattern* loading, its small value merely means that factor 2 makes a very small *unique* contribution to the variance in V1.

For contrast, now consider the **structure loading** for V1 on factor 2 (from Output 2.9) The structure loading reveals that V1 actually demonstrates a correlation with factor 2 of −.43. Why would V1 be negatively correlated with factor 2? Because V1 is directly related to factor 1, and factor 1, in turn, is negatively correlated with factor 2. This negative correlation is illustrated graphically in Figure 2.4. Notice that there is a curved, double-headed arrow that connects factor 1 to factor 2. The arrow is identified with a negative sign. This curved arrow is used to show that factors 1 and 2 are negatively correlated.

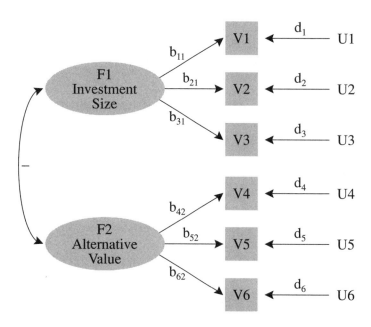

Figure 2.4: Path Model for a 6-Variable, 2-Factor Model, Oblique Factors, Factorial Complexity=1

The path model of Figure 2.4 is identical to that of Figure 2.1, with one exception: a curved, double-headed arrow now connects factors 1 and 2. This means that the factors are now oblique, or correlated. This figure helps demonstrate how a variable could have a moderately large structure loading for a factor, but a small pattern loading. The structure loading for V1 and factor 2 is –.43 because V1 is caused by factor 1, and factor 1 is negatively correlated with factor 2. However, the pattern loading for V1 and factor 2 is much smaller at –.14 because factor 2 has essentially no direct effect on V1.

In summary, you should always review the pattern matrix to determine which groups of variables are measuring a given factor, for purposes of interpreting the *meaning* of that factor. You should then review the structure matrix to get the big picture concerning the simple bivariate relations between variables and factors.

If the structure matrix is so important, then why was it not discussed in the chapter on principal component analysis? Because the pattern matrix and the structure matrix are one and the same in a principal component analysis with an orthogonal rotation. Technically, the loadings of the pattern matrix in principal component analysis can be viewed as regression coefficients, as in common factor analysis. However, remember that the principal components of this analysis are orthogonal, or uncorrelated. Because of this orthogonality, the regression coefficients for the components are equivalent to the correlation between the components and the variables. This is to say that the loadings of the pattern matrix can also be interpreted as correlations between the components and the variables. Hence, there is no difference between a factor pattern matrix and a factor structure matrix in principal component analysis with an orthogonal rotation. This is why only the pattern matrix is printed and interpreted.

Step 5: Creating Factor Scores or Factor-Based Scores

Once the analysis is complete, it is often desirable to assign scores to each subject to indicate where that subject stands on the retained factors. For example, the two factors retained in the present study were interpreted as an investment size factor and alternative value factor. Now, you may want to assign one score to each subject to indicate where that subject stands on the investment size factor, and a different score to indicate where that subject stands on the alternative value factor. With this done, these factor scores could then be used either as predictor variables or as criterion variables in subsequent analyses.

Before discussing the options for assigning these scores, it is necessary to first draw a distinction between factor scores and estimated factor scores. A **factor score** represents a subject's actual standing on an underlying factor. An **estimated factor score**, on the other hand, is merely an estimate of a subject's standing on that underlying factor. In practice, researchers are never able to compute true factor scores for subjects. This is due to a fundamental indeterminancy in factor analysis. In the end, factor scores are estimated by creating linear composites of the observed variables. That is, you compute factor scores by adding together the subjects' (optimally weighted) scores on the observed variables. But remember that the common part of a given variable (that part influenced by the common factor) is inseparately mixed with that variable's unique component. This means that there will always be some error associated with the computation of factor scores, and so it is better to refer to them as *estimated* factor scores.

Estimated factor scores. Broadly speaking, two scoring approaches are available. The more sophisticated approach is to allow PROC FACTOR to compute estimated factor scores. An estimated factor score is a linear composite of the optimally weighted variables under analysis. For example, to calculate a subject's estimated factor score on factor 1, you would use the following equation:

$$F'_1 = b_{11}V_1 + b_{12}V_2 + b_{13}V_3 + \ldots b_{1p}V_p$$

where

F'_1 = the estimated factor score for factor 1
b_{11} = the scoring coefficient for survey question 1, used in creating estimated factor score 1
V_1 = the subject's score on survey question 1
b_{12} = the scoring coefficient for survey question 2, used in creating estimated factor score 1
V_2 = the subject's score on survey question 2
b_{1p} = the scoring coefficient for survey question p (the last question), used in creating estimated factor score 1
V_p = the subject's score on survey question p.

A different equation, with different scoring coefficients, would be used to calculate the subjects' scores on the remaining retained factors. In practice, you do not actually have to create equations such as those appearing here. Instead, these estimated factor scores may be created automatically by requesting the creation of a new data set within the SAS program. This is done by including the OUT= and NFACT= options in the FACTOR statement.

The general form for the NFACT= option is:

```
NFACT=number-of-factors-to-be-retained
```

The general form for the OUT= option is:

```
OUT=name-of-new-SAS-data-set
```

The following SAS program incorporates these options:

```
1       PROC FACTOR    DATA=D1
2                      SIMPLE
3                      METHOD=PRIN
4                      PRIORS=SMC
5                      NFACT=2
6                      ROTATE=PROMAX
7                      ROUND
8                      FLAG=.40
9                      OUT=D2 ;
10         VAR V1-V6 ;
11         RUN;
12
13      PROC CORR    DATA=D2;
14         VAR COMMIT FACTOR1 FACTOR2;
15         RUN;
```

Line 9 of the preceding program asks that an output data set be created and given the name "D2." This name was arbitrary; any name consistent with SAS System requirements would have been acceptable. The new data set named D2 will contain all of the variables contained in the previous data set, as well as new variables named FACTOR1 and FACTOR2. FACTOR1 will contain estimated factors scores for the first retained factor, and FACTOR2 will contain estimates for the second factor. The number of new "FACTOR" variables created will be equal to the number of factors retained by the NFACT= option.

The OUT= option may be used only if the factor analysis has been performed on a raw data set (as opposed to a correlation or covariance matrix). The use of the NFACT= option is also required.

Having created the new estimated factor score variables named FACTOR1 and FACTOR2, you may be interested in seeing how they relate to some of the study's other variables (i.e., variables not analyzed in the factor analysis itself). You may therefore append a PROC CORR statement to your program following the last of the PROC FACTOR statements. In the preceding program, these statements appear on lines 13-15.

These PROC CORR statements request that COMMIT be correlated with FACTOR1 and FACTOR2. COMMIT represents subjects' commitment to the relationship. High scores on this variable indicate that subjects intend to remain in the relationship with their current partners (assume that the variable COMMIT was also measured with the questionnaire, and that scores on COMMIT were keyed as part of data set D1). These PROC CORR statements result in the SAS output that is reproduced here as Output 2.11.

```
                        Correlation Analysis                             7

                3 'VAR' Variables:  COMMIT   FACTOR1   FACTOR2


                          Simple Statistics

Variable          N        Mean      Std Dev         Sum      Minimum      Maximum

COMMIT           50    15.52000      6.67692   776.00000      4.00000     28.00000
FACTOR1          50           0      0.95720           0     -2.25877      1.68987
FACTOR2          50           0      0.88955           0     -1.29220      2.75565


     Pearson Correlation Coefficients / Prob > |R| under Ho: Rho=0 / N = 50

                          COMMIT            FACTOR1            FACTOR2

      COMMIT            1.00000            0.31881           -0.29307
                            0.0             0.0240             0.0389


      FACTOR1           0.31881            1.00000           -0.39458
                         0.0240                0.0             0.0046


      FACTOR2          -0.29307           -0.39458            1.00000
                         0.0389             0.0046                0.0
```

Output 2.11: Correlations between COMMIT and Estimated Factor Score Variables FACTOR1 and FACTOR2

The correlations of interest appear in Output 2.11 below the heading "Pearson Correlation Coefficients...". Look at the first column of coefficients, which is headed "COMMIT". Where this column intersects with the row headed FACTOR1, you can see that FACTOR1 displays a correlation of approximately +.32 with commitment. This makes sense because the first retained factor was interpreted as the investment size factor. It is logical that investment size would be positively correlated with commitment to maintain the relationship. The second estimated factor score variable, FACTOR2, displays a correlation of approximately −.29 with commitment, and this too is logical. The second retained factor was interpreted as alternative value. It makes sense that commitment would decrease as the attractiveness of one's alternatives increases. FACTOR1 and FACTOR2 may now be used as predictor or criterion variables in any other appropriate SAS System procedures.

Factor-based scales. A second (and less sophisticated) approach to scoring involves the creation of factor-based scales. A **factor-based scale** is a variable that estimates subject scores on the underlying factors, but does not use an optimally weighted formula to do this (as was the case with the estimated factor scores created by PROC FACTOR).

Although a factor-based scale can be created in a number of ways, the following method has the advantages of being relatively straightforward and commonly used:

1. To calculate scores on factor-based scale 1, first determine which questionnaire items had high loadings on factor 1.

2. For a given subject, add together that subject's responses to these items. The result is that subject's score on the factor-based scale for factor 1.

3. Repeat these steps to calculate each subject's score on the factor-based scales for remaining retained factors.

Although this may sound like a cumbersome task, it is actually made quite simple through the use of data manipulation statements contained in a SAS program. For example, assume that you have performed the factor analysis on your survey responses, and have obtained the findings reported in this chapter. Specifically, it was found that survey items 1, 2, and 3 loaded on factor 1 (the investment size factor), while items 4, 5, and 6 loaded on factor 2 (the alternative value factor).

You would now like to create two new SAS variables. The first variable, called INVEST, will include each subject's score on the factor-based scale for investment size. The second variable, called ALTERN, will include each subject's score on the factor-based scale for alternative value. Once these variables are created, you can use them as criterion variables or predictor variables in subsequent correlations, multiple regressions, ANOVAs, or other analyses. To keep things simple in the present example, assume that you are interested in determining whether there is a significant correlation between COMMIT and INVEST, and between COMMIT and ALTERN.

At this time, it may be useful to review Appendix A.3, "Working with Variables and Observations in SAS Data Sets," particularly the section on "Creating New Variables from Existing Variables." Such a review should make it easier to understand the data manipulation statements used here.

Assume that earlier statements in the SAS program have already input subject responses to questionnaire items, including subjects' scores on the variable COMMIT. These variables are included in a data set called D2. Here are the subsequent SAS statements that would go on to create a new data set called D3, which would include all of the variables in D2 as well as the newly created, factor-based scales called INVEST and ALTERN.

```
16
17        DATA D3;
18           SET D2;
19
20        INVEST  = (V1 + V2 + V3);
21        ALTERN  = (V4 + V5 + V6);
22
23        PROC CORR    DATA=D3;
24           VAR COMMIT INVEST ALTERN;
25           RUN;
```

Lines 17 and 18 request that a new data set called D3 be created, and that it be set up as a duplicate of existing data set D2. In line 20, the new variable called INVEST is created. For each subject, his or her responses to items 1, 2 and 3 are added together. The result is the subject's score on the factor-based scale for the first factor. These scores are stored in a variable called INVEST. The factor-based scale for the alternative value factor is created on line 21, and these scores are stored in the variable called ALTERN. Lines 23-25 request the correlations between COMMIT, INVEST, and ALTERN be determined.

Fictitious results from the preceding program are reproduced as Output 2.12.

Correlation Analysis   7

3 'VAR' Variables: COMMIT   INVEST   ALTERN

Simple Statistics

Variable	N	Mean	Std Dev	Sum	Minimum	Maximum
COMMIT	50	15.52000	6.67692	776.00000	4.00000	28.00000
INVEST	50	13.36000	4.36479	668.00000	3.00000	21.00000
ALTERN	50	7.68000	3.22895	384.00000	3.00000	18.00000

```
    Pearson Correlation Coefficients / Prob > |R| under Ho: Rho=0 / N = 50

                            COMMIT            INVEST            ALTERN

          COMMIT          1.00000           0.33798          -0.26380
                          0.0               0.0164            0.0642

          INVEST          0.33798           1.00000          -0.30588
                          0.0164            0.0               0.0308

          ALTERN         -0.26380          -0.30588           1.00000
                          0.0642            0.0308            0.0
```

Output 2.12: Correlations between COMMIT and Factor-Based Scales INVEST and ALTERN

You can see that the correlations between COMMIT and the estimated factor scores (FACTOR1 and FACTOR2) discussed earlier are slightly different from the correlations between COMMIT and the factor-based scales (INVEST and ALTERN) presented here. For example, the correlation between COMMIT and FACTOR1 (the estimated factor-score variable for investment size) was approximately .32, while the correlation between COMMIT and INVEST (the factor-based scale for investment size) was slightly higher at approximately .34. These differences are to be expected, as the estimated factor scores (FACTOR1 and FACTOR2) are optimally weighted linear composites, while the factor-based scales (INVEST and ALTERN) are not optimally weighted. In fact, it would be instructive to create a single correlation matrix that includes both the estimated factor scores as well as the factor-based scales. This could be done with the following statements:

```
 1      PROC FACTOR    DATA=D1
 2                     SIMPLE
 3                     METHOD=PRIN
 4                     PRIORS=SMC
 5                     NFACT=2
 6                     ROTATE=PROMAX
 7                     ROUND
 8                     FLAG=.40
 9                     OUT=D2 ;
10          VAR V1-V6 ;
11          RUN;
12
13
14      DATA D3;
15         SET D2;
16
17      INVEST  = (V1 + V2 + V3);
18      ALTERN  = (V4 + V5 + V6);
19
```

```
20      PROC CORR    DATA=D3;
21          VAR COMMIT FACTOR1 FACTOR2 INVEST ALTERN;
22          RUN;
```

This program results in the correlation matrix reproduced here as Output 2.13.

```
                    Correlation Analysis                    7

        5 'VAR' Variables:  COMMIT   FACTOR1   FACTOR2   INVEST   ALTERN


                        Simple Statistics

Variable        N       Mean    Std Dev        Sum     Minimum     Maximum

COMMIT         50    15.52000   6.67692   776.00000    4.00000    28.00000
FACTOR1        50          0    0.95720          0    -2.25877     1.68987
FACTOR2        50          0    0.88955          0    -1.29220     2.75565
INVEST         50    13.36000   4.36479   668.00000    3.00000    21.00000
ALTERN         50     7.68000   3.22895   384.00000    3.00000    18.00000


      Pearson Correlation Coefficients / Prob > |R| under Ho: Rho=0 / N = 50

                COMMIT      FACTOR1     FACTOR2      INVEST      ALTERN

COMMIT         1.00000      0.31881    -0.29307     0.33798     -0.26380
               0.0          0.0240      0.0389      0.0164       0.0642

FACTOR1        0.31881      1.00000    -0.39458     0.99431     -0.32121
               0.0240       0.0        0.0046      0.0001       0.0229

FACTOR2       -0.29307     -0.39458     1.00000    -0.38401      0.99043
               0.0389       0.0046      0.0        0.0059       0.0001

INVEST         0.33798      0.99431    -0.38401     1.00000     -0.30588
               0.0164       0.0001      0.0059      0.0          0.0308

ALTERN        -0.26380     -0.32121     0.99043    -0.30588      1.00000
               0.0642       0.0229      0.0001      0.0308       0.0
```

Output 2.13: Correlations between COMMIT, Estimated Factor Score Variables FACTOR1 and FACTOR2, and Factor-Based Scales INVEST and ALTERN

The correlations of interest appear in the section headed "Pearson Correlation Coefficients...". Remember that FACTOR1 contains the estimated factor scores for investment size, while INVEST is the factor-based scale for investment size. Where the column for FACTOR1 intersects the row for INVEST, you will find a correlation coefficient of approximately .994, meaning that the estimated factor score variable and the factor-based scale for this construct are almost perfectly correlated. In the same way, the correlation of approximately .990 between

FACTOR2 and ALTERN shows that the estimated factor score variable and the factor-based scale for alternative value are also very strongly correlated.

Recoding reversed items prior to analysis. It is generally best to recode any reversed items before conducting any of the analyses described here. In particular, it is essential that reversed items be recoded prior to the program statements that produce factor-based scales. For example, the three questionnaire items designed to assess investment size are once again reproduced here:

```
Please rate each of the following items to indicate the extent to
which you agree or disagree with each statement.  Use a response
scale in which 1=Strongly Disagree and 7=Strongly Agree.

_____   1. I have invested a lot of time and effort in developing
            my relationship with my current partner.
_____   2. My current partner and I have developed interest in a
            lot of fun activities that I would lose if our
            relationship were to end.
_____   3. My current partner and I have developed a lot of mutual
            friendships that I would lose if our relationship were
            to end.
```

None of these items is reversed; with each item, a response of "7" indicates a high level of investment. In the following however, item 1 is a reversed item: with it, a response of "7" indicates a *low* level of investment:

```
_____   1. I have invested very little of my time and effort in
            developing my relationship with my current partner.
_____   2. My current partner and I have developed interest in a
            lot of fun activities that I would lose if our
            relationship were to end.
_____   3. My current partner and I have developed a lot of mutual
            friendships that I would lose if our relationship were
            to end.
```

If you were to perform a factor analysis on responses to these items, the factor loading for item 1 would most likely have a sign that is the opposite of the sign of the loadings for items 2 and 3 (e.g., if items 2 and 3 had positive loadings, item 1 would have a negative loading). This would complicate the creation of a factor-based scale: with items 2 and 3, higher scores indicate greater investment; with item 1, lower scores indicate greater investment. Clearly, you would not want to sum these three items together given the way they are coded presently. First, you need to reverse item 1. Notice how this is done in the following program (assume that the data have already been input in a SAS data set named D1):

```
15      DATA D2;
16          SET D1;
17
18      V1 = 8 - V1;
19
20      INVEST   = (V1 + V2 + V3) ;
21      ALTERN   = (V4 + V5 + V6) ;
22
23      PROC CORR    DATA=D2;
24          VAR COMMIT INVEST ALTERN;
25          RUN;
```

With line 18, you are creating a new version of variable V1. Values on this new version of V1 will be equal to the quantity "8 minus the value of the old version of V1." Therefore, for subjects whose score on the old version of V1 was 1, their value on the new version of V1 will be 7 (because $8 - 1 = 7$); for subjects whose score on the old version of V1 was 7, their value on the new version of V1 will be 1 (because $8 - 7 = 1$); and so forth.

The general form of the formula used when recoding reversed items is:

```
Variable name = constant - variable name ;
```

In this formula, the "constant" is the following quantity:

```
(the number of points on the response scale used with the
questionnaire item + 1)
```

Therefore if you are using the 4-point response scale, the constant is 5; if using a 9-point scale, the constant is 10, and so forth.

If you have prior knowledge about which items will appear as reversed items (with reversed factor loadings) in your results, it is best to place these recoding statements early in your SAS program, before the PROC FACTOR statements. This will make interpretation of the factors a bit more straightforward because it will eliminate significant loadings with opposite signs from appearing on the same factor. In any case, it is essential that the statements that recode reversed items appear before the statements that create any factor-based scales.

Step 6: Summarizing the Results in a Table

In some cases, you may want to prepare a table presenting the rotated factor pattern and factor structure for the variables analyzed. One possible format is presented in Table 2.2.

Table 2.2

Questionnaire Items and Corresponding Factor Loadings from the Rotated Factor Pattern Matrix and Factor Structure Matrix, Decimals Omitted

Factor Pattern		Factor Structure			Questionnaire Item
1	2	1	2		
85	-14	89	-43	1.	I have invested a lot of time and effort in developing my relationship with my current partner.
97	10	93	-23	2.	My current partner and I have developed interest in a lot of fun activities that I would lose if our relationship were to end.
84	-1	84	-30	3.	My current partner and I have developed a lot of mutual friendships that I would lose if our relationship were to end.
-5	73	-30	77	4.	It would be more attractive for me to be involved in a relationship with someone else rather than continue in a relationship with my current partner.
-4	76	-30	78	5.	It would be more attractive for me to be by myself than to continue my relationship with my current partner.
7	79	-20	77	6.	In general, my alternatives to remaining in this relationship are quite attractive.

Note. N = 50.

If the journal format allows for an even larger table, it may be desirable to include an additional column presenting the final communality estimates. The column should be headed "h^2", the symbol for communality. These final communality estimates appear in the SAS output following the factor structure matrix. Table 1.2 from Chapter 1 shows how communalities may be presented in a table.

When many factors are retained, or when the questionnaire items are long or numerous, it probably will not be possible to present the factor loadings, communalities, and questionnaire items all in one table. In these situations, the loadings and communalities should be presented in one table, and the items in a second (or within the text of the paper).

Step 7: Preparing a Formal Description of the Results for a Paper

Although an exploratory factor analysis often involves a good deal of work, it is usually summarized in a brief way in a research paper. For example, the preceding analysis could be summarized as follows:

> Responses to the 6-item questionnaire were subjected to an exploratory factor analysis using squared multiple correlations as prior communality estimates. The principal factor method was used to extract the factors, and this was followed by a promax (oblique) rotation. A scree test suggested two meaningful factors, so only these factors were retained for rotation.
>
> In interpreting the rotated factor pattern, an item was said to load on a given factor if the factor loading was .40 or greater for that factor, and was less than .40 for the other. Using these criteria, three items were found to load on the first factor, which was subsequently labeled the investment size factor. Three items also loaded on the second factor, which was labeled the alternative value factor. Questionnaire items and corresponding factor loadings are presented in Table 2.2.

A More Complex Example:
The Job Search Skills Questionnaire

The results presented in the preceding section were designed to be relatively simple so that the basic concepts of factor analysis would be easier to learn. In conducting actual research, however, the results are seldom so clear cut. Very often you are forced to make somewhat subjective decisions and are forced to choose between more than one interpretable solution. This section illustrates these problems by presenting a somewhat more complex analysis.

Assume that you are now conducting research in the area of college student career development. You have developed an instrument to assess student knowledge and ability in a wide variety of areas related to occupational choice and the job-search process. The instrument consists of 100 items, and the items are divided into 25 scales. Each scale contains 4 items.

Here are the SAS variable names for each scale. To the right of the SAS variable name is the full name for the scale (in italics) and a sample item from the scale (in parentheses). Reviewing the scale names and sample items should make clear what type of knowledge or ability is assessed by each scale.

For example, the first scale is identified with the SAS variable name, VALUES. The full name for this scale is *Clarifying values and interests*, and the sample item is "My ability to describe just what my work-related interests are." Subjects responded to each item on the questionnaire using a 7-point scale in which 1 = Very Bad and 7 = Very Good.

1. VALUES	*Clarifying values and interests* ("My ability to describe just what my work-related interests are").
2. ABILIT	*Identifying work-related abilities and skills* ("My ability to describe just what my strongest work-related skills and abilities are").
3. ASSESS	*Using assessment instruments* ("My knowledge of what specific assessment instruments are available to help me assess my interests").
4. STRAT	*Identifying effective job search strategies* ("My knowledge of which job-search strategies are most effective").
5. EXPER	*Getting job-related experience* ("My knowledge of how I could get relevant job experience in my field before I graduate").
6. ORGCHAR	*Identifying preferred organizational characteristics* ("My ability to clearly describe the exact characteristics an organization should have in order to satisfy my personal preferences").
7. RESOCCU	*Researching potential occupations* ("My knowledge of what specific books and other published sources provide useful information about specific occupations").
8. RESEMPL	*Researching specific employers* ("My ability to collect detailed information on a specific organization just before I have an employment interview there").
9. GOALS	*Setting goals* ("My ability to clearly describe my career goals for the next five years").
10. BARRIER	*Dealing with occupational barriers* ("My knowledge of what types of occupational barriers are likely to stand in my way of getting the job I really want").
11. MOTIVAT	*Staying motivated* ("My ability to maintain a high level of motivation throughout my job search").
12. RESUMES	*Using resumes* ("My ability to write a highly effective resume").
13. LETREC	*Using letters of recommendation* ("My knowledge of what I should do to ensure that my reference writes a very effective letter of recommendation for me").
14. DIRECT	*Using the cover letter/direct mail approach* ("My knowledge of just what should be included in a cover letter used in the direct mail approach").

15. APPLIC   *Completing application forms* ("My ability to complete an application form in such a way as to make the best possible impression on an employer").

16. IDENEMP   *Identifying potential employers* ("My knowledge of exactly what books/references are available to help me identify organizations that might hire me").

17. CARDEV   *Using campus career development services* ("My ability to clearly describe exactly what services are offered by the career development office on this campus").

18. AGENCY   *Using employment agencies* ("My knowledge of how to make the most effective use of an employment agency").

19. FAIRS   *Using job fairs* ("My ability to make the most effective use of a job fair").

20. ADVERT   Responding to advertised job openings ("My knowledge of how to effectively respond to a job advertisement by phone").

21. COUNS   *Using career counselors/consultants* ("My knowledge of how to use career counselors/consultants to make the most of the services they offer").

22. UNADV   *Applying directly for unadvertised jobs* ("My knowledge of the correct way to directly apply for an unadvertised position that I want").

23. NETWORK   *Using the networking approach to job search* ("My knowledge of the most effective ways of producing job leads by asking for help from friends, relatives, past employers, and other contacts").

24. INTERV   *Managing the employment interview process* ("My knowledge of how to respond to tough interview questions").

25. SALARY   *Negotiating salary* ("My ability to successfully negotiate a fair and motivating salary").

Assume that you administered your scale to 258 college students, and obtained usable responses from 220 of these. You determined each student's score on each of the 25 scales, meaning that there are 25 data points for each student. You now want to perform an exploratory factor analysis to identify the factor structure underlying the data. (Note: The data analyzed here are actually from Ruddle, Thompson, and Hatcher, 1993).

Notice that the analysis will be performed on the 25 scale scores, not on the responses to each of the 100 individual questionnaire items. This approach is justifiable only if you have reason to believe that each of the 25 scales assesses just one construct. It would not be appropriate if, for example, items 1 and 2 within scale 1 assessed one construct, and items 3 and 4 within the same scale assessed a different construct. In this latter case, it would be more appropriate to perform a factor analysis using all 100 of the individual items as variables (of course, that analysis would require a very large sample size in order to maintain a good ratio of subjects per variable).

However, assume you have evidence that each scale does in fact assess just one construct. Assume that coefficient alpha exceeds .90 for each scale, and that the item-total correlations are quite high. These findings suggest that the individual scales are unifactorial. Therefore you will use scores on the 25 scales as observed variables in the factor analysis.

The SAS Program

The data analyzed here appear in Appendix B. This is the SAS program (minus the DATA step) to perform an exploratory factor analysis on the data from your study.

```
1     PROC FACTOR    DATA=D1
2                    SIMPLE
3                    METHOD=ML
4                    PRIORS=SMC
5                    NFACT=1
6                    SCREE
7                    ROTATE=PROMAX
8                    ROUND
9                    FLAG=.40    ;
10       VAR VALUES ABILIT ASSESS STRAT EXPER ORGCHAR RESOCCU
11           RESEMPL GOALS BARRIER MOTIVAT RESUMES LETREC
12           DIRECT APPLIC IDENEMP CARDEV AGENCY FAIRS
13           ADVERT COUNS UNADV NETWORK INTERV SALARY ;
14       RUN;
```

In most respects, the preceding program is similar to the other exploratory factor analysis programs discussed earlier in this chapter. The PRIORS= option (line 4) requests that squared multiple correlations again be used as prior communality estimates, and the FLAG= option (line 9) requests that factor loadings whose absolute values exceed .40 be flagged with asterisks. The NFACT= option (line 5) requests that one factor be retained (once again, you have no empirical evidence to expect any specific number of factors at this stage of the analysis; one factor was specified simply as a starting point).

This program differs from the other analyses, however, in that METHOD=ML (line 3) requests that the maximum likelihood method be used to extract the factors. As was mentioned earlier, some researchers prefer this method because it is believed to provide more accurate parameter estimates, and also provides a significance test to help solve the number of factors problem. Because of these advantages, the use of the maximum likelihood method will be demonstrated in this section.

With LINESIZE=80 and PAGESIZE=60 in the OPTIONS statement, the preceding program would produce 7 pages of output. Some of the information appearing on each page is summarized here:

- Page 1 includes simple statistics.

- Page 2 includes prior communality estimates and the eigenvalue table.

- Page 3 includes the scree plot of eigenvalues.

- Page 4 includes the iteration history and the significance tests for the number of factors extracted.

- Page 5 includes the unrotated factor pattern.

- Page 6 includes the final communality estimates.

- Page 7 includes a note reminding you that factor rotation is not possible when the analysis involves just one factor (as this analysis does).

Portions of this output are reproduced on the following pages as Output 2.14, Output 2.15, and Output 2.16.

Determining the Number of Factors to Retain

The scree plot. Because 25 scales were analyzed, you know that 25 factors will be extracted. The eigenvalue table for these factors, along with the scree plot, is reproduced here as Output 2.14.

The SAS System   2

Initial Factor Method: Maximum Likelihood

Prior Communality Estimates: SMC

VALUES	ABILIT	ASSESS	STRAT	EXPER	ORGCHAR	RESOCCU
0.520125	0.423624	0.556893	0.621408	0.584038	0.536296	0.707146

RESEMPL	GOALS	BARRIER	MOTIVAT	RESUMES	LETREC	DIRECT
0.705383	0.561503	0.533967	0.549074	0.664709	0.668555	0.693434

APPLIC	IDENEMP	CARDEV	AGENCY	FAIRS	ADVERT	COUNS
0.526580	0.745961	0.634829	0.559441	0.563656	0.754287	0.699202

UNADV	NETWORK	INTERV	SALARY
0.668733	0.608614	0.635764	0.683456

Preliminary Eigenvalues: Total = 43.2133584 Average = 1.72853434

	1	2	3	4	5
Eigenvalue	35.8841	3.2832	1.6709	1.4372	1.2100
Difference	32.6009	1.6124	0.2336	0.2272	0.3734
Proportion	0.8304	0.0760	0.0387	0.0333	0.0280
Cumulative	0.8304	0.9064	0.9450	0.9783	1.0063

	6	7	8	9	10
Eigenvalue	0.8366	0.6662	0.4418	0.3526	0.3083
Difference	0.1704	0.2245	0.0892	0.0443	0.0880
Proportion	0.0194	0.0154	0.0102	0.0082	0.0071
Cumulative	1.0257	1.0411	1.0513	1.0595	1.0666

	11	12	13	14	15
Eigenvalue	0.2203	0.1997	0.0900	-0.0113	-0.0500
Difference	0.0206	0.1097	0.1013	0.0387	0.0449
Proportion	0.0051	0.0046	0.0021	-0.0003	-0.0012
Cumulative	1.0717	1.0763	1.0784	1.0781	1.0770

	16	17	18	19	20
Eigenvalue	-0.0949	-0.1702	-0.2438	-0.2737	-0.3312
Difference	0.0754	0.0735	0.0300	0.0575	0.0567
Proportion	-0.0022	-0.0039	-0.0056	-0.0063	-0.0077
Cumulative	1.0748	1.0708	1.0652	1.0589	1.0512

	21	22	23	24	25
Eigenvalue	-0.3879	-0.4081	-0.4375	-0.4752	-0.5037
Difference	0.0202	0.0294	0.0376	0.0286	
Proportion	-0.0090	-0.0094	-0.0101	-0.0110	-0.0117
Cumulative	1.0422	1.0328	1.0227	1.0117	1.0000

1 factors will be retained by the NFACTOR criterion.

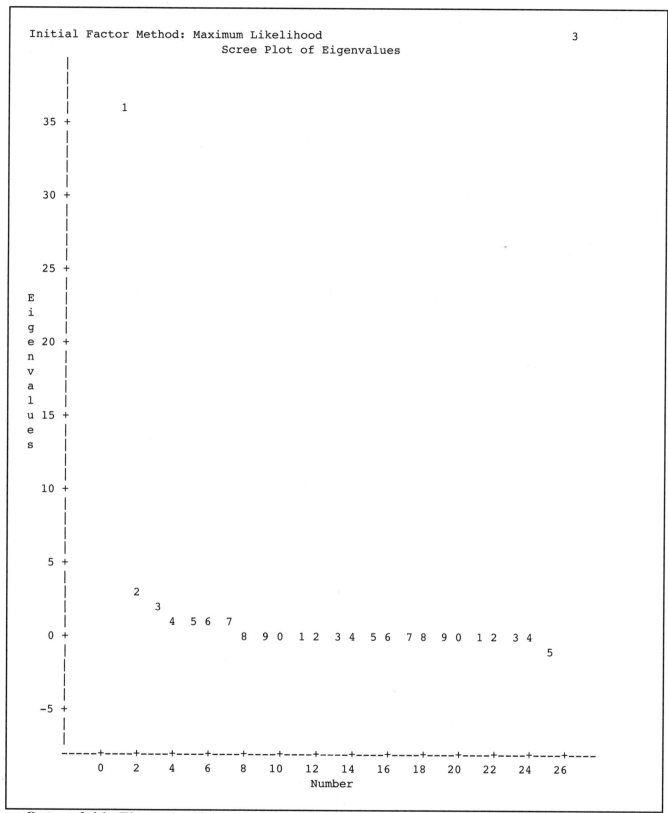

Output 2.14: Eigenvalue Table and Scree Plot from Analysis of Job Search Skills Questionnaire

How many factors should you retain and rotate? Earlier, you used the scree test to help you make this decision. Remember that, with the scree test, you look for a major break in the eigenvalues. You hope that, following this break, the line will begin to "flatten out." The factors that appear before the break are the significant factors to be retained; those appearing on the flat line after the break are assumed to be accounting for only trivial variance, and will not be retained.

In the scree plot of Output 2.14, there is clearly a major break following factor 1. This may mean that this questionnaire is unifactorial; maybe most of the scales measure just one factor, some general job search skills factor. To assess the interpretability of this one-factor model, you will consult the factor pattern to determine which variables display the largest loadings for this factor. Identifying the variables with the highest loadings will help you label the factor.

The interpretation of a one-factor solution is slightly different from the interpretation of multiple-factor models (as have been presented earlier). When only one factor is retained, rotation (whether orthogonal or oblique) is not possible. This actually makes your task easier; when only one factor is retained, it is possible to review the unrotated factor pattern to interpret the factor. The unrotated factor pattern from your one-factor solution is reproduced here as Output 2.15.

```
                 Factor Pattern                        5
                    FACTOR1

            VALUES      47 *
            ABILIT      54 *
            ASSESS      64 *
            STRAT       75 *
            EXPER       70 *
            ORGCHAR     57 *
            RESOCCU     76 *
            RESEMPL     81 *
            GOALS       56 *
            BARRIER     69 *
            MOTIVAT     57 *
            RESUMES     73 *
            LETREC      77 *
            DIRECT      76 *
            APPLIC      64 *
            IDENEMP     82 *
            CARDEV      69 *
            AGENCY      69 *
            FAIRS       68 *
            ADVERT      85 *
            COUNS       78 *
            UNADV       76 *
            NETWORK     75 *
            INTERV      74 *
            SALARY      77 *
NOTE: Printed values are multiplied by 100 and rounded to the nearest integer.
      Values greater than 0.4 have been flagged by an '*'.
```

Output 2.15: Factor Pattern from the One-Factor Solution, Analysis of Job Search Skills Questionnaire

Notice that every variable in Output 2.15 demonstrates a meaningful loading on factor 1 (that is, a loading over .40). This is indicated by the fact that the loading for each variable is flagged with an asterisk. For example, the variable VALUES displays a loading of .47, the variable ABILIT displays a loading of .54, and so forth.

To interpret factor 1 more effectively, it would be helpful to isolate those scales that demonstrate the largest loadings for it. Therefore, you will somewhat arbitrarily choose the value of .70 as a cut off, and construct a table that lists the scales that demonstrate a loading of .70 or greater for factor 1. These scales are listed in Table 2.3.

Table 2.3

Scales with Larger Factor Loadings from Output 2.15, Sorted by Size of Loadings

Factor loading	Variable	Description
.85	ADVERT	Responding to advertised job openings
.82	IDENEMP	Identifying potential employers
.81	RESEMP	Researching specific employers
.78	COUNS	Using career counselors/consultants
.77	LETREC	Using letters of recommendation
.77	SALARY	Negotiating salary
.76	RESOCCU	Researching potential occupations
.76	DIRECT	Using the cover letter/direct mail approach
.76	UNADV	Applying directly for unadvertised jobs
.75	STRAT	Identifying effective job search strategies
.75	NETWORK	Using the networking approach to job search
.74	INTERV	Managing the employment interview process
.73	RESUMES	Using resumes
.70	EXPER	Getting job-related experience

Table 2.3 lists the scales that demonstrated a loading on factor 1 of .70 or greater, and these scales are reordered according to the size of their loadings. Notice that most of the scales that loaded heavily on factor 1 dealt with the nuts-and-bolts tasks associated with the job hunt itself: responding to job openings, learning about potential employers, negotiating salary, and so forth. Because of this, if you ultimately decide that a one-factor solution is best, you will probably define this dimension as a *general job-search skills* factor.

Variance accounted for. Before accepting the one-factor solution as your final solution, you will first review other criteria and consider some alternative solutions. The "Proportion" row of the eigenvalue table from Output 2.14 shows that the first factor accounts for approximately 83% of the common variance. Factor 2 accounts for an additional 8% of the variance: an amount that some researchers would consider a meaningful amount. This information alone would probably warrant exploring a two-factor solution.

The chi-square test. As was mentioned earlier, one of the advantages of the maximum likelihood method of factor extraction is the fact that it provides a chi-square test to help make the number of factors decision. The chi-square test for the current analysis is reproduced in Output 2.16.

```
        Significance tests based on 220 observations:                    4

            Test of H0: No common factors.
                  vs HA: At least one common factor.

            Chi-square = 3891.942    df = 300    Prob>chi**2 = 0.0001

            Test of H0: 1 Factors are sufficient.
                  vs HA: More factors are needed.

            Chi-square = 820.089    df = 275    Prob>chi**2 = 0.0001

       Chi-square without Bartlett's correction = 858.64278687
       Akaike's Information Criterion = 308.64278687
       Schwarz's Bayesian Criterion = -624.6047884
       Tucker and Lewis's Reliability Coefficient = 0.8344510332

                     Squared Canonical Correlations

                              FACTOR1
                              0.965566
```

Output 2.16: Significance Tests for One-Factor Model, Job Search Skills Questionnaire

The test you are most interested in appears below the heading "Test of H0: 1 Factors are sufficient. vs. HA: More factors are needed." This heading is self-explanatory: it tells you that the chi-square statistic tests the null hypothesis that retaining one factor is sufficient. If you obtain a small *p* value for this test (less than .05), you are to reject this null hypothesis and consider the alternative hypothesis that more factors should be retained.

Output 2.16 shows that the obtained value of chi-square for the test was quite large at 820.089 (*df* = 275). To the right of the chi-square statistic and the degrees of freedom, the output provides the entry, "Prob>chi**2 = 0.0001". This is the *p* value for the obtained chi-square statistic. This value was significant at *p* < .0001. Because this obtained *p* value is less than .05, you may reject the null hypothesis that one factor is adequate. This finding can be used as evidence that more factors should be retained; under these circumstances, some researchers would sequentially add more factors to the model until a nonsignificant chi-square value were obtained.

However, use caution not to rely too heavily on the chi-square test in this manner. Under circumstances that are often encountered in applied research, reliance on the chi-square test alone very often causes you to retain too many factors. This is especially likely to happen when the sample is large or there is even a minor misfit between the model and the data (Kim & Mueller, 1978b). For this reason, it is important to use the chi-square test as only one piece of information in making the number-of-factors decision; if the test suggests that additional factors are needed, consult other criteria (e.g., the scree test, proportion of variance accounted for, and interpretability criteria) before making a final decision.

A Two-Factor Solution

So far you have obtained mixed support for a one-factor model. The scree test could be interpreted as supporting the retention of only one factor. On the other hand, the eigenvalue table showed that factor 2 accounts for over 7% of the common variance, and the chi-square test rejected the one-factor model. Combined, these findings justify moving on and exploring the possibility of a two-factor model.

The analysis was therefore repeated, this time specifying NFACT=2. This revised program produced 12 pages of output, some of which is reproduced here as Output 2.17, Output 2.18, and Output 2.19. Some of the information appearing on these 12 pages of output is summarized here:

- Page 1 includes simple statistics.

- Page 2 includes prior communality estimates and the eigenvalue table.

- Pages 2-3 include the iteration history and the significance tests for the number of factors extracted.

- Page 4 includes the unrotated factor pattern.

- Page 5 includes the final communality estimates.

- Page 6 includes the (varimax) rotated factor pattern.

- Page 7 again includes the final communality estimates.

- Page 8 includes the inter-factor correlations.

- Page 9 includes the (promax) rotated factor pattern (standardized regression coefficients).

- Page 10 includes the reference structure (semipartial correlations).

- Page 11 includes the factor structure (correlations).

- Page 12 again includes the final communality estimates.

The rotated factor pattern from the promax rotation is presented as Output 2.17.

```
Rotation Method: Promax                                              9

                    Rotated Factor Pattern (Std Reg Coefs)

                              FACTOR1    FACTOR2

                VALUES         -11         79 *
                ABILIT          15         52 *
                ASSESS          57 *       12
                STRAT           73 *        4
                EXPER           49 *       29
                ORGCHAR          4         73 *
                RESOCCU         65 *       17
                RESEMPL         69 *       17
                GOALS           -1         77 *
                BARRIER         41 *       39
                MOTIVAT          4         72 *
                RESUMES         56 *       24
                LETREC          59 *       25
                DIRECT          73 *        5
                APPLIC          46 *       25
                IDENEMP         80 *        5
                CARDEV          73 *       -4
                AGENCY          73 *       -4
                FAIRS           80 *      -15
                ADVERT          81 *        7
                COUNS           88 *      -11
                UNADV           74 *        5
                NETWORK         59 *       22
                INTERV          60 *       19
                SALARY          77 *        1

NOTE: Printed values are multiplied by 100 and rounded to the nearest integer.
      Values greater than 0.4 have been flagged by an '*'.
```

Output 2.17: Rotated Factor Pattern from Promax Rotation, Two-Factor Solution, Job Search Skills Questionnaire

Remember that the pattern matrix reflects the unique contribution that each factor makes to the variance in a variable, so it is this matrix that you will first use to determine which variables load on which factor. For this analysis, you have flagged any loading over .40 with an asterisk, and will assume the flagged loadings are meaningful.

First, you should read across each row from left to right to see if any variable has a significant loading for more than one factor. These variables need to be identified so that they will not be included in any factor-based scale that you will create later. It turns out that no variables load on both factors.

Next you should read down the first factor to see which variables demonstrated significant loadings for this factor. What do these variables have in common? What general construct do they all seem to be measuring? This process is then repeated in order to interpret factor 2. While doing this, try to determine the way in which factor 1 differs from factor 2: in what way do the variables loading on factor 1 (as a group) tend to differ from those loading on factor 2?

To make this process easier for you, Table 2.4 sorts the scales according to the factors on which they load, and provides brief descriptions for the scales. (In this table, the variables have not been sorted according to the size of their loadings.)

```
Table 2.4
```

Variables Loading on Factors 1 and 2 According to Rotated Factor
Pattern, Two-Factor Solution, Job Search Skills Questionnaire

---

```
Variables loading on factor 1:

 3. ASSESS     Using assessment instruments
 4. STRAT      Identifying effective job search strategies
 5. EXPER      Getting job-related experience
 7. RESOCCU    Researching potential occupations
 8. RESEMPL    Researching specific employers
10. BARRIER    Dealing with occupational barriers
12. RESUMES    Using resumes
13. LETREC     Using letters of recommendation
14. DIRECT     Using the cover letter/direct mail approach
15. APPLIC     Completing application forms
16. IDENEMP    Identifying potential employers
17. CARDEV     Using campus career development services
18. AGENCY     Using employment agencies
19. FAIRS      Using job fairs
20. ADVERT     Responding to advertised job openings
21. COUNS      Using career counselors/consultants
22. UNADV      Applying directly for unadvertised jobs
23. NETWORK    Using the networking approach to job search
24. INTERV     Managing the employment interview process
25. SALARY     Negotiating salary
```

Table 2.4 *continued*

Variables Loading on Factors 1 and 2 According to Rotated Factor
Pattern, Two-Factor Solution, Job Search Skills Questionnaire

---

Variables loading on factor 2:

```
 1. VALUES    Clarifying values and interests
 2. ABILITY   Identifying work-related abilities and skills
 6. ORGCHAR   Identifying preferred organizational characteristics
 9. GOALS     Setting goals
11. MOTIVAT   Staying motivated
```

With factor 1, you can see significant loadings for such variables as STRAT, RESEMPL, DIRECT, ADVERT, COUNS, UNADV, SALARY, and so forth. In general, variables loading on factor 1 seem to deal with a person's ability to perform tasks related to the job search process itself. People who score high on factor 1 tend to be knowledgeable about which job search strategies are effective, how to conduct research on specific employers, how to make use of the services offered by career counselors, how to negotiate salary, and so forth. It therefore seems appropriate to label this the *job search skills* factor.

With factor 2, on the other hand, you can see significant loadings for such scales as VALUES, ABILIT, ORGCHAR, GOALS, and MOTIVAT. People who score high on factor 2 are able to clearly describe just what their work-related values and abilities are. They know what their goals are, and feel that they will be able to stay motivated during their job search. Therefore, you might label factor 2 the *goal clarity and motivation* factor. From the perspective of interpretability (at least) this two-factor solution appears to be acceptable.

Now that you have interpreted the meaning of the factors, it would be useful to know the nature of the relationship between factor 1 and factor 2. For this information, you may turn to the inter-factor correlations provided in the output. These are reproduced as Output 2.18.

```
                    Inter-factor Correlations                    8

                       FACTOR1    FACTOR2

             FACTOR1    100 *       61 *
             FACTOR2     61 *      100 *

NOTE: Printed values are multiplied by 100 and rounded to the nearest integer.
      Values greater than 0.4 have been flagged by an '*'.
```

Output 2.18: Inter-Factor Correlations from Two-Factor Solution, Job Search Skills Questionnaire

The inter-factor correlation of +.61 from Output 2.18 reveals a moderately strong positive correlation between the job search skills factor and the goal clarity and motivation factor. This seems logical; it makes sense that people who have a good deal of self-insight and motivation related to their careers would also have higher levels of the skills necessary to actually find a job. (In fact, this correlation is so high that you might wonder whether you are really justified in interpreting them as two separate factors.)

To understand the big picture concerning the relationship between the two factors, you will now review the factor structure matrix from your two-factor solution. This is reproduced as Output 2.19.

```
Rotation Method: Promax                                              11

                   Factor Structure (Correlations)

                              FACTOR1     FACTOR2

                 VALUES         37          72 *
                 ABILIT         47 *        62 *
                 ASSESS         64 *        46 *
                 STRAT          75 *        48 *
                 EXPER          67 *        59 *
                 ORGCHAR        48 *        75 *
                 RESOCCU        75 *        56 *
                 RESEMPL        79 *        59 *
                 GOALS          46 *        77 *
                 BARRIER        65 *        64 *
                 MOTIVAT        48 *        75 *
                 RESUMES        70 *        58 *
                 LETREC         75 *        61 *
                 DIRECT         76 *        50 *
                 APPLIC         61 *        53 *
                 IDENEMP        83 *        53 *
                 CARDEV         71 *        41 *
                 AGENCY         71 *        41 *
                 FAIRS          71 *        34
                 ADVERT         85 *        57 *
                 COUNS          81 *        42 *
                 UNADV          77 *        50 *
                 NETWORK        73 *        58 *
                 INTERV         72 *        56 *
                 SALARY         78 *        48 *

NOTE: Printed values are multiplied by 100 and rounded to the nearest integer.
      Values greater than 0.4 have been flagged by an '*'.
```

Output 2.19: Factor Structure from Promax Rotation, Job Search Skills Questionnaire

Notice that almost all of the scales are flagged as having significant loadings for both factors. In light of the strong inter-factor correlation reported earlier, this finding makes sense. Remember that in a structure matrix, the loadings represent the correlation between a variable and a factor. Given the strong correlation between factor 1 and factor 2, it makes sense that any variable that loads on factor 1 will also be correlated with factor 2, and that any variable that loads on factor 2 will also be correlated with factor 1. This is why it is necessary to review the structure matrix to fully understand an oblique solution. Reviewing only the pattern matrix would not reveal how strongly most variables are related to both factors.

A Four-Factor Solution

To illustrate that it is often possible to obtain more than one interpretable solution from a factor analysis, a four-factor solution is now reviewed. Consider the eigenvalue table from page 2 of the current output, reproduced once again here as Output 2.20.

```
      Preliminary Eigenvalues:  Total = 43.2133584  Average = 1.72853434

                         1          2          3          4          5
      Eigenvalue     35.8841     3.2832     1.6709     1.4372     1.2100
      Difference     32.6009     1.6124     0.2336     0.2272     0.3734
      Proportion      0.8304     0.0760     0.0387     0.0333     0.0280
      Cumulative      0.8304     0.9064     0.9450     0.9783     1.0063

                         6          7          8          9         10
      Eigenvalue      0.8366     0.6662     0.4418     0.3526     0.3083
      Difference      0.1704     0.2245     0.0892     0.0443     0.0880
      Proportion      0.0194     0.0154     0.0102     0.0082     0.0071
      Cumulative      1.0257     1.0411     1.0513     1.0595     1.0666

                        11         12         13         14         15
      Eigenvalue      0.2203     0.1997     0.0900    -0.0113    -0.0500
      Difference      0.0206     0.1097     0.1013     0.0387     0.0449
      Proportion      0.0051     0.0046     0.0021    -0.0003    -0.0012
      Cumulative      1.0717     1.0763     1.0784     1.0781     1.0770

                        16         17         18         19         20
      Eigenvalue     -0.0949    -0.1702    -0.2438    -0.2737    -0.3312
      Difference      0.0754     0.0735     0.0300     0.0575     0.0567
      Proportion     -0.0022    -0.0039    -0.0056    -0.0063    -0.0077
      Cumulative      1.0748     1.0708     1.0652     1.0589     1.0512

                        21         22         23         24         25
      Eigenvalue     -0.3879    -0.4081    -0.4375    -0.4752    -0.5037
      Difference      0.0202     0.0294     0.0376     0.0286
      Proportion     -0.0090    -0.0094    -0.0101    -0.0110    -0.0117
      Cumulative      1.0422     1.0328     1.0227     1.0117     1.0000
```

Output 2.20: Eigenvalue Table from Analysis of Job Search Skills Questionnaire

Clearly, there is a large break following factor 1, another (much smaller) break following factor 2, and from that point on the eigenvalues seem to flatten out. On the basis of these breaks in the eigenvalues alone, it would be difficult to justify rotating four factors.

Still, some have argued that retaining and rotating too few factors has a more serious negative effect on the factor structure than rotating too many, and that it is probably best to err in the direction of overfactoring if you are to err at all (Cattell, 1952, 1958; Rummel, 1970). In fact, one of Cattell's (1958) proposed solutions to the number-of-factors problem is to retain enough factors to account for 99% of the variance. With the preceding eigenvalue table, you can see that this would involve retaining the first four factors. This can be seen by reviewing the figures in the "Cumulative" row of Output 2.20. Notice that factors 1–4 (combined) account for approximately 98% of the common variance in the data set, while factors 1–5 (combined) account for approximately 101% of the common variance. If you were to heed Cattell's recommendation, you would therefore retain and interpret factors 1–4.

As an illustration, the results of this four-factor solution is presented here. Output 2.21 provides the factor pattern matrix resulting from a promax rotation of four factors.

```
             Rotated Factor Pattern (Std Reg Coefs)
               FACTOR1    FACTOR2    FACTOR3    FACTOR4
    VALUES       -3       77 *       -2         -3
    ABILIT       12       47 *       17         -8
    ASSESS       38       12          2         24
    STRAT        52 *      2         15         15
    EXPER        24       32         -1         34
    ORGCHAR      10       70 *        2         -3
    RESOCCU      74 *     16        -12          6
    RESEMPL      78 *     12          2         -6
    GOALS        -5       78 *       -2         11
    BARRIER      29       38          5         14
    MOTIVAT       2       69 *        9         -2
    RESUMES       3       15         71 *        3
    LETREC       22       18         50 *        4
    DIRECT       15       -5         72 *        9
    APPLIC       22       19         37         -1
    IDENEMP      66 *      1         11         12
    CARDEV        2        3          1         89 *
    AGENCY       48 *     -8         23         13
    FAIRS        31      -13         21         41 *
    ADVERT       65 *      2         22          5
    COUNS        51 *    -10          7         41 *
    UNADV        75 *      1          4          0
    NETWORK      49 *     21          0         17
    INTERV       55 *     11         25        -10
    SALARY       90 *     -4         -3         -5
NOTE: Printed values are multiplied by 100 and rounded to the nearest integer.
      Values greater than 0.4 have been flagged by an '*'.
```

Output 2.21: Rotated Factor Pattern from Promax Rotation, Four-Factor Solution, Job Search Skills Questionnaire

As usual, you should begin by reviewing the rows of the factor pattern matrix (from left to right) to identify any variables with significant loadings for more than one factor. This process identifies one problem variable: COUNS loads on both factors 1 and 4. This variable will therefore not be used in interpreting the factors.

To make it easier to interpret the meaning of these four factors, Table 2.5 groups together the scales according to the factors on which they load.

Table 2.5

Variables Loading on Factors 1, 2, 3, and 4 According to Rotated
Factor Pattern, Four-Factor Solution, Job Search Skills Questionnaire

---

Variables loading on factor 1:

4.	STRAT	Identifying effective job search strategies
7.	RESOCCU	Researching potential occupations
8.	RESEMPL	Researching specific employers
16.	IDENEMP	Identifying potential employers
18.	AGENCY	Using employment agencies
20.	ADVERT	Responding to advertised job openings
22.	UNADV	Applying directly for unadvertised jobs
23.	NETWORK	Using the networking approach to job search
24.	INTERV	Managing the employment interview process
25.	SALARY	Negotiating salary

Variables loading on factor 2:

1.	VALUES	Clarifying values and interests
2.	ABILIT	Identifying work-related abilities and skills
6.	ORGCHAR	Identifying preferred organizational characteristics
9.	GOALS	Setting goals
11.	MOTIVAT	Staying motivated

Variables loading on factor 3:

12.	RESUMES	Using resumes
13.	LETREC	Using letters of recommendation
14.	DIRECT	Using the cover letter/direct mail approach

Variables loading on factor 4:

17.	CARDEV	Using campus career development services
19.	FAIRS	Using job fairs

According to Table 2.5, the scales that loaded on factor 1 all seem to deal with finding and following-up on job leads. People who score high on this factor feel that they understand the best job-search strategies, are able to research employers and identify organizations that might hire them, are able to respond effectively to advertised and unadvertised job openings, are able to manage the interview process, and are able to successfully negotiate a good salary. This might therefore be labelled the *finding and pursuing job leads* factor.

Factor 2, on the other hand, should look familiar at this point. People who score high on this factor know what their work-related values and abilities are, and feel that they are able to set goals and stay motivated throughout their job search. Clearly, this factor is similar to the *goal clarity and motivation* factor observed with the two-factor solution.

Three variables loaded on factor 3. Students scoring high on this factor believe that they are able to work effectively with resumes and letters of recommendation, and are able to use the cover letter/direct mail approach to job search. This factor could be labelled the *using resumes and letters* factor.

Only two variables loaded exclusively on factor 4. Subjects who scored high on this factor knew how to use the campus career services office and also knew how to make effective use of job fairs (events that are typically coordinated by campus career development offices). This factor could be labelled the *using campus career services* factor.

This four-factor solution proved to be fairly interpretable. For example, with each factor, all of the scales loading on that factor seemed to be measuring the same underlying construct (i.e., all scales loading on factor 1 seemed to be measuring the *finding and pursuing job leads* construct). In addition, the scales that loaded on different factors did seem to be measuring conceptually different constructs (i.e, the *finding and pursuing job leads* factor seemed to be conceptually different from the *goal clarity and motivation* factor).

Unfortunately, the four-factor solution is less than satisfactory in the sense that one of its factors (factor 4) displayed meaningful loadings for less than three variables (after eliminating a factorially complex variable). If you believe that the nature of job search skills is best described with this four-factor model, you may want to improve on these results by creating additional scales that could be expected to load on factor 4. This would be desirable because with self-report instruments (such as this job search skills questionnaire), the more variables that are used to assess each construct, the more reliable the instrument tends to be.

After adding new items or new scales to the questionnaire, it could be administered to a new sample of subjects. A factor analysis could then be performed on the responses to this revised questionnaire to see if a more acceptable four-factor solution is obtained.

Conclusion

The point made in the preceding paragraph is important in that it describes the role that exploratory factor analysis ideally should play in an ongoing program of research. In many situations, factor analytic research should be regarded as an iterative process in which you begin with some a priori ideas regarding the nature of the factors to be investigated (hopefully based on theory and prior research), and then identify a number of variables that can be expected to measure these factors. Performing an exploratory factor analysis on the collected data will often teach you something that was not previously known; perhaps a five-factor model emerges when a three-factor model was expected; perhaps variables expected to load on factor 1 instead load on factor 4. These results should encourage you to return to the relevant literature, revise the initial model, and perhaps even find new ways of measuring the constructs of interest. A program of research that includes a number of exploratory factor analyses of different data sets, perhaps using improved measures at each step, stands the best chance of discovering the true nature of the factor structure that is responsible for covariation between the observed variables.

Appendix: Assumptions Underlying Exploratory Factor Analysis

As with principal component analysis, a factor analysis is performed on a matrix of correlations, and this means that the data should satisfy the assumptions for the Pearson correlation coefficient. These assumptions are briefly reviewed here.

- **Interval-level measurement**. All analyzed variables should be assessed on an interval or ratio level of measurement.

- **Random sampling**. Each subject will contribute one score on each observed variable. These sets of scores should represent a random sample drawn from the population of interest.

- **Linearity**. The relationship between all observed variables should be linear.

- **Normal distributions**. Each observed variable should be normally distributed. Variables that demonstrate marked skewness or kurtosis may be transformed to approximate normality (see Rummel, 1970).

- **Bivariate normal distribution**. Each pair of observed variables should display a bivariate normal distribution; e.g., they should form an elliptical scattergram when plotted. However, the Pearson correlation coefficient is robust against violations of this assumption when the sample size is greater than 25.

When the maximum likelihood method is used to extract factors, the output provides a significance test for the null hypothesis that the number of factors retained in the current analysis is sufficient to explain the observed correlations. The following assumption should be met for the probability value associated with this test to be valid:

- **Multivariate normality**. The measurements obtained from subjects should follow an approximate multivariate normal distribution.

References

Cattell, R. B. (1952). *Factor analysis: An introduction and manual for the psychologist and social scientist.* New York: Harper & Row.

Cattell, R. B. (1958). Extracting the correct number of factors in factor analysis. *Educational and Psychological Measurement, 18,* 791-837.

Cattell, R. B. (1966). The scree test for the number of factors. *Multivariate Behavior Research , 1 ,* 245-276.

Hatcher, L. & Stepanski, E. (1994). *A step-by-step approach to using the SAS system for univariate and multivariate statistics.* Cary, NC: SAS Institute Inc.

Kim, J. O. & Mueller, C. W. (1978a). *Introduction to factor analysis: What it is and how to do it.* Beverly Hills, CA: Sage.

Kim, J. O. & Mueller, C. W. (1978b). *Factor analysis: Statistical methods and practical issues.* Beverly Hills, CA: Sage.

Loehlin, J. C. (1987). *Latent variable models.* Hillsdale, NJ: Lawrence Erlbaum Associates.

Ruddle, K., Thompson, J., & Hatcher, L. (1993). *Development of the job search skills inventory: A measure of job search abilities and knowledge.* Paper presented at the Carolinas Psychology Conference, North Carolina State University, Raleigh, NC.

Rummel, R. J. (1970). *Applied factor analysis.* Evanston, IL: Northwestern University Press.

Rusbult, C.E. (1980). Commitment and satisfaction in romantic associations: A test of the investment model. *Journal of Experimental Social Psychology, 16,* 172-186.

SAS Institute Inc. (1989). *SAS/STAT users guide, version 6, fourth edition, volume 1.* Cary, NC: SAS Institute Inc..

Spector, P. E. (1992). *Summated rating scale construction: An introduction.* Newbury Park, CA: Sage.

Stevens, J. (1986). *Applied multivariate statistics for the social sciences.* Hillsdale, NJ: Lawrence Erlbaum Associates.

Chapter 3

ASSESSING SCALE RELIABILITY WITH COEFFICIENT ALPHA

Overview. This chapter shows how to use PROC CORR to compute the coefficient alpha reliability index for a multiple-item scale. It reviews basic issues regarding the assessment of reliability, and describes the circumstances under which a measure of internal consistency reliability is likely to be high. Fictitious questionnaire data are analyzed to demonstrate how you can use the results of PROC CORR to perform an item analysis, thereby improving the reliability of the scale.

Introduction: The Basics of Scale Reliability

You may compute coefficient alpha when you have administered a multiple-item, summated rating scale to a group of subjects and want to determine the internal consistency reliability of the scale. The items constituting the scale may be scored dichotomously (scored as "right" or "wrong"), or they may use a multiple-point rating format (e.g., subjects may respond to each item using a 7-point "agree-disagree" rating scale).

This chapter shows how to use the SAS System's PROC CORR to easily compute coefficient alpha for the types of scales that are often used in social science research. However, this chapter will not show how to actually *develop* a multiple-item scale for use in research. To learn about recommended approaches for creating summated rating scales, see Spector (1992).

Example of a summated rating scale. A **summated rating scale** usually consists of a short list of statements, questions, or other items that a subject responds to. Very often, the items that constitute the scale are statements, and subjects indicate the extent to which they agree or disagree with each statement by circling or checking some response on a rating scale (perhaps a 7-point rating scale in which 1 = "Strongly Disagree" and 7 = "Strongly Agree"). The scale is called a *summated* scale because the researcher typically sums the response numbers circled by a subject to create an overall score on the scale for that subject. These scales are often referred to as **Likert scales** or **Likert-type scales**.

For example, imagine that you are interested in measuring job satisfaction in a sample of employees. To do this, you might develop a 10-item scale that includes items such as "In general, I am satisfied with my job." Employees respond to these items using a 7-point response format in which 1 = "Strongly Disagree" and 7 = "Strongly Agree."

You administer this scale to 200 employees and compute a job satisfaction score for each individual by summing his or her responses to the 10 items. Scores may range from a low of 10 (if the employee circled "Strongly Disagree" for each item) to a high of 70 (if the employee circled "Strongly Agree" for each item). Given the way these scores were created, higher scores indicate higher levels of job satisfaction. With the job satisfaction scale now developed and administered to a sample, you hope to use it as a predictor variable or criterion variable in research. However, the people who later read about your research are going to have questions about the psychometric properties of your scale; at the very least, they will want to see empirical evidence that the scale is reliable. This chapter discusses the meaning of scale reliability, and shows how to use the SAS System to obtain an index of internal consistency reliability for summated rating scales.

True scores and measurement error. Most observed variables assessed in the social sciences (such as scores on your job satisfaction scale) actually consist of two components: a **true score** component (which indicates where the subject actually stands on the variable of interest), along with a **measurement error** component. Almost all observed variables in the social sciences contain at least some measurement error, even the variables that seem to be objectively measured.

For example, imagine that you assess the observed variable "Age" in a group of subjects by asking the subjects to write down their age in years. To a large extent, this observed variable (what the subjects wrote down) is influenced by the true score component: To a large extent, what they write will be influenced by how old they actually are. Unfortunately, however, this observed variable will also be influenced by measurement error: some subjects will write down the wrong age because they don't know how old they are; some will write the wrong age because they don't want the researcher to know how old they are; some will write the wrong age because they didn't understand the question. In short, it is likely that there will not be a perfect correlation between the observed variable (what the subjects write down) and their true scores on the underlying construct (their actual age).

And remember that the preceding "Age" variable was relatively objective and straightforward. If a question such as this is going to be influenced by measurement error, imagine how much more error is likely to be displayed by responses to subjective questionnaire items (such as the items that constitute your job satisfaction scale).

Underlying constructs versus observed variables. In conducting research, it is also useful to draw a distinction between underlying constructs versus observed variables. An **underlying construct** is the hypothetical variable that you actually want to measure. For example, in the job satisfaction study just described, you wanted to measure the underlying construct of job satisfaction in a group of employees. The **observed variable**, on the other hand, consists of the measurements that you actually obtained. In that example, the observed variable consisted of scores on the 10-item measure of job satisfaction. These scores may or may not be a good measure of the underlying construct.

Reliability defined. With this foundation laid, it is now possible to provide some definitions. Technically, a **reliability coefficient** may be defined as the percent of variance in an observed variable that is accounted for by true scores on the underlying construct. For example, imagine that in the study just discussed, you were able to obtain two scores for the 200 employees in the sample: their observed scores on the job satisfaction questionnaire, and their true scores on the underlying construct of job satisfaction. Assume that you compute the correlation between these two variables. The square of this correlation coefficient represents the reliability of your job satisfaction scale; it is the percent of variance in observed job satisfaction scores that is accounted for by true scores on the underlying construct of job satisfaction.

The preceding was a technical definition for reliability, and this definition is of little use in practice because it is generally not possible to obtain true scores on a variable. For this reason, reliability is usually defined in practice in terms of the **consistency** of the scores that are obtained on the observed variable; an instrument is said to be reliable if it is shown to provide consistent scores upon repeated administration, upon administration by alternate forms, and so forth. A variety of methods of estimating scale reliability are actually used in practice.

Test-retest reliability. For example, assume that you administer your measure of job satisfaction to a group of 200 employees at two points in time: once in January, and again in March. If the instrument is indeed reliable, you would expect that the subjects who displayed high scores in January will also tend to display high scores in March, and that those who displayed low scores in January will also display low scores in March. These results will support the test-retest reliability of the scale. Test-retest reliability is assessed by administering the same instrument to the same sample of subjects at two points in time, and computing the correlation between the two sets of scores.

Internal consistency reliability. One problem with the test-retest reliability procedure is the time that it requires. What if you do not have time to perform two administrations of the scale? In such situations, you are likely to turn to reliability indices that may be obtained with only one administration. In research that involves the use of questionnaire data, the most popular of these are the internal consistency indices of reliability. Briefly, **internal consistency** is the extent to which the individual items that constitute a test correlate with one another or with the test total. In the social sciences, one of the most widely-used indices of internal consistency reliability is coefficient alpha (Cronbach, 1951).

Coefficient Alpha

Formula. Coefficient alpha is a general formula for scale reliability based on internal consistency. It provides the lowest estimate of reliability that can be expected for an instrument.

The formula for coefficient alpha is as follows:

$$r_{xx} = \left(\frac{N}{N - 1} \right) \left(\frac{S^2 - \Sigma S_i^2}{S^2} \right)$$

where

r_{xx} = Coefficient alpha.

N = Number of items constituting the instrument.

S^2 = Variance of the summated scale scores (for example, assume that you compute a total score for each subject by summing each subject's responses to the items that constitute the scale; the variance of this total score variable would be S^2).

ΣS_i^2 = The sum of the variances of the individual items that constitute this scale.

When will coefficient alpha be high? Other factors held constant, coefficient alpha will be high to the extent that many items are included in the scale, and the items that constitute the scale are highly correlated with one another.

To understand why coefficient alpha is high when the items are highly correlated with one another, consider the second term in the preceding formula:

$$\left(\frac{s^2 - \Sigma s_i^2}{s^2} \right)$$

This term shows that the variance of the summated scales scores is (essentially) divided by itself to compute coefficient alpha. However, the combined variance of the individual items is first subtracted from this variance before the division is performed. This part of the equation shows that, if combined variance of the individual items is a small value, then coefficient alpha will be a relatively larger value.

This is important because (once again, with other factors held constant), the stronger the correlations between the individual items, the smaller the term ΣS_i^2 will be. This is why coefficient alpha for a given scale is likely to be large to the extent that the variables constituting that scale are strongly correlated with one another.

Assessing Coefficient Alpha with PROC CORR

Imagine that you have conducted research in the area of prosocial behavior, and have developed an instrument designed to measure two separate underlying constructs: acquaintance helping and financial giving. **Acquaintance helping** refers to prosocial activities performed to help coworkers, relatives, and friends. **Financial giving**, on the other hand, refers to giving money to charities or panhandlers (see Chapter 1: "Principal Component Analysis," for a more detailed description of these constructs). In the following questionnaire, items 1–3 were designed to assess acquaintance helping, and items 4–6 were designed to assess financial giving.

```
Instructions:  Below are a number of activities that people
sometimes engage in.  For each item, please indicate how
frequently you have engaged in this activity over the preceding
six months.  Make your rating by circling the appropriate number
to the left of the item, and use the following response format:

    7 = Very Frequently
    6 = Frequently
    5 = Somewhat Frequently
    4 = Occasionally
    3 = Seldom
    2 = Almost Never
    1 = Never
```

```
1 2 3 4 5 6 7     1.    Went out of my way to do a favor for a
                        coworker.

1 2 3 4 5 6 7     2.    Went out of my way to do a favor for a
                        relative.

1 2 3 4 5 6 7     3.    Went out of my way to do a favor for a
                        friend.

1 2 3 4 5 6 7     4.    Gave money to a religious charity.

1 2 3 4 5 6 7     5.    Gave money to a charity not associated with
                        a religion.

1 2 3 4 5 6 7     6.    Gave money to a panhandler.
```

Assume that you have administered this 6-item questionnaire to 50 subjects. For the moment, we are concerned only with the reliability of the scale that includes items 1-3 (the items that assess acquaintance helping).

Let us further assume that you have made a mistake in assessing the reliability of this scale. Assume that you erroneously believed that the acquaintance helping construct was assessed by items 1-4 (whereas, in reality, of course, the construct was assessed by items 1-3). It will be instructive to see what you learn when you mistakenly include item 4 in the analysis.

General form. Here is the general form for the SAS statements that estimate coefficient alpha reliability for a summated rating scale:

```
PROC CORR    DATA=data-set-name    ALPHA    NOMISS;
   VAR  list-of-variables;
   RUN;
```

In the preceding program, the ALPHA option requests that coefficient alpha be computed for the group of variables included in the VAR statement. The NOMISS option is required to compute coefficient alpha. The VAR statement should list only the variables (items) that constitute the scale in question. You must perform a separate CORR procedure for each scale whose reliability you want to assess.

A 4-item scale. Here is an actual program, including the DATA step, that you could use to analyze some fictitious data from your study. Only a few sample lines of data appear here; the complete data set appears in Appendix B.

```
 1       DATA D1;
 2          INPUT    #1    @1    (V1-V6)    (1.)  ;
 3       CARDS;
 4       556754
 5       567343
 6       777222
 7       .
 8       .
 9       .
10       767151
11       455323
12       455544
13       ;
14       PROC CORR    DATA=D1    ALPHA    NOMISS;
15          VAR V1 V2 V3 V4;
16          RUN;
```

The results of this analysis appear as Output 3.1. Page 1 of these results provides the means, standard deviations, and other descriptive statistics that you should review to verify that the analysis proceeded as expected. Page 2 provides the results pertaining to the reliability of the scale.

```
                        The SAS System                          1

                     Correlation Analysis

        4 'VAR' Variables:  V1        V2        V3       V4


                     Simple Statistics

Variable        N       Mean     Std Dev        Sum     Minimum     Maximum

V1             50     5.18000     1.39518    259.00000     1.00000     7.00000
V2             50     5.40000     1.10657    270.00000     3.00000     7.00000
V3             50     5.52000     1.21622    276.00000     2.00000     7.00000
V4             50     3.64000     1.79296    182.00000     1.00000     7.00000
```

2

```
                           Correlation Analysis

                        Cronbach Coefficient Alpha

                 for RAW variables         :  0.490448
                 for STANDARDIZED variables:  0.575912


                    Raw Variables                 Std. Variables

     Deleted    Correlation                    Correlation
     Variable   with Total       Alpha         with Total        Alpha
     ----------------------------------------------------------------------
     V1          0.461961       0.243936        0.563691        0.326279
     V2          0.433130       0.318862        0.458438        0.420678
     V3          0.500697       0.240271        0.546203        0.342459
     V4         -0.037388       0.776635       -0.030269        0.773264


      Pearson Correlation Coefficients / Prob > |R| under Ho: Rho=0 / N = 50

                    V1              V2              V3              V4

     V1          1.00000         0.49439         0.71345        -0.10410
                 0.0             0.0003          0.0001          0.4719

     V2          0.49439         1.00000         0.38820         0.05349
                 0.0003          0.0             0.0053          0.7122

     V3          0.71345         0.38820         1.00000        -0.02471
                 0.0001          0.0053          0.0             0.8648

     V4         -0.10410         0.05349        -0.02471         1.00000
                 0.4719          0.7122          0.8648          0.0
```

Output 3.1: Simple Statistics and Coefficient Alpha Results for Analysis of Scale that Includes Items 1-4, Prosocial Behavior Study

On page 2 of Output 3.1, to the right of the heading "Cronbach Coefficient Alpha for RAW variables" you see that the reliability coefficient for the scale that includes items 1–4 is only .490448, which rounds to .49 (reliability for raw variables is normally reported in published articles).

But how large must a reliability coefficient be to be considered acceptable? A widely used rule of thumb of .70 has been suggested by Nunnally (1978): For scales used in research, reliability coefficients less than .70 are generally seen as inadequate. However, you should remember that this is only a rule of thumb, and the social science literature does sometimes report studies employing variables with coefficient alpha reliabilities under .70 (and sometimes even under .60!).

The coefficient alpha of .49 reported in Output 3.1 is not acceptable; obviously, it should be possible to significantly improve the reliability of this scale. But how?

In some situations, the reliability of a multiple-item scale is improved by dropping from the scale those items that demonstrate poor item-total correlations. An **item-total correlation** is the correlation between an individual item, and the sum of the remaining items that constitute the scale. If an item-total correlation is small, this may be seen as evidence that the item is not measuring the same construct that is measured by the other items in the scale. This may mean that you should drop an item demonstrating a small item-total correlation from the scale.

Consider Output 3.1. Under the "Correlation with Total" heading, you can see that items 1–3 each demonstrate reasonably strong correlations with the sum of the remaining items on the scale. However, item V4 demonstrates an item-total correlation of approximately –.04. This suggests that item V4 is probably not assessing the same construct that is assessed by items V1–V3.

In Output 3.1 under the "Alpha" heading you find an estimate of what alpha would be if a given variable (item) were deleted from the scale. To the right of "V4", PROC CORR estimates that alpha would be approximately .78 if V4 were deleted (this value appears where the row headed "V4" intersects with the column headed "Alpha" in the "Raw Variables" section). This makes sense, because variable V4 demonstrates a correlation with the remaining scale items of only –.04. Obviously, you could substantially improve this scale by removing the item that is clearly not measuring the same construct that is assessed by the other items.

A 3-item scale. Output 3.2 reveals the results of PROC CORR when coefficient alpha is requested for just variables V1–V3 (this is done by specifying only V1–V3 in the VAR statement).

```
                          The SAS System                              3

                        Correlation Analysis

             3 'VAR' Variables:   V1        V2        V3


                         Simple Statistics

Variable        N       Mean     Std Dev        Sum     Minimum     Maximum

V1             50     5.18000    1.39518    259.00000   1.00000     7.00000
V2             50     5.40000    1.10657    270.00000   3.00000     7.00000
V3             50     5.52000    1.21622    276.00000   2.00000     7.00000
```

```
                                                                    4

                        Correlation Analysis

                     Cronbach Coefficient Alpha

               for RAW variables        :   0.776635
               for STANDARDIZED variables:  0.773264


                  Raw Variables                    Std. Variables

   Deleted      Correlation                      Correlation
   Variable     with Total        Alpha          with Total        Alpha
   ------------------------------------------------------------------------
   V1            0.730730        0.557491          0.724882        0.559285
   V2            0.480510        0.828202          0.476768        0.832764
   V3            0.657457        0.649926          0.637231        0.661659


      Pearson Correlation Coefficients / Prob > |R| under Ho: Rho=0 / N = 50

                          V1              V2              V3

             V1        1.00000         0.49439         0.71345
                       0.0             0.0003          0.0001

             V2        0.49439         1.00000         0.38820
                       0.0003          0.0             0.0053

             V3        0.71345         0.38820         1.00000
                       0.0001          0.0053          0.0
```

Output 3.2: Simple Statistics and Coefficient Alpha Results for Analysis of Scale that Includes Items 1-3, Prosocial Behavior Study

Page 4 of Output 3.2 provides a raw-variable coefficient alpha of .78 for the three variables included in this analysis (this value appears to the right of the heading "Cronbach Coefficient Alpha for RAW variables"). This coefficient exceeds the minimum value of .70 recommended by Nunnally (1978). Obviously, the acquaintance-helping scale demonstrates a much higher level of reliability with item V4 deleted.

Summarizing the Results

Summarizing the results in a table. Researchers typically report the reliability of a scale in a table that reports simple descriptive statistics for the study's variables, such as means, standard deviations, and intercorrelations. In these tables, coefficient alpha reliability estimates are usually reported on the diagonal of the correlation matrix, within parentheses. Such an approach appears in Table 3.1.

Table 3.1

Means, Standard Deviations, Intercorrelations, and Coefficient Alpha Reliability Estimates for the Study's Variables

Variables	Mean	SD	1	2	3
1. Authoritarianism	13.56	2.54	(.90)		
2. Acquaintance helping	15.60	3.22	.37	(.78)	
3. Financial giving	12.55	1.32	.25	.53	(.77)

Note: N = 200. Reliability estimates appear on the diagonal.

In the preceding table, information for the authoritarianism variable is presented in both the row and the column that is headed "1". Where the row headed "1" intersects with the column headed "1", you will find the coefficient alpha reliability index for the authoritarianism scale. You can see that this index is .90. In the same way, you can find coefficient alpha for acquaintance helping where row 2 intersects with column 2 (alpha = .78), and you can find coefficient alpha for financial giving where row 3 intersects with column 3 (alpha = .77).

Preparing a formal description of the results for a paper. When reliability estimates are computed for a relatively large number of scales, it is common to report them in a table (such as Table 3.1), and make only a global reference to them within the text of the paper. For example, within the section on instrumentation, you might indicate:

Coefficient alpha reliability estimates (Cronbach, 1951) all exceeded .70, and are reported on the diagonal of Table 3.1.

When reliability estimates are computed for only a small number of scales, it is possible to instead report these estimates within the body of the text itself. Here is an example of how this might be done:

```
     Scale reliability was assessed by calculating
coefficient alpha (Cronbach, 1951).  Reliability estimates
were .90, .78, and .77 for the authoritarianism,
acquaintance helping, and financial giving scales,
respectively.
```

Conclusion

Assessing scale reliability with coefficient alpha (or some other reliability index) should be one of the first tasks you complete when conducting questionnaire research; if the scales you use are not reliable, there is no point performing additional analyses. You can often improve a single scale demonstrating poor reliability by deleting items with poor item-total correlations, according to the procedures just discussed. When several scales on a questionnaire display poor reliability, you may be better advised to perform a principal component analysis or an exploratory factor analysis on responses to all items on the questionnaire, to determine which items tend to group together empirically. If many items load on each retained factor, and if the factor pattern obtained from such an analysis displays simple structure, chances are good that the resulting scales will demonstrate adequate internal consistency reliability estimates.

References

Cronbach, L. J. (1951). Coefficient alpha and the internal structure of tests. *Psychometrika, 16,* 297-334.

Nunnally, J. (1978). *Psychometric theory.* New York: McGraw-Hill.

Spector, P. E. (1992). *Summated rating scale construction: An introduction.* Newbury Park, CA: Sage.

Chapter 4

PATH ANALYSIS WITH MANIFEST VARIABLES

Overview. This chapter shows how to use the CALIS procedure to test path models with manifest (observed) variables. It introduces important terms and concepts in path analysis, and describes the necessary conditions for performing the analysis. The chapter shows how to prepare a program figure that specifies the parameters to be estimated, and how to turn this figure into a SAS program that tests the model of interest. You are shown how to prepare the PROC CALIS statement and related statements, how to interpret the results of the analysis, how to modify a model to achieve a better fit, and how to summarize the results for a paper. This chapter deals only with the testing of recursive (unidirectional) models.

Introduction: The Basics of Path Analysis

Path analysis can be used to test theoretical models that specify causal relationships between a number of observed variables. Path analysis determines whether your theoretical model successfully accounts for the actual relationships observed in the sample data. The output of the CALIS procedure provides indices that indicate whether the model, as a whole, fits the data, as well as significance tests for specific causal paths. When a model provides a relatively poor fit to the data, additional results from PROC CALIS can be used to modify the model and improve its fit.

This chapter deals only with causal models in which all variables are **manifest** (observed) variables. It does not deal with path models that specify causal relationships between **latent** (unobserved) variables (such models are often called "LISREL-type" models). For guidelines on how to test these latent-variable models, see Chapter 5, "Developing Measurement Models with Confirmatory Factor Analysis" and Chapter 6, "Path Analysis with Latent Variables" of this text.

Some Simple Path Diagrams

When studying complex phenomena, it often becomes clear that a given outcome variable of interest is actually influenced by a variety of other variables; few outcome variables of any importance are causally determined by just one variable. For example, imagine that you are a theorist in industrial psychology who believes that an employee's work performance (the outcome variable of interest) is influenced by the following four variables:

• The employee's level of intelligence

• The employee's level of work motivation

- Work place norms

- Supervisory support

A very simple path diagram for this hypothetical causal model is presented in Figure 4.1.

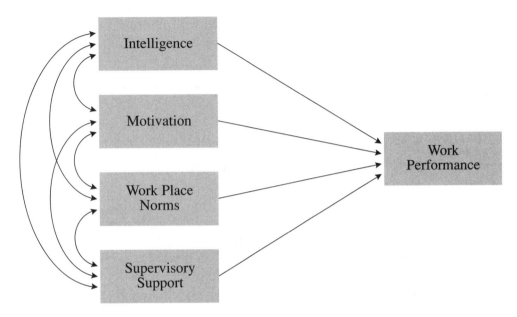

Figure 4.1: Path Diagram: A Simple Model of the Determinants of Work Performance

Intelligence, motivation, work place norms, and supervisory support are all **antecedent variables** within this framework, as they are clearly predicted to precede and have a causal effect on work performance. Similarly, work performance is the **consequent variable** in the model, as it is said to be affected by these antecedent variables. In the path analysis literature, many refer to antecedent variables as **independent variables**, and consequent variables as **dependent variables** (although these terms obviously do not have the same meaning in path analysis research, which is correlational in nature, as they have in experimental research).

The boxes of Figure 4.1 are connected to one another by means of straight, single-headed arrows and curved, double-headed arrows. In the path analysis literature, **straight, single-headed arrow** is generally used to represent a unidirectional causal path in a path diagram. The arrow originates at the variable exerting the causal influence (the independent variable), and the arrow points toward the variable being affected (the dependent variable). For example, the straight, single-headed arrow from intelligence to work performance represents the prediction that intelligence has a causal effect on performance.

In contrast, a **curved, double-headed arrow** connecting two variables represents a simple covariance, or correlation, between the variables. A curved arrow connecting two variables means that the two variables are expected to covary, but that no hypothesis is made regarding any causal influence between them. For example, the two-headed, curved arrow connecting intelligence to motivation means that you makes no hypothesis as to which variable influences which. Perhaps intelligence causes motivation, perhaps motivation causes intelligence, perhaps each has some causal influence on the other, or perhaps their correlation is spurious, due to the influence of some shared but unmeasured variable.

The model presented in Figure 4.1 suggests that each of the four antecedent variables is expected to have a direct effect on work performance; notice that each arrow goes directly from the independent variable of interest to work performance. This is the simplest possible type of path model. However, most models in the social science literature also predict that some variables have *indirect* effects on other variables. Figure 4.2 provides a model that includes indirect effects.

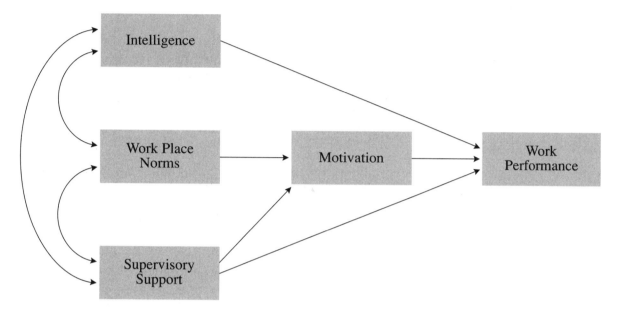

Figure 4.2: Path Diagram: A More Complex, Recursive Model of the Determinants of Work Performance

Figure 4.2 includes the same four variables discussed earlier, but they are arranged in a somewhat different causal sequence. Most notably, worker motivation is now viewed as a **mediator variable**: a variable that mediates, or conveys, the effect of an antecedent variable onto a consequent variable. Notice a single-headed arrow goes from work place norms to motivation, and that a separate single-headed arrow goes from motivation to work performance. This indicates that norms have only an indirect effect on performance; norms influence performance by first influencing motivation. This same idea can be expressed by saying that motivation completely mediates the effect of norms on performance. It is clear that norms are not expected to have a direct effect on performance, as there is no path going directly from norms to performance.

A variable may have both direct and indirect effects on a consequent variable; Figure 4.2 illustrates such a relationship between supervisory support and work performance. A single-headed arrow goes directly from support to performance, indicating the predicted direct effect. However, a path also goes from support to motivation, and a second path goes from motivation to performance. This indicates that support should also indirectly affect performance by first affecting motivation.

It may be possible to defend the causal relationships predicted in Figure 4.2, since variables such as intelligence, motivation, work place norms, and supervisory support should have an effect on work performance. However, these hypothetical path models are used for illustrative purposes only. There is no well-developed theory that posits precisely the relations displayed in the figure, and no claim regarding its validity is made here.

Important Terms Used in Path Analysis

Endogenous versus exogenous variables. In path analysis, a distinction is made between endogenous variables and exogenous variables. An **endogenous variable** is one whose variability is predicted to be causally affected by other variables in the model. Any variable that has a straight, single-headed arrow pointing at it is an endogenous variable. In Figure 4.2, work performance is clearly an endogenous variable, as it is directly influenced by intelligence, motivation, and supervisory support. Motivation is also an endogenous variable; it is affected by work place norms and support.

Exogenous variables, on the other hand, are constructs that are influenced only by variables that lie outside of the causal model. Exogenous variables do not have any straight, single-headed arrows pointing at them. In Figure 4.2, intelligence, norms, and support are all exogenous variables. These three variables are connected by curved arrows (indicating that they are expected to covary), but no single-headed arrows point toward them. This means that the theorist makes no predictions about what influences intelligence, norms, or supervisory support. In most models, exogenous variables will affect other variables, but by definition, exogenous variables are never affected by other variables in the model.

Notice the straight arrow that runs from motivation to work performance. Does this mean that motivation is an exogenous variable? No, because there are also two arrows that point toward motivation. Anytime a single-headed arrow points at a variable, that variable is an endogenous variable.

Most of the figures in this chapter will follow the convention of having causal paths run from left to right. In general, exogenous variables will be presented toward the left side of the figure, and endogenous variables will be presented toward the middle and right.

Manifest versus latent variables. A **manifest variable** is one that is directly measured or observed in the course of an investigation, while a **latent variable** is a hypothetical construct that is not directly measured or observed. For example, "scores on the Weschler Adult Intelligence Scale (WAIS)" is a manifest variable; it is possible to directly determine exactly where each subject stands on this variable. On the other hand, "intelligence" may be thought of as a latent variable; it is a construct that is presumed to exist, although it cannot be directly observed. You may give your subjects an intelligence test, but you know that the resulting IQ scores are only estimates of where they stand on the underlying construct of intelligence. You hope that the underlying latent variable of intelligence will influence where the subjects stand on the manifest variable of WAIS scores.

What do rectangles and ovals represent in path diagrams? In this text, manifest variables are represented as squares or rectangles in path diagrams (notice the rectangles in Figures 4.1 and 4.2). A different procedure related to path analysis (often called "structural equation modeling" or "covariance structure modeling") includes both manifest and latent variables. In those analyses, latent variables are represented by circles or ovals in path diagrams. This chapter deals only with path models containing manifest variables; Chapters 5 and 6 will discuss causal models with latent variables.

Recursive versus nonrecursive models. A **recursive path model** is one in which causation flows in only one direction. In a recursive model, a consequent variable never exerts causal influence (either directly or indirectly) on an antecedent variable that first exerts causal influence on it. In other words, recursive models are unidirectional. The model presented in Figure 4.2 is recursive; notice that causation flows only in a left-to-right direction.

In contrast, in a **nonrecursive path model**, causation may flow in more than one direction, and a variable may have a direct or indirect effect on another variable that preceded it in the causal chain. For example, a model may be nonrecursive because it predicts reciprocal causation between two variables. This is illustrated in Figure 4.3, which reveals reciprocal causation between the work performance and motivation variables. Here, motivation is predicted to affect performance which, in turn, is expected to affect motivation.

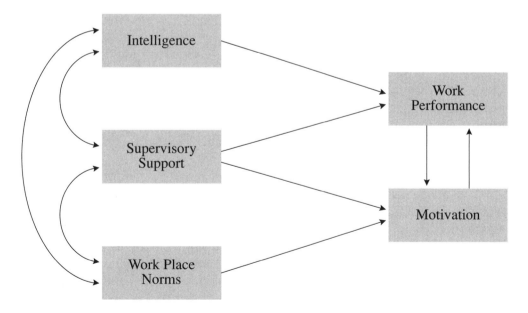

Figure 4.3: Path Diagram: Reciprocal Causation in a Nonrecursive Model

Models may also be nonrecursive because they contain a feedback loop. For example imagine a model in which variable A affects variable B, variable B affect variable C, and variable C, in turn, affects variable A. Variables A, B, and C can be said to constitute a feedback loop, and the model that contains this loop is therefore nonrecursive.

Until recently, it was important to make a distinction between recursive and nonrecursive models because the most popular procedures for performing path analysis (multiple regression procedures) did not lend themselves easily to the testing of nonrecursive models. However, either type of model can be analyzed with PROC CALIS.

NOTE: This chapter deals only with recursive models. This chapter will cover only the analysis of recursive (unidirectional) causal models. This will prepare you to deal with a wide variety of research problems that are commonly encountered in the social sciences. It will also serve as a good foundation for readers who want to move on to the somewhat more complex nonrecursive models. References for the analysis of nonrecursive causal models are provided later in the chapter.

Necessary Conditions for Path Analysis

The use of path analysis assumes that a number of requirements have been met concerning the nature of the data as well as the theoretical model itself. Some important assumptions associated with the analysis of the simple recursive models to be covered in this chapter are listed here.

1. **Interval- or ratio-level measurement.** All endogenous variables should be assessed on an interval- or ratio-level of measurement. Exogenous variables should also be on an interval- or ratio-level, although exogenous variables may be assessed at a nominal level if they are dummy-coded. Alternative procedures for situations in which these assumptions are violated have been discussed elsewhere (e.g., Joreskog & Sorbom, 1989), but will not be covered in this chapter. Although it has been argued that, in some cases, ordinal-level data may be treated as interval-level data without serious problems (e.g., Asher, 1988), this issue remains controversial.

2. **Minimal number of values.** Endogenous variables should be continuous and should assume a minimum of four values (Bentler & Chou, 1987).

3. **Normally distributed data.** Although parameter estimates may be correct with non-normal data, the statistical tests used with PROC CALIS (such as the model chi-square test and significance tests for path coefficients) assume a multivariate normal distribution. It has been argued, however, that the maximum likelihood and generalized-least squares estimation procedures appear to be fairly robust against moderate violations of this assumption (Anderson & Gerbing, 1988; Joreskog & Sorbom, 1989). When data are markedly non-normal, you should consider data transformations and/or the removal of outliers.

4. **Linear and additive relationships.** Relationships between variables should be linear and additive (that is, relationships between independent and dependent variables should not be curvilinear or interactive).

5. **Absence of multicollinearity.** Variables should be free of multicollinearity. Multicollinearity is a condition in which one or more variables exhibit very strong correlations (perhaps above .80) with one another.

6. **Absence of measurement error.** All independent (antecedent) variables should be measured without error. This means that any independent manifest variable that is analyzed should be a perfectly reliable indicator of the underlying construct that it is intended to measure. Given the relatively high levels of measurement error associated with variables typically studied in the social sciences, this may be the most frequently violated assumption in the use of path analysis. The following two chapters will show how this problem can be minimized through the analysis of causal models with multiple indicators and latent variables.

7. **Inclusion of all nontrivial causes.** All known nontrivial causes of a model's endogenous variables should be included in the model as independent variables. For example, if you perform a path analysis in which "work performance" is to be a dependent variable, the model must specify as independent variables all constructs that are known to have nontrivial effects on work performance (and, needless to say, these variables must be assessed in the study itself). If important antecedent variables are omitted, the path coefficients for the remaining antecedents are likely to be biased. If this requirement is met, the model is said to be **self-contained**, and all residual terms in the model should be noncorrelated (the meaning of "residual terms" is discussed later).

8. **Overidentified model.** To be tested for goodness of fit, the causal model must be overidentified. The meaning of "just-identified" versus "overidentified" versus "underidentified" models is discussed later.

10. **Minimal number of observations.** Path analysis is a large-sample procedure; it is best that the analysis is based on at least 200 subjects (although results based on fewer subjects have certainly been reported in the literature). In addition, there should be a ratio of at least 5 subjects for each parameter to be estimated. The total number of parameters is the sum of the

   - path coefficients

   - variances

   - covariances to be estimated.

Along with the preceding requirements, James, Mulaik, and Brett (1982, pp. 56-57) describe a number of "conditions pertaining to the appropriateness of theoretical models for confirmatory analysis." These prerequisites are useful for determining whether your conceptual model is sufficiently well developed for testing with path analysis or related confirmatory procedures. Their prerequisites include the requirement that you make a formal statement of theory in terms of a structural model, that you provide a theoretical rationale for each causal hypothesis, along with other requirements. You are encouraged to use this valuable source to verify that your model is sufficiently developed prior to conducting the confirmatory procedures discussed in this chapter.

Overview of the Analysis

Although it is relatively easy to perform path analysis using PROC CALIS, this process must be divided into a number of steps with important decisions to be made at each step. Here is an overview of the process:

Preparing the program figure. It is recommended that you always prepare a detailed figure that describes the predicted relationships between all variables, and identifies the parameters to be estimated. This program figure guides you when writing the SAS program. If carefully prepared, the program figure will make writing the program a much simpler task.

Preparing the SAS program. The analyses will be conducted using PROC CALIS, a SAS procedure that can be used for path analysis, confirmatory factor analysis, structural equation modeling with latent variables, and other purposes. The CALIS program will include statements to represent the model displayed in your program figure.

Interpreting the results. The output of PROC CALIS provides a significance test for the null hypothesis that your theoretical model fits the data, along with a number of descriptive goodness of fit statistics. It also provides estimates and significance tests for parameters such as path coefficients, variances, and covariances. If there is not a good fit between model and data, you may use modification indices to determine how the model should be changed. Any revised model can then be estimated to determine if it provides a good fit.

Example 1: A Path-Analytic Investigation of the Investment Model

The investment model (Rusbult, 1980) will again be used, this time to illustrate how path analysis may be conducted with PROC CALIS. Remember that the investment model identifies variables that are believed to affect satisfaction and commitment in romantic relationships. One conceptualization of the model holds that **commitment** to a romantic relationship is determined by the following three constructs:

- **Satisfaction** with the relationship

- The size of one's **investments** in the relationship (e.g., the amount of personal time and resources put into the relationship)

• **Alternative value**, or the attractiveness of one's alternatives to the relationship

Satisfaction, in turn, is said to be predicted by **rewards**, or the positive features that one associates with the relationship, and **costs**, or the negative features associated with the relationship.

It is possible to use some of the fundamental concepts of the investment model to develop the causal model presented in Figure 4.4. The paths presented in this figure represent the prediction that commitment is determined by satisfaction, investment size, and alternative value, and that satisfaction is determined by rewards and costs.

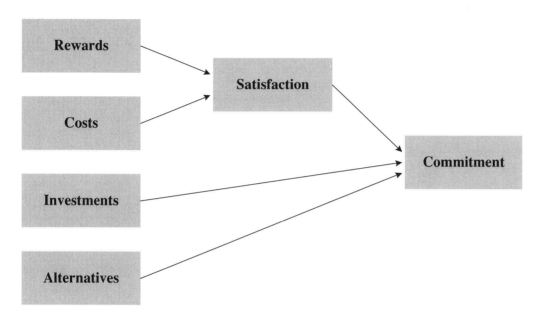

Figure 4.4: The Basic Causal Model

NOTE: The model presented in Figure 4.4 is merely based on the investment model, and should not necessarily be interpreted as an accurate representation of the model as it was originally developed by Rusbult (1980). Note that the data analyses reported in this book are fictitious and are used only to illustrate statistical procedures. These results should not be viewed as valid tests of the investment model, or of any other theoretical framework.

Overview of the Rules for Performing Path Analysis

The following sections of this chapter will show you how to prepare a program figure and how to write the SAS program that will test the path model presented in Figure 4.4. These sections will present 14 rules that you can use as guidelines in preparing program figures and writing SAS programs that will perform a path analysis. The 14 rules are summarized here:

Rule 1: In general, only exogenous variables are allowed to have covariances.

Rule 2: A residual term must be identified for each endogenous variable in the model.

Rule 3: Exogenous variables do not have residual terms.

Rule 4: Variances should be estimated for every exogenous variable in the model, including residual terms.

Rule 5: In most cases, covariances should be estimated for every possible pair of manifest exogenous variables; covariances are not estimated for endogenous variables.

Rule 6: For simple recursive models, covariances should not be estimated for residual terms.

Rule 7: One equation should be created for each endogenous variable, with that variable's name to the left of the equals sign.

Rule 8: Variables that have a direct effect on that endogenous variable are listed to the right of the equals sign.

Rule 9: Exogenous variables, including residual terms, are never listed to the left of the equals sign.

Rule 10: To estimate a path coefficient for a given independent variable, a unique path coefficient name should be created for the path coefficient associated with that independent variable.

Rule 11: The last term in each equation should be the residual (disturbance) term for that endogenous variable; this E (or D) term will have no name for its path coefficient.

Rule 12: To *estimate* a parameter, create a name for that parameter.

Rule 13: To *fix* a parameter at a given numerical value, insert that value in the place of the parameter's name.

Rule 14: To *constrain* two or more parameters to be equal, use the same name for those parameters.

The fact that these guidelines are assigned numbers (e.g., Rule 1, Rule 2) is not meant to indicate any hierarchy of importance. The numbers were merely assigned consecutively so that it will be possible to easily refer to specific rules in this text.

Obviously, it is not expected that the preceding rules will make any sense to you at this point if you are just learning path analysis for the first time. However, as you read the following sections and get some experience in conducting path analyses on your own, these guidelines will become second nature to you. They are listed here so that they will be easy to locate for future reference.

Preparing the Program Figure

The PROC CALIS program that performs a path analysis is somewhat more complicated than the SAS program that performs, say, multiple regression. This is because the PROC CALIS program must include a relatively large number of statements that represent your causal model as a series of structural equations. One error in any statement, and your results will likely be incorrect.

Fortunately, the likelihood of making errors can be greatly reduced by first preparing a program figure. This **program figure** will display all of the important features of your causal model. Among other things, it will identify the endogenous and exogenous variables, it will identify the parameters (such as path coefficients) to be estimated, and it will indicate which variables are free to covary. If you do a careful job of preparing the program figure, writing the PROC CALIS program will be quite easy.

This section discusses the steps to follow in preparing a program figure. A later section will show how the figure is then translated into a SAS program.

Step 1: Drawing the Basic Causal Model

When preparing a program figure, try to adhere to the convention of placing the antecedent variables toward the left side of the figure, and the consequent variables to the right. Draw straight, single-headed arrows to indicate the predicted causal connections. The basic causal model to be tested here was presented in Figure 4.4. Notice that the four exogenous variables are lined up on the left side of that figure. The reasons for this will become clear later.

Step 2: Assigning Short Variable Names to the Manifest Variables

In Figure 4.5, short variable names have been assigned to the variables of the path diagram. The conventions for naming variables and error terms in this chapter are based largely on those developed by Bentler (1989) for the EQS structural equations program. With this system, manifest variables are represented by the letter "V" followed by a number. This text uses the convention of starting with the last dependent variable in the causal chain and naming it V1, then tracing backward (right to left), assigning the names V2, V3, etc. to variables that appear earlier in the causal chain. These short variable names are written just above the variable's actual name in the path diagram. Following these conventions, Figure 4.5 shows that commitment has been named V1, satisfaction has been named V2, and so forth.

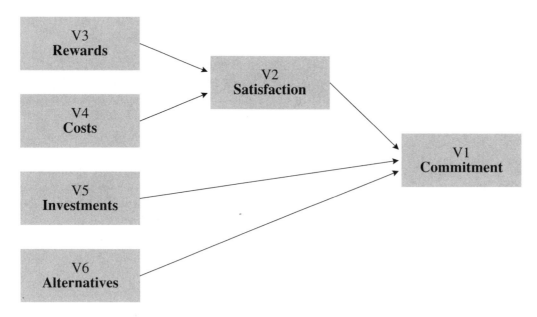

Figure 4.5: Assigning Short Variables Names to the Manifest Variables

Step 3: Identifying Covariances between Exogenous Variables

Remember that exogenous variables are those constructs that do not have any single-headed arrows pointing at them. In the present model, the exogenous variables are Rewards, Costs, Investments, and Alternatives.

In Figure 4.6, curved, double-headed arrows are used to connect all four of these exogenous variables. This indicates that the four exogenous variables are expected to covary (be correlated). This also means that you will ultimately prepare a program that estimates the covariances between these variables.

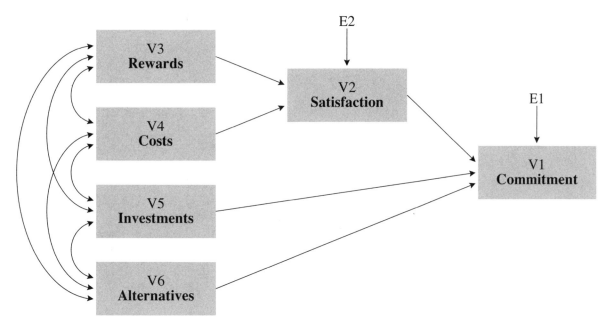

Figure 4.6: Identifying Covariances between Exogenous Variables, and the Residual Terms for Endogenous Variables

In preparing the program figure for your model, you should identify any variables that are expected to covary in the same manner, by connecting them with curved, double-headed arrows. However, you should heed the following rule:

> **Rule 1:** In general, only exogenous variables are allowed to have <u>covariances</u>.

With the relatively simple causal models discussed in this chapter, an endogenous variable is never allowed to have a covariance with any other variable. This means that if a straight, single-headed arrow points at a variable, there should not be a curved, double-headed arrow pointing at that same variable.

With the simple causal models discussed in this chapter, every exogenous variable is usually allowed to covary with every other exogenous variable in the model. That is, you should estimate a covariance for every possible pair of exogenous variables. However, this is not a strict rule. If you have good reason to believe that two exogenous variables should not be correlated, it may be acceptable to not estimate a covariance for this pair of constructs. For example, you may have strong theoretical reasons to believe that two exogenous variables may be uncorrelated, and you want to test a model that reflects this prediction. However, in the vast majority of cases, it will be appropriate to estimate covariances for every pair of exogenous variables.

Step 4: Identifying Residual Terms for Endogenous Variables

After identifying the covariances for the exogenous variables, you should identify the residual terms for the endogenous variables, in accordance with Rule 2:

> **Rule 2:**   A residual term must be identified for each endogenous variable in the model.

In other references, the residual term is sometimes referred to as the **error term** or **disturbance term.** The **residual term** for a variable represents all the factors that influence variability in that variable, but are not included as antecedent variables in the model.

For example, the path diagram in Figure 4.6 predicts that Commitment will be affected by Satisfaction, Investments, and Alternatives. However, it is highly unlikely that these three antecedent variables will account for 100% of the variability in Commitment. Therefore, Commitment is expected to also be affected by a residual term. This residual term may represent causal effects on the dependent variable due to omitted independent variables, random shocks, or specifications errors in the equation (James, Mulaik, & Brett, 1982).

Using Bentler's (1989) conventions, residual terms are represented by the letter E (for Error term). In Figure 4.6, the residual term for commitment is given the short name E1 because the short name for the commitment variable is V1; the numerical suffix for the V and E terms ("1" in this case) should always match. The figure shows that commitment (V1) is expected to be affected by the residual term E1, while satisfaction (V2) is expected to be affected by its residual term, E2.

The remaining variables are exogenous variables, and therefore are not given residual terms, in accordance with Rule 3:

> **Rule 3:** Exogenous variables do not have residual terms.

Step 5: Identifying Variances to Be Estimated

Rule 4 indicates which variance terms should be estimated:

> **Rule 4:**   Variances should be estimated for every exogenous variable in the model, including exogenous manifest variables as well as residual terms.

This text recommends that you identify every parameter to be estimated by placing a question mark (?) in the appropriate location in your program figure. More specifically, you should use the symbol "VAR?" to identify each variance to be estimated. Place this symbol directly under the name of that manifest variable or residual term. For example, Figure 4.7 shows that the VAR? symbol is placed directly under the word "Rewards" in the rectangle representing that variable. Similar symbols appear in the rectangles for the Costs, Investments, and Alternatives variables. Because these are all exogenous variables, their variances must be estimated.

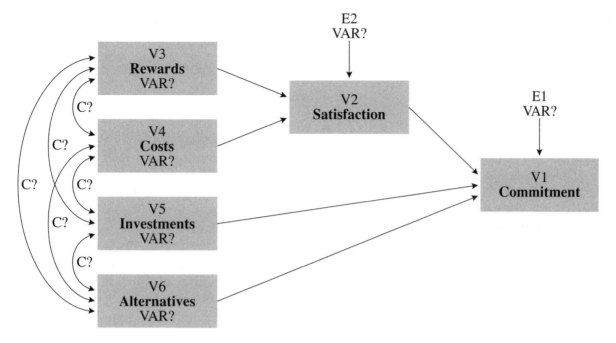

Figure 4.7: Identifying the Variances and Covariances to Be Estimated

When identifying variances to be estimated, do not forget the residual terms! The residual terms of Figure 4.7 do not have any single-headed arrows pointing toward them, and this means that they are also exogenous variables. They therefore must have their variances estimated. Figure 4.7 shows that the VAR? symbol is also placed just under the short names for residual terms E1 and E2.

Step 6: Identifying Covariances to Be Estimated

Figure 4.7 uses the symbol "C?" to identify the covariances that are to be estimated. In accordance with Rule 5, these covariances will usually involve all manifest exogenous variables:

> **Rule 5:** In most cases, covariances should be estimated for every possible pair of manifest exogenous variables; covariances are not estimated for endogenous variables.

Notice that a curved, double-headed arrow connects the rewards variable (at the top of Figure 4.7) with the alternatives variable (at the bottom of the figure). This curved arrow is identified with the C? symbol, indicating that the SAS program should estimate the covariance between these two variables. The remaining curved arrows in the figure connect the remaining pairs of manifest exogenous variables, and each of these is identified with the C? symbol, meaning that a total of six covariances is to be estimated for this causal model.

However, notice that no curved arrow connects either E1 or E2 with any other variable. This is in accordance with Rule 6:

> **Rule 6:** For simple recursive models, covariances should not be estimated for residual terms.

Rule 6 does not necessarily apply to other types of models. For example, residual terms are sometimes allowed to covary in a time-series design in which the same variable is measured at more than one point in time. Also, in models with reciprocal causation, the residual terms for the two variables involved in the reciprocal relationship may be allowed to covary. But for the relatively simple models to be discussed in this chapter, Rule 6 will generally hold.

Step 7: Identifying the Path Coefficients to Be Estimated

Each straight, single-headed arrow in your figure represents a causal path. For each of these, the SAS program will estimate a **path coefficient:** a number that represents the amount of change in a dependent variable that is associated with a one-unit change in a given independent variable, while holding constant the other independent variables. Path coefficients represent the size of the effect that a given independent variable has on a dependent variable.

This text uses the symbol "P?" to identify path coefficients that are to be estimated. In Figure 4.8, each path has been identified with this symbol. After the standardized path coefficients have been estimated by the SAS program, you will review their relative size to determine which independent variables had the stronger effects on the dependent variables. The program results will also provide statistics testing the null hypothesis that a given path coefficient is zero in the population.

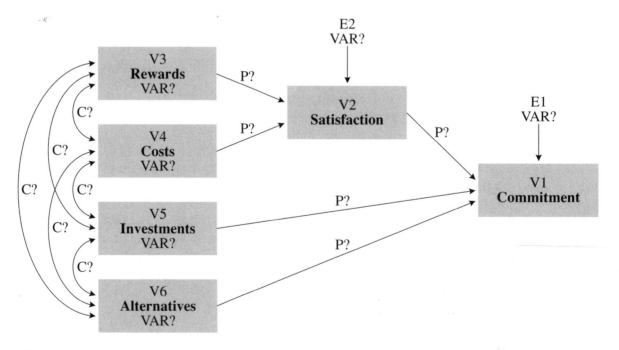

Figure 4.8: Identifying the Path Coefficients to Be Estimated

Step 8: Verifying that the Model Is Overidentified

The identification problem is one of the most important concepts to understand in causal modeling. Because identification is less of a problem when testing simple recursive path models (such as those presented in this chapter), this section will merely introduce some basic issues. Identification becomes a more serious problem when estimating nonrecursive causal models or models with latent variables; for users interested in testing these more complicated models, this section also lists references that provide a more in-depth treatment of the topic.

To understand identification, it is necessary to first understand that a causal model may be represented as a system of functional equations. For example, the path diagram presented in Figure 4.8 predicts that the endogenous variable, Satisfaction (V2), will be causally determined by Rewards (V3), Costs (V4), and its residual term (E2). A functional equation representing this part of the model could take the following form:

$$V_2 = p_{23}\ V_3 + p_{24}\ V_4 + E_2$$

where p_{23} represents the path coefficient for the effect of V_3 on V_2, and p_{24} represents the effect of V_4 on V_2. This equation includes two unknowns: the two path coefficients. In performing the analysis, PROC CALIS will estimate values for these two path coefficients. A typical path model (such as that represented in Figure 4.8) will be represented by a number of equations, and will usually include a variety of unknown parameters to be estimated, including variances, covariances, and path coefficients.

It is highly desirable that a model not be **underidentified** prior to estimation. Technically, a system is underidentified when it includes fewer linearly independent equations than unknowns (Asher, 1988). When a model is underidentified, an infinite number of solutions can be generated to solve for its parameters. For example, if PROC CALIS were used to estimate an underidentified model, performing the analysis with one set of starting values might generate one set of parameter estimates, while running the analysis a second time with a different set of starting values might generate a radically different set of parameter estimates (different values for the same path coefficients, etc.). Obviously, results obtained from the analysis of an underidentified model are completely meaningless.

Parameter estimates such as path coefficients are meaningful only if they are obtained from the estimation of an **identified** model. A model may be identified either by being just-identified or overidentified. A **just-identified** model is one in which there are exactly as many linearly independent equations as unknowns (some texts refer to just-identified models as **exactly-identified** or **saturated** models). Although a just-identified model has the advantage of allowing the estimation of just one unique set of parameters (for a given sample), it has the disadvantage of not allowing any tests for goodness of fit (this is because just-identified models always provide a perfect fit to the data). It is for this reason that researchers typically need to be sure that their models are overidentified.

A model is said to be **overidentified** when it includes more equations than unknowns. As with a just-identified model, the estimation of an overidentified model will result in only one set of parameter estimates for a given sample of data. Overidentified models, however, have an additional desirable property: they can be tested for overall goodness of fit.

Fortunately, simple recursive models such as those covered in this chapter are always just-identified or overidentified (Asher, 1988, pp. 54-55). "Simple" means manifest-variable models without nonrecursive relationships, in which residual terms are uncorrelated with other terms. When dealing with recursive models with uncorrelated residuals, the model may be said to be just identified if every variable in the model is related to every other variable by either a curved, two-headed arrow, or a causal path. Figure 4.9 displays an example of a just-identified model. Notice that every variable is interrelated with every other variable, either through a causal path or a covariance.

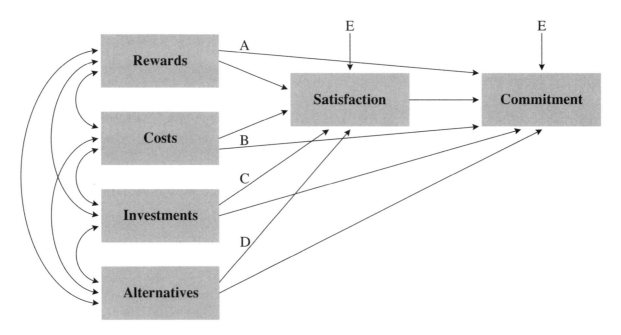

Figure 4.9: A Just-Identified, Fully Recursive Model

To create this just-identified model, four new paths were added to the investment model presented earlier. Paths A and B predict that Rewards and Costs, respectively, will affect Commitment, and paths C and D predict that Investments and Alternatives will affect Satisfaction.

In performing path analysis, it is customary to statistically test the "fit" between the model and data; the better the fit, the stronger the support for the theory. However, it is important to remember that a just-identified model cannot be tested for goodness of fit. This is because a just-identified model always fits the data perfectly. The reason for this should become clear when you consider that the data that are analyzed in path analysis consist simply of a correlation matrix (or a variance-covariance matrix). This matrix provides the correlations between every manifest variable. Now consider the just-identified model presented in Figure 4.9. Note that every variable is predicted to be related to every other variable in some way. It comes as no surprise that such a complex model is able to account for the correlations in the original correlation matrix. A more valuable model would be a simpler model: one that did not contain so many interconnections between the variables, but was still able to account for the observed correlations. Such a model would be an overidentified model.

A just-identified model becomes overidentified when you place restrictions on certain model parameters. A number of different types of restrictions may be imposed; for example, it is possible to constrain two path coefficients to take on the same value. But by far, the most common approach is to fix certain path coefficients to take on a value of zero. Fixing a path at zero has the same effect as eliminating that path from the model.

Figure 4.10 shows the model that resulted from fixing paths A, B, C, and D (from Figure 4.9) at zero. These paths have been eliminated, with the result that the path model in Figure 4.10 is now overidentified. Because of this, the model may now be tested for goodness of fit.

In summary, remember that a recursive model may be tested for goodness of fit only if certain restrictions are placed on the just-identified model, and that this is usually achieved by eliminating some of the possible causal paths.

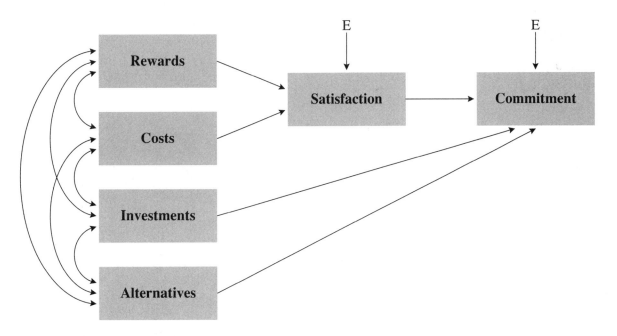

Figure 4.10: An Overidentified Model

For more complex models such as nonrecursive models or models with latent variables, the identification problem can be more troublesome. These models *can* be underidentified, and any researcher who estimates an underidentified model may unknowingly obtain meaningless parameter estimates.

How can you be sure that one of these more complex path models is overidentified? Long (1983a, p. 44) argues that "...the most effective way to demonstrate that a model is identified is to show that through algebraic manipulations of the model's covariance equations each of the parameters can be solved in terms of the population's variances and covariances of the observed variables." When the variables are standardized, solving the equations for parameters is equivalent to solving the path equations, and Duncan (1975) and Kenny (1979) are useful sources describing the necessary techniques. Long (1983b, pp. 66-72) demonstrates this approach to proving identification. Additional approaches for proving identification in manifest-variable models are provided by Asher (1988) and Berry (1985).

A particularly useful guide to model identification is provided by Bollen (1989), who describes a number of necessary and/or sufficient conditions for identification. Especially helpful is the way the text provides separate summaries of identification rules for path models with manifest variables, confirmatory factor analysis models, and structural equation models with latent variables. Several of the procedures described in the text require the use of matrix algebra.

The procedures just discussed can be time-consuming, but are necessary if you are to be sure that a nonrecursive or latent-variable model is identified. Fortunately, the CALIS procedure itself can often detect an underidentified model. When it does, it will print equations showing the linear relationships among the relevant parameter estimates (see "Identification of Models" in the chapter "Introduction to Structural Equations with Latent Variables" in the *SAS/STAT users guide*).

In addition to checking log and output files for this warning sign, you should also routinely verify that the number of data points in the analysis is larger than the number of parameters to be estimated; when this is not the case, the model is not identified. The number of data points may be calculated with the following equation:

```
Number of data points = (p ( p + 1 )) / 2
```

where p = the number of manifest variables being analyzed. For example, in the analysis of the investment model, six variables are being analyzed. Inserting this in the formula results in the following:

```
Number of data points = (6 (6 + 1)) / 2
                      = (6 (7)) / 2
                      = (42) / 2
                      = 21
```

Once it has been established that the current analysis involves 21 data points, it is necessary to determine the number of parameters to be estimated. This will be equal to the sum of the number of

- path coefficients,

- variances

- covariances to be estimated.

This will be easy to determine, as you have already identified these parameters in your program figure. This program figure is again reproduced as Figure 4.11. Remember that the symbol VAR? was used to represent each variance to be estimated, C? was used to represent each covariance, and P? was used to represent each path coefficient. By referring to Figure 4.11, you can see that the analysis will estimate 6 variances (for E1, E2, V3, V4, V5, and V6), 6 covariances (between variables V3, V4, V5, and V6), and 5 paths. The total number of parameters to be estimated is therefore equal to 17.

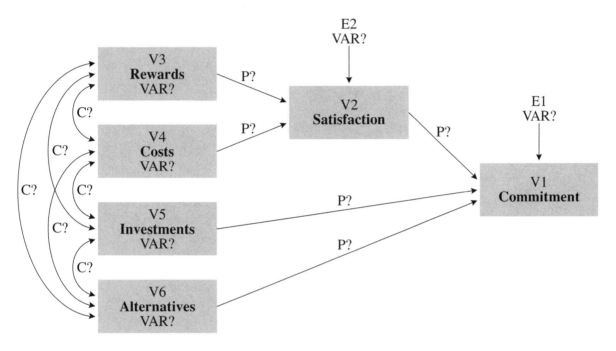

Figure 4.11: The Completed Program Figure

It has already been said that one criterion for overidentification is that the number of data points should exceed the number of parameters to be estimated. Because the present analysis involves 21 data points but only 17 parameters, you may conclude that the model presented in Figure 4.11 meets this criterion.

Is it necessary to actually determine the number of data points and parameter estimates in this way prior to each analysis? Technically, the answer is no, because this information is also provided in the printed output of PROC CALIS. On page 3 of the output resulting from the initial analysis (to be presented later) PROC CALIS provides the number of data points to the left of the word "Informations", and provides the number of parameter estimates to the right of the word "Parameters". Despite this, it is good practice to calculate these values prior to gathering data, to verify that your model is in fact testable.

As a final test for identification, you may want to repeat your analyses several times, using very different starting values for parameter estimates each time. If PROC CALIS arrives at the same final parameter estimates each time, it is likely (but not certain) that your model is identified.

Important: Remember that the last two procedures (counting parameters and informations and conducting analyses with differing starting values) are necessary but not sufficient conditions for demonstrating identification. This means that, if your model fails to pass these tests, they are clearly underidentified, but if they do pass these tests it does not conclusively prove that they are identified. The only way to be sure is to use one of the more time-consuming approaches such as those recommended by Asher (1988), Berry (1985), Bollen (1989), or Long (1983a, 1983b). Remember that these more complicated approaches are necessary only with more complex causal models such as nonrecursive models, models with correlated residuals, or latent-variable models. Simple recursive systems without correlated residuals will always be overidentified as long as some of the causal paths have been fixed at zero.

Preparing the SAS Program

Overview

The PROC CALIS program that analyzes a simple recursive path model (such as the one described here) is longer than most SAS programs, but is not especially complicated or difficult to understand. Preparing the program figure in advance will greatly facilitate the writing of this program, and the instructions provided here should be helpful in leading you step-by-step through the various PROC CALIS statements. Here is the entire PROC CALIS program, including the DATA step, that was used to analyze the model presented in Figure 4.11:

```
1       DATA D1(TYPE=CORR) ;
2          INPUT _TYPE_ $ _NAME_ $ V1-V6 ;
3             LABEL
4                V1 ='COMMITMENT'
5                V2 ='SATISFACTION'
6                V3 ='REWARDS'
7                V4 ='COSTS'
8                V5 ='INVESTMENTS'
9                V6 ='ALTERNATIVES' ;
10      CARDS;
11      N    .    240     240     240     240     240     240
12      STD  .   2.3192  1.7744  1.2525  1.4086  1.5575  1.8701
13      CORR V1  1.0000    .       .       .       .       .
14      CORR V2   .6742  1.0000    .       .       .       .
15      CORR V3   .5501   .6721  1.0000    .       .       .
16      CORR V4  -.3499  -.5717  -.4405  1.0000    .       .
17      CORR V5   .6444   .5234   .5346  -.1854  1.0000    .
18      CORR V6  -.6929  -.4952  -.4061   .3525  -.3934  1.0000
19        ;
20      PROC CALIS   COVARIANCE CORR RESIDUAL MODIFICATION ;
21          LINEQS
22             V1 = PV1V2 V2 + PV1V5 V5 + PV1V6 V6 + E1,
```

```
23              V2 = PV2V3 V3 + PV2V4 V4              + E2;
24        STD
25            E1 = VARE1,
26            E2 = VARE2,
27            V3 = VARV3,
28            V4 = VARV4,
29            V5 = VARV5,
30            V6 = VARV6;
31        COV
32            V3 V4 = CV3V4,
33            V3 V5 = CV3V5,
34            V3 V6 = CV3V6,
35            V4 V5 = CV4V5,
36            V4 V6 = CV4V6,
37            V5 V6 = CV5V6;
38        VAR  V1 V2 V3 V4 V5 V6 ;
39        RUN;
```

handwritten annotations: for V_1 V_2 } endogenous V. (pointing to lines 25–26)
exogenous V. (pointing to COV section)

Lines 1–19 of the preceding program constitute the DATA step; the data are input as a correlation matrix. Line 20 includes the PROC CALIS statement that calls up the procedure that will perform the analyses. The equations included in the LINEQS statement (lines 21–23) identify the nature of the predicted causal relationships between the model's variables (e.g., which variables have causal effects on the model's endogenous variables). Lines 24 through 30 include the STD statement, which identifies the variances to be estimated, while the subsequent COV statement (lines 31-37) indicates which covariances are to be estimated. Finally, the VAR statement on line 38 indicates which variables are to be selected for analysis. Each of these sections will be discussed in detail in the following section, which is divided into five subsections. First, the options for data input are described. Next, the conventions and options to be used with the PROC CALIS, LINEQS, STD, and COV statements are reviewed. Once these statements are completed, your path model will have been converted into a series of structural equations, and will be ready for analysis.

The Data Input Step

With PROC CALIS, data may be input either as raw data or as a correlation or covariance matrix. Appendix A.2, "Data Input," provides instruction on how to input these different types of data.

Inputting raw data. The advantage of inputting raw data is that this format allows use of the KURTOSIS option with PROC CALIS. This option computes a number of univariate and multivariate measures of kurtosis, and identifies the observations that make the greatest contribution to kurtosis. This can be used to identify multivariate outliers that possibly should be removed from the data set. The disadvantage associated with raw data is that the analysis may take longer to run, especially with large samples. This extra computer time can add up, as it is often necessary to perform several analyses before a satisfactory model is developed.

Inputting a correlation or covariance matrix. Inputting a correlation or covariance matrix has the advantage of usually taking less computer time, but the disadvantage of not allowing for the computation of measures of kurtosis or the identification of outliers. In many cases, a good compromise is to use both methods; the first runs should involve the analysis of raw data (if available) to assess kurtosis and identify outliers. Once outliers have been eliminated, the correlation or covariance matrix based on the resulting sample can be analyzed in subsequent runs. PROC CORR can be used to create the correlation matrix, if this is desired. It is usually desirable to specify the NOMISS option when computing these correlations, as this assures that every correlation will be based on the same set of observations. For example, the correlation matrix needed for the current model could have been produced with the following statements (the NOPROB option suppresses printing of *p* values for the correlations):

```
   PROC CORR    DATA=D1 NOMISS NOPROB ;
      VAR V1 V2 V3 V4 V5 V6;
      RUN;
```

Problems with large differences in standard deviations. With PROC CALIS, problems can result if the standard deviations for some of the variables are much larger than the standard deviations for other variables. For example, if the largest standard deviation in an analysis is 150 for variable V1, and the smallest is 1.3 for variable V2, the CALIS procedure may encounter difficulty in estimating the model. This may be noted on the SAS output itself, with a message that not all parameters are identified. Near-zero standard errors for parameter estimate *t* tests are another warning sign.

When inputting raw data, these problems can be avoided by rescaling variables so that they are all on approximately the same scale. For example, if everyone's score on V1 were divided by 100, the standard deviation for that variable would decrease from 150 to 1.5, which is comparable to the standard deviation of 1.3 for V2. Despite the new standard deviation, all correlations between the variables would remain the same.

When data are input in the form of a correlation matrix with standard deviations, it is not necessary to actually divide everyone's score by some constant. When you encounter a problem with standard deviations of this nature, simply move the decimal point on the troublesome standard deviation to achieve a standard deviation of the desired magnitude.

The PROC CALIS Statement

The CALIS procedure can perform a variety of structural equation analyses, including confirmatory factor analysis and covariance structural modeling with latent variables. The models tested with these procedures are often called "LISREL-type" models because the LISREL program (Joreskog & Sorbom, 1989) was the first widely available program that was capable of performing these analyses. However, this chapter will focus only on how the CALIS procedure may be used to perform path analysis with *manifest* variables. Confirmatory factor analysis and path analysis with latent variables are covered in Chapter 5 and Chapter 6 of this text, respectively.

The general form for the PROC CALIS statement is as follows:

```
PROC CALIS    options ;
```

The words PROC CALIS invoke the procedure, and these should be followed by at least one blank space and a list of options (if desired), with the name of each option separated by at least one blank space. The statement ends with a semicolon.

The preceding program uses the following PROC CALIS statement. This statement requests the COVARIANCE, CORR, RESIDUAL, and MODIFICATION options.

```
    ✓ PROC CALIS    COVARIANCE CORR RESIDUAL MODIFICATION ;
```

The COVARIANCE option requests that the analysis be performed on the covariance matrix rather than the correlation matrix. If your input step includes both a correlation matrix and standard deviations, the SAS System will use these to create a covariance matrix. Without this option, the analysis is performed on the correlation matrix by default, which is generally less desirable. Analyses performed on the correlation matrix are more likely to produce invalid standard errors for parameter estimates, which means that the significance tests for path coefficients (and other parameters) may be inaccurate. The model chi-square test (described later) is generally the same regardless of which matrix is analyzed.

The CORR option prints the correlation or covariance matrix that is analyzed, along with the predicted model matrix. The latter matrix is sometimes referred to as the **reproduced** matrix, and will also be discussed later.

The RESIDUAL option prints the absolute and normalized residual matrix, which is helpful in assessing the fit of the model and identifying possible areas requiring modification. A few words of explanation should clarify why this matrix is important.

After a path analysis has been performed on a covariance matrix, it is possible to use the resulting path coefficients to calculate the elements of a reproduced covariance matrix. If your path model provides a good fit to the data, the reproduced covariance matrix should be nearly identical to the original covariance matrix (the one that was analyzed).

By subtracting each coefficient in the reproduced matrix from the corresponding coefficient in the original matrix, it is possible to create a residual matrix. If your model provides a good fit to the data, then the elements of the residual matrix should be zero or very near to zero. A large element in the residual matrix may indicate a specification error in your model. For example, if the matrix displays a residual of .50 at the intersection of V1 and V2, then your model may be doing a poor job describing the relationship between these two variables. Perhaps your model includes a path between the two variables that should be deleted. Or perhaps it does not include a path between them that should be added. In summary, the residual matrix allows you to assess the model's fit, and helps identify specification errors. (In practice, it is generally easier to actually review the *normalized* residual matrix, as the elements of the residual matrix are not standardized in any meaningful way; more on this later).

Finally, the MODIFICATION option requests a number of modification indices that can be useful in identifying changes that would improve your model's fit. Often, your original model will not provide a satisfactory fit to the data, and it will be necessary to change it so as to better reflect the actual relationships between model variables. The MODIFICATION option prints results of the **Lagrange multiplier test**, which identifies paths or covariances that possibly should be added to the model. It may also print results of the **Wald test**, which helps identify paths and covariances that possibly should be deleted from the model. Although these indices can be quite valuable in developing an improved model, it is emphasized that any changes made to the model should be driven by your understanding of theory and research in the area being investigated, not simply by the results of these statistical tests. This point will be elaborated in a later section of this chapter titled "Modifying the Model."

The PROC CALIS chapter in the *SAS/STAT users guide* (1989) describes many additional options that may be used with the procedure. A few that may be particularly useful in social science research are presented here:

ALL
> prints all optional output.

CORR
> prints the original correlation (or covariance) matrix to be analyzed, along with the predicted model correlation (or covariance) matrix. Comparing these two matrices helps in identifying potential model modifications.

COVARIANCE or COV
> causes PROC CALIS to analyze the covariance matrix rather than the correlation matrix (analyzing the correlation matrix is default). In order to analyze the covariance matrix, it is necessary to input either a correlation matrix that includes standard deviations (as in the preceding program), a covariance matrix, a TYPE=SSCP data set, or raw data.

DATA=data-set-name
> specifies the input data set to be analyzed.

FCONV=p
> specifies the relative function convergence criterion. By default, this value (p) is the smallest real value that can be represented by your computer. For most problems, the optimization technique used by PROC CALIS requires the repeated computation of two values: the function value (optimization criterion) and the gradient vector (first-order partial derivatives).

GCONV=p
> specifies the absolute gradient convergence criterion. By default, this value (p) is equal to $1 E - 3$. Smaller values may be specified to obtain more precise parameter estimates, but this will significantly increase the time required for computation.

KURTOSIS or KU

> prints coefficients of univariate kurtosis and skewness along with various coefficients of multivariate kurtosis. This option also prints the numbers of the observations that make the greatest contribution to normalized multivariate kurtosis. The KURTOSIS option can help identify outliers, and should be requested during the first run. The data set must be a raw data set, however; it cannot be requested if the input data set is a correlation or covariance matrix.

MAXITER=n

> specifies the maximum number of iterations in the optimization process, where n = the number of iterations. For the default optimization process, the default number of iterations is 50.

METHOD=name

> specifies the method of parameter estimation. Maximum likelihood estimation is the default. Here are the names for the various optimization techniques that may be requested:

> > GLS
> >
> > > requests generalized least-squares parameter estimation and requires a nonsingular correlation matrix. This method performs a statistical test of the goodness of fit of the model to the data, but assumes multivariate normality of all variables and independence of observations.
> >
> > LSGLS
> >
> > > requests unweighted least-squares estimation followed by generalized least-squares estimation.
> >
> > LSML
> >
> > > requests unweighted least-squares parameter estimation followed by normal-theory maximum-likelihood estimation.
> >
> > ML
> >
> > > requests normal-theory maximum-likelihood parameter estimation. Requires a nonsingular correlation matrix. This method performs a statistical test of the goodness of fit of the model to the data, but assumes multivariate normality of all variables and independence of observations. This is the default method.
> >
> > NONE
> >
> > > requests that no estimation method be used.
> >
> > ULS
> >
> > > requests unweighted least-squares parameter estimation.

MODIFICATION or MOD

> requests Lagrange multiplier indices, along with univariate and multivariate Wald test indices.

PRIVEC or PVEC

    requests that parameter estimates, standard errors, the gradient, and *t* values be printed in vector form.

RESIDUAL

    requests the absolute and normalized correlation (or covariance) matrix be printed, along with the rank order of the largest residuals and a bar chart of the residuals.

SIMPLE or S

    requests means, standard deviations, skewness, and univariate kurtosis of manifest variables.

SUMMARY or PSUM

    requests that only the fit assessment table be printed.

TOTEFF or TE

    requests that total effects, indirect effects, and latent variable regression score coefficients be printed.

The LINEQS Statement

PROC CALIS provides a number of ways for describing the path model to be analyzed. This can be done using RAM-style input, COSAN-style input, LINEQS-style input, or FACTOR-style input. This chapter will discuss only the LINEQS approach to model specification. The LINEQS statement is used to identify the variables that have direct effects on the endogenous variables in the path model. This is done with a series of equations, with a separate equation for each endogenous variable. This chapter will present a system of notation to use with the LINEQS statement that is largely based upon the system of notation developed by Bentler (1989) for the EQS progam.

Here is the general form for the LINEQS statement:

```
LINEQS
    v = p v + p v + p v ..... + e,
    v = p v + p v + p v ..... + e,
    v = p v + p v + p v ..... + e ;
```

  where

        v = manifest variables
        p = path coefficients
        e = residual term for corresponding endogenous variables

Although the preceding displays the general form for three equations, any number of equations are actually possible. Also, any number of independent variables (to the right of the equals sign) is also possible.

To make this a bit more concrete, here is the LINEQS statement for the present path model.

```
LINEQS
    V1 = PV1V2 V2 + PV1V5 V5 + PV1V6 V6 + E1,
    V2 = PV2V3 V3 + PV2V4 V4              + E2;
```

The LINEQS statement begins with the word LINEQS, and ends with a semicolon. Each equation in the statement must be separated by a comma. The preceding statement includes two equations. The first equation identifies the variables that have a direct effect on the endogenous variable V1. This is indicated by the fact that "V1" appears to the left of the equals sign in the first equation. The second equation identifies the variables that have a direct effect on V2.

The names of the independent variables that have direct effects on an endogenous variable appear to the right of the equals sign in the equation for that variable. In the current equations, you can see that V1 is predicted to be affected by V2, V5, V6, and its residual term, E1. At the same time, V2 is predicted to be affected by V3 and V4, along with its residual term E2.

To the immediate left of each independent variable is the name for that variable's path coefficient. In the present case, the path coefficient for V2 is given the name "PV1V2," the coefficient for V5 is given the name "PV1V5," and the coefficient for V6 is given the name "PV1V6." The conventions used in naming these coefficients will be discussed shortly.

The SAS System requires that certain conventions be used when creating names for some of the variables included in the LINEQS statement. For example, it requires that names for the residual terms of manifest variables begin with the letter E (for Error term), that names for latent factors begin with the letter F, and names for the disturbance terms of these factors begin with the letter D (latent factors and their disturbance terms are not relevant to manifest-variable path analysis, so the F and D terms will not be discussed in this chapter). This chapter will recommend some additional conventions to simplify your task.

Naming manifest variables. In the previous section on preparing the program figure, this text has already advised that the manifest variables be named using string variables. Each variable name consists of the prefix V and a numerical suffix. Thus, you have variable names such as V1, V2, etc. Technically, manifest variables actually can be given any name, as long as the name adheres to the usual SAS System conventions (e.g., begins with a letter, no longer than eight letters or numbers). However, the conventions recommended here will help create more meaningful names for path coefficients, as discussed later.

Naming residual terms. Similarly, the earlier section advised that residual terms be given names such as E1 and E2. The numerical suffix should always match the suffix of the corresponding manifest variable (so that E1 is the residual term for V1, E2 for V2, and so forth).

Naming path coefficients. Technically, path coefficients may be given any name that meets the usual SAS System conventions. This text, however, will recommend a system similar to that developed by Bentler (1989). The use of the present system results in path coefficient names such a PV1V2, PV1V5, and PV2V4. With this convention, the name for a path coefficient

- begins with the prefix P (to identify this at the name of a Path coefficient)

- continues with the name of the dependent variable being affected (such as V1 or V2)

- concludes with the name of the independent variable where the path originates (such as V4, V5, or V6).

Thus, the name PV1V5 tells you that this is the name of the path coefficient for the path to V1 from V5. The name PV2V4 indicates that this is the name of the coefficient for the path to V2 from V4. This approach has the advantage of making each coefficient's name both unique and meaningful. For example, path coefficients sometimes appear in sections of output (such as with the Wald test results) that provide no independent information to indicate which variables are associated with that coefficient. By using this system, you will be able to identify the relevant variables simply by looking at the path coefficient's name.

Identifying the variables to be included in each equation. It is now possible to show how the program figure (developed earlier) can be used to help you more easily construct the equations to be included in the LINEQS statement. That program figure for the investment model is reproduced here as Figure 4.12.

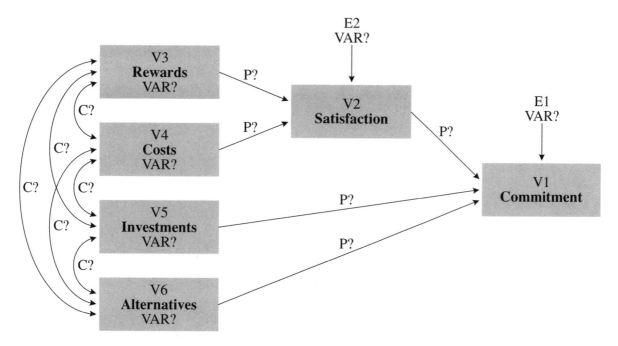

Figure 4.12: The Completed Program Figure

The process begins by determining how many equations should be included in the LINEQS statement (there will be one equation for each endogenous variable). Remember that a variable is an endogenous variable if any straight single-headed arrow points at it. Figure 4.12 shows two endogenous variables: V1 and V2. One equation will be created for each of these.

Next, review the figure to determine which variables should be listed as independent variables for a given endogenous variable. In this context, **independent variables** are those variables that have arrows pointing directly at the endogenous variable of interest. For V1, the independent variables are V2, V5, and V6. With these independent variables identified, you are now ready to write the equation for V1, while adhering to Rules 7, 8, and 9 (these rules pick up where Rule 6 left off in the previous section on "Preparing the Program Figure").

Rule 7: One equation should be created for each endogenous variable, with that variable's name to the left of the equals sign.

Rule 8: Variables that have a direct effect on that endogenous variable are listed to the right of the equals sign.

Rule 9: Exogenous variables, including residual terms, are never listed to the left of the equals sign.

In compliance with Rule 7, your equation for V1 would therefore begin with:

```
V1 =
```

In accordance with Rule 8, the following independent variable names would then be added:

```
V1 =        V2   +      V5   +         V6
```

Next, Rule 10 makes provision for path coefficient names:

> **Rule 10:** To estimate a path coefficient for a given independent variable, a unique path coefficient name should be created for the path coefficient associated with that independent variable.

Thus, the following path coefficient names would be added to the equation:

```
V1 = PV1V2 V2 + PV1V5 V5 + PV1V6 V6
```

Finally, Rule 11 indicates how each equation should end:

> **Rule 11:** The last term in each equation should be the residual term for that endogenous variable; this E term will have no name for its path coefficient.

The completed equation for V1 takes the following form:

```
V1 = PV1V2 V2 + PV1V5 V5 + PV1V6 V6 + E1,
```

Notice that given the conventions for creating path coefficient names, just looking at PV1V2 tells you that this coefficient represents the strength of the effect of V2 on V1. Notice also that the equation ends with a comma, since it will be followed by the equation for V2.

Following the same rules, it is now possible to create the equation for V2. Because the program figure shows that it is affected by V3 and V4, the equation will take on the following form:

```
V2 = PV2V3 V3 + PV2V4 V4 + E2;
```

The preceding equation ends with a semicolon, as it is the last equation.

The program figure tells you which variables should be included in each LINEQS equation. You need only find the endogenous variable of interest and determine which variables have direct effects on it, as indicated by straight, single-headed arrows. Figure 4.12 shows that V1 (Commitment) is affected by V2, V5, and V6, along with its residual term, E1. Similarly, you can see that V2 (Satisfaction) is affected by V3 and V4, along with its residual term, E2. It is in this way that the figure tells you which terms should be included in the equations for V1 and V2.

Once again, the full LINEQS statement appears as follows:

```
LINEQS
  V1 = PV1V2 V2 + PV1V5 V5 + PV1V6 V6 + E1,
  V2 = PV2V3 V3 + PV2V4 V4               + E2;
```

Is it really necessary to line up the residual terms? The above statement was prepared so that the residual variables (E1 and E2) are lined up vertically. This alignment is not required by the CALIS procedure, but is recommended to minimize the chance of errors. Experience teaches that it is easy to forget a residual term for a given equation if they are not lined up in this way!

Estimating, fixing, and constraining paths. The CALIS procedure estimates three different types of parameters:

- path coefficients, which represent the amount of change in a dependent variable that is associated with a one-unit change in the relevant independent variable, while holding constant the effects of the remaining independent variables

- variances, which represent the variability in exogenous variables

- covariances, which represent the covariation between pairs of exogenous variables.

All three types of parameters either may be estimated, fixed, or constrained. The following rules explain how this is done in the LINEQS statement, and this section provides specific examples of how path coefficients either may be estimated, fixed, or constrained.

> **Rule 12:** To *estimate* a parameter, create a name for that parameter.

When the CALIS procedure estimates a path coefficient, it simply determines the "optimal" value for that coefficient, in much the same way that PROC REG determines the optimal value for regression coefficients in a multiple regression equation. Whenever your LINEQS statement includes a name for a coefficient, the procedure's output will provide a numerical estimate of that coefficient. For example, consider the following equation:

```
V2 = PV2V4 V4 + E2;
```

The preceding equation asks the CALIS procedure to estimate just one coefficient, the one representing the effect of V4 on V2. The output of this fictitious program may indicate that this path coefficient is equal to 0.2345.

> **Rule 13:** To *fix* a parameter at a given numerical value, insert that value in the place of the parameter's name.

For example, the following equation fixes the path for the effect of V4 on V2 at .50, while estimating the coefficient for the effect of V5 on V2:

```
V2 = .50 V4 + PV2V5 V5 + E2;
```

While it is unlikely that researchers new to path analysis will want to fix path coefficients at specific numbers, it is nonetheless easy to do, as the preceding rule suggests.

It is sometimes desirable to fix the value of a path coefficient at 1.00, and this is especially easy to do with PROC CALIS. You may have noticed that there are no path coefficient names to represent the effects of the E terms on the endogenous variables. This is because these coefficients are usually fixed so that they are equal to 1.00. With PROC CALIS, leaving off the name of a path coefficient just before the independent variable's name has the effect of fixing that coefficient at 1.00. In the following equation, both the path for the E term, as well as the path for the V4 term, are fixed at 1.00 (the path coefficient for V5 is again estimated):

```
V2 = V4 + PV2V5 V5 + E2;
```

Be warned that this convention applies only to fixing *path coefficients* at 1.00. It will not apply to variances or covariances, as you will see later.

When you fix a path coefficient at zero, this has the effect of eliminating the corresponding path from the model. In practice, it is not actually necessary to include a zero (0) in the equation followed by the name of the relevant independent variable. To fix that path at zero, merely omit from the equation the name of the relevant independent variable. For example, consider the following equation:

```
V2 = PV2V4 V4 + E2;
```

This equation indicates that V2 is affected only by V4; the paths from all other variables to V2 are automatically fixed at zero.

Finally, Rule 14 indicates how two or more paths may be constrained to be equal to each other:

 Rule 14: To *constrain* two or more parameters to be equal, use the same name for those parameters.

Again, it is unlikely that researchers learning path analysis at an introductory level will need to do this, but it is a useful fact to know for future purposes. The two (or more) path coefficients may be given any name that complies with the usual SAS System conventions. However, this text recommends that the coefficients be given a "PEQ" prefix (for "Paths constrained to be EQual"), followed by a numerical suffix such as PEQ1 or PEQ2. For example, assume that you want the path from V4 to V2 to be constrained to be equal to the path from V5 to V1. This means that, in the printed output, these two paths will take on exactly the same value (whatever that will be). Therefore, both paths will be given exactly the same name: PEQ1. The following LINEQS statements show how this constraint may be requested:

```
LINEQS
    V1 = PV1V2 V2 + PEQ1 V5 + E1,
    V2 = PV2V3 V3 + PEQ1 V4 + E2;
```

Starting values for path coefficients. It is usually unnecessary to provide the CALIS procedure with starting values for path coefficients or any other parameters to be estimated. However, if you provide starting figures that are close to the parameter estimates that the procedure will ultimately arrive at, this may allow the program to converge more quickly and thus save computing time. In addition, it has been mentioned that one test to determine whether a model is identified involves running the program several different times with different starting values.

When used, the starting value for a given path coefficient should appear in parentheses to the immediate right of that coefficient. Here, .55 is used as the starting value for the coefficient named PV1V2, while .3 is used as the starting value for the coefficient named PV2V3:

```
LINEQS
    V1 = PV1V2 (.55) V2 + PV1V5 V5 + PV1V6 V6 + E1,
    V2 = PV2V3 (.3)  V3 + PV2V4 V4            + E2;
```

The STD Statement

The primary purpose of the STD statement is to identify the variables whose variances are to be estimated in the analysis. Here, you will see that the statement can also be used to fix or constrain variances, although this will not frequently be necessary in path analysis.

The STD statement follows a format very similar to the LINEQS statement. It begins with the letters STD, and ends with a semicolon. Each equation in the statement is separated by a comma. This is the STD statement used in the present program:

```
STD
    E1 = VARE1,
    E2 = VARE2,
    V3 = VARV3,
    V4 = VARV4,
    V5 = VARV5,
    V6 = VARV6;
```

These lines request that the variance be estimated for the residual terms E1 and E2, along with the exogenous manifest variables V3, V4, V5, and V6. This is consistent with Rule 4 (provided earlier) which stated that the variance should be estimated for every exogenous variable in the model, including both exogenous manifest variables as well as residual terms.

Fortunately, by completing the program figure in advance, the variances to be estimated have already been identified. This was done in Figure 4.12 by inserting the symbol VAR? just below the name of every exogenous variable (again, including the residual variables). The program figure shows that variances should be estimated for V3, V4, V5, V6, E1, and E2.

Naming variance estimates. A variance is a parameter, just like a path coefficient. Like path coefficients, actual numerical estimates for these variances will be calculated by the CALIS procedure. Also like path coefficients, a variance must be given a name to be estimated. Technically, any name that conforms to SAS System conventions for variable names can be used. This text, however, recommends an approach that results in meaningful names that facilitate understanding the output of the procedure. With this system, you will begin the name with a VAR prefix (to identify this as the name of a VARiance estimate), and conclude the name with the name of the variable whose variance is being estimated (such as V3 or E2).

This results in variance estimate names such as VARV3 and VARE2. With this system, you need only observe the name VARV3 to realize that this is the variance estimate for V3, for example.

Estimating, fixing, and constraining variances. An earlier section of this chapter provided rules for estimating, fixing, and constraining parameters. This section will demonstrate how these rules apply to variance estimates.

To estimate a variance for an exogenous variable, you need only create an equation within the STD statement that contains the variable's name to the left of the equals sign, and the variance estimate's name to the right. The following statement requests variance estimates for exogenous variables V3 and E2:

```
STD
   V3 = VARV3,
   E2 = VARE2;
```

To fix a variance at a specific numerical value, provide that numerical value in place of the variance estimate name. Here, the variance of V3 is fixed at .93, while the variance of E2 is estimated:

```
STD
   V3 = .93,
   E2 = VARE2;
```

To constrain two or more variance estimates to be equal, use the same variance estimate name for those variables. This text advises that the name for this estimate begin with the VAREQ prefix (for "VARiances constrained to be EQual"), and end with a numerical suffix, so that the resulting name takes on the form of VAREQ1, VAREQ2, etc. Here, the variances for V3 and V4 are constrained to be equal; the name for the resulting variance estimate is VAREQ1:

```
STD
   V3 = VAREQ1,
   V4 = VAREQ1,
   V5 = VARV5,
   E2 = VARE2;
```

Starting values for variance estimates. Starting values are usually not required, but may result in less computing time if they are accurate. If used, starting values should appear in parentheses to the immediate right of the variance estimate name, but before the comma or semicolon. Here, VARV3 is given a starting value of 1.3, while VARE2 is given a starting value of .65:

```
STD
   V3 = VARV3 (1.3),
   E2 = VARE2 (.65);
```

The COV Statement

The COV statement is used to identify pairs of variables that are expected to covary (be correlated). The statement begins with the letters COV, and ends with a semicolon. Commas are used to separate the equations included in the statement. Within each equation, the pairs of covarying variables are presented to the left of the equals sign, and the name for the corresponding covariance estimate appears to the right. The COV statement used in the present program is:

```
COV
   V3  V4  =  CV3V4,
   V3  V5  =  CV3V5,
   V3  V6  =  CV3V6,
   V4  V5  =  CV4V5,
   V4  V6  =  CV4V6,
   V5  V6  =  CV5V6;
```

The preceding statement requests that covariances be estimated for every possible combination of the four manifest exogenous variables displayed in Figure 4.12. It will be noted that no endogenous variables appear in the statement. This is consistent with Rule 5 (presented earlier) which stated that covariances are normally estimated for pairs of exogenous variables, but not endogenous variables.

Naming covariance estimates. As with variances, covariances must be given a name if they are to be estimated. To create meaningful names, this text advises that the name begin with the C prefix (to identify this as the name of a Covariance estimate), and conclude with the names of the two variables that are covarying. In this way, for example, the name CV3V4 is the name for the covariance between V3 and V4. This approach has the advantage of resulting in highly meaningful names. You need only see the name CV5V6 to know that it represents the covariance between V5 and V6.

Note: Be consistent when referring to these covariance estimates. The name given to a covariance estimate must always be typed the same way each time it appears in the program. For example, if it appears once as CV3V4, and later as CV4V3, errors will result. To avoid this confusion, the variable name with the lower numerical value should always appear first, and the variable with the higher value should always appear last. That is, the covariance name should be created as CV3V4, not as CV4V3.

Estimating, fixing, and constraining covariances. The general rules developed earlier for estimating, fixing, and constraining parameters will apply to covariances as well. To estimate a covariance, the two variables expected to covary are listed on the left side of the equals sign (separated by at least one blank space), and the name for that covariance is listed to the right. To illustrate, the following COV statement requests that the covariance between V4 and V5 be estimated:

```
COV
    V4 V5 = CV4V5;
```

Which covariances should be estimated? You have already identified these by preparing your program figure in advance. In Figure 4.12, manifest variables V3, V4, V5, and V6 have all been interconnected by curved, double-headed arrows, indicating that they are all expected to covary. Each of these arrows has been identified with the C? symbol. There will be one equation in the COV statement for every C? symbol in the figure.

All manifest exogenous variables are usually allowed to covary in a path analysis. Although they are manifest variables, the residual terms (E variables) are usually not allowed to covary. This can be seen in the COV statement used for this program:

```
COV
    V3 V4 = CV3V4,
    V3 V5 = CV3V5,
    V3 V6 = CV3V6,
    V4 V5 = CV4V5,
    V4 V6 = CV4V6,
    V5 V6 = CV5V6;
```

To fix a covariance at a specific numerical value, provide that numerical value in place of the covariance estimate name. Here, the covariance between V4 and V5 is fixed at .45, and the variance between V4 and V6 and between V5 and V6 is estimated:

```
COV
    V4 V5 = .45,
    V4 V6 = CV4V6
    V5 V6 = CV5V6;
```

Fixing a covariance at zero has the effect of eliminating any curved, double-headed arrow between the relevant variables. To fix a covariance at zero, however, it is not necessary to actually list the relevant variables to the left of the equals sign and a zero to the right of the sign; simply leaving the relevant pair of variables out of the COV statement has the effect of fixing their covariances at zero. For example, the variables V1 and V2 are not listed in the preceding COV statement. This means that the covariance between V1 and V2 will be fixed at zero.

To constrain two or more covariance estimates to be equal, use the same covariance estimate name for those pairs of variables. This approach involves beginning the common name with the

CEQ prefix (for "Covariances constrained to be EQual"), and concluding with a numerical suffix, so that the resulting name takes on the form of CEQ1, CEQ2, and so forth. Here, the covariance between V4 and V5 is constrained to be equal to the covariance between V4 and V6:

```
COV
    V4 V5 = CEQ1,
    V4 V6 = CEQ1,
    V5 V6 = CV5V6;
```

Starting values for covariance estimates. Starting values for covariance estimates are optional. If used, starting values should appear in parentheses to the immediate right of the covariance estimate name, but before the comma or semicolon. Here, starting values are provided for all covariance estimates in the current program:

```
COV
    V3 V4 = CV3V4 (.50),
    V3 V5 = CV3V5 (.40),
    V3 V6 = CV3V6 (.40),
    V4 V5 = CV4V5 (.50),
    V4 V6 = CV4V6 (.75),
    V5 V6 = CV5V6 (.20);
```

The VAR Statement

The VAR statement identifies the manifest variables to be analyzed in the path analysis. If your data set includes variables that will not be analyzed, then using the VAR statement can save computer time. Technically, the VAR statement made no difference in the present program because all variables in the data set were analyzed. However, it was included just as a matter of good practice.

The general form for the VAR statement is

```
VAR   list-of-manifest-variables-to-be-analyzed ;
```

For the current program, the following VAR statement was used:

```
VAR   V1 V2 V3 V4 V5 V6 ;
```

Interpreting the Results of the Analysis

When the analysis has been completed, the SAS System creates two new files. The log file contains lines from the original SAS program along with notes and error messages, and the output file contains the actual results of the analysis (the output file is sometimes called the lis file). As always, the log file should be reviewed first to check for errors or warnings. Next, the output file should be reviewed for signs of other problems; the following section shows how to do this.

Making Sure that the SAS Output File Looks Right

The SAS output file contains the results of the analyses. Specifying LINESIZE=80 and PAGESIZE=60 in the OPTIONS statement will cause the preceding program to produce 13 pages of output. Here is a brief description of the information contained on each page (in the interest of brevity, some of the information appearing in the output file has not been listed here):

- Page 1 includes a list of the endogenous variables and exogenous variables specified in the LINEQS statement.

- Page 2 includes the general form of the structural equations specified in the LINEQS statement.

- The top of page 3 provides general information about the estimation problem, including

    - the number of observations

    - the number of variables

    - the amount of independent information in the data matrix (number of "data points")

    - the number of parameters to be estimated.

  Below this are some univariate statistics for the manifest variables and the correlation (or covariance) matrix to be analyzed.

- Page 4 includes a vector of initial parameter estimates.

- Page 5 includes the iteration history.

- Page 6 includes the predicted model correlation (or covariance) matrix and the residual matrix.

- Page 7 includes the normalized residual matrix and a bar chart displaying the distribution of normalized residuals.

- Page 8 includes a variety of goodness of fit indices, to be discussed later.

- Page 9 includes equations containing parameter estimates, their approximate standard errors, and *t* values. These equations correspond to those constructed in the LINEQS statement. This page also provides estimates of variances and covariances, along with corresponding approximate standard errors and *t* values.

- Page 10 includes equations containing the standardized parameter estimates. Below these are estimates of the endogenous variable variances, and the R^2 values for each endogenous variable. The correlations among exogenous variables are also presented.

- Pages 11-13 include the modification indices (Lagrange multiplier and Wald test results).

The output itself is reproduced here as Output 4.1 through Output 4.11. Several pages of the output must be reviewed to verify that the program ran as expected. First, the structural equations printed on pages 1 and 2 should be reviewed to verify that the LINEQS statements were written correctly. The information at the bottom of page 1 indicates that your program has specified two endogenous variables (V1 and V2) and six exogenous variables (V3, V4, V5, V6, E1, and E2). This is consistent with your model.

```
                         The SAS System                                1

          Covariance Structure Analysis: Pattern and Initial Values

              LINEQS Model Statement
          -------------------------------
              Matrix        Rows & Cols         Matrix Type
     Term   1-------------------------------------------------------------
              1    _SEL_        6        8      SELECTION
              2    _BETA_       8        8      EQSBETA         IMINUSINV
              3    _GAMMA_      8        6      EQSGAMMA
              4    _PHI_        6        6      SYMMETRIC



        Number of endogenous variables = 2

Manifest:      V1          V2



        Number of exogenous variables = 6

Manifest:      V3          V4          V5          V6
Error:         E1          E2
```

```
        Covariance Structure Analysis: Pattern and Initial Values       2

                       Manifest Variable Equations
                          Initial Estimates

     V1       =    .    *V2    +   .    *V5    +   .    *V6    + 1.0000 E1
                        PV1V2            PV1V5            PV1V6

     V2       =    .    *V3    +   .    *V4    + 1.0000 E2
                        PV2V3            PV2V4
```

```
            Variances of Exogenous Variables
         -----------------------------------------
         Variable     Parameter        Estimate
         -----------------------------------------
         V3           VARV3                    .
         V4           VARV4                    .
         V5           VARV5                    .
         V6           VARV6                    .
         E1           VARE1                    .
         E2           VARE2                    .


          Covariances among Exogenous Variables
             ---------------------------------
                Parameter          Estimate
             ---------------------------------
          V4    V3    CV3V4                .
          V5    V3    CV3V5                .
          V5    V4    CV4V5                .
          V6    V3    CV3V6                .
          V6    V4    CV4V6                .
          V6    V5    CV5V6                .
```

Output 4.1: PROC CALIS Output Pages 1-2, Analysis of Initial Model for Investment Model Study

The equations at the top of output page 2 (under the heading "Manifest Variable Equations Inititial Estimates") correspond to the equations in the LINEQS statements. Parameters to be estimated are identified by an asterisk ("*"), and the information on this page shows that five path coefficients (e.g., PV1V2, PV1V5) are to be estimated. The number 1.0000 appears before E1 and E2, indicating that these paths are fixed at 1. Finally, the tables on the lower half of the page can be reviewed to verify that you have correctly indicated which variances and covariances are to be estimated.

The top of page 3 from Output 4.2 provides general information about the estimation problem. It shows that the analysis was based on 240 observations and 6 variables. The number "21" appearing to the left of "Informations" reveals the number of data points associated with the analysis (the amount of independent information in the data matrix). Earlier, it was shown that this value may be calculated with the following equation:

$$\texttt{Number of data points = (p (p + 1)) / 2}$$

where p = number of manifest variables being analyzed. Because p = 6 in this analysis, the resulting number of data points is 21. Earlier, it was also said that a necessary (but not sufficient) condition for model identification is that the number of data points must exceed the number of parameters to be estimated. The top of this page shows that only 17 parameters are to be estimated, so this condition has been met.

```
                        The SAS System                              3

         Covariance Structure Analysis: Maximum Likelihood Estimation

              240 Observations      Model Terms        1
                6 Variables         Model Matrices     4
               21 Informations      Parameters        17

          VARIABLE              Mean             Std Dev

          V1                      0          2.319200000    COMMITMENT
          V2                      0          1.774400000    SATISFACTION
          V3                      0          1.252500000    REWARDS
          V4                      0          1.408600000    COSTS
          V5                      0          1.557500000    INVESTMENTS
          V6                      0          1.870100000    ALTERNATIVES


                              Covariances

                   V1                 V2                 V3

     V1       5.378688640        2.774460073        1.597929380    COMMITMENT
     V2       2.774460073        3.148495360        1.493699236    SATISFACTION
     V3       1.597929380        1.493699236        1.568756250    REWARDS
     V4      -1.143062109       -1.428918323       -0.777161596    COSTS
     V5       2.327672038        1.446482895        1.042880974    INVESTMENTS
     V6      -2.658230605       -1.643224854       -0.951208132    ALTERNATIVES


                   V4                 V5                 V6

     V1      -1.143062109        2.327672038       -2.658230605    COMMITMENT
     V2      -1.428918323        1.446482895       -1.643224854    SATISFACTION
     V3      -0.777161596        1.042880974       -0.951208132    REWARDS
     V4       1.984153960       -0.406748040        0.928563558    COSTS
     V5      -0.406748040        2.425806250       -1.145848607    INVESTMENTS
     V6       0.928563558       -1.145848607        3.497274010    ALTERNATIVES
                 Determinant = 26.8 (Ln = 3.288)

           Set covariances of exogenous manifest variables:
                         V3 V4 V5 V6

        Some initial estimates computed by two-stage LS method.
```

Output 4.2: PROC CALIS Output Page 3, Analysis of Initial Model for Investment Model Study

Page 3 of Output 4.2 also provides univariate statistics such as means and standard deviations for the variables being analyzed. These should be checked to verify that no errors have been made when inputting the data.

Finally, page 5 of Output 4.3 provides the iteration history, and this page should be reviewed to verify that the program converged. In the present case, you can see that convergence was achieved after only 3 iterations. If the model is overidentified and the program has been written correctly but still has not converged within the allowed number of iterations, it may be necessary to provide better starting values and/or increase the allowed number of iterations with the MAXITER option in the PROC CALIS statement.

```
                         The SAS System                              5

           Covariance Structure Analysis: Maximum Likelihood Estimation

                       Levenberg-Marquardt Minimization
                            Algorithm for Hessian= 1
                            Maximum Iterations= 50
                          Maximum Function Calls= 125
                    Maximum Absolute Gradient Criterion= 0.001
                Number of Estimates= 17 Lower Bounds= 0 Upper Bounds= 0
   Minimization Start: Active Constraints= 0 Criterion= 0.154
   Maximum Gradient Element= 0.019 Radius= 0.034

    Iter  nfun  act    mincrit   maxgrad   difcrit    lambda rhoratio
      1     2    0     0.15416   0.003915  0.00008        0   1.1994
      2     3    0     0.15415   0.001355  7.959E-6       0   1.3115
      3     4    0     0.15415   0.000368  9.207E-7       0   1.3401
   Minimization Results: Iterations= 3 Function Calls= 4 Derivative Calls= 4
   Active Constraints= 0 Criterion= 0.154 Maximum Gradient Element= 0.000368
   Radius= 4.46E-6

   NOTE: Convergence criterion satisfied.
```

Output 4.3: PROC CALIS Output Page 5, Analysis of Initial Model for Investment Model Study

The final note on page 5 tells you that the convergence criterion is satisfied. For most problems, the optimization technique used by PROC CALIS requires the repeated computation of two values: the function value (optimization criterion) and the gradient vector (first-order partial derivatives). The FCONV= option used in the PROC CALIS statement may be used to specify the relative function convergence criterion. The default for FCONV= is the smallest real value that can be represented by your computer. In the same way, the GCONV= option specifies the absolute gradient convergence criterion. By default, this value is equal to $1 E - 3$. Smaller values may be specified to obtain more precise parameter estimates, but this will increase significantly the time required for computation.

Because the program did converge, it is now appropriate to review the remainder of the SAS output file to assess the fit of your model.

Assessing the Fit between Model and Data

When conducting path analysis, you will usually begin with some theoretical model that describes the causal connections between a set of observed variables. You then obtain data from a sample, usually in the form of a correlation matrix or, as in this case, a covariance matrix. A theoretical model provides a good fit to the data when it successfully accounts for the observed covariances in this matrix.

A number of procedures and statistics have been developed to assess the extent to which a model fits the data, and you will usually refer to several of these in the course of the analysis. This is because there is no single index of goodness of fit that is universally accepted; each index provides somewhat different information. This section describes a few indices that should be particularly useful to you if you are learning about path analysis for the first time. Consistent with the format used throughout this text, these indices are introduced by leading you through a number of structured steps.

Step 1: Reviewing the residual matrix and normalized residual matrix. When your original covariance matrix is analyzed, it is possible to use the resulting path coefficients to create a reproduced covariance matrix, or a predicted model matrix as it is referred to in the SAS output. If the theoretical model successfully accounts for the actual causal relationships between the variables, this reproduced covariance matrix should be nearly identical to the original covariance matrix.

An easy way to determine the similarity between the two matrices is to subtract each element of the predicted model matrix from the corresponding element of the original covariance matrix, and use the resulting differences to form a residual matrix. If the model provides a good fit to the data, each element of this residual matrix should be zero or near zero. Large elements in the residual matrix may indicate a specification error in the theoretical model. For example, if a large residual appears at the intersection of V1 and V2 in the residual matrix, then the model probably does not successfully account for the relationship between these two variables. Perhaps your theoretical model predicts that V2 has a causal effect on V1, but in reality this causal path should be eliminated. On the other hand, it is possible that your model predicts no relationship between the two variables, while in reality they should be connected by a causal path. Other specification problems are also possible, including the presence of correlated residual terms or an incorrect ordering of variables. A later section will discuss this concept of model modification in more detail.

When path analysis is performed on a correlation matrix (rather than a covariance matrix), the elements of the residual matrix itself may be meaningfully interpreted. Any residual whose absolute value exceeds some specified value is considered problematic, and may indicate a specification error in the model. How large may a residual be before it is considered "too large?" The value of .05 has evolved as a widely used rule of thumb for identifying large residuals (Billings and Wroten, 1978), although some researchers do use a more liberal criterion such as .10.

When path analysis is performed on a covariance matrix (rather than a correlation matrix), the elements of the residual matrix are not standardized in any meaningful way, making it difficult to determine just how large a residual should be to be considered too large. To address this problem, PROC CALIS also prints a normalized residual matrix. Elements in this matrix whose absolute value exceed 2.00 should be considered large. Below this table, PROC CALIS also prints the largest normalized residuals in descending order according to size. This allows you to identify quickly the most problematic residuals.

For the present analysis, the normalized residual table and the rank order of largest normalized residuals appear on output page 7. These results are reproduced here as Output 4.4.

```
                            The SAS System                              7

           Covariance Structure Analysis: Maximum Likelihood Estimation

                          Normalized Residual Matrix

           V1          V2          V3          V4          V5          V6

V1      0.9114      1.5705      0.1604      0.2661      0.8505     -0.7733   COMMITMENT
V2      1.5705      0.0000      0.0000      0.0000      2.6563     -2.3946   SATISFACTION
V3      0.1604      0.0000      0.0000      0.0000      0.0000      0.0000   REWARDS
V4      0.2661      0.0000      0.0000      0.0000      0.0000      0.0000   COSTS
V5      0.8505      2.6563      0.0000      0.0000      0.0000      0.0000   INVESTMENTS
V6     -0.7733     -2.3946      0.0000      0.0000      0.0000      0.0000   ALTERNATIVES
                 Average Normalized Residual = 0.4563
            Average Off-diagonal Normalized Residual = 0.5781

                 Rank Order of 8 Largest Normalized Residuals

        V5,V2       V6,V2       V2,V1       V1,V1       V5,V1       V6,V1       V4,V1
       2.6563      -2.3946      1.5705      0.9114      0.8505     -0.7733      0.2661

                                  V3,V1
                                 0.1604
```

```
            Distribution of Normalized Residuals
              (Each * represents 1 residuals)

   -2.50000 -    -2.25000    1    4.76% | *
   -2.25000 -    -2.00000    0    0.00% |
   -2.00000 -    -1.75000    0    0.00% |
   -1.75000 -    -1.50000    0    0.00% |
   -1.50000 -    -1.25000    0    0.00% |
   -1.25000 -    -1.00000    0    0.00% |
   -1.00000 -    -0.75000    1    4.76% | *
   -0.75000 -    -0.50000    0    0.00% |
   -0.50000 -    -0.25000    0    0.00% |
   -0.25000 -           0    0    0.00% |
          0 -     0.25000   14   66.67% | **************
    0.25000 -     0.50000    1    4.76% | *
    0.50000 -     0.75000    0    0.00% |
    0.75000 -     1.00000    2    9.52% | **
    1.00000 -     1.25000    0    0.00% |
    1.25000 -     1.50000    0    0.00% |
    1.50000 -     1.75000    1    4.76% | *
    1.75000 -     2.00000    0    0.00% |
    2.00000 -     2.25000    0    0.00% |
    2.25000 -     2.50000    0    0.00% |
    2.50000 -     2.75000    1    4.76% | *
```

Output 4.4: PROC CALIS Output Page 7, Analysis of Initial Model for Investment Model Study

For the current analysis, the "Rank Order of the 8 Largest Normalized Residuals" table shows two residuals whose absolute value exceeds 2.00: the normalized residual for V5 and V2 is 2.66, and the residual for V6 and V2 is –2.39. This indicates that there are problems with your theoretical model. Nonetheless, you may still proceed to review some additional goodness of fit measures before considering any modifications to the model.

Step 2: Reviewing the chi-square test. Perhaps the most widely reported goodness of fit index used in path analysis is the chi-square test. The *SAS/STAT users guide* says that this statistic provides a "test of the specified model versus the alternative that the data are from a multivariate normal distribution with unconstrained covariance matrix" (1989, *volume 1*, p. 139). In short, the chi-square statistic provides a test of the null hypothesis that the theoretical model fits the data. If the null hypothesis is correct, then the obtained chi-square value should be small, and the *p* value associated with the chi-square should be relatively large. The *p* value (probability value) associated with the test indicates the likelihood of obtaining a chi-square value this large or larger if the null hypothesis were true (i.e., if the model fits the data).

Remember that the null hypothesis in path analysis is a hypothesis of "good fit," and think about what this means in practical terms. For example, if you have developed a theoretical model and hope to obtain support for it (as most often will be the case), then you hope *not* to reject this null hypothesis. You hope that the chi-square value will be small (near zero), and the *p* value will be large (at a minimum it should be larger than the standard *p* cutoff value of .01 or .05; the closer it is to 1.00, the better).

The chi-square test for the present analysis is reproduced here as Output 4.5. This section of the PROC CALIS output provides a large number of goodness of fit indices, some of which will be discussed here.

```
                         The SAS System                              8

        Covariance Structure Analysis: Maximum Likelihood Estimation

        Fit criterion . . . . . . . . . . . . . . . . . . .       0.1542
        Goodness of Fit Index (GFI) . . . . . . . . . . .         0.9544
        GFI Adjusted for Degrees of Freedom (AGFI)  . . .         0.7609
        Root Mean Square Residual (RMR) . . . . . . . . .         0.2227
        Chi-square = 36.8420        df = 4        Prob>chi**2 = 0.0001
        Null Model Chi-square:      df = 15                     672.7238
        RMSEA Estimate  . . . . . .  0.1854   90%C.I.[0.1334, 0.2423]
        Bentler's Comparative Fit Index . . . . . . . . .         0.9501
        Normal Theory Reweighted LS Chi-square  . . . . .        34.2185
        Akaike's Information Criterion  . . . . . . . . .        28.8420
        Consistent Information Criterion  . . . . . . . .        10.9195
        Schwarz's Bayesian Criterion  . . . . . . . . . .        14.9195
        McDonald's (1989) Centrality. . . . . . . . . . .         0.9339
        Bentler & Bonett's (1980) Non-normed Index. . . .        0.8128
        Bentler & Bonett's (1980) Normed Index. . . . . .        0.9452
        James, Mulaik, & Brett (1982) Parsimonious Index.        0.2521
        Z-Test of Wilson & Hilferty (1931) . . . . . . .         4.8864
        Bollen (1986) Normed Index Rho1 . . . . . . . . .        0.7946
        Bollen (1988) Non-normed Index Delta2 . . . . . .        0.9509
        Hoelter's (1983) Critical N . . . . . . . . . . .            63
```

Output 4.5: PROC CALIS Output Page 8, Analysis of Initial Model for Investment Model Study

The chi-square test appears on the fifth line down from the top of the table, and takes this form:

```
Chi-square = 36.8420          df = 4          Prob>chi**2 = 0.0001
```

Remember that, if the model provides a good fit, you expect to see a small value of chi-square and a large *p* value. In the present analysis, chi-square was 36.84 with 4 degrees of freedom, which was highly significant. Because it was significant, you are forced to reject your null hypothesis of good model fit. In other words, this test did not support your model.

How are the degrees of freedom determined? For a simple recursive path analytic model without correlated residuals (such as the one considered here), the degrees of freedom are equal to the number of data points (or "informations") used in the analysis, minus the number of parameters to be estimated. Page 3 of the SAS output (Output 4.2) indicated that your analysis involved 21 informations and 17 parameters; hence 21 minus 17 equals the 4 degrees of freedom associated with this test.

It is also useful to think of the degrees of freedom as being equal to the number of restrictions that are placed on the data. Remember that, if your model were a fully recursive, or saturated, model, then no restrictions would have been placed on any of the relationships. Every variable would be connected to every other variable by either a causal path or a double-headed arrow. However, you did impose some restrictions in order to create the model tested here. Specifically, you fixed the following four paths to be zero: the path from V3 to V1, the path from V4 to V1, the path from V5 to V2, and the path from V6 to V2. Fixing these paths to be equal to zero had the same effect as eliminating them from your path model. Imposing these four restrictions resulted in four degrees of freedom, and made it possible to assess your model with the chi-square test.

Although the chi-square test is a useful index, it is generally accepted that it should be interpreted with caution and supplemented with other goodness of fit indices. This is because the chi-square test can be influenced by factors in addition to the validity of the theoretical model; these factors include departures from multivariate normality, sample size, and even the complexity of the model.

1. First, the chi-square test requires that the data demonstrate multivariate normality. If the data are leptokurtic (peaked), a well fitting model is more likely to be incorrectly rejected. If the data are platykurtic (flat), the analysis is likely to fail to reject a poorly fitting model (Anderson & Gerbing, 1988). When data demonstrate high or low kurtosis, they should be transformed and/or a search for outliers should be conducted.

2. To obtain reliable results, it has been recommended that the sample should include at least 100 observations, and that the sample size should be at least 5 to 20 times the number of parameters being estimated (*SAS/STAT users guide* [1989], p. 140). Unfortunately, when path analysis is performed on a very large sample, the chi-square statistic attains such a level of power that it may demonstrate statistical significance when there are only relatively trivial differences between the predicted and actual covariance matrices. In this way, the chi-square test can result in the rejection of a model that appears to fit the data quite well (Mulaik, James, Alstine, Bennett, Lind, & Stillwell, 1989).

Because the chi-square test is frequently not valid in applied settings, it has been recommended that it be treated as a general goodness of fit index, but not as a statistical test in the strictest sense (Joreskog & Sorbom, 1989). For these reasons, many researchers supplement the chi-square test with a number of other stand-alone goodness of fit indices. In some cases, these indices may reveal a relatively good fit even when the chi-square test suggests rejection of the model. Some of these additional fit indices are discussed here.

Step 3: Reviewing the non-normed index and comparative fit index. Bentler and Bonett's (1980) normed-fit index (NFI) has been proposed as an alternative to the chi-square test. Values on this index may range from 0 to 1, with values over .9 indicative of an acceptable fit of the model to the data. This index may be viewed "as the percentage of observed-measure covariation explained by a given measurement or structural model (compared with an overall, null model...that solely accounts for the observed measure variances)" (Anderson and Gerbing, 1988, p. 421). Although the NFI has the advantage of being easily interpreted, it has the disadvantage of sometimes underestimating goodness of fit in small samples.

A variation on the NFI is the non-normed fit index (NNFI, Bentler & Bonett, 1980). The NNFI has been shown to better reflect model fit at all sample sizes (Bentler, 1989; Anderson & Gerbing, 1988; Marsh, Balla, & McDonald, 1988). NNFI values over .9 are also viewed as desirable, although, unlike the NFI, the NNFI may assume values below 0 and above 1.

Finally, Bentler's (1989) comparative fit index (CFI) is similar to the NNFI in that it provides an accurate assessment of fit regardless of sample size. In addition, the CFI tends to be more precise than the NNFI in describing comparative model fit (Bentler, 1989). Values of the CFI will always lie between 0 and 1, with values over .9 indicating a relatively good fit.

The NFI, NNFI, and CFI appear in Output 4.5, in the same table that contained the chi-square statistic. The CFI appears on the 7th line of this table, the NNFI appears on the 13th line, and the NFI on the 14th. These entries are reproduced once again here:

```
        Bentler's Comparative Fit Index . . . . . . . . .     0.9501

        Bentler & Bonett's (1980) Non-normed Index. . . .     0.8128

        Bentler & Bonett's (1980) Normed Index. . . . . .     0.9452
```

These three indices provide mixed signals concerning your model's fit. Although the CFI and NFI suggest an acceptable fit with indices at .95 and .95 respectively, the NNFI indicates an inadequate fit at .81.

Step 4: Reviewing R^2 values for the endogenous variables. Remember that path analysis is often conducted because you want to identify the variables that determine variability in the endogenous, or dependent, variables. Although the preceding indices may reflect the goodness of fit between model and data, they do not necessarily reflect the extent to which the independent variables in the model account for variability in the dependent variables.

Fortunately, however, the CALIS procedure also produces R^2 values for all endogenous variables included in the model. These R^2 values indicate the percent of variance in the endogenous variables that is accounted for by their direct antecedents. As in multiple regression, R^2 values may range from 0 to 1 with larger values indicating a greater percent of variance accounted for.

In the present analysis, R^2 values appear on page 10 of the SAS output, in the table titled "Variances of Endogenous Variables." This table is reproduced as part of Output 4.6.

Covariance Structure Analysis: Maximum Likelihood Estimation

Manifest Variable Equations

```
V1       =       0.4456*V2    +  0.5132*V5    -  0.3827*V6    +  1.0000 E1
Std Err          0.0553 PV1V2    0.0646 PV1V5    0.0536 PV1V6
t Value          8.0632          7.9461          -7.1407


V2       =       0.7387*V3    -  0.4308*V4    +  1.0000 E2
Std Err          0.0686 PV2V3    0.0610 PV2V4
t Value          10.7632         -7.0592
```

Variances of Exogenous Variables

Variable	Parameter	Estimate	Standard Error	t Value
V3	VARV3	1.568756	0.143207	10.954
V4	VARV4	1.984154	0.181128	10.954
V5	VARV5	2.425806	0.221445	10.954
V6	VARV6	3.497274	0.319256	10.954
E1	VARE1	1.930730	0.176251	10.954
E2	VARE2	1.429455	0.130491	10.954

Covariances among Exogenous Variables

	Parameter	Estimate	Standard Error	t Value
V4 V3	CV3V4	-0.777162	0.124443	-6.245
V5 V3	CV3V5	1.042881	0.142786	7.304
V5 V4	CV4V5	-0.406748	0.144029	-2.824
V6 V3	CV3V6	-0.951208	0.163187	-5.829
V6 V4	CV4V6	0.928564	0.180293	5.150
V6 V5	CV5V6	-1.145849	0.202038	-5.671

Covariance Structure Analysis: Maximum Likelihood Estimation   10

Equations with Standardized Coefficients

```
V1       =       0.3548*V2    + 0.3587*V5    -  0.3211*V6    + 0.6236 E1
                 PV1V2           PV1V5           PV1V6


V2       =       0.5214*V3    - 0.3420*V4    +  0.6738 E2
                 PV2V3           PV2V4
```

```
            Variances of Endogenous Variables
--------------------------------------------------------
   Variable          Estimate          R-squared
--------------------------------------------------------
    1   V1           4.965570           0.611177
    2   V2           3.148495           0.545988


          Correlations among Exogenous Variables
          ------------------------------------------
            Parameter              Estimate
          ------------------------------------------
       V4   V3    CV3V4            -0.440500
       V5   V3    CV3V5             0.534600
       V5   V4    CV4V5            -0.185400
       V6   V3    CV3V6            -0.406100
       V6   V4    CV4V6             0.352500
       V6   V5    CV5V6            -0.393400
```

Output 4.6: Output Pages 9-10, Analysis of Initial Model for Investment Model Study

In the present theoretical model, V1 (Commitment) was said to be directly determined by V2 (Satisfaction), V5 (Investments), and V6 (Alternatives). The preceding output shows that these three variables accounted for 61% of the variance in Commitment. Most researchers would consider this to be a relatively large percent of the variance (although the figure is probably somewhat inflated since the same method was used to assess both independent and dependent variables). In your model, V2 (Satisfaction) was predicted to be directly affected by V3 (Rewards) and V4 (Costs). These variables accounted for 55% of the variance in Satisfaction, again a substantial proportion of the variance.

Step 5: Reviewing significance tests for path coefficients and covariances. The chi-square test, the NFI, the NNFI, and the CFI reflect the overall fit of the model to the data, and in this case have not provided strong, consistent support for your model. However, even if they had indicated a relatively good overall fit, it would still be necessary to inspect specific features of the model to see if any of these features failed to receive support.

For example, the present model predicts that satisfaction will have a significant effect on commitment. To test this prediction, it is necessary to test the significance of the path coefficient that represents the effect of satisfaction on commitment. If this coefficient is significantly different from zero, the prediction receives support.

Significance tests for path coefficients appeared on page 9 of the present SAS output, under the heading "Manifest Variable Equations." The second equation in this table, the equation for V2 (Satisfaction), is reproduced again in Output 4.7.

```
V2        =      0.7387*V3      -    0.4308*V4    +    1.0000 E2
Std Err          0.0686 PV2V3        0.0610 PV2V4
t Value          10.7632             -7.0592
```

Output 4.7: Manifest Variable Equation for Variable V2 (Satisfaction), Analysis of Initial Model for Investment Model Study

The first line of this output provides the manifest variable equation itself. V2 (Satisfaction) is on the left of the equals sign because it is the endogenous variable being predicted, and V3 (Rewards) and V4 (Costs) are on the right because they are the independent variables.

A given path coefficient appears just before the short name for the predictor variable. The preceding output therefore shows that the path coefficient for V3 is .73, and the coefficient for V4 is –.43 (notice the minus sign in the equation). The signs of these coefficients are as you would expect. It makes sense that Rewards would be positively related to Satisfaction, and that Costs would be negatively related.

Notice that the path coefficient for E2 (the residual term) is 1.00. This is because this coefficient was fixed at 1.00 in the SAS program.

Given the definition for path coefficients, the preceding values show that for a one-unit increase in V3 there is an increase of .7387 units in V2 while holding constant the effects of the other independent variables. It also shows that for a one-unit increase in V4 there is a decrease of .4308 units in V2 while holding constant the effects of the other independent variables.

Just below each path coefficient, the CALIS procedure prints that coefficient's standard error. The standard error for V3 is .0686. These standard errors will be valid only if the data are multivariate normal and the sample is adequately large (as discussed earlier in reference to the chi-square test). In addition, the standard error estimates may be incorrect if the analysis is based on a correlation matrix rather than a covariance matrix, even if the sample is relatively large (*SAS/STAT users guide, volume 1*, 1989).

What does it mean if my standard errors are small? Before interpreting the *t* tests for the path coefficients, the corresponding standard errors should be reviewed to determine if any of them are excessively small (i.e., close to zero). This sometimes reflects an estimation problem that results when one parameter is linearly dependent on other parameters, and can result in invalid tests. In the current output, none of the standard errors are excessively small.

The *t* test for each path coefficient appears just below the standard error for that coefficient. This value of *t* is determined by dividing the appropriate path coefficient by its standard error. If the appropriate assumptions are met, these values may be used to test the null hypothesis that the corresponding path coefficient is equal to zero in the population. Because these *t* tests are equivalent to large-sample z tests, they are statistically significant at the $p < .05$ level whenever their absolute value exceeds 1.96 (two-tailed test). In other words, any path coefficient may be viewed as being statistically different from zero if the absolute value of its *t* statistic is greater than 1.96. The path coefficient is significant at the .01 level if *t* exceeds 2.58, and is significant at the .001 level if *t* exceeds 3.30.

Because the *t* values for V3 and V4 are 10.7632 and −7.0592, respectively, it is clear that both of these paths are significant. Nonetheless, even a significant path coefficient must be viewed with great caution if it appears within the context of a model whose overall fit is questionable (as is the case with the present model).

It should be remembered that the path coefficients on this page are in "raw score" form, and are not standardized. **Standardized path coefficients** are usually more desirable for publication in articles, because, with these coefficients, all variables are on the same unit of measurement (all variables have a standard deviation of 1.00). You may compare the relative size of standardized path coefficients to determine which independent variable has the largest effect on the dependent variable.

The standardized coefficients from the current analysis are again reproduced here in Output 4.8. You can see that the standardized path coefficient for the effect of V2 on V1 is .35. This means that there is an increase of .35 standard deviations in V1 for an increase of 1 standard deviation in V2, while holding constant the effect of the other independent variables. The correlations among the exogenous variables also appear on this page of output (but are not reproduced here).

```
                          The SAS System                              10

          Covariance Structure Analysis: Maximum Likelihood Estimation

                Equations with Standardized Coefficients

    V1     =    0.3548*V2    + 0.3587*V5    - 0.3211*V6    + 0.6236 E1
                  PV1V2          PV1V5          PV1V6

    V2     =    0.5214*V3    - 0.3420*V4    + 0.6738 E2
                  PV2V3          PV2V4
```

Output 4.8: Equations with Standardized Path Coefficients, Analysis of Initial Model for Investment Model Study

How do I test the significance of the correlations between exogenous variables?
Earlier, Output 4.6 provided *t* tests for each covariance specified in the model (under the heading "Covariances among Exogenous Variables"). These statistics test the null hypothesis that the covariance (or correlation) between a given set of variables is zero in the population. Although the statistical significance of covariances is usually of less interest to researchers than the significance of path coefficients, these tests are still useful for obtaining a more complete understanding of how model variables are related to one another.

Characteristics of an Ideal Fit

As the preceding sections suggest, assessing the adequacy of a path model is no simple matter; there are a number of statistical tests and goodness of fit indices that must be consulted. However, it is possible to simplify things somewhat by summarizing the characteristics that would be displayed by a model that demonstrates an "ideal" fit to the data:

- The absolute values of entries in the normalized residual matrix should not exceed 2.00.

- The *p* value associated with the model chi-square test should exceed .05; the closer to 1.00, the better.

- The comparative fit index (CFI) and the non-normed fit index (NNFI) should both exceed .9; the closer to 1.00, the better.

- The R^2 value for each endogenous variable should be relatively large, compared to what typically is obtained in research with these variables.

- The absolute value of the *t* statistics for each path coefficient should exceed 1.96, and the standardized path coefficients should be nontrivial in magnitude (i.e., absolute values should exceed .05).

Remember that a model does not necessarily have to display all of these characteristics to be considered acceptable; in fact, the literature contains many studies reporting acceptable models that fail to demonstrate one or more of the preceding traits. In particular, a nonsignificant chi-square value is normally not essential, as long as the value of chi-square is not very large in relation to the degrees of freedom (an informal rule of thumb used by some researchers is to view the value of chi-square as being acceptable as long as the ratio of chi-square to its degrees of freedom is less than 2). Nonetheless, the preceding provides a useful standard against which models may be compared, and you can have more confidence in a path model that does demonstrate these characteristics.

Modifying the Model

When researchers perform path analysis, they usually hope that the analysis will reveal a good fit between model and data: that the residuals will be small, the chi-square statistic will be nonsignificant, and the other characteristics of an "ideal fit" will be displayed. If such a fit is obtained, then the model has survived an attempt at disconfirmation, and gains some support. The analysis terminates at that point.

But what if the theoretical model does not demonstrate a good fit? In most cases (in fact, in all cases with which we are familiar), the researcher then begins modifying the model so that it better accounts for the observed relationships between the variables. For example, if the theoretical model predicts no path from variable X to variable Y, but there actually appears to be a rather strong relationship between these variables, the researcher may add the path, revise the PROC CALIS program to reflect this change, resubmit the program, and ascertain whether the revised model now demonstrates an acceptable fit.

In actual practice, you typically should rely on a number of modification indices to determine how the model should be changed. For example, the CALIS procedure provides a modification index called the Lagrange multiplier which estimates the extent to which the model chi-square statistic would decrease (improve) if a given parameter were freed (e.g., if a given path were added to the model). In most cases, you will modify models by

- making the single modification that results in the greatest improvement in overall model fit

- analyzing the fit of the revised model

- reviewing the modification indices that are produced by that analysis to identify the next change that would most improve the fit of the revised model, if necessary.

In this way, the model is modified, one parameter at a time, until an acceptable fit is obtained.

Problems Associated with Model Modification

Unfortunately, many researchers who modify their initial models in the manner just described run the risk of arriving at a final model which, although it may fit the data, is nonetheless invalid: a model that will not generalize to other samples or to the population of interest.

This is because these model modifications very often capitalize on chance characteristics of the sample data. When researchers perform path analysis, they normally hope that their final model will represent the nature of the relationships between variables in the population, not just in the sample that they have drawn from the population. Unfortunately, the relationships observed in the sample data will usually differ from those existing in the population, because no sample is perfectly representative of the population; some differences will exist simply due to sampling error. Therefore, when researchers make many modifications in order to better fit the sample data, chances are good that the resulting model will fit data only from that specific sample; it will not generalize to other samples or to the population.

The modifications just described have been labeled **data-driven modifications**, because it is the characteristics of the sample data, and not primarily the researcher's theory, that are determining what modifications will be made. MacCallum, Roznowski, and Necowitz (1992) have identified a worst-case scenario in which data-driven modifications are particularly likely to lead to models that will not generalize, and, although their discussion deals with covariance structure models with latent variables (as opposed to manifest variables), the concerns raised are relevant to path analysis as well. MacCallum et al. concluded that data-driven model modifications are especially likely to lead to nongeneralizable models under the following circumstances:

- **When the sample is small**
    MacCallum (1986) has shown that data-driven model modifications based on samples of 100 observations almost invariably lead to poor outcomes, and that the outcomes are only mediocre with samples of 300.

- **When many modifications are made**
    When models are modified, the first few changes typically result in relatively large improvements in fit, with each successive change resulting smaller improvements. In addition, when the analysis is performed on data from a large sample, the largest discrepancies between model and data are more likely to be stable (common to most samples drawn from the population); smaller discrepancies are more likely to be unstable (unique to the current sample). For this reason, only the first few modifications made to the model have a reasonable chance of leading to a model that will generalize. If many modifications are made, the latter changes are more likely to be capitalizing on chance properties of the sample data.

- **When the modification is not interpretable or justified according to theory or prior research**
    The literature often warns that modifications may be made to path models only if they can be justified in light of existing theory or prior research (e.g., Joreskog & Sorbom, 1989; Pedhazur, 1982; Sorbom, 1989). It is bad science to implement a given modification simply because it improves the model's fit: the change must be interpretable. You must be able to explain why adding a given path (for example) makes sense in light of what is already known concerning this phenomenon. Despite these frequent warnings, MacCallum et al. (1992) found in a review of the literature that very few researchers actually bother to justify their model modifications on substantive grounds.

Recommendations for Modifying Models

Sooner or later, almost every researcher who performs path analysis will have to confront these problems associated with model modification, since it is relatively rare that an initial theoretical model demonstrates a good fit. Most researchers are forced to make some model modifications, but they can minimize the dangers associated with this process by adhering to the following recommendations, most of which are provided by MacCallum et al. (1992):

- **Use large samples**

    Some references suggest that path analysis may be performed with as few as 100-150 subjects, but these sample sizes are likely to lead to unstable final models if the initial analysis is followed by data-driven model modifications. MacCallum et al. (1992) found that some level of model stability was achieved with relatively large samples ($N = 800$-$1,200$), and ideally you should attempt to obtain samples of this size if many model modifications are anticipated. With smaller samples, the article describing the study should clearly acknowledge the potential unreliability of the final model.

- **Make few modifications**

    As was previously mentioned, only the first few modifications are likely to result in a model that reflects population relationships; subsequent changes are likely to be sample specific.

- **Make only changes that can be meaningfully interpreted**

    Before making a change, consider whether the modification really makes sense in terms of relevant theory or prior research. And never make such a change without discussing it in the research report, carefully summarizing your justification for the modification.

- **Follow a parallel specification search procedure**

    A **specification search** is simply the search for changes that will improve the model's fit. MacCallum et al. (1992) recommend that researchers begin with two independent samples and conduct a separate specification search in each sample. If both searches lead to the same set of modifications, more confidence can be placed in the stability of the final model. As an added advantage, this procedure also allows for a double cross-validation of the final model.

- **Compare alternative a priori models**

    Rather than begin with a single model and then perform modifications until a final model fits the data, it is often possible to begin with several competing models, perform a single analysis on each, and determine which of these provide the best fit. For example, assume that Theory A states that both attitudes and intentions directly affect behavior, but Theory B states that attitudes affect behavior only indirectly, by first influencing intentions (which then directly affect behavior). These alternative models may be tested in a single study. This would involve assessing all variables in a single large sample, then conducting one analysis of the data to test Theory A, and a separate analysis to test Theory B (and Theories C and D, if they exist). The results of these analyses could then be evaluated in terms of overall fit, interpretability, and other criteria to determine which theory obtained the most support. Such an approach is possible both in confirmatory studies based on specific, well-developed theories, as well as in exploratory studies when theory development is still in its earliest stages.

- **Carefully describe the limitations of your study**

    Most causal modeling studies reported in the literature today are single-sample studies in which a series of data-driven modifications are made to arrive at a better-fitting model. When such an approach is followed, it is important that the published research report warn the reader that this approach to model development can result in models that do not generalize to the population or to other samples. The report should state that the model must be considered tentative until it successfully survives additional tests in new samples.

Modifying the Present Model

Having reviewed the problems associated with model modifications, you may now turn your attention back to your path analysis of the investment model. Because your sample is not large enough to divide into the two samples needed for the parallel specification search procedure, and because we are familiar with no alternative theory specifying the predicted relationships between commitment, satisfaction, rewards, costs, investments and alternatives, neither of these preferred approaches to your analysis is possible. Instead, you will review the modification indices and other results obtained from the first analysis to see if any changes can be made to improve the fit of the original model. To minimize the negative consequences of following this admittedly less desirable approach, you will make as few changes as are necessary, will make only changes that are substantively interpretable, and will discuss the limitations of this approach in your research report.

The normalized residual matrix. The modification procedure should begin by reviewing the original covariance matrix, the predicted model matrix, and the normalized residual matrix. If your theoretical model provides a good fit to the data, the predicted model matrix should be nearly identical to the original covariance matrix. By subtracting each element of the predicted model matrix from the original covariance matrix, a residual matrix is formed. In an ideal solution, the elements of the residual matrix should be zero or near zero; a large residual for a given pair of variables indicates that the model does a poor job of accounting for the relationship between those two variables. Because the residual matrix is not standardized, it is generally easier to actually review the normalized residual matrix. Entries in this matrix whose absolute values exceed 2.00 should be viewed as problematic. These matrices are again reproduced as Output 4.9.

```
                        The SAS System                              3
                         Covariances

                V1                V2                V3

  V1      5.378688640       2.774460073       1.597929380    COMMITMENT
  V2      2.774460073       3.148495360       1.493699236    SATISFACTION
  V3      1.597929380       1.493699236       1.568756250    REWARDS
  V4     -1.143062109      -1.428918323      -0.777161596    COSTS
  V5      2.327672038       1.446482895       1.042880974    INVESTMENTS
  V6     -2.658230605      -1.643224854      -0.951208132    ALTERNATIVES


                V4                V5                V6

  V1     -1.143062109       2.327672038      -2.658230605    COMMITMENT
  V2     -1.428918323       1.446482895      -1.643224854    SATISFACTION
  V3     -0.777161596       1.042880974      -0.951208132    REWARDS
  V4      1.984153960      -0.406748040       0.928563558    COSTS
  V5     -0.406748040       2.425806250      -1.145848607    INVESTMENTS
  V6      0.928563558      -1.145848607       3.497274010    ALTERNATIVES
                Determinant = 26.8 (Ln = 3.288)
```

```
   Covariance Structure Analysis: Maximum Likelihood Estimation     6

                      Predicted Model Matrix

                V1                V2                V3

  V1      4.965570083       2.310228984       1.564799549    COMMITMENT
  V2      2.310228984       3.148495360       1.493699236    SATISFACTION
  V3      1.564799549       1.493699236       1.568756250    REWARDS
  V4     -1.200791663      -1.428918323      -0.777161596    COSTS
  V5      2.104821555       0.945639792       1.042880974    INVESTMENTS
  V6     -2.417734982      -1.102725355      -0.951208132    ALTERNATIVES


                V4                V5                V6

  V1     -1.200791663       2.104821555      -2.417734982    COMMITMENT
  V2     -1.428918323       0.945639792      -1.102725355    SATISFACTION
  V3     -0.777161596       1.042880974      -0.951208132    REWARDS
  V4      1.984153960      -0.406748040       0.928563558    COSTS
  V5     -0.406748040       2.425806250      -1.145848607    INVESTMENTS
  V6      0.928563558      -1.145848607       3.497274010    ALTERNATIVES
                Determinant = 31.26 (Ln = 3.442)
```

```
                          Residual Matrix

                  V1                V2                V3

V1       0.4131185573      0.4642310894      0.0331298303      COMMITMENT
V2       0.4642310894      0.0000000000      0.0000000000      SATISFACTION
V3       0.0331298303      0.0000000000      0.0000000000      REWARDS
V4       0.0577295538      0.0000000000      0.0000000000      COSTS
V5       0.2228504825      0.5008431029      0.0000000000      INVESTMENTS
V6      -.2404956231      -.5404994989      0.0000000000      ALTERNATIVES


                  V4                V5                V6

V1       0.0577295538      0.2228504825     -.2404956231      COMMITMENT
V2       0.0000000000      0.5008431029     -.5404994989      SATISFACTION
V3       0.0000000000      0.0000000000      0.0000000000      REWARDS
V4       0.0000000000      0.0000000000      0.0000000000      COSTS
V5       0.0000000000      0.0000000000      0.0000000000      INVESTMENTS
V6       0.0000000000      0.0000000000      0.0000000000      ALTERNATIVES
                  Average Absolute Residual = 0.1178
              Average Off-diagonal Absolute Residual = 0.1373
```

```
         Covariance Structure Analysis: Maximum Likelihood Estimation      7

                          Normalized Residual Matrix

           V1       V2       V3       V4       V5       V6

V1      0.9114   1.5705   0.1604   0.2661   0.8505  -0.7733   COMMITMENT
V2      1.5705   0.0000   0.0000   0.0000   2.6563  -2.3946   SATISFACTION
V3      0.1604   0.0000   0.0000   0.0000   0.0000   0.0000   REWARDS
V4      0.2661   0.0000   0.0000   0.0000   0.0000   0.0000   COSTS
V5      0.8505   2.6563   0.0000   0.0000   0.0000   0.0000   INVESTMENTS
V6     -0.7733  -2.3946   0.0000   0.0000   0.0000   0.0000   ALTERNATIVES
                  Average Normalized Residual = 0.4563
              Average Off-diagonal Normalized Residual = 0.5781

            Rank Order of 8 Largest Normalized Residuals

       V5,V2      V6,V2      V2,V1      V1,V1      V5,V1      V6,V1      V4,V1
      2.6563    -2.3946     1.5705     0.9114     0.8505    -0.7733     0.2661

                                V3,V1
                                0.1604
```

Output 4.9: Original Covariance Matrix, Predicted Model Matrix, Residual Matrix, and Normalized Residual Matrix, Analysis of Initial Model for Investment Model Study

Because the normalized residual matrix is symmetric, it is necessary only to review that part of the matrix which appears below the diagonal; identical values appear above the diagonal. The rank order of the largest normalized residuals appears just below this matrix, and it is usually necessary to consult only this summary to find potentially problematic residuals.

Inspection of the matrix headed "Normalized Residual Matrix" reveals two elements whose absolute values exceed 2.00. The residual appearing at the row for V5 and the column for V2 is approximately 2.66, and the residual appearing at the row for V6 and the column for V2 is approximately −2.39.

To better understand the meaning of these residuals, you will now compare corresponding elements of the predicted model matrix and the original covariance matrix. The matrix headed "Predicted Model Matrix" (from page 6 of Output 4.9) shows that your theoretical model predicts a covariance between V2 (Satisfaction) and V5 (Investments) of just .9456. However, the matrix headed "Covariances" (from page 3 of Output 4.9) shows that the *actual* covariance between these variables is higher at approximately 1.4465. Apparently, V2 and V5 are more strongly related than your theoretical model would suggest.

Similarly, the predicted model matrix shows that your model predicts a covariance between V2 (Satisfaction) and V6 (Alternatives) of −1.1027. However, the covariance matrix shows that the actual covariance between these variables is stronger at −1.6432. Again, the actual relationship between V2 and V6 is stronger than your model would predict. It is too early to make any decisions yet, but these results suggest that it may be necessary to add a causal path connecting V2 to V5, V2 to V6, or both.

In a way, this is unfortunate because it suggests that you may have to add new paths to the model. In general, the two types of changes that researchers most frequently make when modifying a path model include freeing parameters to be estimated (e.g., adding a new causal path or covariance), or constraining parameters to be zero (e.g., deleting an existing path or covariance). Bentler and Chou (1987) argue that, of these two options, the first (adding new paths or covariances) may be somewhat more likely to capitalize on chance characteristics of the sample data and lead to an inaccurate model.

2. **The significance of the causal paths.** Because it is generally more desirable to drop nonsignificant paths than to add new paths, you will next review the path coefficient estimates to see if any of these are not statistically significant. If any prove to be nonsignificant, it may be useful to re-estimate the model without these paths before proceeding.

The significance tests for these equations appear under the heading "Manifest Variable Equations" on page 9 of the SAS output. These equations are again reproduced in Output 4.10.

```
                              The SAS System                              9

             Covariance Structure Analysis: Maximum Likelihood Estimation

                         Manifest Variable Equations

    V1       =      0.4456*V2    +   0.5132*V5    -   0.3827*V6    +  1.0000 E1
    Std Err         0.0553 PV1V2     0.0646 PV1V5     0.0536 PV1V6
    t Value         8.0632           7.9461          -7.1407


    V2       =      0.7387*V3    -   0.4308*V4    +  1.0000 E2
    Std Err         0.0686 PV2V3     0.0610 PV2V4
    t Value        10.7632          -7.0592
```

Output 4.10: Manifest Variable Equations, Analysis of Initial Model for Investment Model Study

Earlier, it was said that path coefficients are statistically significant at the .05 level if the absolute values of their *t* statistics exceed 1.96. These results show that all five paths included in the present model are statistically significant, and therefore should probably not be dropped from the model.

For example, consider the first equation, for the endogenous variable V1 (Commitment). The path coefficient for independent variable V2 (Satisfaction) is .4456, and the *t* statistic for this path coefficient is 8.0632. Because this *t* is larger than the standard cutoff of 1.96, you know that it is significant at *p* < .05. A review of the remaining terms in this manifest variable equation shows that all of the included independent variables (V2, V5, and V6) demonstrated significant path coefficients in the prediction of V1. In the same way, the two independent variables predicted to affect V2 (Satisfaction) also demonstrated significant path coefficients. In other words, these results provide no empirical reason for dropping any of the causal paths from your theoretical model.

The modification indices. The MODIFICATION option included in the program's PROC CALIS statement causes two types of modification indices to be printed: the Lagrange multiplier test and the Wald test.

First, the **Lagrange multiplier test** estimates the decrease in the overall chi-square that would result from freeing a parameter to be estimated. In practical terms, this means that the test estimates an improvement in chi-square that would result from adding a new causal path (or a new covariance) to your theoretical model. Normally, one hopes that this test will identify new paths (or covariances) that should be added to the model and result in a significant decrease in chi-square. A *p* value is also printed with this statistic; if the *p* value is less than .05, you may assume that adding this path to the model will result in a statistically significant improvement in the model chi-square.

How do I perform a chi-square difference test? The Lagrange multiplier provides an approximate test regarding the change in model fit that results from adding a new path or covariance. A more accurate test can be obtained by performing a **chi-square difference test**. This involves

- estimating the original model

- estimating the revised model in which the new path has been added

- calculating the difference between the two resulting chi-square values.

The resulting chi-square difference statistic also has a chi-square distribution, and can be tested for statistical significance. This is done by referring to a table of critical values of chi-square. You merely need to look up the critical value of chi-square for the degrees of freedom associated with this test. The degrees of freedom for the chi-square difference test is equal to the difference between degrees of freedom associated with the two models (these degrees of freedom are listed on the "Chi-square" line of the table that provides goodness of fit statistics on page 8 of the output).

In the present case, your original model had 4 degrees of freedom, while the revised model with the new path had 3 degrees of freedom. Therefore, the degrees of freedom associated with the chi-square difference test would be $4 - 3$, or 1 degree of freedom. A table reporting critical values of chi-square shows that the critical value at 1 degree of freedom is 3.84 ($p < .05$). This means that, if the chi-square difference statistic is greater than 3.84, then adding the new path to the original model resulted in a significant improvement in the model's fit.

The MODIFICATION option also requests that the **Wald test** be conducted. This test estimates the extent to which chi-square would change if a currently free parameter were fixed at zero. In practical terms, this means that the test estimates the change in chi-square that would result from deleting a path (or covariance) that exists in the current model. Normally, one hopes that this test will identify unimportant paths or covariances that may be deleted with only a small and nonsignificant increase in chi-square. A p value is also printed for this test.

What if my output doesn't contain the Wald test results? Wald test results are printed only if it is possible to drop paths or covariances from a model without a significant decrease in chi-square. The output for the current analysis does not include any Wald test results; this means it is probably not possible to drop any of these parameters without significantly hurting the model's fit.

The Lagrange multiplier tests requested by the MODIFICATION option are printed on pages 11-13 of the SAS output. These indices are computed for three matrices:

- the **phi matrix**, which consists of the residual terms paired with manifest exogenous variables or other residual terms

- the **gamma matrix**, whose rows consist of endogenous variables and columns consist of manifest exogenous variables

- the **beta matrix**, which consists of the two endogenous variables.

Below each of these matrices, the CALIS procedure prints the largest Lagrange multipliers ranked in descending order. The Lagrange multiplier tests for the current analysis are reproduced here as Output 4.11.

```
                        The SAS System                              11

            Covariance Structure Analysis: Maximum Likelihood Estimation

             Lagrange Multiplier and Wald Test Indices _PHI_[6:6]
                            Symmetric Matrix
                  Univariate Tests for Constant Constraints
             ------------------------------------------------
             |  Lagrange Multiplier   or   Wald Index     |
             ------------------------------------------------
             |  Probability  | Approx Change of Value |
             ------------------------------------------------


                        V3                    V4                    V5

     V3        120.000 [VARV3]       39.002 [CV3V4]        53.345 [CV3V5]


     V4         39.002 [CV3V4]      120.000 [VARV4]         7.975 [CV4V5]


     V5         53.345 [CV3V5]        7.975 [CV4V5]       120.000 [VARV5]


     V6         33.977 [CV3V6]       26.526 [CV4V6]        32.165 [CV5V6]


     E1          0.412                0.494                 0.364
                 0.521    0.061       0.482    0.082        0.546   -0.170

     E2         31.233                2.694                16.207
                 0.000   -0.780       0.101    0.642        0.000    0.395
```

	V6	E1		E2	
V3	33.977 [CV3V6]	0.412		31.233	
		0.521	0.061	0.000	-0.780
V4	26.526 [CV4V6]	0.494		2.694	
		0.482	0.082	0.101	0.642
V5	32.165 [CV5V6]	0.364		16.207	
		0.546	-0.170	0.000	0.395
V6	120.000 [VARV6]	0.065		9.374	
		0.799	-0.148	0.002	-0.382
E1	0.065		120.000 [VARE1]	0.000	
	0.799	-0.148		0.992	0.001
E2	9.374		0.000	120.000 [VARE2]	
	0.002	-0.382	0.992	0.001	

Covariance Structure Analysis: Maximum Likelihood Estimation   12

Rank order of 9 largest Lagrange multipliers in _PHI_

E2 : V3	E2 : V5	E2 : V6
31.2330 : 0.000	16.2068 : 0.000	9.3735 : 0.002

E2 : V4	E1 : V4	E1 : V3
2.6936 : 0.101	0.4938 : 0.482	0.4124 : 0.521

E1 : V5	E1 : V6	E2 : E1
0.3637 : 0.546	0.0646 : 0.799	0.000107 : 0.992

Lagrange Multiplier and Wald Test Indices _GAMMA_[2:4]
General Matrix
Univariate Tests for Constant Constraints

```
-------------------------------------------
| Lagrange Multiplier  or  Wald Index    |
-------------------------------------------
| Probability | Approx Change of Value |
-------------------------------------------
```

	V3		V4		V5	V6
V1	0.200		0.327		63.141 [PV1V5]	50.990 [PV1V6]
	0.654	0.049	0.568	0.046		
V2	115.846 [PV2V3]		49.833 [PV2V4]		24.415	17.582
					0.000 0.290	0.000 -0.194

```
          Rank order of 4 largest Lagrange multipliers in _GAMMA_

          V2 : V5                 V2 : V6                 V1 : V4
     24.4154 : 0.000         17.5821 : 0.000          0.3269 : 0.568

                                  V1 : V3
                             0.2004 : 0.654
```

```
          Covariance Structure Analysis: Maximum Likelihood Estimation         13

            Lagrange Multiplier and Wald Test Indices _BETA_[2:2]
                              General Matrix
                    Identity-Minus-Inverse Model Matrix
                 Univariate Tests for Constant Constraints
            -----------------------------------------------
            | Lagrange Multiplier  or  Wald Index   |
            -----------------------------------------------
            | Probability | Approx Change of Value |
            -----------------------------------------------

                              V1                      V2

          V1          sing              65.015 [PV1V2]
                        .         .

          V2        18.493                    sing
                     0.000   0.237             .       .

      Rank order of 1 largest Lagrange multipliers in _BETA_

                              V2 : V1
                         18.4930 : 0.000
```

Output 4.11: Output pages 11-13, Analysis of Initial Model for Investment Model Study

The interpretation of the modification indices is easiest if you have at hand a figure displaying the model being estimated. Therefore, the initial model being tested in the current analysis is again reproduced here as Figure 4.13.

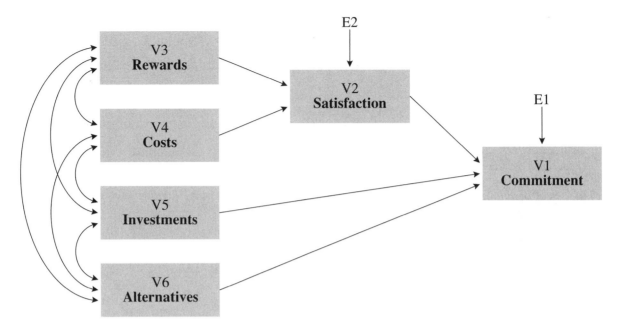

Figure 4.13: The Initial Model, Investment Model Study

Remember that the Lagrange multiplier estimates the improvement in chi-square that would result from *adding* a given path or covariance to the model. At the top of page 12 of Output 4.13, you can see that PROC CALIS has ranked the largest Lagrange multipliers from that matrix. These appear under the heading "Rank order of 9 largest Lagrange multipliers in _PHI_." The largest Lagrange multiplier is for the E2-V3 entry. The observed value of the Lagrange multiplier is 31.23, which means that the model chi-square should decrease by a value of about 31.23 if you were to add a covariance connecting E2 (the residual term for Satisfaction) and V3 (Rewards). To the right of the Lagrange multiplier is the *p* value of 0.000, which means that this would result in a significant decrease in chi-square ($p < .001$).

Such a modification would be questionable, however. One of the assumptions of path analysis is that the residual terms should be uncorrelated with other residual terms and other model variables. There are exceptions to this rule, as in repeated measures designs, but such an exception would not apply to the current study. Closer inspection shows that all of the ranked Lagrange multipliers in this matrix involve relationships either between residual terms and manifest variables or between the residual terms themselves. Because it is preferable that your model not include residuals correlated with any other variables in the model, you will not consider any of these modifications.

How do I know whether the Lagrange multiplier test refers to adding a *covariance* as opposed to adding a *causal path*? Earlier, it was said that the Lagrange multiplier test estimates the improvement in chi-square that would result from adding either a covariance or a path between two variables. If the test involves two exogenous variables, then the test clearly deals with adding a covariance between the variables, since exogenous variables may only be connected by a covariance. If the test involves an exogenous variable and an endogenous variable, then the test clearly deals with adding a path from the exogenous variable to the endogenous variable (remember that exogenous variables are almost never connected to endogenous variables with a covariance). Finally, if the test involves two endogenous variables, then it clearly deals with adding a causal path, as endogenous variables are almost never connected by covariances.

But the user must remember that there may be two Lagrange multipliers computed for each pair of endogenous variables in the beta matrix. One test estimates the model improvement that would result from adding a path from endogenous variable A to endogenous variable B, and the second estimates the improvement that would result from adding a path from variable B to variable A. In the beta matrix, the rows represent the variables when considered as dependent variables, while the columns represent the variables when considered as independent variables. In the current beta matrix, for example, where the row for V2 intersects with the column for V1, you can see a Lagrange multiplier estimate of 18.493. This means that the Lagrange multiplier for adding a path from V1 to V2 is 18.493. Where the row for V1 intersects with the column for V2, there is no Lagrange multiplier test; this is because the path from V2 to V1 is already included in your model. If this path were not already in the model, you would find the Lagrange multiplier for adding a path from V2 to V1 at this intersection.

The gamma matrix appears on the bottom half of page 12 in Output 4.11, and the ranked Lagrange multipliers from this matrix appear under the heading "Rank order of 4 largest Lagrange multipliers in _GAMMA_." The largest value is for the V2–V5 entry at 24.42. This value is significant at $p < .001$. Earlier it was noted that the V2-V5 entry in the normalized residual matrix was the largest entry of that matrix as well. Comparing the original covariance matrix to the predicted model matrix suggested that the actual relationship between V2 and V5 is stronger than was predicted by your theoretical model. The results of the Lagrange multiplier test suggest that your model may be significantly improved by adding a path from the independent variable V5 (investments) to the dependent variable V2 (satisfaction). But now for the important question: will such a change be interpretable? Can it be justified on the basis of theory and research in the area?

Given that I (the author) am an expert on neither interpersonal attraction nor the investment model, an honest response would be "I don't know." However, for the sake of demonstration, assume that adding this causal path can be justified on substantive grounds and proceed with the analysis.

The careful reader may have noted that the V2-V5 entry was not the only entry demonstrating a significant Lagrange multiplier; the Lagrange multiplier for the V2-V6 entry was 17.58, which was also significant at $p < .001$. In addition, the single Lagrange multiplier from the beta matrix estimates that adding a path from V1 (commitment) to V2 (satisfaction) will result in an improvement in the model chi-square of 18.493. Despite this, however, it is more conservative to make only one change to the model at a time, and re-estimate the model after each change. Among other reasons, this is because it is possible that adding just the one path from V5 to V2 will lead to an acceptable model, and may make it unnecessary to add the other paths suggested by the Lagrange multiplier tests. Because you hope to make as few changes as possible, you will revise your SAS program so that it includes only the new path from V5 to V2, and add the additional paths only if the results of the new model indicate that they are needed. The new model is presented as revised model 1 in Figure 4.14.

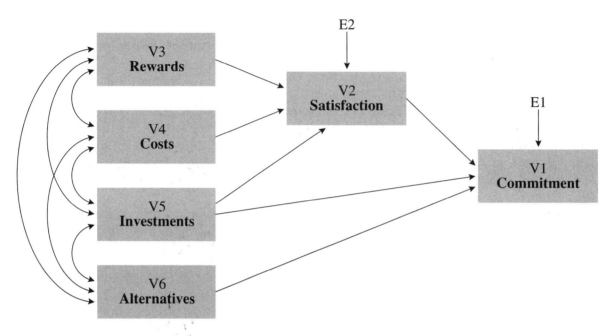

Figure 4.14: Revised Model 1, Developed by Adding a Path from V5 (Investments) to V2 (Satisfaction)

It might also be noted that it is not necessarily best to always choose the path with the largest Lagrange multiplier as the path that must be added to the model. If a path with a smaller value makes better sense in terms of theory or prior research, it may be added instead.

Creating revised model 1. To change the initial theoretical model, it was necessary only to revise one line (line 23) in the LINEQS statement of the PROC CALIS program. Specifically, the equation for the endogenous variable V2 was modified so that V5 now appears as a cause of V2. Since the variables that were originally exogenous variables are still exogenous variables, and since all of the previous endogenous variables are still endogenous variables, it was not necessary to change the STD or COV statements. Part of the revised program appears here.

```
20      PROC CALIS   COVARIANCE CORR RESIDUAL MODIFICATION ;
21         LINEQS
22            V1 = PV1V2 V2 + PV1V5 V5 + PV1V6 V6 + E1,
23            V2 = PV2V3 V3 + PV2V4 V4 + PV2V5 V5 + E2;
24         STD
25            E1 = VARE1,
26            E2 = VARE2,
```

```
27              V3 = VARV3,
28              V4 = VARV4,
29              V5 = VARV5,
30              V6 = VARV6;
31          COV
32              V3 V4 = CV3V4,
33              V3 V5 = CV3V5,
34              V3 V6 = CV3V6,
35              V4 V5 = CV4V5,
36              V4 V6 = CV4V6,
37              V5 V6 = CV5V6;
38          VAR  V1 V2 V3 V4 V5 V6 ;
39          RUN;
```

This SAS program also produced an output file 13 pages long. Some of these results are reproduced here as Output 4.12 through Output 4.15.

Adding the path from investments to satisfaction did result in an improvement in the revised model's fit over the original model's fit. With the first analysis, two entries in the normalized residual matrix exceeded 2.00 in absolute value; with the present analysis, no normalized residuals exceeded this level (although the normalized residual for the V6-V2 entry did come close at -1.67). Most of the residuals in the residual matrix were zero or near zero, with the average absolute residual being .0438. The normalized residual matrix for the revised model is presented in Output 4.12.

```
                          The SAS System                              7

          Covariance Structure Analysis: Maximum Likelihood Estimation

                       Normalized Residual Matrix

           V1        V2        V3        V4        V5        V6

V1     0.2726    0.4697    0.1575    0.2597    0.0000   -0.5320    COMMITMENT
V2     0.4697    0.0000    0.0000    0.0000    0.0000   -1.6683    SATISFACTION
V3     0.1575    0.0000    0.0000    0.0000    0.0000    0.0000    REWARDS
V4     0.2597    0.0000    0.0000    0.0000    0.0000    0.0000    COSTS
V5     0.0000    0.0000    0.0000    0.0000    0.0000    0.0000    INVESTMENTS
V6    -0.5320   -1.6683    0.0000    0.0000    0.0000    0.0000    ALTERNATIVES
                  Average Normalized Residual = 0.16
            Average Off-diagonal Normalized Residual = 0.2058

              Rank Order of 6 Largest Normalized Residuals

          V6,V2     V6,V1     V2,V1     V1,V1     V4,V1     V3,V1
         -1.6683   -0.5320    0.4697    0.2726    0.2597    0.1575
```

```
                Distribution of Normalized Residuals
                  (Each * represents 1 residuals)

        -1.75000 -   -1.50000    1    4.76%  | *
        -1.50000 -   -1.25000    0    0.00%  |
        -1.25000 -   -1.00000    0    0.00%  |
        -1.00000 -   -0.75000    0    0.00%  |
        -0.75000 -   -0.50000    1    4.76%  | *
        -0.50000 -   -0.25000    0    0.00%  |
        -0.25000 -          0    0    0.00%  |
               0 -    0.25000   16   76.19%  | ***************
         0.25000 -    0.50000    3   14.29%  | ***
```

Output 4.12: Output Page 7, Analysis of Revised Model 1 for Investment Model Study

The goodness of fit indices on page 8 of the new output were also indicative of a better model fit (see Output 4.13). The comparative fit index (CFI) appears to the right of "Bentler's Comparative Fit Index," and you can see that the observed value for the CFI was approximately .99. Similarly, the non-normed fit index (NNFI) appears to the right of "Bentler & Bonett's (1980) Non-normed Index," and you can see that the NNFI was .9377. In short, both the CFI and the NNFI exceeded the desired .9 level, suggesting an adequate fit.

```
                        The SAS System                              8

    Covariance Structure Analysis: Maximum Likelihood Estimation

    Fit criterion . . . . . . . . . . . . . . . . . . .     0.0469
    Goodness of Fit Index (GFI) . . . . . . . . . . .       0.9849
    GFI Adjusted for Degrees of Freedom (AGFI)  . . .       0.8945
    Root Mean Square Residual (RMR) . . . . . . . . .       0.1019
    Chi-square = 11.2008       df = 3       Prob>chi**2 = 0.0107
    Null Model Chi-square:     df = 15                    672.7238
    RMSEA Estimate . . . . . .  0.1070  90%C.I.[0.0452, 0.1770]
    Bentler's Comparative Fit Index . . . . . . . .         0.9875
    Normal Theory Reweighted LS Chi-square  . . . . .      10.9693
    Akaike's Information Criterion  . . . . . . . . .       5.2008
    Consistent Information Criterion  . . . . . . . .      -8.2411
    Schwarz's Bayesian Criterion  . . . . . . . . . .      -5.2411
    McDonald's (1989) Centrality. . . . . . . . . .         0.9831
    Bentler & Bonett's (1980) Non-normed Index. . . .       0.9377
    Bentler & Bonett's (1980) Normed Index. . . . . .       0.9834
    James, Mulaik, & Brett (1982) Parsimonious Index.      0.1967
    Z-Test of Wilson & Hilferty (1931) . . . . . . .        2.2979
    Bollen (1986) Normed Index Rho1 . . . . . . . . .       0.9168
    Bollen (1988) Non-normed Index Delta2 . . . . . .       0.9878
    Hoelter's (1983) Critical N . . . . . . . . . . .          168
```

Output 4.13: Output Page 8, Analysis of Revised Model 1 for Investment Model Study

Unfortunately, however, the model chi-square statistic of 11.20 (at *df*= 3) was still statistically significant at *p* < .05, suggesting that the null hypothesis of good model fit can be rejected at the .05 level of confidence. This may not necessarily mean that your model is unacceptable, as the chi-square statistic is known to be very sensitive to seemingly trivial differences between model and data, and the CFI and NNFI do suggest that the present model may be acceptable. For the time being, you will proceed with the analysis.

In your analysis of the original model, the Lagrange multiplier test suggested that adding the path from Investments to Satisfaction would result in a significant improvement in model chi-square. But remember that the Lagrange multiplier test merely provides an *estimate* of the improvement in model chi-square. Since the revised model has now been estimated, it is possible to directly determine whether the model improvement was significant by performing the chi-square difference test discussed earlier. This process involves determining the difference between the chi-square values for the two models, and testing this difference for significance.

The chi-square for the initial model was 36.8420, while the chi-square for the revised model was 11.2008. The difference chi-square is therefore equal to

```
36.8420 - 11.2008 = 25.6412
```

The degrees of freedom for the test are obtained by subtracting the degrees of freedom for the revised model (3) from the degrees of freedom for the original model (4), resulting in 1 degree of freedom for the difference test. A table of chi-square indicates that the critical value of chi-square with 1 degree of freedom is 3.841 (*p* < .05), and since 25.6412 is larger than 3.841, this difference is clearly significant. In fact, this improvement in chi-square is significant at the *p* < .001 level, since the critical value of chi-square is 10.827 at that level of significance (*df* = 1).

Output 4.14 displays path coefficients and other parameter estimates for revised model 1, and this information provides additional evidence of a fairly good fit:

- All of the path coefficients on output page 9 are significant. These appear under the heading "Manifest Variable Equations," and you can see that the *t* values for all path coefficients exceed 1.96 in absolute value.

- The standardized path coefficients on page 10 all exceed .05 in absolute value, indicating they are not trivial in size. These appear under the heading "Equations with Standardized Coefficients."

- The R^2 values for the two endogenous variables on page 10 are substantial in size. These appear in the table headed "Variances of the Endogenous Variables," under the heading "R-squared."

The SAS System   9

Covariance Structure Analysis: Maximum Likelihood Estimation

Manifest Variable Equations

```
V1        =      0.4456*V2    + 0.5131*V5    - 0.3827*V6    + 1.0000 E1
Std Err          0.0608 PV1V2   0.0697 PV1V5   0.0535 PV1V6
t Value          7.3240         7.3568        -7.1519


V2        =      0.5358*V3    - 0.4508*V4    + 0.2903*V5    + 1.0000 E2
Std Err          0.0758 PV2V3   0.0580 PV2V4   0.0557 PV2V5
t Value          7.0686        -7.7762         5.2135
```

Variances of Exogenous Variables

Variable	Parameter	Estimate	Standard Error	t Value
V3	VARV3	1.568756	0.143207	10.954
V4	VARV4	1.984154	0.181128	10.954
V5	VARV5	2.425806	0.221445	10.954
V6	VARV6	3.497274	0.319256	10.954
E1	VARE1	1.930733	0.176251	10.954
E2	VARE2	1.284036	0.117216	10.954

Covariances among Exogenous Variables

		Parameter	Estimate	Standard Error	t Value
V4	V3	CV3V4	-0.777162	0.124443	-6.245
V5	V3	CV3V5	1.042881	0.142786	7.304
V5	V4	CV4V5	-0.406748	0.144029	-2.824
V6	V3	CV3V6	-0.951208	0.163187	-5.829
V6	V4	CV4V6	0.928564	0.180293	5.150
V6	V5	CV5V6	-1.145849	0.202038	-5.671

Covariance Structure Analysis: Maximum Likelihood Estimation   10

Equations with Standardized Coefficients

```
V1        =      0.3452*V2    + 0'.3488*V5    - 0.3124*V6    + 0.6065 E1
                 PV1V2          PV1V5           PV1V6


V2        =      0.3782*V3    - 0.3578*V4    + 0.2549*V5    + 0.6386 E2
                 PV2V3          PV2V4           PV2V5
```

```
                    Variances of Endogenous Variables
          ------------------------------------------------
          Variable           Estimate          R-squared
          ------------------------------------------------
             1    V1          5.248075           0.632106
             2    V2          3.148495           0.592175


                 Correlations among Exogenous Variables
            --------------------------------------------
                   Parameter            Estimate
            --------------------------------------------
            V4   V3    CV3V4            -0.440500
            V5   V3    CV3V5             0.534600
            V5   V4    CV4V5            -0.185400
            V6   V3    CV3V6            -0.406100
            V6   V4    CV4V6             0.352500
            V6   V5    CV5V6            -0.393400
```

Output 4.14: Output Pages 9-10, Analysis of Revised Model 1, Investment Model Study

Lagrange multiplier test results for the current analysis are provided in Output 4.15. A possible problem appears on output page 12, where the rank order of the largest Lagrange multiplier indices from the gamma matrix is presented. You can see that the Lagrange multiplier for the V2-V6 entry is statistically significant at 10.4351 ($p < .001$). This would suggest that adding a path from V6 (Alternatives) to V2 (Satisfaction) could result in a significant improvement in the model's fit. For the sake of demonstration, you will assume that the addition of such a path is theoretically interpretable, and will revise the SAS program to reflect this change. The resulting model with the new path from V6 to V2 appears as revised model 2 in Figure 4.15.

```
                       The SAS System                                11

          Covariance Structure Analysis: Maximum Likelihood Estimation

            Lagrange Multiplier and Wald Test Indices _PHI_[6:6]
                             Symmetric Matrix
                 Univariate Tests for Constant Constraints
            --------------------------------------------
            |  Lagrange Multiplier  or  Wald Index   |
            --------------------------------------------
            |  Probability  | Approx Change of Value  |
            --------------------------------------------
```

	V3	V4	V5
V3	120.000 [VARV3]	39.002 [CV3V4]	53.345 [CV3V5]
V4	39.002 [CV3V4]	120.000 [VARV4]	7.975 [CV4V5]
V5	53.345 [CV3V5]	7.975 [CV4V5]	120.000 [VARV5]
V6	33.977 [CV3V6]	26.526 [CV4V6]	32.165 [CV5V6]
E1	0.380 0.538　0.056	0.514 0.473　0.085	0.325 0.568　−0.155
E2	10.435 0.001　−1.567	10.435 0.001　1.242	10.435 0.001　−1.210

	V6	E1	E2
V3	33.977 [CV3V6]	0.380 0.538　0.056	10.435 0.001　−1.567
V4	26.526 [CV4V6]	0.514 0.473　0.085	10.435 0.001　1.242
V5	32.165 [CV5V6]	0.325 0.568　−0.155	10.435 0.001　−1.210
V6	120.000 [VARV6]	0.062 0.804　−0.142	10.435 0.001　−0.382
E1	0.062 0.804　−0.142	120.000 [VARE1]	0.018 0.894　0.020
E2	10.435 0.001　−0.382	0.018 0.894　0.020	120.000 [VARE2]

Covariance Structure Analysis: Maximum Likelihood Estimation   12

Rank order of 9 largest Lagrange multipliers in _PHI_

E2 : V4	E2 : V6	E2 : V5
10.4351 : 0.001	10.4351 : 0.001	10.4351 : 0.001

E2 : V3	E1 : V4	E1 : V3
10.4351 : 0.001	0.5140 : 0.473	0.3800 : 0.538

E1 : V5	E1 : V6	E2 : E1
0.3254 : 0.568	0.0617 : 0.804	0.0178 : 0.894

```
            Lagrange Multiplier and Wald Test Indices _GAMMA_[2:4]
                             General Matrix
                  Univariate Tests for Constant Constraints
            ---------------------------------------------
            | Lagrange Multiplier  or  Wald Index   |
            ---------------------------------------------
            | Probability | Approx Change of Value |
            ---------------------------------------------

                   V3                V4                V5                V6

      V1     0.179             0.336             54.122 [PV1V5]     51.150 [PV1V6]
             0.672   0.043     0.562   0.047

      V2   49.965 [PV2V3]    60.469 [PV2V4]     27.180 [PV2V5]     10.435
                                                                   0.001   -0.146


          Rank order of 3 largest Lagrange multipliers in _GAMMA_

                 V2 : V6              V1 : V4              V1 : V3
             10.4351 : 0.001      0.3360 : 0.562       0.1789 : 0.672




            Lagrange Multiplier and Wald Test Indices _BETA_[2:2]
                             General Matrix
                  Identity-Minus-Inverse Model Matrix
                  Univariate Tests for Constant Constraints
            ---------------------------------------------
            | Lagrange Multiplier  or  Wald Index   |
            ---------------------------------------------
            | Probability | Approx Change of Value |
            ---------------------------------------------

                            V1                V2

             V1          sing              53.641 [PV1V2]
                          .        .

             V2         3.826              sing
                        0.050   0.132       .        .
```

```
      Covariance Structure Analysis: Maximum Likelihood Estimation      13

          Rank order of 1 largest Lagrange multipliers in _BETA_

                            V2 : V1
                          3.8255 : 0.050
```

Output 4.15: Output Pages 11-13, Analysis of Revised Model 1, Investment Model Study

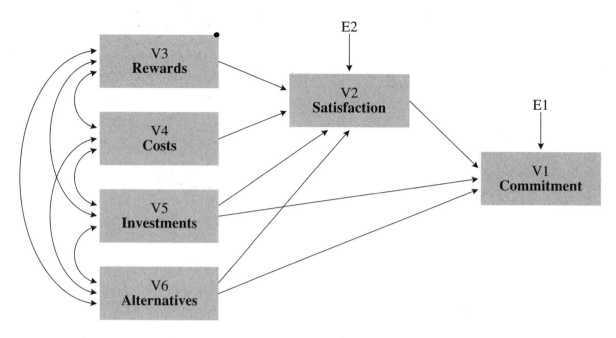

Figure 4.15: Revised Model 2, Developed by Adding a Path from V6 (Alternatives) to V2 (Satisfaction)

Creating revised model 2. Here is the revised SAS program which includes the new path from V6 to V2. It was necessary to change only the LINEQS statement for the endogenous variable V2 to make this modification; all other aspects of the program remain the same.

```
20      PROC CALIS    COVARIANCE CORR RESIDUAL MODIFICATION ;
21        LINEQS
22            V1 = PV1V2 V2 + PV1V5 V5 + PV1V6 V6          + E1,
23            V2 = PV2V3 V3 + PV2V4 V4 + PV2V5 V5 + PV2V6 V6 + E2;
24        STD
25            E1 = VARE1,
26            E2 = VARE2,
27            V3 = VARV3,
28            V4 = VARV4,
29            V5 = VARV5,
30            V6 = VARV6;
31        COV
32            V3 V4 = CV3V4,
33            V3 V5 = CV3V5,
34            V3 V6 = CV3V6,
35            V4 V5 = CV4V5,
36            V4 V6 = CV4V6,
```

```
37            V5 V6 = CV5V6;
38            VAR  V1 V2 V3 V4 V5 V6 ;
39            RUN;
```

The SAS output created by this program was again 13 pages long, parts of which are reproduced here as Output 4.16 and Output 4.17. Output 4.16 presents the predicted model matrix, residual matrix, and normalized residual matrix for this solution.

```
                              The SAS System                               6

                Covariance Structure Analysis: Maximum Likelihood Estimation

                           Predicted Model Matrix

                     V1              V2              V3

      V1       5.378860563     2.774460073     1.564855116     COMMITMENT
      V2       2.774460073     3.148495360     1.493699236     SATISFACTION
      V3       1.564855116     1.493699236     1.568756250     REWARDS
      V4      -1.200981573    -1.428918323    -0.777161596     COSTS
      V5       2.327672038     1.446482895     1.042880974     INVESTMENTS
      V6      -2.658230605    -1.643224854    -0.951208132     ALTERNATIVES


                     V4              V5              V6

      V1      -1.200981573     2.327672038    -2.658230605     COMMITMENT
      V2      -1.428918323     1.446482895    -1.643224854     SATISFACTION
      V3      -0.777161596     1.042880974    -0.951208132     REWARDS
      V4       1.984153960    -0.406748040     0.928563558     COSTS
      V5      -0.406748040     2.425806250    -1.145848607     INVESTMENTS
      V6       0.928563558    -1.145848607     3.497274010     ALTERNATIVES
                   Determinant = 26.86 (Ln = 3.291)


                            Residual Matrix

                     V1              V2              V3

      V1       -.0001719231    0.0000000000    0.0330742636     COMMITMENT
      V2       0.0000000000    0.0000000000    0.0000000000     SATISFACTION
      V3       0.0330742636    0.0000000000    0.0000000000     REWARDS
      V4       0.0579194638    0.0000000000    0.0000000000     COSTS
      V5       0.0000000000    0.0000000000    0.0000000000     INVESTMENTS
      V6       0.0000000000    0.0000000000    0.0000000000     ALTERNATIVES
```

	V4	V5	V6	
V1	0.0579194638	0.0000000000	0.0000000000	COMMITMENT
V2	0.0000000000	0.0000000000	0.0000000000	SATISFACTION
V3	0.0000000000	0.0000000000	0.0000000000	REWARDS
V4	0.0000000000	0.0000000000	0.0000000000	COSTS
V5	0.0000000000	0.0000000000	0.0000000000	INVESTMENTS
V6	0.0000000000	0.0000000000	0.0000000000	ALTERNATIVES

Average Absolute Residual = 0.004341
Average Off-diagonal Absolute Residual = 0.006066

Covariance Structure Analysis: Maximum Likelihood Estimation   7

Normalized Residual Matrix

	V1	V2	V3	V4	V5	V6	
V1	-.00035	0.00000	0.15529	0.25779	0.00000	0.00000	COMMITMENT
V2	0.00000	0.00000	0.00000	0.00000	0.00000	0.00000	SATISFACTION
V3	0.15529	0.00000	0.00000	0.00000	0.00000	0.00000	REWARDS
V4	0.25779	0.00000	0.00000	0.00000	0.00000	0.00000	COSTS
V5	0.00000	0.00000	0.00000	0.00000	0.00000	0.00000	INVESTMENTS
V6	0.00000	0.00000	0.00000	0.00000	0.00000	0.00000	ALTERNATIVES

Average Normalized Residual = 0.01969
Average Off-diagonal Normalized Residual = 0.02754

Rank Order of 3 Largest Normalized Residuals

V4,V1	V3,V1	V1,V1
0.2578	0.1553	-0.000350

Output 4.16: Output Pages 6-7, Analysis of Revised Model 2, Investment Model Study

The values in the residual matrix are all zero or near zero; the average absolute residual is only 0.004341. In addition, the normalized residual matrix displays no normalized residual greater than 2.00 (in fact, it displays none that even comes close to 1.00). These results are all indicative of a good model fit.

The goodness of fit statistics are presented in Output 4.17. At last you have obtained a nonsignificant model chi-square value of 0.5765 ($df = 2$, $p = .75$). The large p value means that you cannot reject the null hypothesis of good model fit.

```
                        The SAS System                         8

        Covariance Structure Analysis: Maximum Likelihood Estimation

        Fit criterion . . . . . . . . . . . . . . . . . . .    0.0024
        Goodness of Fit Index (GFI) . . . . . . . . . . .      0.9992
        GFI Adjusted for Degrees of Freedom (AGFI)  . . .      0.9916
        Root Mean Square Residual (RMR) . . . . . . . . .      0.0146
        Chi-square = 0.5765        df = 2       Prob>chi**2 = 0.7496
        Null Model Chi-square:     df = 15                     672.7238
        RMSEA Estimate  . . . . . . . . . .  0.0000  90%C.I.[., .]
        Bentler's Comparative Fit Index . . . . . . . .        1.0000
        Normal Theory Reweighted LS Chi-square  . . . . .      0.5757
        Akaike's Information Criterion  . . . . . . . . .     -3.4235
        Consistent Information Criterion  . . . . . . . .    -12.3848
        Schwarz's Bayesian Criterion  . . . . . . . . . .    -10.3848
        McDonald's (1989) Centrality. . . . . . . . . .        1.0030
        Bentler & Bonett's (1980) Non-normed Index. . . .      1.0162
        Bentler & Bonett's (1980) Normed Index. . . . . .      0.9991
        James, Mulaik, & Brett (1982) Parsimonious Index.      0.1332
        Z-Test of Wilson & Hilferty (1931) . . . . . . .      -0.6850
        Bollen (1986) Normed Index Rho1 . . . . . . . . .      0.9936
        Bollen (1988) Non-normed Index Delta2 . . . . . .      1.0021
        Hoelter's (1983) Critical N . . . . . . . . . . .      2485
```

Output 4.17: Output Page 8, Analysis of Revised Model 2, Investment Model Study

The Lagrange multiplier test from the analysis of revised model 1 suggested that adding the V2-V6 path would result in a significant improvement in chi-square. But did it? This can be assessed by again performing a chi-square difference test. Revised model 1 demonstrated a model chi-square value of 11.2008 with 3 degrees of freedom. In the present analysis, the model chi-square for revised model 2 is 0.5765 with 2 degrees of freedom. The chi-square difference value is therefore

```
    11.2008 - 0.5765 = 10.6243
```

with difference test degrees of freedom equal to

```
    3 - 2 = 1
```

The critical value of chi-square with 1 degree of freedom is 3.841 for $p < .05$, and 6.635 for $p < .01$, so it is clear that adding the V2–V6 path did result in a significant improvement in the model's fit.

Other information from the SAS output (some of which is not reproduced here) provided even more evidence that the addition of the new V2-V6 path resulted in a model with a very good fit to the data. The CFI for the new model was 1.00, and NNFI was 1.02. All path coefficients were statistically significant, and all standardized coefficients exceeded .05 in absolute value.

Values of R^2 for the two endogenous variables each exceeded .60. Some of the Lagrange multiplier values did attain statistical significance (such as the value for the V1-V3 relationship), but adding any of these paths would have resulted in only trivial decreases in model chi-square. In short, revised model 2, in which the path from Investments to Satisfaction and the path from Alternatives to Satisfaction had been added to the original model, provided an excellent fit to the data.

Choosing a final model. At this point, three models have been investigated in this study:

- the initial investment model (as illustrated in Figure 4.13)

- revised model 1, in which a path from investments to satisfaction had been added to the initial model (as illustrated in Figure 4.14)

- revised model 2, in which a path from alternatives to satisfaction had been added to revised model 1 (as illustrated in Figure 4.15).

In an actual analysis, you would have to decide at this point which of these will be presented as the final model in the research report.

The decision will not necessarily be an easy one. On empirical grounds, it is difficult to justify accepting the initial model as the study's final model. The NNFI was less than .9, the absolute value of two normalized residuals exceeded 2.00, and the Lagrange multiplier tests indicated that the model could be improved substantially by the addition of at least one new path.

Also on purely empirical grounds, it would be easy to justify revised model 2 as the study's final model. This model (which included two new paths from Investments and Alternatives to Satisfaction) demonstrated all five characteristics of an ideal fit:

- Normalized residuals were less than 2.00.

- The model chi-square was nonsignificant.

- The CFI and NNFI each exceeded .9.

- R^2 values were large.

- All path coefficients were statistically significant and nontrivial in size.

In addition, the chi-square difference test showed that revised model 2 demonstrated a fit to the data that was significantly superior to that of revised model 1.

On the negative side, revised model 2 can hardly be called a parsimonious model. A **parsimonious** path model is one that can account for covariation in the data with a minimal number of paths. In revised model 2, however, nearly every variable is interconnected to every other variable by means of a path or a curved, double-headed arrow. The model's lack of parsimony is reflected in the fact that it has only two degrees of freedom. A more parsimonious model would have more degrees of freedom. Finally, the admonitions of MacCallum et al. (1992) against making many model modifications would seem to argue in favor of revised model 1 and against model 2.

Your final decision should be based on the preceding considerations, along with your knowledge of theory and prior research in the area. If there are strong substantive reasons to support both of the new causal paths included in revised model 2, then that may be the more defensible model to retain as the study's final model (provided that you clearly acknowledge the dangers inherent in data-driven model modifications in the research report). The author's knowledge concerning this area of research is not adequate to make an informed decision, however, so you will (somewhat arbitrarily) choose revised model 1 as your final model.

Preparing a Formal Description of the Analysis and Results for a Paper

The results section of a path analytic study can be much longer than the results sections of studies that use simpler statistical procedures such as PROC ANOVA. This is because, in most path analytic studies, the initial model fails to provide an adequate fit to the data and must be modified. The research report must describe how these modifications were made, and this is usually done in the results section.

Preparing Figures and Tables

The figures. It will be much easier for a reader to understand your article if you organize the text around a few figures and tables. First, one figure should illustrate your initial theoretical model. In most cases, this figure will appear in the introduction section of the paper in which you state the hypotheses to be tested in the study. An example of such a figure for the present study is presented as Figure 4.16.

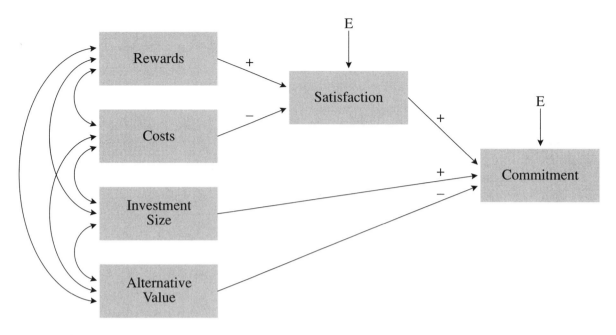

Figure 4.16: The Initial Theoretical Model (for the Published Article)

This figure follows the same conventions for path diagrams used throughout this chapter, with a few minor exceptions. For example, notice that each of the causal paths have been identified with either a positive or negative sign. These indicate whether you predict a positive or negative relationship between a given pair of variables (e.g., this figure predicts a positive relationship between Satisfaction and Commitment, but a negative relationship between Alternative value and Commitment). The use of these signs is optional, however; some researchers simply describe the nature of the predicted relationships in the text of their papers.

Later, in the results section, many papers then provide an additional figure that illustrates modifications made to the initial figure, or perhaps the final version of the model, after modifications (if any) have been completed. Examples of this are presented as Figures 4.17 and 4.18.

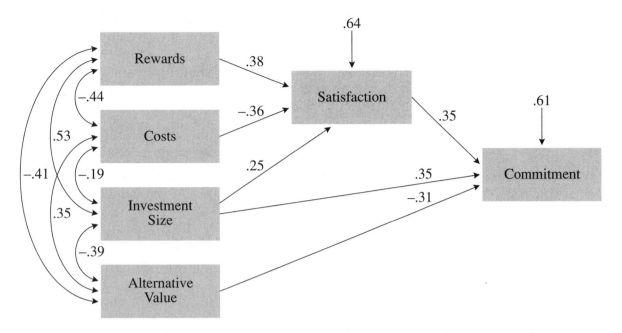

Figure 4.17: Revised Model 1, Standardized Path Coefficients Appear on Single-Headed Straight Arrows; Correlations Appear on Double-Headed Curved Arrows; All Path Coefficients Significant at *p* < .05 or Lower

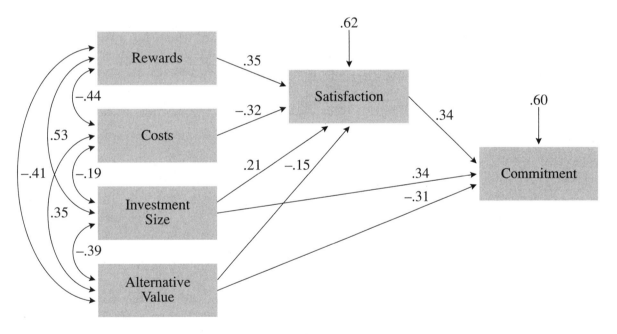

Figure 4.18: Revised Model 2, Standardized Path Coefficients Appear on Single-Headed Straight Arrows; Correlations Appear on Double-Headed Curved Arrows; All Path Coefficients Significant at *p* < .05 or Lower

These figures present revised models 1 and 2 from the present study. Notice that the positive and negative signs from the previous figure have now been replaced with the standardized path coefficients obtained in the analyses. Be sure to get these coefficients from the SAS output table titled "Equations with Standardized Coefficients" (such as the one presented in Output 4.14). The table titled "Manifest Variable Equations" (such as the one presented in Output 4.14) will present nonstandardized coefficients, which usually are not presented in figures such as this. The "Equations with Standardized Coefficients" table will also present the residual terms needed for the figure (in the case of revised model 1, the residual term for satisfaction is .64, and the residual term for commitment is .61). The residual term for a given endogenous variable is equal to the square root of the quantity $(1-R^2)$ for that endogenous variable.

In the present case, the revised figures also provide correlations between the model's exogenous variables. The correlation between two variables is located next to the curved two-headed arrow that connects those variables. Technically, this information is optional, since it already appears in the table of means, standard deviations, and intercorrelations (to be discussed below). In fact, the path models in many papers do not even bother to include the curved, two-headed arrows that indicate the correlations between the study's exogenous variables, as these correlations are usually assumed.

The tables. In most research papers published in the social sciences (including most path-analytic studies), one of the first tables presented should provide simple descriptive statistics for the study's variables, including the means, standard deviations, and intercorrelations. This will allow subsequent researchers to repeat your analyses if they like. An example of such a table is presented as Table 4.1.

Table 4.1

Means, Standard Deviations, Reliability Estimates (in parentheses), and Intercorrelations for the Investment Model Study

Measure	M	SD	1	2	3	4	5	6
1. Commitment	7.42	2.32	(81)					
2. Satisfaction	7.88	1.77	67	(92)				
3. Rewards	7.92	1.25	55	67	(77)			
4. Costs	3.74	1.41	-35	-57	-44	(71)		
5. Investment size	6.55	1.50	64	52	53	-19	(86)	
6. Alternative value	3.18	1.87	-69	-50	-41	35	-39	(85)

Note. $\underline{N}$ = 240. Coefficient alpha reliability estimates are reported in parentheses. Decimals are omitted from correlations and reliability estimates.

An additional table that will prove useful to your readers is one that summarizes the goodness of fit indices obtained for the different path models. An example of such a table is presented as Table 4.2 here.

Table 4.2

Goodness of Fit Indices for Various Models, Investment Model Study

Model	Chi-square	df	p	NFI	NNFI	CFI
Null model	672.72	15	.001	0.000	--	--
Initial model	36.84	4	.001	.945	.813	.950
✓ Revised model 1 [a]	11.20	3	.011	.983	.938	.988
Revised model 2 [b]	0.58	2	.750	.999	1.016	1.000

Note. N = 240. NFI = Normed fit index; NNFI = non-normed fit index; CFI = comparative fit index.

[a] Identical to the initial model, except that a path from Investment size to Satisfaction was added. [b] Identical to revised model 1, except that a path from Alternative value to Satisfaction was added.

The columns of Table 4.2 present the goodness of fit statistics referred to in this chapter:

- the chi-square statistic, along with degrees of freedom and *p* values associated with that test

- the normed fit index (NFI)

- the non-normed fit index (NNFI)

- the comparative fit index (CFI).

Table 4.2 presents goodness of fit indices for four models, although only the last three of these models will be of major interest to the reader. The first model, called the "null model" in Table 4.2, represents a hypothetical path model in which none of the variables are related to any of the other variables. This null model chi-square is useful because, in small samples, it may show that the null, uncorrelated-variables model fits the data as well as your theoretical model. If this is the case, your theoretical model receives little support. In large samples, the chi-square for the null model can be used as a baseline, and the chi-square values obtained for the other models can be compared against it. If your theoretical model achieves a large reduction in chi-square compared to the null model (while taking into account the degrees of freedom), then this theoretical model gains support.

Here, Table 4.2 reveals a null model chi-square of 672.72 which, at 15 degrees of freedom, indicates a very poor fit for the uncorrelated variables model ($p < .001$). Your initial and revised models achieve dramatic decreases in chi-square, compared to this null model (while taking into account the degrees of freedom). This provides some nominal support for your models.

Preparing Text

There is a great deal of variability in the way that path analyses are described in published research articles. To a large extent, the way the results section is written will depend on factors such as the hypotheses being tested and the number and nature of the models being compared. If you are just learning path analysis, you should review several published path analytic studies to learn about the different options that are available. The following represents just one approach that could be used to discuss the results of the present study. For the sake of completeness, the following example is somewhat lengthy; for many journals, you will have to describe your analyses and results in a more concise manner.

>   Path analysis was performed to test the theoretical
> model presented in Figure 4.16. All analyses were
> conducted using the SAS System's CALIS procedure. These
> analyses used the maximum likelihood method of parameter
> estimation, and all analyses were performed on the
> variance-covariance matrix.
>   Goodness of fit indices for the various models are
> presented in Table 4.2. The chi-square statistic included
> in this table provides a test of the null hypothesis that
> the reproduced covariance matrix has the specified model
> structure, i.e., that the model "fits the data."
>   Table 4.2 also provides three additional goodness of
> fit indices: the normed fit index, or NFI (Bentler &
> Bonett, 1980), the non-normed fit index, or NNFI (Bentler &
> Bonett, 1980), and the comparative fit index, or CFI
> (Bentler, 1989). The NFI may range in value from 0 to 1,
> where 0 represents the goodness of fit associated with a
> "null" model (one specifying that all variables are
> uncorrelated), and 1 represents the goodness of fit
> associated with a "saturated" model (a model with 0 degrees
> of freedom that perfectly reproduces the original
> covariance matrix). The NNFI and CFI are variations on the
> NFI that have been shown to be less biased in small samples

(Bentler, 1989). Values on the NFI, NNFI, and CFI over .9 indicate an acceptable fit between model and data.

The "initial" model of Table 4.2 is this study's theoretical model, as presented in Figure 4.16. Estimation of this model revealed a significant model chi-square value, χ^2 (4, $\underline{N}$ = 240) = 36.84, $\underline{p}$ < .001. Although values on the NFI and CFI exceeded .9, the NNFI value for this model was only .813, indicating that the fit between model and data could probably be substantially improved. We therefore rejected the original model, and attempted to identify modifications that would improve the model's fit.

First, the path coefficients were reviewed to see if any of the paths in the initial model should be deleted. The $\underline{t}$ values for all path coefficients proved to be statistically significant ($\underline{p}$ < .05), and all standardized path coefficients exceeded .32 in absolute magnitude, indicating that they were meaningful in size (Billings & Wroten, 1978). For these reasons, none of the existing paths were eliminated from the initial model.

The largest residual in the normalized residual matrix was for the relationship between the Satisfaction and Investment Size variables. In addition, a Lagrange multiplier test (Bentler, 1989) suggested that the model could be significantly improved by adding a path from Investment Size to Satisfaction. Adding such a path would be consistent with the prediction from cognitive dissonance theory (Festinger, 1957), that individuals often adjust their attitudes (i.e., become more satisfied) so that their attitudes will be consistent with their behaviors (i.e., the behavior of investing time and effort in a relationship). Because its addition could be justified on theoretical grounds, a path from Investment Size to Satisfaction was added to the initial model. The resulting model, called "revised model 1," was then re-estimated.

The goodness of fit indices for revised model 1 are also presented in Table 4.2. By comparing the chi-square statistic for the initial model to the chi-square statistic for revised model 1, it was possible to perform a chi-square difference test to determine whether the addition of the new path resulted in a significant improvement in the model's fit. This difference test was computed as 36.84 - 11.20 = 25.64. With $\underline{df}$ = 1, the chi-square difference statistic of 25.64 was significant ($\underline{p}$ < .001), indicating that revised model 1 provided a superior fit to the data.

Path coefficients for revised model 1 are presented in Figure 4.17. All coefficients were significant at p .05 or lower. The analysis revealed R^2 values of .63 for commitment and .59 for satisfaction.

Table 4.2 shows that the model chi-square for revised model 1 was statistically significant ($p < .02$). Under some circumstances, a significant model chi-square can suggest a poor model fit. On the other hand, the NFI, NNFI and CFI for this model all exceeded .9, indicative of an acceptable fit. Because of these inconsistencies, the results of the analysis of revised model 1 were reviewed to see if the model's fit could be further improved.

None of the normalized residuals from model 1 exceeded 2.00 in absolute magnitude, although the largest normalized residual was −1.67 for the Alternative Value-Satisfaction relationship. This same relationship also demonstrated a significant Lagrange multiplier test (10.44, $p < .001$). We therefore modified revised model 1 by adding a path from Alternative Value to Satisfaction. The resulting model was called revised model 2.

Goodness of fit indices in Table 4.2 show that revised model 2 provided a very good fit to the data. The model chi-square statistic was nonsignificant, χ^2 (2, $N = 240$) = 0.58, $p = .750$, and the NFI, NNFI, and CFI all exceeded .99. In addition, a chi-square difference test revealed that adding the path from Alternative Value to Satisfaction resulted in a significant improvement in the model's fit, χ^2 (1, $N = 240$) = 10.62, $p < 01$.

The standardized path coefficients for revised model 2 are presented in Figure 4.18. All of these coefficients are meaningful in absolute magnitude, and t tests revealed that all were significant at $p < .05$. The analysis revealed R^2 values of .64 for Commitment and .61 for Satisfaction.

Somewhere in the discussion section of the paper, you may want to evaluate the three models and justify acceptance of one of them as the study's final model. This would be difficult in the present study, because there are positive features to recommend both revised model 1 and revised model 2. However, assume that you have chosen the more conservative approach of making relatively few modifications to a model, and have selected revised model 1 as the study's final model. Part of the discussion section may proceed as follows:

Both revised model 1 and revised model 2 provided, at the very least, a minimally acceptable fit to the data. Revised model 2 provided the best fit, with a near-zero model chi-square value. In addition, revised model 2 demonstrated very high values--in excess of .99--on the NFI, NNFI, and CFI.

Nonetheless, it is argued here that revised model 1 should be tentatively accepted as the "final" model identified by this investigation. This conclusion rests on three arguments.

First, although revised model 1 demonstrated a significant model chi-square test, this statistic does not provide a valid test of model fit in most applied situations, and should be viewed more as a general goodness of fit index rather than as a statistical test (Joreskog & Sorbom, 1989). In support of revised model 1, the NNFI exceeded .9, and the NFI and CFI were close to 1, indicating a good-to-superior fit between model and data.

Second, revised model 1 has the advantage of incorporating only one modification to the initial model, whereas revised model 2 incorporates two changes. This makes model 2 less desirable, as MacCallum, Roznowski, & Necowitz, (1992) have advised that specification searches are less likely to lead to inaccurate models when investigators make only a relatively small number of modifications.

Finally, MacCallum et al. (1992) have also advised that changes be made only when they are theoretically meaningful. In our view, the path from Investment Size to Satisfaction in revised model 1 is clearly interpretable within the framework of cognitive dissonance theory; the path from Alternative Value to Satisfaction in revised model 2 is less interpretable, and hence more suspect.

Still, the revised model that has been "accepted" as the study's final model (model 1) is itself of admittedly questionable validity, since it results from data-driven modifications made to a rejected initial model, and is based on a single sample of only moderate size. It is therefore possible that the model will not generalize to other samples or to the population. Future path analytic

studies should be conducted to test the validity of this
model, preferably by comparing the three models
investigated in this study as a priori models, to determine
which provides the best fit to the data in new samples.

Example 2: Path Analysis of a Model Predicting Victim Reactions to Sexual Harassment

The next example comes from the field of women's studies. In recent years there has been increased interest in research dealing with sexual harassment in the work place, and some researchers have sought to identify the variables that determine whether women will report instances of sexual harassment (e.g., Brooks & Perot, 1991).

One approach to studying this phenomenon might involve a type of role-playing task. Women could be asked to read a brief scenario in which a fictional woman was being sexually harassed by her supervisor. The subjects would be asked to imagine how they would feel if they were the individual being harassed. They could then complete questionnaires that assess a number of their attitudes, beliefs, and intentions related to the harassment. The purpose of the study would be to identify the attitudes, beliefs, and other variables that influence the subject's intention to report the harassment to authorities.

Figure 4.19 displays a path model identifying the predicted determinants of a subject's intention to report sexual harassment (although some of the variables in the model were inspired by Brooks & Perot [1991], this model was constructed for purposes of illustration only, and should not be regarded as a serious theory). The model includes the following variables:

- the subject's **intention to report** the harassment to a higher-level manager at the organization. With this variable, higher scores indicate greater intention to report.

- the **expected outcomes** of reporting the harassment, where higher scores indicate stronger belief that reporting the harassment will result in positive results for the victim.

- **feminist ideology**, where higher scores indicate pro-feminist, egalitarian attitudes about sex roles.

- **seriousness of the offense**, where higher scores reveal stronger belief that the woman in the scenario experienced a serious form of harassment.

- **victim marketability**, where higher scores indicate stronger belief that the woman in the scenario could easily find another good job if she were to leave her current position.

- **victim's age**, in this case the actual age of the subject who is role-playing the victim.

- **normative expectations**, where higher scores reflect stronger belief that the victim's family, friends, and coworkers would support her if she reported the harassment.

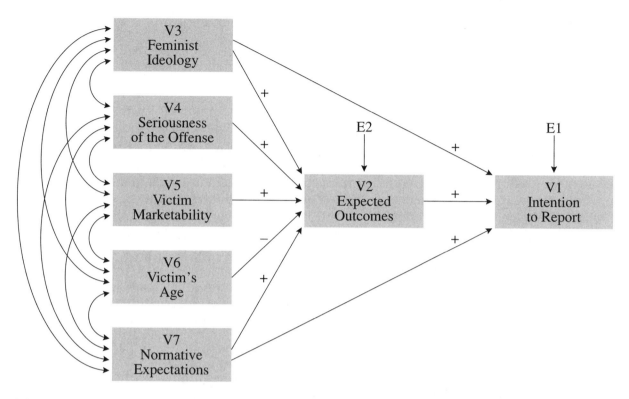

Figure 4.19: Model 1, The Predicted Determinants of the Intention to Report Sexual Harassment (This Model Predicts Both Direct and Indirect Effects of Feminist Ideology on Intention to Report)

The causal paths displayed in the figure are identified by "+" and "–" signs to indicate whether positive or negative causal relationships are predicted. The model includes two endogenous variables: intention to report (V1), and expected outcomes (V2). As a whole, the model makes the following predictions:

- There will be direct, positive causal paths from expected outcomes, feminist ideology, and normative expectations to intention to report.

- There will be direct, positive causal paths from feminist ideology, seriousness of the offense, victim marketability, and normative expectations to expected outcomes.

- there will be a direct, negative path from victim's age to expected outcomes.

Comparing Alternative Models

To make things more interesting, imagine that a controversy has erupted among scholars who study sexual harassment. The controversy focuses on the relationship between feminist ideology and intention to report, and scholars are divided into three schools of thought on this issue. The first group of scholars believes that feminist ideology is a very important determinant of the intention to report. They believe that feminist ideology has both direct and indirect effects on these intentions. The direct effect of feminist ideology is represented by the single path that runs from V3 to V1 in Figure 4.19. The first group of scholars believe that women who score high on feminist ideology will be more likely to report harassment because they will view it as a personal obligation as a feminist to take this action. Hence, there is a direct causal path between these two variables in the figure.

The first group of scholars believe that feminist ideology will also have an indirect effect on intentions through its effect on expected outcomes. They argue that, if a woman is a feminist, she will be more likely to believe that reporting harassment will lead to positive outcomes such as punishment for the perpetrator, increased consciousness about harassment in the work place, and so forth. From this perspective, feminist ideology has a positive effect on expected outcomes (represented by the path from V3 to V2), and expected outcomes, in turn, have a positive effect on intention to report (represented by the path from V2 to V1). In this way feminist ideology has an indirect effect on intentions. (Of course, any discussion of feminist ideology having an indirect effect on intention assumes that the relationship between expected outcomes and intentions is significant and of meaningful size; if this latter relationship were zero or near zero, then feminist ideology could not have an indirect effect on intentions through expected outcomes, no matter how strong the relationship between ideology and expected outcomes.)

The second group of scholars does not believe that feminist ideology is as important as the first group believes. The second group believes that feminist ideology has only indirect effects on intentions. The path model that represents the predictions of this group is presented as Figure 4.20.

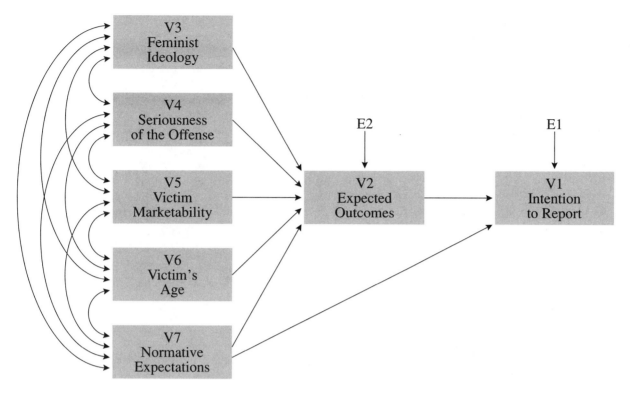

Figure 4.20: Model 2, Predicting Only Indirect Effects of Feminist Ideology on Intention to Report

The path model promoted by the second group is identical to that proposed by the first group, except that the direct path from feminist ideology to intentions has been deleted. Feminist ideology is now predicted to have only an indirect influence on intentions, through its effect on expected outcomes.

Finally, a third group of scholars is convinced that feminist ideology has absolutely no effect on the intention to report harassment. They believe that the intention to report is directly or indirectly determined by the other variables presented in the preceding model, but that feminist ideology does not influence these intentions in any way. The model that represents this last school of thought is presented in Figure 4.21.

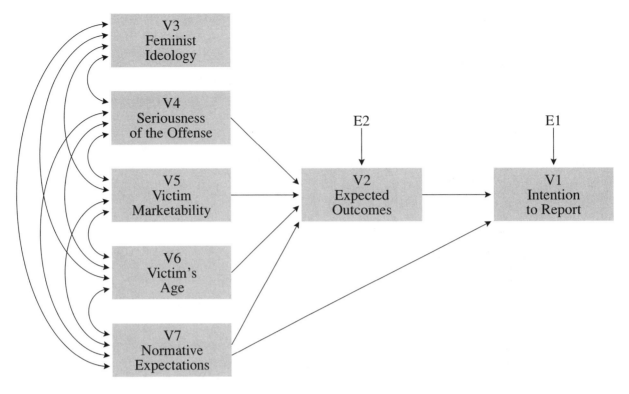

Figure 4.21: Model 3, Predicting No Effects of Feminist Ideology on Intention to Report

Notice that Figure 4.21 is identical to the Figure 4.20, except that the causal path from feminist ideology to expected outcomes has been deleted.

To the author's knowledge, there is no controversy of this nature in the field of women's studies, but if there were it would present an advantageous state of affairs for researchers performing path analysis, as it would present an opportunity to test alternative a priori models. Earlier, it was noted that it is generally safer to test alternative theory-based models and identify the one with the best fit to the data than it is to begin with a single model and make a number of data-driven modifications to achieve a better fit, as the latter approach is less likely to result in a model that will generalize (MacCallum et al., 1992).

Evaluating these three models will involve conducting three analyses. In the first, a PROC CALIS program will be written to test the first model, to be referred to here as the **direct and indirect effects model**. The program will then be modified to eliminate the direct path from feminist ideology to intentions, and the resulting model, to be called the **indirect effects model**, will then be re-estimated. If dropping this path does not result in a significant decrease in the fit between model and data, this will provide support for the indirect effects position. Finally, the program will be modified once more to eliminate the path from feminist ideology to expected outcomes, producing a **no effects model**. If dropping this path does not result in a significant decrease in model fit, this will provide support for the last model.

The SAS Program

The PROC CALIS program that would analyze the direct and indirect effects model is presented here (the data analyzed here are from Murphy, Walters, and Hatcher, 1993).

```
 1    DATA D1(TYPE=CORR) ;
 2      INPUT _TYPE_ $ _NAME_ $ V1-V7;
 3      LABEL
 4          V1 ='REPORT'
 5          V2 ='EXPECTED_OUTCOMES'
 6          V3 ='FEMINIST'
 7          V4 ='SERIOUSNESS'
 8          V5 ='MARKETABILITY'
 9          V6 ='AGE'
10          V7 ='NORMS' ;
11    CARDS;
12    N       .    202    202    202    202    202    202    202
13    STD     . 2.0355 1.4500 0.4393 2.1873 2.7433 4.0513 1.0552
14    CORR  V1  1.0000    .      .      .      .      .      .
15    CORR  V2   .4815  1.0000   .      .      .      .      .
16    CORR  V3  -.0306   .0014  1.0000  .      .      .      .
17    CORR  V4   .1458   .1683   .1148  1.0000  .      .      .
18    CORR  V5   .0479   .1939   .0128   .0599  1.0000  .      .
19    CORR  V6  -.0302  -.1165   .1479   .1061  -.0998  1.0000  .
20    CORR  V7   .3952   .3700   .0512   .2486   .1275   .0606  1.0000
21    ;
22    PROC CALIS   COVARIANCE CORR RESIDUAL MODIFICATION ;
23      LINEQS
24          V1 = PV1V2 V2 + PV1V3 V3 + PV1V7 V7                        + E1,
25          V2 = PV2V3 V3 + PV2V4 V4 + PV2V5 V5 + PV2V6 V6 + PV2V7 V7 + E2;
26      STD
27          E1 = VARE1,
28          E2 = VARE2,
29          V3 = VARV3,
30          V4 = VARV4,
31          V5 = VARV5,
32          V6 = VARV6,
33          V7 = VARV7;
34      COV
35          V3 V4 = CV3V4,
36          V3 V5 = CV3V5,
37          V3 V6 = CV3V6,
38          V3 V7 = CV2V7,
39          V4 V5 = CV4V5,
40          V4 V6 = CV4V6,
41          V4 V7 = CV4V7,
42          V5 V6 = CV5V6
43          V5 V7 = CV5V7,
44          V6 V7 = CV6V7;
45      VAR  V1 V2 V3 V4 V5 V6 V7 ;
46      RUN;
```

The indirect effects model (in which the direct path from feminist ideology to intention to report has been deleted) would be identical to the preceding program, except that V3 (feminist ideology) would be dropped as an independent variable in the LINEQS statement for V1 (intentions). The resulting LINEQS statements would be as follows:

```
23    LINEQS
24        V1 = PV1V2 V2                    + PV1V7 V7                              + E1,
25        V2 = PV2V3 V3 + PV2V4 V4 + PV2V5 V5 + PV2V6 V6 + PV2V7 V7 + E2;
```

Similarly, the no effects model (in which the path from ideology to expected outcomes has been deleted) would be identical to the preceding (modified) program, except that V3 (ideology) would be dropped as an independent variable in the LINEQS statement for V2 (outcomes). This is how the resulting LINEQS statements would appear:

```
23    LINEQS
24        V1 = PV1V2 V2                     + PV1V7 V7                             + E1,
25        V2 =                  PV2V4 V4 + PV2V5 V5 + PV2V6 V6 + PV2V7 V7 + E2;
```

All other aspects of the PROC CALIS program remain the same for the three analyses. In all three, V3 (feminist ideology) remains an exogenous variable that is allowed to covary with the other exogenous variables. The only aspect of the model that was modified involves its relationship with intentions and expected outcomes.

Results of the Analysis

Some of the results of the three analyses are presented in Table 4.3. Notice that, in contrast to Table 4.2, Table 4.3 presents the models in ascending order of complexity, with the null (uncorrelated variables) model at the bottom, and the model predicting both direct and indirect effects at the top. Either way of organizing this table is acceptable.

Table 4.3

Goodness of Fit Indices for Various Models, Sexual Harassment Study

Model	Chi-square	df	p	NFI	NNFI	CFI
Direct & indirect effects	1.20	3	.753	.991	1.107	1.000
Indirect effects	1.75	4	.782	.987	1.101	1.000
No effects	1.76	5	.882	.987	1.116	1.000
Null	137.89	21	--	0.000	--	--

Note. N = 202. NFI = Normed fit index; NNFI = non-normed fit index; CFI = comparative fit index.

The goodness of fit indices for the direct and indirect effects model seem to suggest a very good fit to the data. The model chi-square is very small, the *p* value very large, and the NFI, NNFI, and CFI values are as near to perfect as possible. So is the model acceptable?

Not quite. The *t* tests for path coefficients (under the heading "Manifest Variable Equations" from Output 4.18) reveal three nonsignificant parameters:

- the path coefficient for the effect of feminist ideology (V3) on intention to report (V1)

- the coefficient for the effect of feminist ideology on expected outcomes (V2)

- the coefficient for the effect of seriousness of the offense (V4) on expected outcomes.

As you can see, it is quite possible for goodness of fit indices to be high even when the model contains nonsignificant paths.

```
                          The SAS System                              9

            Covariance Structure Analysis: Maximum Likelihood Estimation

                      Manifest Variable Equations

V1       =      0.5441*V2    - 0.2046*V3    + 0.4901*V7    + 1.0000 E1
Std Err         0.0897 PV1V2   0.2754 PV1V3   0.1234 PV1V7
t Value         6.0651        -0.7427         3.9707
```

```
V2      = - 0.0282*V3    + 0.0604*V4    + 0.0698*V5    - 0.0473*V6
Std Err     0.2136 PV2V3   0.0440 PV2V4   0.0342 PV2V5   0.0233 PV2V6
t Value    -0.1321          1.3738         2.0450        -2.0325


          + 0.4658*V7    + 1.0000 E2
Std Err     0.0910 PV2V7
t Value     5.1166
```

                      Variances of Exogenous Variables

| | | | Standard | |
Variable	Parameter	Estimate	Error	t Value
V3	VARV3	0.192984	0.019203	10.050
V4	VARV4	4.784281	0.476054	10.050
V5	VARV5	7.525695	0.748835	10.050
V6	VARV6	16.413032	1.633158	10.050
V7	VARV7	1.113447	0.110792	10.050
E1	VARE1	2.948511	0.293388	10.050
E2	VARE2	1.720336	0.171180	10.050

                     Covariances among Exogenous Variables

| | | | | Standard | |
	Parameter		Estimate	Error	t Value
V4	V3	CV3V4	0.110309	0.068051	1.621
V5	V3	CV3V5	0.015426	0.084800	0.182
V5	V4	CV4V5	0.359425	0.422945	0.850
V6	V3	CV3V6	0.263223	0.126584	2.079
V6	V4	CV4V6	0.940195	0.626986	1.500
V6	V5	CV5V6	-1.109170	0.785858	-1.411
V7	V3	CV3V7	0.023734	0.032658	0.727
V7	V4	CV4V7	0.573778	0.167336	3.429
V7	V5	CV5V7	0.369078	0.205321	1.798
V7	V6	CV6V7	0.259061	0.301335	0.860

Output 4.18: Output Page 9, Analysis of Direct and Indirect Effects Model, Sexual Harassment Study

The problems with these three paths were also revealed in the Wald tests included in the SAS output. When a path model contains paths or covariances that may be dropped without causing a significant decrease in fit, these are often identified by the Wald test. The Wald tests for the current analysis appear in Output 4.19.

```
                    Stepwise Multivariate Wald Test
----------------------------------------------------------------------
                Cumulative Statistics            Univariate Increment
Parameter    Chi-Square    D.F.      Prob       Chi-Square      Prob
----------------------------------------------------------------------
PV2V3         0.017448       1      0.8949       0.017448      0.8949
CV3V5         0.050538       2      0.9750       0.033090      0.8557
CV3V7         0.553629       3      0.9070       0.503091      0.4781
PV1V3         1.105287       4      0.8934       0.551658      0.4576
CV6V7         1.682101       5      0.8911       0.576814      0.4476
CV4V5         2.392062       6      0.8803       0.709961      0.3995
PV2V4         4.262029       7      0.7492       1.869967      0.1715
CV4V6         6.241035       8      0.6203       1.979005      0.1595
CV3V4         7.837242       9      0.5506       1.596208      0.2064
CV5V6        10.628933      10      0.3871       2.791691      0.0948
CV5V7        13.625284      11      0.2544       2.996351      0.0835
CV3V6        16.716227      12      0.1606       3.090943      0.0787
PV2V6        20.502565      13      0.0834       3.786338      0.0517
```

Output 4.19: Output Page 13, Wald Test Results, Sexual Harassment Study

On the right side of this output, under the heading "Univariate Increment," the Wald test estimates the change in model chi-square that would result from deleting a given parameter from the model. For example, the first line of the output provides Wald test estimates for deleting the parameter named "PV2V3". This, of course, is the path coefficient for the path from V3 (feminist ideology) to V2 (expected outcomes). The Wald test estimates that the model chi-square would change by only 0.017448 if PV2V3 were deleted. This is a very small change in chi-square, and suggests that you may safely drop this unnecessary path without significantly hurting the model's overall fit. In general, the first parameter listed in the Wald test table is the parameter that could be dropped with the least harm to the model's fit. The "Cumulative Statistics" on the left side of the table estimate how much chi-square would change if entire groups of variables were dropped from the model.

Remember that your study began by comparing the predictions of three groups of scholars who disagreed regarding the importance of feminist ideology in your model. The school proposing a direct and indirect effects model basically predicted that the model must include the path from ideology to intentions, that dropping this path would seriously harm the model's ability to account for relationships in the data. You can test this prediction directly by performing a chi-square difference test on chi-square values from Table 4.3. The model chi-square for the direct and indirect effects model was 1.20, while the chi-square for the indirect effects model was 1.75. The chi-square difference was therefore

$$1.75 - 1.20 = 0.55$$

The degrees of freedom for this test is equal to the difference between the degrees of freedom for the two models, or

```
4-3 = 1
```

The critical value of chi-square ($p < .05$) is 3.84, and your obtained difference chi-square was far below that at 0.55. Therefore, you may conclude that there is not a significant difference between the fit of the two models. In other words, deleting the path from feminist ideology to intentions did not significantly hurt the model's fit to the data. Because the model without the path is the more parsimonious (and therefore the more desirable) of the two, you tentatively accept it, and reject the direct and indirect effects model.

But your analysis is not complete. Is it possible to also drop the path from ideology to expected outcomes without significantly hurting the model's fit? To determine this, you compute a second chi-square difference test, this one comparing the indirect effects model to the no effects model.

Using the appropriate chi-square values from Table 4.3, this chi-square difference is calculated as

```
1.76 - 1.75 = 0.01
```

This obtained value is also far below the critical chi-square value of 3.84, so you know that you may drop this path without causing a significant decrease in the model's fit. Because the no effects model is the more parsimonious of the two, you tentatively accept it over the indirect effects model.

In short, if you are to believe these results, feminist ideology appears to play no role whatsoever in influencing a victim's intention to report harassment; It neither affects these intentions directly, nor does it affect them indirectly by first affecting expected outcomes. An interesting feature of these results is the fact that the model still provides a good fit to the data even after deleting these two paths. Table 4.3 shows that the model chi-square for the no effects model is still quite small and nonsignificant, and the NFI, NNFI, and CFI are still very high.

Unfortunately for you, however, the model is still not quite perfect, as the output from the analysis of the no effects model (not reproduced here) shows that the path from seriousness of the offense to expected outcomes is still nonsignificant. If this were an actual investigation, your specification search would most likely continue at this point.

Important: Compare only nested models. It is hoped that this example has been useful for showing how the chi-square difference test can be used to make comparisons between competing a priori models. However, remember that these difference tests can only be used to compare **nested models**. For example, Model B is said to be nested within Model A if it is identical to that model, with the exception that some of the paths in Model A have been deleted in Model B.

In this sense, you can see that the indirect effects model was nested within the direct and indirect effects model, and that the no effects model was nested within both of the other models. To date, there is no widely accepted method for comparing models that are not nested (Bentler & Chou, 1987).

Conclusion: Learning More about Path Analysis

The best way to learn about path analysis is to *do it*. Find a good text on path analysis, or perhaps some published research articles reporting path-analytic studies, and replicate the analyses that they contain. But be forewarned: authors do not always indicate whether they analyzed the correlation matrix or the covariance matrix, so it may be necessary to perform the analysis both ways in order to obtain results that match the published results.

Now that you have had a concise introduction to the analysis of simple recursive models, you should be ready for texts that provide a more in-depth treatment of path-analytic procedures. This is particularly important if you want to test more complex models such as nonrecursive models with reciprocal causation or feedback loops, or time-series designs with repeated measures on some variables. Introductions to these and other issues are provided by Asher (1988), Berry (1985), Bollen (1989), Duncan (1975), James et al. (1982), Kenny (1979), and Long (1983a, 1983b).

Finally, the following two chapters build on the current chapter by introducing causal models with latent variables. Analyzing path models with latent variables has many important advantages over the analysis of path models with manifest variables, and the increasing availability of software capable of analyzing these models (such as PROC CALIS) represents one of the most exciting recent developments in the social sciences. Those interested in learning about this powerful new research procedure are encouraged to read on.

References

Anderson, J.C. & Gerbing, D.W. (1988). Structural equation modeling in practice: A review and recommended two-step approach. *Psychological Bulletin, 103,* 411-423.

Asher, H.B. (1988). *Causal modeling, second edition.* Beverly Hills, CA: Sage.

Bentler, P.M. (1989). *EQS structural equations program manual.* Los Angeles: BMDP Statistical Software.

Bentler, P.M. & Bonett, D.G. (1980). Significance tests and goodness-of-fit in the analysis of covariance structures. *Psychological Bulletin, 88,* 588-606.

Bentler, P.M. & Chou, C.P. (1987). Practical issues in structural modeling. *Sociological Methods & Research, 16,* 78-117.

Berry, W. D. (1985). *Nonrecursive causal models.* Beverly Hills, CA: Sage.

Billings, R.S. & Wroten, S.P. (1978). Use of path analysis in industrial/organizational psychology: Criticisms and suggestions. *Journal of Applied Psychology, 63,* 677-688.

Bollen, K.A. (1989). *Structural equations with latent variables.* New York: John Wiley & Sons.

Brooks, L. & Perot, A.R. (1991). Reporting sexual harassment: exploring a predictive model. *Psychology of Women Quarterly, 15,* 31-47.

Duncan, O.D. (1975). *Introduction to structural equation models.* New York: Academic.

Festinger, L. (1957). *A theory of cognitive dissonance.* Stanford, CA: Stanford University Press.

James, L.R., Mulaik, S.A., & Brett, L.M. (1982). *Causal analysis: Assumptions, models, and data.* Beverly Hills, CA: Sage.

Joreskog, K.G. & Sorbom, D. (1989). *LISREL 7: A guide to the program and applications.* Chicago: SPSS Inc.

Kenny, D.A. (1979). *Correlation and causality.* New York: John Wiley & Sons.

Long, J. S. (1983a). *Confirmatory factor analysis: A preface to LISREL.* Beverly Hills, CA: Sage.

Long, J. S. (1983b). *Covariance structure models: An introduction to LISREL.* Beverly Hills, CA: Sage.

MacCallum, R.C. (1986). Specification searches in covariance structure modeling. *Psychological Bulletin, 100,* 107-120.

MacCallum, R.C., Roznowski, M., & Necowitz, L.B. (1992). Model modifications in covariance structure analysis: The problem of capitalization on chance. *Psychological Bulletin, 111,* 490-504.

Marsh, H.W., Balla, J.R., & McDonald, R.P. (1988). Goodness-of-fit indexes in confirmatory factor analysis: The effect of sample size. *Psychological Bulletin, 103,* 391-410.

Mulaik, S.A., James, L.R., Alstine, J.V., Bennett, N., Lind, S., & Stilwell, D. (1989). Evaluation of goodness-of-fit indices for structural equation models. *Psychological Bulletin, 105,* 430-445.

Murphy, L., Walters, K., & Hatcher, L. (1993). *Preferred strategies for dealing with sexual harassment: A role-playing experiment.* Paper presented at the Carolinas Psychology Conference, North Carolina State University, Raleigh, NC.

Pedhazur, E. (1982). *Multiple regression in behavioral research.* New York: Holt.

Rusbult, C.E. (1980). Commitment and satisfaction in romantic associations: A test of the investment model. *Journal of Experimental Social Psychology, 16,* 172-186.

SAS Institute Inc. (1989). *SAS/STAT users guide, version 6, fourth edition, volume 1.* Cary, NC: SAS Institute Inc.

Sorbom, D. (1989). Model modification. *Psychometrika, 54,* 371-384.

Chapter 5

DEVELOPING MEASUREMENT MODELS WITH CONFIRMATORY FACTOR ANALYSIS

Overview. This chapter introduces a two-step process for testing causal models with latent variables. The current chapter (Chapter 5) shows how to develop acceptable measurement models with confirmatory factor analysis; the following chapter (Chapter 6) shows how to modify these measurement models in order to perform path analysis with latent variables. This chapter introduces basic terminology in latent-variable analyses, and describes the necessary conditions for performing these analyses. It shows how to prepare the program figure for a confirmatory factor analysis, how to translate this figure into the corresponding CALIS program, how to interpret the output, and how to use this output to develop a measurement model that provides a better fit to the data. It describes a number of procedures for assessing the reliability, convergent validity, and discriminant validity of the model's constructs and indicators.

Introduction: A Two-Step Approach to
Path Analysis with Latent Variables

One of the most important recent advances in social science research has been the development of software capable of testing causal models with latent (unobserved) variables. These models have been referred to as structural equation models, covariance structure models, latent-variable models, and causal models with unmeasured variables. They are perhaps most frequently referred to as "LISREL-type" models, as many people associate them with the LISREL program (for LInear Structural RELations). LISREL was the first widely available software that made possible the analysis of causal models with latent variables (Joreskog & Sorbom, 1989).

This chapter and the following chapter (Chapter 6) show how the SAS System's PROC CALIS can be used to test these latent-variable models. These chapters follow a two-step approach recommended by Anderson and Gerbing (1988). With this approach, the first step involves using confirmatory factor analysis to develop an acceptable measurement model. When you test a measurement model, you look for evidence that your indicator variables really are measuring the underlying constructs of interest, and that your measurement model demonstrates an acceptable fit to the data. This measurement model does not specify any causal relationships between the latent constructs of interest; at this stage of the analysis, each latent variable is allowed to correlate freely with every other latent variable. The current chapter focuses exclusively on how to estimate measurement models, how to assess their reliability and validity, and how to modify them (when necessary) to achieve a better fit.

Chapter 6, "Path Analysis with Latent Variables," then builds on this foundation by showing how these measurement models can be modified so that they predict specific causal relationships between the latent variables. When you test these modified measurement models, you are performing path analysis with latent variables. Among other things, performing this type of path analysis allows you to test hypotheses that certain latent constructs have causal effects on other latent constructs.

You should therefore view Chapters 5 and 6 as a two-part introduction to causal models with latent variables. Chapter 5 shows how to develop adequate measurement models, and Chapter 6 shows how to test the (theoretical) causal models of interest. Even if you are interested only in the topic of path analysis with latent variables (from Chapter 6), you will probably still need to read the current chapter first to obtain the necessary background information (unless, of course, you are already familiar with PROC CALIS and latent-variable models).

All of the structural models discussed in this chapter and the next are **recursive** models. This means that none of the variables that constitute the structural portion of the models will be involved in feedback loops or reciprocal causation; the structural models discussed here will demonstrate only unidirectional causal flow. Nonetheless, once the basic principles discussed here have been learned, you will be well prepared to move on to more advanced texts that discuss nonrecursive models. A list of helpful references is provided at the end of Chapter 6.

Important: This chapter builds on material presented in Chapter 4, "Path Analysis with Manifest Variables." That chapter introduced basic terminology and concepts, and showed how to write programs for the CALIS procedure. With only a few exceptions, most of that introductory material will not be repeated here. It is assumed that you have completed the preceding chapter before beginning this one.

A Model of the Determinants of Work Performance

This chapter begins with an illustration of a theoretical path model that includes only manifest variables (similar to the path models discussed in the preceding chapter). It will then show how a path model with latent variables differs from the manifest-variable case.

Figure 5.1 provides a model predicting the causal relationships between a number of variables related to work performance. The model includes two endogenous variables: Work Performance, which is predicted to be directly determined by Intelligence, Motivation, and Supervisory Support, and Motivation, which is said to be directly determined by Work Place Norms and Supervisory Support. The model includes three exogenous variables (Intelligence, Work Place Norms, and Supervisory Support) which are expected to covary.

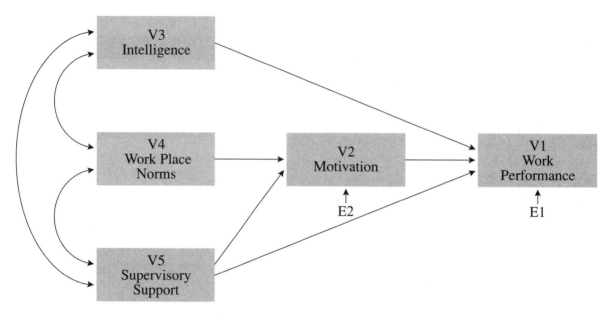

Figure 5.1: A Causal Model with Manifest Variables

The Manifest Variable Model

This figure displays the manifest variable model that was presented in the preceding chapter. You will remember that **manifest variables** are variables that are directly observed, such as scores on an intelligence test or on a motivation scale. Manifest variables are sometimes referred to as **observed variables**, **measured variables**, or **indicator variables**. With the type of path analysis discussed in the preceding chapter, all variables in the path model were manifest variables.

The Latent Variable Model

Figure 5.2, on the other hand, presents a path-analytic model with latent variables. In the figure, the latent variables are represented by ovals, and the manifest variables by rectangles. **Latent variables** are sometimes referred to as **unobserved variables**, **unmeasured variables**, or **latent factors**. Notice that the latent variables in the figure are identified by short names such as F1,

F2, etc., because the F stands for latent Factor. A latent variable is a **hypothetical construct**: a variable that cannot be directly observed. The existence of a latent variable can only be inferred by the way that it influences certain manifest variables that can be directly observed (more on this shortly).

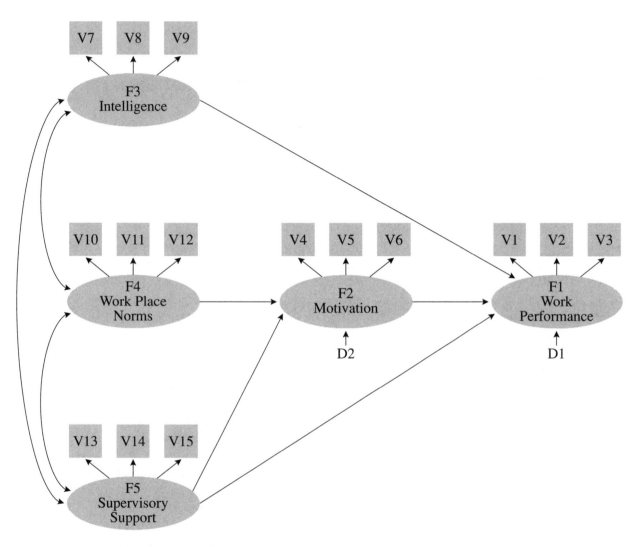

Figure 5.2: A Causal Model with Latent Variables

The path connecting the latent variables (ovals) in Figure 5.2 predicts exactly the same causal relationships between latent variables that were predicted to exist between the observed variables in the previous manifest variable model; Work Performance is directly determined by Intelligence, Motivation, and Supervisory Support, while motivation is directly determined by Work Place Norms and Supervisory Support.

In the model, Work Performance is also influenced by a disturbance term, D1 (the "D" stands for Disturbance, and the "1" corresponds to the "1" in F1). A **disturbance term** for a latent variable has the same interpretation as a residual term, or error term, for a manifest variable (from the preceding chapter). A disturbance term represents "causal effects on the endogenous variables due to such things as omitted causal variables, random shocks, and misspecifications of equations" (James, Mulaik, & Brett, 1982, p. 163). For example, you would expect a large disturbance term for the work performance construct if other constructs that have important effects on work performance have been left out of your model. If you have left out only those causal variables that have a trivial effect on work performance, the disturbance term for work performance should be relatively small.

Basic Concepts in Latent-Variable Analyses

Latent Variables versus Manifest Variables

The system presented in Figure 5.2 has been referred to as a latent variable model, and it is important to understand the difference between latent variables and manifest variables. Earlier, a latent variable was described as a hypothetical construct that is not directly observed, but whose existence is inferred from the way that it influences manifest variables.

For example, consider the latent variable Intelligence (F3) in Figure 5.2. The arrows in the figure suggest that this latent factor is expected to affect three manifest variables, labelled here as V7, V8, and V9. Assume that these three manifest variables are three different tests of intelligence. For example, V7 may be subjects' scores on the Wechsler Adult Intelligence Scale, V8 may be scores on the Stanford Binet intelligence test, and V9 may be scores on some other test of intelligence. If there really is some underlying construct that could reasonably be labelled "intelligence" (and if these three tests are valid), you would expect to see certain results when your data are analyzed using PROC CALIS. For example, the path coefficients for the path going from F3 (Intelligence) to V7, V8, and V9 should be relatively large and statistically significant (later, you will see that these path coefficients are usually referred to as factor loadings). These findings, along with others, would reassure you that there really is an underlying intelligence variable, and your manifest variables really are measuring it.

Notice that the path model in Figure 5.2 identifies five different latent F variables, and also indicates which indicator (manifest) variables are expected to measure each. Variables V1 through V3 measure work performance, V4 through V6 measure motivation, and so forth. It is important to remember that the indicator variables (the variables represented by rectangles) are the only variables that you actually gather during the study. Although the hypothetical constructs (the variables represented by ovals) are the variables that you are most interested in, they are never measured directly.

Choosing Indicator Variables

A wide variety of variables may be used as indicators, as long as they are assessed on an interval-ratio-scale, and meet certain other conditions to be discussed later (in the section titled "Necessary Conditions for Confirmatory Factor Analysis and Path Analysis with Latent Variables"). For example, in the present study it would be possible to use objective measures as indicators for some of the latent variables. In measuring the latent variable Work Performance, the indicator V1 could be the units produced by the subjects per hour, V2 could be the quality of the units produced by the subjects, and V3 could be the number of times per month each subject exceeds his or her production quota.

Subjective measures may also be used as indicator variables. In measuring motivation, V4 could be the subjects' own responses to a self-report questionnaire assessing work motivation, V5 could be ratings of the subjects' level of motivation provided by their direct supervisors, and V6 could be ratings of the subjects' motivation provided by coworkers.

In designing a study, great care should be used in deciding which indicator variables will measure each latent variable. Ideally, the indicators already should have been shown to be reliable and valid through previous research. In many studies, indicator variables are composite scores on standardized tests such as the Wechsler Adult Intelligence Scale, or on carefully developed scales that assess attitudes or beliefs such as the Job Descriptive Index, a measure of job satisfaction (Smith, Kendall, & Hulin, 1969). Using previously developed instruments is usually the preferred approach, as these instruments will generally have known psychometric characteristics.

In some studies, however, individual items on a questionnaire may serve as indicator variables. This is generally less desirable, as individual items tend to display lower levels of reliability than composite scores on carefully developed tests and scales. However, given the constraints commonly encountered in research (e.g., well developed measures of your constructs do not exist; your subjects don't have the time to take the Wechsler, the Stanford Binet, plus thirteen other instruments), many researchers in the social sciences will find themselves using individual items on questionnaires as indictor variables.

Regardless of what type of indicator variable you use, remember that you will be able to perform meaningful tests of your causal model only if these indicators show certain empirical properties. Specifically, it is essential that a group of indicators chosen to measure the same latent construct show a high level of **convergent validity**. They must all clearly be measuring the same underlying construct (in practical terms, this means that the indicators should be moderately or strongly correlated with one another). What is more, groups of variables that are intended to measure different latent constructs should display **discriminant validity**. A group of indicators intended to assess one latent variable (say, F1) should not at the same time be measuring a different latent variable (say, F2). In practical terms, this means that if V1 through V3 are measuring F1, and V4 through V6 are measuring F2, V1 through V3 should not show strong correlations with V4 through V6.

Later, you will learn how to assess the convergent and discriminant validity of your measurement model in a systematic way. However, these procedures will bring only disappointment if you have been careless in the initial selection of your indicator variables.

The Confirmatory Factor Analytic Approach

The procedures to be discussed in this chapter represent a special type of confirmatory factor analysis (CFA). These procedures differ in some important ways from those discussed in the earlier chapter on exploratory factor analysis. With exploratory factor analysis, you are often unsure of the number of factors being measured, and use the results of the analysis to help solve the number-of-factors problem. With the procedures to be discussed here, however, you must not only have a good idea regarding the number of factors (latent variables) being assessed, but you must also know which manifest variables load on which factors.

A more important distinction involves the relationship between the latent factors. With exploratory procedures, the latent variables are generally all allowed to be correlated (as with oblique solutions), or are forced to be uncorrelated (as with orthogonal solutions). With the procedures covered here and in Chapter 6, however, you are allowed to test hypotheses that certain factors have causal effects on other factors. It is this feature that makes PROC CALIS such a powerful tool for testing hypotheses in social science research.

The Measurement Model versus the Structural Model

You should think of the model presented in Figure 5.2 as actually consisting of two components. The measurement portion of the model, generally referred to as the **measurement model**, describes the relationships between the latent factors and their indicator variables. The measurement model is said to provide a good fit to the data if V1 through V3 are actually doing a good job of measuring work performance (F1), if V4 through V6 are actually doing a good job of measuring motivation (F2), and so forth. In this chapter, when manifest variables are used as measures of latent factors, they will be referred to as **indicator variables** or **measurement variables** (these terms are necessary as later it will be shown that manifest variables are not always indicator variables).

On the other hand, the structural portion of the model (usually referred to as the **structural model** or the **causal model**) describes the predicted causal relationships between the constructs of central theoretical interest. In Figure 5.2, the structural model consists of F1, F2, F3, F4, and F5, as well as the paths that connect them. In this chapter, the variables that constitute this structural model will be called **structural variables** (this term is necessary as it will later be shown that structural variables are not always latent factors; sometimes they are also manifest variables).

Standard versus Nonstandard Models

The model displayed in Figure 5.2 is sometimes called a multiple-indicator model, because multiple manifest variables are used as indicators of each latent factor. The term **standard model** is used here to refer to systems in which all variables constituting the structural portion of the model are latent factors with multiple indicators.

Unfortunately, it is not always possible to obtain multiple measures for each of the constructs included in the structural model. For example, consider the model presented in Figure 5.3. Here, the construct of work performance is no longer a latent factor with multiple indicators. Instead, a single manifest variable, V1, is used to assess work performance (perhaps you had access to only a single measure of "units produced per hour" in conducting the study). Similarly, the construct of supervisory support is also measured by a single indicator. The remaining constructs of motivation, intelligence, and work place norms are all latent factors with multiple indicators. This type of causal model, in which some structural variables are single indicator variables and other structural variables are latent factors with multiple indicators, will be referred to here as a **nonstandard model** (Bentler, 1989).

The current chapter focuses on the analysis of standard models only. However, Chapter 6 shows how to analyze both standard and nonstandard models.

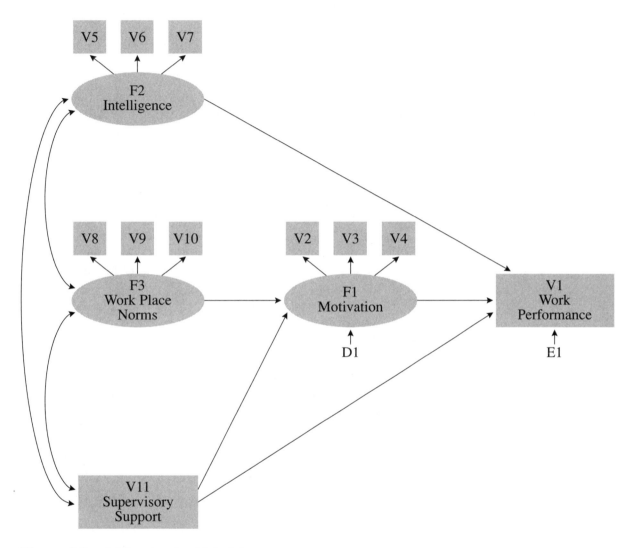

Figure 5.3: A Nonstandard Model

Advantages of Path Analysis with Latent Variables

Path analysis with latent variables has at least two important advantages over path analysis with only manifest variables. First, as was mentioned earlier, the latent-variable approach allows researchers to assess the convergent and discriminant validity of their measures. If the proposed measurement model fares well with regard to the tests to be discussed here, it obtains support for the construct validity of its manifest variables (James et al., 1982). This provides evidence that you really are studying the hypothetical constructs of interest. This is important, because many studies published in the social sciences offer no evidence concerning the construct validity of their variables.

Second, latent-variable analyses provide the opportunity to work with perfectly reliable causes and effects within the structural model. In the preceding chapter, it was mentioned that path analysis with manifest variables assumes that all variables are measured without error. This is to say the manifest variables are assumed to be perfectly reliable indicators of the underlying constructs they are intended to measure. Needless to say, this assumption is often not justified with the types of measures used in the social sciences.

For example, assume that you were to perform the work performance study described here using the standard single-indicator path analytic approach described in the last chapter. Further assume that you use full-scale IQ scores on the Wechsler Adult Intelligence Scale (WAIS) as your single measure of the hypothetical construct of intelligence. Although the WAIS is a carefully developed and relatively reliable instrument, its reliability is not perfect. When the WAIS is administered to a group of people, you assume that their scores are going to be influenced, in large part, by the underlying construct of intelligence. Unfortunately, you also assume that their scores on the IQ test will not be a perfect representation of their underlying level of intelligence, because some of the variability in scores will be due to errors of measurement (either random or systematic). This creates a problem for you because you do not really want to study the relationship between WAIS full-scale IQ scores and work performance; you want to study the relationship between the underlying construct of *intelligence* and work performance. And because the IQ scores provide an imperfect measure of intelligence, the path coefficients that you obtain from a single-indicator path analysis will likely be biased (see, for example, James et al. [1982] and Netemeyer, Johnston, and Burton [1990]).

Path analysis with latent variables circumvents this difficulty by excluding the unwanted part of the manifest variables (the error variance) from the definition of the latent factors, and modeling this unwanted part separately. This is illustrated in Figure 5.4, in which residual terms (indicated by the symbol E for Error term) have been added to the model. For illustrative purposes, consider the manifest variables V7 through V9, and assume that these variables are three different tests of intelligence. The paths leading from the latent construct F3 (Intelligence) to V7, V8, and V9 represent the prediction that variability in scores on these three tests will be influenced, in part, by the hypothetical construct of intelligence. However, notice that a causal arrow also points from the residual term E7 to V7. This represents the assumption that variability in V7 will also be influenced by factors in addition to intelligence such as sources of random and systematic error. This is that component of V7 that is not shared in common with V8 and V9, and so it is modeled separately. In the same way, E8 represents errors of measurement for V8, and E9 represents errors of measurement for V9.

As a result, the latent variable F3 consists only of variance that is shared in common between V7, V8, and V9. Because the error components have been modeled separately, the latent variable F3 is perfectly reliable, and is therefore well suited for path-analytic procedures.

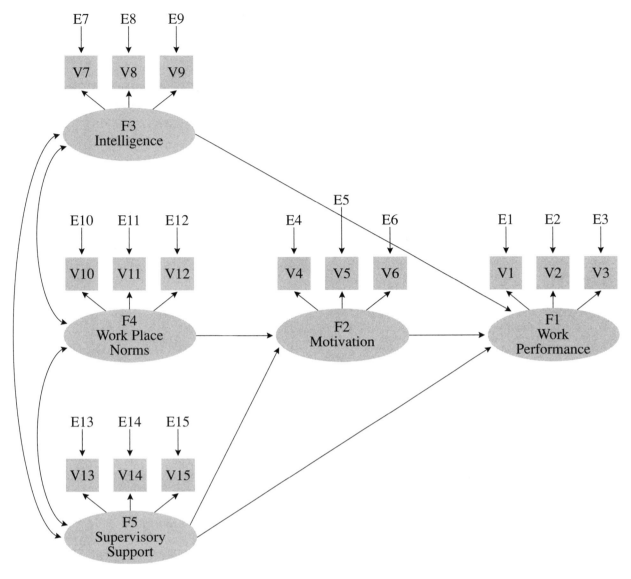

Figure 5.4: A Path Diagram with Residual Terms for Manifest Variables Identified

Necessary Conditions for Confirmatory Factor Analysis and Path Analysis with Latent Variables

The preceding chapter presented a number of requirements that must be met in order to perform path analysis with manifest variables, and many of those conditions apply to latent-variable analyses as well. These are briefly summarized here (most of the details explained in the preceding chapter will not be repeated here):

1. **Interval- or ratio-level measurement for all indicator variables.**

2. **Minimal number of values**. Indicator variables should be continuous and should assume a minimum of four values.

3. **Normally distributed data**.

4. **Linear and additive relationships.** This applies only to the types of analyses discussed in this text. More advanced texts show how to test models with nonlinear relationships, and CALIS is capable of testing these models as well.

5. **Absence of multicollinearity**.

6. **Inclusion, within the model, of all nontrivial causal variables.**

7. **Overidentified model.**

8. **Minimal number of observations.** For the analyses discussed here, a minimally acceptable number of observations would be the larger of 150 observations or 5 observations per parameter to be estimated. Larger samples are always preferable, and if many model modifications are to be made, substantially larger samples are required to arrive at a model that will successfully generalize to other samples (see MacCallum, Roznowski, and Necowitz [1992]).

In addition to the preceding, the analysis of latent-variable models also requires that the following conditions be met:

9. **At least three indicator variables per latent factor.** Technically, a latent factor may be assessed with just two indicators under certain conditions. However, models with only two indicator variables per factor often exhibit problems with identification and convergence, so it is recommended that each latent variable be assessed with at least three indicators (Anderson & Gerbing, 1988; Bentler & Chou, 1987; Lomax, 1982). In practice, researchers are well advised to have at least four or five indicators for each latent factor, as it is often necessary to drop some of the indicators in order to arrive at a well fitting measurement model.

10. **A maximum of 20-30 indicator variables.** One of the limitations of this model involves the maximum number of indicator variables that can be effectively studied. Bentler and Chou (1987) caution that it is easy to become too grandiose when developing structural models, and advise that researchers who lack a great deal of knowledge about the variables of interest should work only with relatively small data sets of perhaps 20 indicator variables or less. Larger data sets often result in large chi-square values and an inability to fit your model to the data. Note that this limitation also affects the maximum number of latent factors that may be effectively studied. Researchers who play it safe by measuring each latent variable with 4 indicators each will be limited to investigating a total of just 5 latent factors (assuming that the study includes just 20 manifest variables).

Having listed the necessary conditions for the analysis, it is important to once again emphasize that the procedures discussed in this chapter are **confirmatory** procedures. They are most appropriate for situations in which you have a fairly good understanding of the phenomena under investigation, and are poorly suited for the early stages of a program of research.

For example, assume that you wish to test the model of work performance presented in Figure 5.2. Ideally, you have begun your program of research with a thorough review of the literature pertaining to all constructs in the model. You either located previously developed instruments to measure these constructs, or, if necessary, you developed your own. You administered all instruments to subjects in a pilot study and performed exploratory factor analyses to discover the number of factors assessed, how the factors intercorrelate, and how the variables load on the factors. In the course of doing this, you likely discovered that some of your instruments did not measure the factors as expected, and had to be replaced.

Through this series of exploratory studies, you eventually arrived at a set of acceptable measures and a measurement model that provided a good fit to the data. Having done all of these things, you are now ready to administer the instruments to a new sample, and perform the confirmatory analyses to be covered here.

The preceding emphasizes the importance of developing a reasonably good measurement model before obtaining data from additional subjects for confirmatory analyses. Many researchers omit this crucial step, and learn too late that they cannot test the structural model of interest because their measurement model provides a poor fit to the data. In this situation, some researchers may be tempted to perform exploratory factor analyses on their data to discover the factor structure, revise their model to reflect this structure, and then perform confirmatory analyses on the same data set to test the revised theoretical model. However, performing both exploratory and confirmatory analyses on the same data set is likely to capitalize on chance characteristics of the sample, and may lead to a final model that will not generalize to other samples or to the population (for more details on this issue, review the section on "Modifying the Model" from the preceding chapter).

Example: The Investment Model

In previous chapters, this text has made frequent references to a theory of relationship commitment called the investment model (Rusbult, 1980). In the preceding chapter, it was noted that one possible interpretation of the investment model predicts that **commitment** to a relationship (e.g., the intention to remain in the relationship) is determined by

- satisfaction with the relationship

- the size of personal investments (e.g., time, energy) put into the relationship

- the attractiveness of one's alternatives to the relationship (e.g., the attractiveness of other potential partners).

Satisfaction, in turn, is expected to be determined by the rewards experienced in the relationship (e.g., the good things associated with it) as well as the costs experienced (e.g., hardships, unpleasant things).

The Theoretical Model

Figure 5.5 presents a latent-variable path model that illustrates these theoretical relationships. The structural portion of the model consists of the ovals (the latent variables) and the paths that connect them. You can see that Commitment (F1) is predicted to be directly determined by Satisfaction (F2), Investments (F5), and Alternatives (F6), while Satisfaction (F2) is expected to be determined by Rewards (F3) and Costs (F4).

The measurement portion of the model consists of the causal arrows that go from the latent variables to the manifest variables that measure them (the manifest variables are represented by rectangles). For example, manifest indicators V1, V2, V3, and V4 are predicted to measure F1 (Commitment), indicators V5, V6, and V7 are predicted to measure F2 (Satisfaction), and so forth.

Ideally, each of the indicator variables should be a different composite scale of known reliability and validity. For example, under ideal circumstances, variables V1 through V4 would be four different composite scales, each consisting of multiple items, and all designed to measure relationship commitment. A less desirable approach would be to have V1 through V4 simply be four different questions on the same scale. The latter is less desirable because individual items tend to be less reliable than composites. Nonetheless, individual items are often used as manifest variables in latent-variable analyses, and researchers who use this method are advised, at the very least, to use a response format with the items that includes a relatively large number of response categories, perhaps seven or nine.

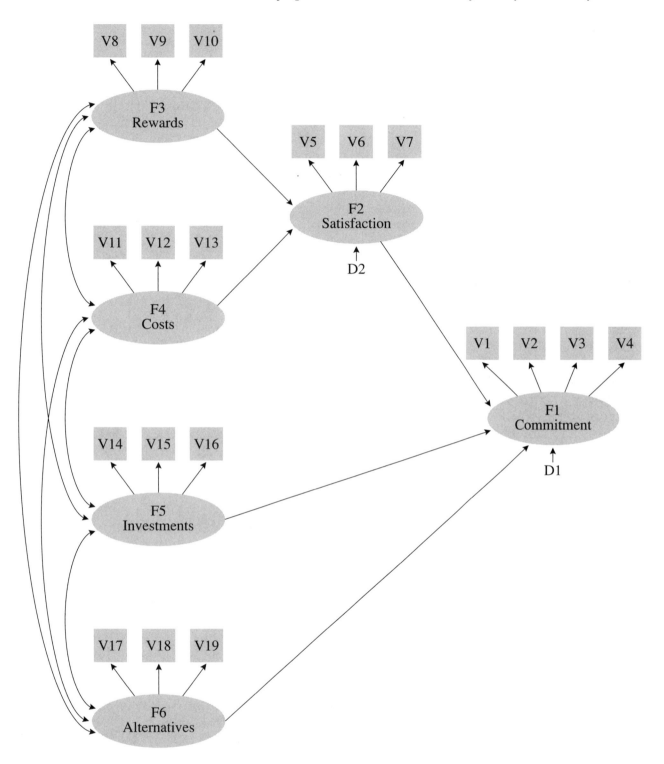

Figure 5.5: The Theoretical Model to Be Tested

Research Method and Overview of the Analysis

For purposes of illustration, assume that you have developed a 19-item instrument to assess the six constructs constituting the investment model. All questionnaire items used a 9-point response format. Items 1 through 4 assessed subject commitment to their current relationship, items 5 through 7 assessed satisfaction, and so forth. The questionnaire was administered to 247 subjects involved in romantic relationships, and usable responses were obtained from 240 of these. (It is again emphasized that the results presented here are fictitious, and should not be viewed as legitimate tests of the investment model.)

In large part, the analyses reported in this chapter and the next will follow a two-phase procedure recommended by Anderson and Gerbing (1988). With this approach, you begin by developing a measurement model that provides an acceptable fit to the data. In this phase, each latent F variable is allowed to covary with every other latent F variable, and the analysis is essentially a confirmatory factor analysis. If the initial measurement model is inadequate, variables are reassigned or deleted in order to attain a better fit.

Once an acceptable measurement model has been developed, the analysis moves to a second phase in which the theoretical model itself is tested. This is done by fixing at zero the covariances between some of the F variables in the measurement model, and replacing some other covariances with unidirectional causal paths so that the relationships between the latent variables come to reflect the causal relations to be tested. Testing the resulting theoretical model for goodness of fit allows a simultaneous test of the measurement model developed in phase one, along with the structural model that is of primary substantive interest. If the theoretical model survives this test, support for the theory is obtained. If the theoretical model does not survive, it may be modified to attain a better fit.

Testing the Fit of the Measurement Model
from the Investment Model Study

Confirmatory factor analysis (CFA) is used to test the fit of the measurement model, and PROC CALIS can perform this analysis. In many respects, this CFA program will be similar to the program used to perform path analysis with manifest variables, as illustrated in the preceding chapter. Among other things, the program will include one functional equation for each manifest variable. These equations will define each V variable in terms of the latent factor (F variable) that it is believed to measure, as well as its residual term (E term).

As was mentioned in the preceding chapter, many of the conventions used here (e.g., representing latent factors with the letter F, representing observed variables with the letter V) are largely based on the conventions developed by Bentler (1989) for the EQS structural equations program.

Preparing the Program Figure

The typical CALIS program to perform this confirmatory factor analysis is longer than the typical program that performs a manifest-variable path analysis because the CFA usually involves more variables, and therefore more equations. Writing this program will usually be much easier if you first prepare a **program figure**: a figure that identifies the latent variables and their indicators, residual terms, and all parameters to be estimated. This section shows how to prepare a program figure following the same procedure used in the last chapter. Because the general steps followed in preparing a program figure were described in detail in that chapter (in the section titled "Preparing the Program Figure"), they will be covered in a more abbreviated form here.

The previous chapter presented a list of rules to guide you in preparing your program figures, and most of these rules will also apply to the analysis of latent variable models (there will be a few exceptions to these rules, and they will be discussed where appropriate). The rules are reproduced again here for purposes of reference:

Rule 1: In general, only exogenous variables are allowed to have covariances.

Rule 2: A residual term must be identified for each endogenous variable in the model.

Rule 3: Exogenous variables do not have residual terms.

Rule 4: Variances should be estimated for every exogenous variable in the model, including residual terms.

Rule 5: In most cases, covariances should be estimated for every possible pair of manifest exogenous variables; covariances are not estimated for endogenous variables.

Rule 6: For simple recursive models, covariances should not be estimated for residual terms.

Rule 7: One equation should be created for each endogenous variable, with that variable's name to the left of the equals sign.

Rule 8: Variables that have a direct effect on that endogenous variable are listed to the right of the equals sign.

Rule 9: Exogenous variables, including residual terms, are never listed to the left of the equals sign.

Rule 10: To estimate a path coefficient for a given independent variable, a unique path coefficient name should be created for the path coefficient associated with that independent variable.

Rule 11: The last term in each equation should be the residual (disturbance) term for that endogenous variable; this E (or D) term will have no name for its path coefficient.

Rule 12: To estimate a parameter, create a name for that parameter.

Rule 13: To fix a parameter at a given numerical value, insert that value in the place of the parameter's name.

Rule 14: To constrain two or more parameters to be equal, use the same name for those parameters.

Step 1: Drawing the basic confirmatory factor model. Figure 5.6 presents the basic measurement model to be tested in this phase of the study. As with the theoretical model presented earlier, the model consists of five latent variables, or factors: Commitment (F1), Satisfaction (F2), Rewards (F3), Costs (F4), Investments (F5), and Alternatives (F6). F1 has four manifest indicator variables, represented by rectangles. Remember that these indicator variables are responses to questionnaire items. The remaining factors have three indicators each. Each factor is connected to every other factor by a curved, two-headed arrow, meaning that every factor is allowed to covary with every other factor. This is a situation analogous to the oblique solution discussed in the chapter on exploratory factor analysis (see the section titled "Orthogonal versus Oblique Models" in Chapter 2).

Notice that each indicator variable is predicted to load on only one factor; this is to say that none of the indicators are **complex variables** (measuring more than one latent variable). Notice also that there are no covariances between any of the indicators. This is because only exogenous variables are allowed to have covariances (you know that indicator variables are endogenous variables because a straight, one-headed arrow points to each of them).

Figure 5.6: The Initial Measurement Model

Step 2: Identifying residual terms for endogenous variables. Rule 2 states that a residual term must be identified for each endogenous variable in the model, and you know that all of the indicator variables are endogenous variables (because they are affected by the F variables). Therefore, a residual term must be created for each indicator. This is illustrated in Figure 5.7.

This figure follows the same conventions used in the last chapter, in that the names for the residual terms begin with the letter "E" and end with the same numerical suffix as is used with the short name for the corresponding indicator. For example, the residual for V1 is E1, the residual for V2 is E2, and so forth.

Figure 5.7 illustrates all of the causal effects in the confirmatory factor model, and you can see that these causal effects are relatively simple. Each indicator is affected only by the underlying common factor (F) on which it loads, along with its residual term. That is to say, V1 is affected only by F1 and E1, V5 is affected only by F2 and E5, and so forth.

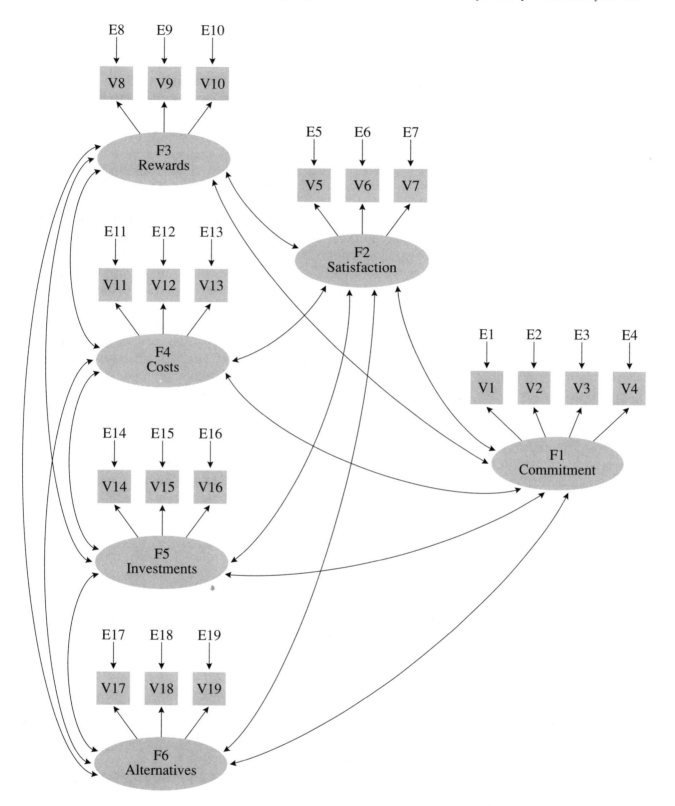

Figure 5.7: The Initial Measurement Model, Including Residual Terms for Endogenous Variables

Step 3: Identifying all parameters to be estimated. Figure 5.8 illustrates all of the parameters to be estimated in the CFA. Do not be intimidated by the seeming complexity of the figure; the rules presented in the last chapter taught virtually everything you need to know to determine which parameters should be estimated. Only a few new concepts will be taught here. Only three types of parameters are estimated in the analysis: variances for exogenous variables, covariances between latent factors (F variables), and factor loadings.

Begin with the variances. Rule 4 states that variances should be estimated for every exogenous variable in the model, including residual terms. Consistent with this, notice that the symbol "VAR?" appears below all of the E residual terms in the model (here, any symbol ending with a "?" indicates a parameter to be estimated; in this case "VAR?" represents a variance parameter to be estimated). This means that the variance will be estimated for each E term in the model. Notice that there are no VAR? symbols under the V terms, because all manifest variables in this CFA model are endogenous variables, and variances are not estimated for endogenous variables.

This figure introduces a new concept, and an exception to one of the last chapter's rules. Notice that all of the latent factors (the F variables) are exogenous variables. You can tell this because an exogenous variable is a variable that is not affected by a straight, single-headed arrow. Rule 4 stated that variances should be estimated for every exogenous variable in the model, but this will not apply to the F variables in confirmatory factor analysis. This is because there exists a basic indeterminancy involving the variance of the F variables and the factor loadings for the manifest indicators that measure those F variables. Because it is a hypothetical construct (rather than a real-world manifest variable), an F variable has no established metric, or scale. This is the **scale indeterminancy problem**. Left unaddressed, this problem makes it impossible to distinguish between situations in which the factor has a large variance and the factor loadings are small, and situations in which the factor variance is small and the loadings are large.

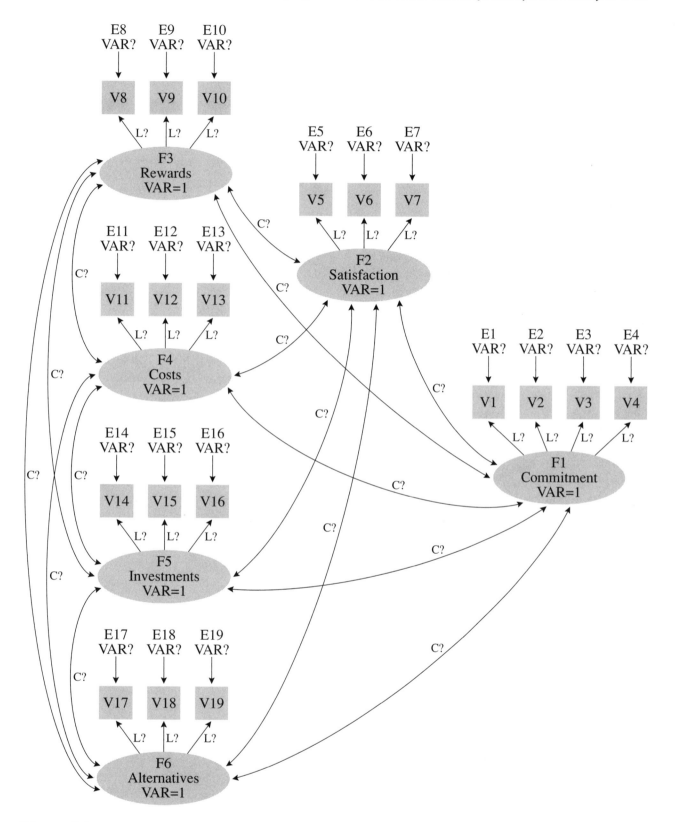

Figure 5.8: The Initial Measurement Model, after Identifying All Parameters to Be Estimated (Completed Program Figure for Confirmatory Factor Analysis)

To solve this problem, you can give all factors unit variances by fixing their variances at 1. This establishes a scale for the F variables, and helps ensure that the model is identified. This is illustrated in the program figure (Figure 5.8) by indicating "VAR=1" below the long name of each F variable. For example, "VAR=1" is listed below "Commitment" in the oval for F1, below "Satisfaction" in the oval for F2, and so forth. Later, this chapter will show how these factor variances are actually set at 1 in the CALIS program itself. This convention for dealing with the scale indeterminancy problem is summarized in Rule 15:

> **Rule 15:** In confirmatory factor analysis, the variances of the latent F variables are usually fixed at 1.

Having identified the variances to be estimated or fixed, you now turn to the covariances. Rule 5 states that covariances are usually estimated for all possible pairs of manifest exogenous variables. The counterpart to this rule in CFA is that covariances should be estimated for every possible pair of latent factors. In Figure 5.8, this is illustrated by the curved, two-headed arrows that connect all of the F variables. The "C?" symbol on each curved arrow represents the covariance estimate.

Notice that covariances are listed only for the latent F variables. There are no covariances between the V variables because these are all endogenous variables, and endogenous variables cannot have covariances. There are no covariances for the residual terms either, because residual terms are not allowed to covary (except in special models such as time-series or nonrecursive models).

Finally, it is necessary to estimate factor loadings for the model, and these are represented by the symbol "L?" in Figure 5.8. In this model, factor loadings are basically just path coefficients for the paths leading from a factor to an indicator variable. For example, the "L?" symbol appears on the causal arrow from F1 (Commitment) to V1, one of its indicator variables. If the path coefficient (or factor loading) for this path is relatively large and significantly different from zero, it means that V1 is doing a good job of measuring F1. You can see that factor loadings are estimated for every causal path leading from a factor to an indicator variable.

Step 4: Verifying that the model is overidentified. The preceding chapter discussed a number of necessary but not sufficient procedures that can be used to determine whether a model is identified. These included

- verifying that the number of data points in the analysis is larger than the number of parameters to be estimated

- checking the SAS log and output carefully to see if PROC CALIS discovered an identification problem

- repeating the analysis several times using different starting values, and verifying that the same parameter estimates are obtained with each run.

Along with these reminders, it is again noted that researchers are more likely to run into identification problems if they

- measure a latent variable with less than three indicators

- fail to establish the metric of a factor (for example, by failing to fix its variance at 1) or

- analyze nonrecursive structural models.

The last chapter stated that recursive path models with manifest variables will always be either just-identified or overidentified. In contrast, causal models with latent variables may be just-identified, overidentified, or underidentified. Remember that the procedures listed previously cannot conclusively prove that a model is identified, and that the results obtained from the analysis of an underidentified model are meaningless. Worse, if PROC CALIS fails to detect an underidentified model, it will still produce seemingly interpretable results that you may go on to publish (and may later have to retract!).

It is possible to demonstrate that your model is overidentified by showing mathematically that the parameters of the model's covariance equations can be solved in terms of population variances and covariances; Long (1983a, 1983b) illustrates this approach. Bollen (1989) provides a particularly useful discussion of the identification problem and approaches for proving identification. Bollen's text discusses a number of necessary and/or sufficient conditions for identification in path analysis models with manifest variables, confirmatory factor analysis models, and path analysis models with latent variables. Some of the procedures do require the use of matrix algebra, but an appendix to the book provides a helpful review of basic definitions and operations in matrix algebra. Bollen's text is highly recommended.

Preparing the SAS Program

The DATA step. Since the DATA step was given extensive coverage in the preceding chapter, only a few central points will be repeated here. First, when possible it is desirable to perform at least the first few analyses on raw data, as this allows use of the KURTOSIS option. This option prints various indices of skewness and kurtosis, and helps you identify observations that make large contributions to kurtosis. In this way, you may identify outliers that could lead to biased parameter estimates if not deleted.

It is also possible to input the data as a correlation or covariance matrix. When creating such a matrix from raw data, you should generally use a listwise procedure for deleting observations with missing data (e.g., use the NOMISS option with PROC CORR when creating the matrix). If inputting a correlation matrix, the standard deviations should also be input if available. This allows PROC CALIS to transform the correlation matrix into a covariance matrix, and this is desirable because the procedures discussed in this chapter generally should be performed on the covariance matrix, not a correlation matrix.

Table 5.1 presents the standard deviations and intercorrelations for the manifest variables analyzed in the present study (as before, these data are fictitious). Readers who want to replicate the analyses reported here may analyze this data set by inputting it as a TYPE=CORR data set.

It is important to input the standard deviations along with the correlations, so that the analysis may be performed on the covariance matrix. Because the last chapter showed how to input a TYPE=CORR data set, instructions for managing the DATA step of the SAS program will not be repeated here.

Table 5.1

Standard Deviations and Intercorrelations for Manifest Variables, Investment Model Study

											Intercorrelations								
	S.D.	V1	V2	V3	V4	V5	V6	V7	V8	V9	V10	V11	V12	V13	V14	V15	V16	V17	V18	V19
V1	2.486	1.000																		
V2	2.909	.734	1.000																	
V3	2.724	.819	.786	1.000																
V4	2.926	.672	.732	.751	1.000															
V5	1.926	.514	.362	.496	.471	1.000														
V6	2.113	.534	.346	.452	.452	.713	1.000													
V7	2.056	.522	.345	.507	.546	.720	.764	1.000												
V8	1.417	.346	.293	.341	.294	.337	.375	.285	1.000											
V9	1.408	.209	.147	.167	.214	.251	.306	.357	.390	1.000										
V10	1.724	.349	.241	.287	.236	.282	.351	.304	.506	.492	1.000									
V11	2.595	.051	.082	.005	−.038	−.161	−.166	−.117	−.091	−.055	−.063	1.000								
V12	2.691	−.040	.013	−.057	−.090	−.189	−.150	−.190	−.102	−.036	−.018	.714	1.000							
V13	2.360	−.029	−.012	−.066	−.023	−.110	−.101	−.083	−.043	.055	−.033	.379	.403	1.000						
V14	2.102	.559	.428	.581	.485	.433	.451	.470	.305	.345	.333	.003	−.005	.043	1.000					
V15	2.219	.434	.322	.454	.424	.472	.418	.415	.205	.231	.241	−.037	.007	−.062	.595	1.000				
V16	1.874	.375	.326	.431	.311	.253	.256	.225	.157	.151	.140	.093	.022	.054	.457	.410	1.000			
V17	2.001	−.141	−.075	−.145	−.275	−.300	−.282	−.305	−.198	−.200	−.188	.226	.216	.051	−.220	−.256	−.065	1.000		
V18	1.966	−.135	−.184	−.154	−.266	−.192	−.184	−.204	−.158	−.133	−.243	.132	.119	.107	−.144	−.149	−.074	.529	1.000	
V19	2.185	−.167	−.142	−.145	−.327	−.234	−.209	−.251	−.288	−.298	−.220	.147	.142	.076	−.179	−.202	−.803	.460	.572	1.000

Note: N = 240.

Overview of the PROC CALIS program. The CALIS program that performs confirmatory factor analysis is very similar to the program that performs path analysis with manifest variables. It is somewhat longer (especially the LINEQS statement) because there are more parameters to estimate, but the basic principles for writing the statements remain the same. Here is the complete program (minus the DATA step) for performing CFA on the model presented in Figure 5.8:

```
1      PROC CALIS  COVARIANCE  CORR  RESIDUAL  MODIFICATION ;
2         LINEQS
3            V1  = LV1F1  F1 + E1,
4            V2  = LV2F1  F1 + E2,
5            V3  = LV3F1  F1 + E3,
6            V4  = LV4F1  F1 + E4,
7            V5  = LV5F2  F2 + E5,
8            V6  = LV6F2  F2 + E6,
9            V7  = LV7F2  F2 + E7,
```

```
10              V8  = LV8F3   F3 + E8,
11              V9  = LV9F3   F3 + E9,
12              V10 = LV10F3  F3 + E10,
13              V11 = LV11F4  F4 + E11,
14              V12 = LV12F4  F4 + E12,
15              V13 = LV13F4  F4 + E13,
16              V14 = LV14F5  F5 + E14,
17              V15 = LV15F5  F5 + E15,
18              V16 = LV16F5  F5 + E16,
19              V17 = LV17F6  F6 + E17,
20              V18 = LV18F6  F6 + E18,
21              V19 = LV19F6  F6 + E19;
22        STD
23              F1 = 1,
24              F2 = 1,
25              F3 = 1,
26              F4 = 1,
27              F5 = 1,
28              F6 = 1,
29              E1-E19 = VARE1-VARE19;
30        COV
31              F1 F2 = CF1F2,
32              F1 F3 = CF1F3,
33              F1 F4 = CF1F4,
34              F1 F5 = CF1F5,
35              F1 F6 = CF1F6,
36              F2 F3 = CF2F3,
37              F2 F4 = CF2F4,
38              F2 F5 = CF2F5,
39              F2 F6 = CF2F6,
40              F3 F4 = CF3F4,
41              F3 F5 = CF3F5,
42              F3 F6 = CF3F6,
43              F4 F5 = CF4F5,
44              F4 F6 = CF4F6,
45              F5 F6 = CF5F6;
46        VAR   V1-V19 ;
47        RUN;
```

The PROC CALIS statement. The PROC CALIS statement, which requests the CALIS procedure, appears on line 1 of the preceding program. Any options desired for the analysis are listed in this statement, separated by at least one space. The PROC CALIS statement in the preceding program includes the COVARIANCE option, which requests that the analysis be performed on the covariance matrix rather than the correlation matrix, the CORR option, which requests that CALIS print the correlation or covariance matrix that is analyzed, along with the predicted (or reproduced) matrix, the RESIDUAL option, which requests the residual and normalized residual correlation or covariance matrices, and the MODIFICATION option, which requests the Lagrange Multiplier and Wald test modification indices. The CORR, RESIDUAL, and MODIFICATION options are useful for determining how the model should be modified if it does not demonstrate an adequate fit.

A number of other frequently used options for PROC CALIS are discussed in the preceding chapter. The complete list of PROC CALIS options may be found in Chapter 14 of the *SAS/STAT users guide, volume 1* (1989).

The LINEQS statement. The LINEQS statement is used to indicate which manifest variables load on which latent factors. This is done with a series of equations, and there is a separate equation for each manifest variable. Here is the general form for the LINEQS statements that appear in a confirmatory factor analysis:

```
LINEQS
    v = l  f + e,
    v = l  f + e,
    v = l  f + e;
```

where

    v = manifest variable (indicator variable)
    l = coefficient name for the factor loading
    f = factor that the manifest variable loads on
    e = residual term for corresponding manifest variable.

Although the preceding provides the general form for three equations, any number of equations is actually possible. The program provided earlier includes 19 equations: one for each indicator variable.

For a concrete illustration, consider the first seven equations and the last three equations from your program:

```
2          LINEQS
3              V1  = LV1F1  F1 + E1,
4              V2  = LV2F1  F1 + E2,
5              V3  = LV3F1  F1 + E3,
6              V4  = LV4F1  F1 + E4,
7              V5  = LV5F2  F2 + E5,
8              V6  = LV6F2  F2 + E6,
9              V7  = LV7F2  F2 + E7,
               .
               .
               .
```

```
19              V17 = LV17F6 F6 + E17,
20              V18 = LV18F6 F6 + E18,
21              V19 = LV19F6 F6 + E19;
```

The LINEQS statement begins with the word LINEQS and ends with a semicolon. Each equation in the statement is separated by a comma. These equations are prepared following the same conventions discussed in the previous chapter.

The preceding LINEQS equations reflect the causal relationships depicted in Figure 5.8. Figure 5.8 shows that only the indicator variables are endogenous variables in the confirmatory factor analysis model, so only these indicators (e.g., V1, V2, V3) appear to the left of the equals sign in the LINEQS equations. Figure 5.8 also shows that each V variable is affected by a single latent factor (an F variable) and a single residual term (an E variable). This is reflected in the LINEQS equations as well: V1 is affected by F1 and E1; V2 is affected by F1 and E2, and so forth.

Figure 5.8 shows which variables load on which factors. If an arrow goes from a factor to an indicator variable, that indicator loads on that factor. The loadings illustrated in Figure 5.8 are also reflected in the LINEQS equations. If an equation for a V variable includes a certain F variable, it means that the V variable loads on that factor. It is in this way that the following statements show that V1 through V4 load on F1, while V5 through V7 load on F2:

```
3               V1  = LV1F1  F1 + E1,
4               V2  = LV2F1  F1 + E2,
5               V3  = LV3F1  F1 + E3,
6               V4  = LV4F1  F1 + E4,
7               V5  = LV5F2  F2 + E5,
8               V6  = LV6F2  F2 + E6,
9               V7  = LV7F2  F2 + E7,
```

Notice that, although a group of variables may share a common factor, each variable still has its own unique residual term. In other words, V1 has the residual term E1, V2 has E2, and so forth.

The only parameters that are estimated by these LINEQS equations are the factor loadings. In this case, the factor loading coefficients are equivalent to path coefficients for the paths that lead from an F variable to a V variable. For example, PROC CALIS may estimate that the factor loading for the path from F1 to V1 is .89, or that the coefficient for the path from F1 to V2 is .65.

Rule 12 from the preceding chapter said that, in order to estimate a parameter, you must create a name for that parameter. This means that you must generate names for the factor loadings that you want to estimate in this analysis. Technically, you may create any names for these coefficients that you like, as long as they comply with the usual SAS System rules for variable names. To create more meaningful names, however, this text recommends a system similar to that used in Chapter 4 to name path coefficients.

Specifically, this text recommends that each factor loading be given a name that begins with the letter "L" (for Loading). Next comes the short name for the indicator variable being affected, and last comes the short name for the underlying factor involved. In this way, LV1F1 is the name for the factor loading of V1 on F1, LV17F6 is the name for the factor loading of V17 on F6, and so forth. Review the names of the factor loadings included in the preceding program to verify that these names accurately reflect the relationships between factors and variables portrayed in Figure 5.8.

This system is largely based on the system developed by Bentler (1989) for the EQS program. The advantage of this system is that it gives each factor loading a name that is both unique and meaningful. For example, if you saw the parameter name LV14F5 among the parameter names listed in the Wald test, you would immediately know that this is the factor loading for manifest variable V14 on latent factor F5.

Again consider the LINEQS statements for indicators V1 through V7:

```
3               V1  = LV1F1  F1 + E1,
4               V2  = LV2F1  F1 + E2,
5               V3  = LV3F1  F1 + E3,
6               V4  = LV4F1  F1 + E4,
7               V5  = LV5F2  F2 + E5,
8               V6  = LV6F2  F2 + E6,
9               V7  = LV7F2  F2 + E7,
```

The previous chapter noted that dependent variables are always listed to the left of the equals sign, and that the independent variables that have direct causal effects on these variables are listed to the right of the sign. It was also said that you place a parameter name to the immediate left of an independent variable in order to estimate the parameter for that independent variable. The preceding equations reflect these conventions; placing the name for a factor loading to the left of an F variable in the preceding statements means that a factor loading will be estimated to describe the relationship between the V variable and that F variable.

But notice that each of the preceding statements also includes a second independent variable: the indicator variable's E term (residual term). Shouldn't you create a name for the path coefficient that represents the relationship between each E term and its corresponding V variable? The answer is no. As with path analysis, the paths for these E terms are fixed at 1. You accomplish this by simply leaving the names for these path coefficients out of the equation, as this automatically fixes the path from the E term to the V variable at 1.

The STD statement. The STD statement is used to indicate which variables' variances are to be estimated, and which variables' variances are to be fixed. In the last chapter, you learned that variances are estimated or fixed for all of the exogenous variables in a model, but are never estimated for endogenous variables. A glance at Figure 5.8 reveals two types of exogenous variables in your confirmatory path model: the F variables and the E variables. This means that the F and E variable will be listed in the STD statement. Since all manifest indicator variables are endogenous variables, none of the V terms will appear here.

Here is the STD statement for the current program:

```
22          STD
23              F1 = 1,
24              F2 = 1,
25              F3 = 1,
26              F4 = 1,
27              F5 = 1,
28              F6 = 1,
29              E1-E19 = VARE1-VARE19;
```

The STD statement begins with the letters STD, and ends with a semicolon. The equations that constitute the statement are each separated by a comma.

In Figure 5.8, you wrote "VAR=1" below the long name of each of the latent factors. This was to indicate that the variance of each factor was to be fixed at 1 (this was necessary to solve the scale indeterminancy problem). In the CALIS program, these variances are actually fixed at 1 by listing the short name of the F variable to the left of the equals sign, and the number "1" to the right of the sign. This is done in lines 23-28 of the program.

Figure 5.8 shows that the symbol "VAR?" appears below the short name of each of the E terms, indicating that the variance for each of these terms is to be estimated. Remember that, in order to estimate any parameter, it is necessary to include a name for that parameter in the PROC CALIS program. This is done in line 29 of your program. The short names for the E terms (E1-E19) appear to the left of the equals sign, and the names for the corresponding variance parameter estimates (VARE1-VARE19) appear to the right of the sign. This single equation will result in variance estimates for all 19 residual terms.

The COV statement. The COV statement is used to identify pairs of variables that are expected to covary. The statement begins with the letters COV and ends with a semicolon, and each equation in it is separated by a comma.

If two variables are expected to covary, the short names for these variables are listed to the left of the equals sign, and the name for the covariance estimate appears to the right. Here is the COV statement for the present program:

```
30          COV
31              F1 F2 = CF1F2,
32              F1 F3 = CF1F3,
33              F1 F4 = CF1F4,
34              F1 F5 = CF1F5,
35              F1 F6 = CF1F6,
36              F2 F3 = CF2F3,
37              F2 F4 = CF2F4,
```

```
38            F2  F5  =  CF2F5,
39            F2  F6  =  CF2F6,
40            F3  F4  =  CF3F4,
41            F3  F5  =  CF3F5,
42            F3  F6  =  CF3F6,
43            F4  F5  =  CF4F5,
44            F4  F6  =  CF4F6,
45            F5  F6  =  CF5F6;
```

In confirmatory factor analysis, all latent factors generally are allowed to covary. This means that covariances must be estimated for every possible pair of F variables, and the preceding equations do just this.

To ensure that you do not miss any covariances, you should follow the general format presented here: lines 31-35 list the pairings of variable F1 with variables F2 through F6; lines 36-39 list the pairings of F2 with F3 through F6; lines 40-42 list the pairings of F3 with F4 though F6, and so forth. Following this systematic approach assures that all possible pairings of the F variables will be listed.

The names for the covariance estimates appear to the right of the equals sign. Here, this text uses the convention presented in the last chapter, in which the name for the estimate begins the letter "C" (for Covariance), followed by the short names for the two F variables involved in the covariance. As usual, covariances are not estimated for any residual terms.

The VAR statement. The VAR statement appears as line 46 of the program. It is reproduced here:

```
46        VAR  V1-V19 ;
```

The VAR statement begins the letters VAR and ends with a semicolon. It contains a list of the subset of variables to be analyzed by PROC CALIS. This statement is optional, but will usually save computer time when it is used.

Making Sure That the SAS Log and Output Files Look Right

When the analysis is completed, the SAS System creates two new files. The log file contains lines of the original SAS program along with notes, warnings, and error messages, and the output file (or lis file) contains the actual results of the analysis. Both files should be reviewed to verify that the analysis was performed in the desired manner.

Specifically, the log file should be inspected for any notes, warnings, or error messages that indicate a problem in the analysis. Toward the end of the log, look for the statement "Convergence criterion satisfied."

If you have specified LINESIZE=80 and PAGESIZE=60 in the OPTIONS statement, the preceding program would produce an output file 37 pages long. Here is a brief description of some of the information contained in this output:

- Page 1 includes a list of the endogenous variables and exogenous variables specified in the LINEQS statement.

- Pages 2-3 include the general form of the structural equations specified in the LINEQS statement.

- The top of page 4 provides general information about the estimation problem, including

    - the number of observations

    - the number of variables

    - the amount of independent information in the data matrix (number of "data points")

    - the number of parameters to be estimated.

- Below this, page 4 presents some univariate statistics for the manifest variables, and this is followed by part of the covariance matrix to be analyzed.

- Page 5 provides the remainder of the covariance matrix to be analyzed.

- Pages 6-7 include a vector of initial parameter estimates.

- Page 8 includes the iteration history.

- Pages 9-13 include the predicted model covariance matrix, the residual matrix, and the normalized residual matrix.

- The bottom of page 13 includes a bar chart displaying the distribution of normalized residuals.

- Page 14 includes a variety of goodness of fit indices, to be discussed later.

- Pages 15-17 include equations containing parameter estimates (factor loadings), their approximate standard errors, and t values. These equations correspond to those constructed in the LINEQS statement. These are followed by estimates of variances and covariances, along with corresponding approximate standard errors and t values.

- Pages 18-19 include equations containing the standardized parameter estimates (factor loadings).

- Page 19 also includes estimates of the endogenous variable variances, and the R^2 values for each endogenous variable. The correlations among exogenous variables are also presented here.

- Pages 20-37 include modification indices (Lagrange multiplier and Wald test results).

Before reviewing the substantive results of the analysis (e.g., the goodness of fit indices, the factor loadings), you should routinely review the first pages of the output to verify that the program was executed as expected. The first eight pages of the current output are reproduced here as Output 5.1.

```
                         The SAS System                              1

            Covariance Structure Analysis: Pattern and Initial Values

              LINEQS Model Statement
            --------------------------------
              Matrix        Rows & Cols         Matrix Type
    Term    1------------------------------------------------------
              1    _SEL_        19      44     SELECTION
              2    _BETA_       44      44     EQSBETA         IMINUSINV
              3    _GAMMA_      44      25     EQSGAMMA
              4    _PHI_        25      25     SYMMETRIC



         Number of endogenous variables = 19

    Manifest:      V1        V2        V3        V4        V5        V6
                   V7        V8        V9        V10       V11       V12
                   V13       V14       V15       V16       V17       V18
                   V19



         Number of exogenous variables = 25

    Latent:        F1        F2        F3        F4        F5        F6
    Error:         E1        E2        E3        E4        E5        E6
                   E7        E8        E9        E10       E11       E12
                   E13       E14       E15       E16       E17       E18
                   E19
```

```
Covariance Structure Analysis: Pattern and Initial Values          2
                 Manifest Variable Equations
                     Initial Estimates
           V1      =      .    *F1      + 1.0000 E1
                             LV1F1

           V2      =      .    *F1      + 1.0000 E2
                             LV2F1

           V3      =      .    *F1      + 1.0000 E3
                             LV3F1

           V4      =      .    *F1      + 1.0000 E4
                             LV4F1

           V5      =      .    *F2      + 1.0000 E5
                             LV5F2

           V6      =      .    *F2      + 1.0000 E6
                             LV6F2

           V7      =      .    *F2      + 1.0000 E7
                             LV7F2

           V8      =      .    *F3      + 1.0000 E8
                             LV8F3

           V9      =      .    *F3      + 1.0000 E9
                             LV9F3

           V10     =      .    *F3      + 1.0000 E10
                             LV10F3

           V11     =      .    *F4      + 1.0000 E11
                             LV11F4

           V12     =      .    *F4      + 1.0000 E12
                             LV12F4

           V13     =      .    *F4      + 1.0000 E13
                             LV13F4

           V14     =      .    *F5      + 1.0000 E14
                             LV14F5

           V15     =      .    *F5      + 1.0000 E15
                             LV15F5

           V16     =      .    *F5      + 1.0000 E16
                             LV16F5

           V17     =      .    *F6      + 1.0000 E17
                             LV17F6

           V18     =      .    *F6      + 1.0000 E18
                             LV18F6
```

```
       Covariance Structure Analysis: Pattern and Initial Values          3

            V19      =       .      *F6      + 1.0000 E19
                                   LV19F6

                   Variances of Exogenous Variables
            ----------------------------------------
            Variable      Parameter         Estimate
            ----------------------------------------
            F1                              1.000000
            F2                              1.000000
            F3                              1.000000
            F4                              1.000000
            F5                              1.000000
            F6                              1.000000
            E1            VARE1                    .
            E2            VARE2                    .
            E3            VARE3                    .
            E4            VARE4                    .
            E5            VARE5                    .
            E6            VARE6                    .
            E7            VARE7                    .
            E8            VARE8                    .
            E9            VARE9                    .
            E10           VARE10                   .
            E11           VARE11                   .
            E12           VARE12                   .
            E13           VARE13                   .
            E14           VARE14                   .
            E15           VARE15                   .
            E16           VARE16                   .
            E17           VARE17                   .
            E18           VARE18                   .
            E19           VARE19                   .


                 Covariances among Exogenous Variables
            ------------------------------------
                  Parameter              Estimate
            ------------------------------------
            F2    F1    CF1F2                    .
            F3    F1    CF1F3                    .
            F3    F2    CF2F3                    .
            F4    F1    CF1F4                    .
            F4    F2    CF2F4                    .
            F4    F3    CF3F4                    .
            F5    F1    CF1F5                    .
            F5    F2    CF2F5                    .
            F5    F3    CF3F5                    .
            F5    F4    CF4F5                    .
            F6    F1    CF1F6                    .
            F6    F2    CF2F6                    .
            F6    F3    CF3F6                    .
            F6    F4    CF4F6                    .
            F6    F5    CF5F6                    .
```

```
      Covariance Structure Analysis: Maximum Likelihood Estimation          4

               240 Observations      Model Terms          1
                19 Variables         Model Matrices        4
               190 Informations      Parameters           53

              VARIABLE              Mean           Std Dev

              V1                       0         2.486000000
              V2                       0         2.909000000
              V3                       0         2.724000000
              V4                       0         2.926000000
              V5                       0         1.929000000
              V6                       0         2.113000000
              V7                       0         2.056000000
              V8                       0         1.417000000
              V9                       0         1.408000000
              V10                      0         1.724000000
              V11                      0         2.595000000
              V12                      0         2.691000000
              V13                      0         2.360000000
              V14                      0         2.102000000
              V15                      0         2.219000000
              V16                      0         1.874000000
              V17                      0         2.001000000
              V18                      0         1.966000000
              V19                      0         2.185000000


                              Covariances

           V1         V2         V3         V4         V5         V6         V7

V1     6.180196   5.308122   5.546157   4.888152   2.464884   2.805058   2.668055
V2     5.308122   8.462281   6.228355   6.230589   2.031349   2.126764   2.063412
V3     5.546157   6.228355   7.420176   5.985788   2.606280   2.601627   2.839476
V4     4.888152   6.230589   5.985788   8.561476   2.658444   2.794552   3.284657
V5     2.464884   2.031349   2.606280   2.658444   3.721041   2.906172   2.855537
V6     2.805058   2.126764   2.601627   2.794552   2.906172   4.464769   3.319067
V7     2.668055   2.063412   2.839476   3.284657   2.855537   3.319067   4.227136
V8     1.218841   1.207762   1.316229   1.218966   0.921153   1.122795   0.830305
V9     0.731560   0.602093   0.640510   0.881639   0.681724   0.910382   1.033461
V10    1.495767   1.208643   1.347803   1.190484   0.937818   1.278627   1.077541
V11    0.329010   0.619006   0.035344  -0.288533  -0.805927  -0.910217  -0.624232
V12   -0.267593   0.101766  -0.417826  -0.708648  -0.981087  -0.852912  -1.051212
V13   -0.170142  -0.082383  -0.424290  -0.158823  -0.500768  -0.503655  -0.402729
V14    2.921095   2.617099   3.326718   2.982969   1.755710   2.003128   2.031205
V15    2.394132   2.078533   2.744228   2.752945   2.020373   1.959896   1.893340
V16    1.747037   1.777178   2.200158   1.705314   0.914581   1.013699   0.866912
V17   -0.701403  -0.436568  -0.790355  -1.610105  -1.157979  -1.192328  -1.254787
V18   -0.659809  -1.052313  -0.824729  -1.530169  -0.728143  -0.764365  -0.824588
V19   -0.907129  -0.902575  -0.863031  -2.090612  -0.986278  -0.964933  -1.127582
```

```
           Covariance Structure Analysis: Maximum Likelihood Estimation        5

                                  Covariances

             V8          V9          V10         V11         V12         V13         V14

V1       1.218841    0.731560    1.495767    0.329010   -0.267593   -0.170142    2.921095
V2       1.207762    0.602093    1.208643    0.619006    0.101766   -0.082383    2.617099
V3       1.316229    0.640510    1.347803    0.035344   -0.417826   -0.424290    3.326718
V4       1.218966    0.881639    1.190484   -0.288533   -0.708648   -0.158823    2.982969
V5       0.921153    0.681724    0.937818   -0.805927   -0.981087   -0.500768    1.755710
V6       1.122795    0.910382    1.278627   -0.910217   -0.852912   -0.503655    2.003128
V7       0.830305    1.033461    1.077541   -0.624232   -1.051212   -0.402729    2.031205
V8       2.007889    0.778103    1.236111   -0.334617   -0.388941   -0.143797    0.908453
V9       0.778103    1.982464    1.194277   -0.200957   -0.136401    0.182758    1.021068
V10      1.236111    1.194277    2.972176   -0.281848   -0.083507   -0.012206    1.206741
V11     -0.334617   -0.200957   -0.281848    6.734025    4.985966    2.321072    0.016364
V12     -0.388941   -0.136401   -0.083507    4.985966    7.241481    2.559356   -0.028282
V13     -0.143797    0.182758   -0.012206    2.321072    2.559356    5.569600    0.213311
V14      0.908453    1.021068    1.206741    0.016364   -0.028282    0.213311    4.418404
V15      0.644586    0.721725    0.921959   -0.213057    0.041799   -0.324684    2.775281
V16      0.416907    0.398427    0.452309    0.452262    0.110945    0.238823    1.800191
V17     -0.561413   -0.563482   -0.648548    1.173526    1.163093    0.240840   -0.925342
V18     -0.440160   -0.368161   -0.823620    0.673434    0.629570    0.496454   -0.595085
V19     -0.891690   -0.916791   -0.828727    0.833501    0.834937    0.391902   -0.822124


                    V15         V16         V17         V18         V19

        V1       2.394132    1.747037   -0.701403   -0.659809   -0.907129
        V2       2.078533    1.777178   -0.436568   -1.052313   -0.902575
        V3       2.744228    2.200158   -0.790355   -0.824729   -0.863031
        V4       2.752945    1.705314   -1.610105   -1.530169   -2.090612
        V5       2.020373    0.914581   -1.157979   -0.728143   -0.986278
        V6       1.959896    1.013699   -1.192328   -0.764365   -0.964933
        V7       1.893340    0.866912   -1.254787   -0.824588   -1.127582
        V8       0.644586    0.416907   -0.561413   -0.440160   -0.891690
        V9       0.721725    0.398427   -0.563482   -0.368161   -0.916791
        V10      0.921959    0.452309   -0.648548   -0.823620   -0.828727
        V11     -0.213057    0.452262    1.173526    0.673434    0.833501
        V12      0.041799    0.110945    1.163093    0.629570    0.834937
        V13     -0.324684    0.238823    0.240840    0.496454    0.391902
        V14      2.775281    1.800191   -0.925342   -0.595085   -0.822124
        V15      4.923961    1.704946   -1.136696   -0.650021   -0.979400
        V16      1.704946    3.511876   -0.243742   -0.272637   -0.339859
        V17     -1.136696   -0.243742    4.004001    2.081068    2.011205
        V18     -0.650021   -0.272637    2.081068    3.865156    2.457146
        V19     -0.979400   -0.339859    2.011205    2.457146    4.774225
                   Determinant = 143562421 (Ln = 18.782)
```

```
        Covariance Structure Analysis: Maximum Likelihood Estimation      6
      Some initial estimates computed by instrumental variable method.
                        Vector of Initial Estimates

            LV1F1        1     2.27362   Matrix Entry: _GAMMA_[1:1]
            LV2F1        2     2.27595   Matrix Entry: _GAMMA_[2:1]
            LV3F1        3     2.50694   Matrix Entry: _GAMMA_[3:1]
            LV4F1        4     2.49029   Matrix Entry: _GAMMA_[4:1]
            LV5F2        5     1.64547   Matrix Entry: _GAMMA_[5:2]
            LV6F2        6     1.76356   Matrix Entry: _GAMMA_[6:2]
            LV7F2        7     1.81496   Matrix Entry: _GAMMA_[7:2]
            LV8F3        8     1.07470   Matrix Entry: _GAMMA_[8:3]
            LV9F3        9     0.82712   Matrix Entry: _GAMMA_[9:3]
            LV10F3      10     1.21539   Matrix Entry: _GAMMA_[10:3]
            LV11F4      11     2.24076   Matrix Entry: _GAMMA_[11:4]
            LV12F4      12     2.18215   Matrix Entry: _GAMMA_[12:4]
            LV13F4      13     1.14459   Matrix Entry: _GAMMA_[13:4]
            LV14F5      14     1.78930   Matrix Entry: _GAMMA_[14:5]
            LV15F5      15     1.55315   Matrix Entry: _GAMMA_[15:5]
            LV16F5      16     1.04368   Matrix Entry: _GAMMA_[16:5]
            LV17F6      17     1.58313   Matrix Entry: _GAMMA_[17:6]
            LV18F6      18     1.27275   Matrix Entry: _GAMMA_[18:6]
            LV19F6      19     1.55041   Matrix Entry: _GAMMA_[19:6]
            CF1F2       20     0.62157   Matrix Entry: _PHI_[2:1]
            CF1F3       21     0.44343   Matrix Entry: _PHI_[3:1]
            CF2F3       22     0.53535   Matrix Entry: _PHI_[3:2]
            CF1F4       23    -0.02315   Matrix Entry: _PHI_[4:1]
            CF2F4       24    -0.22628   Matrix Entry: _PHI_[4:2]
            CF3F4       25    -0.09097   Matrix Entry: _PHI_[4:3]
            CF1F5       26     0.69541   Matrix Entry: _PHI_[5:1]
            CF2F5       27     0.63909   Matrix Entry: _PHI_[5:2]
            CF3F5       28     0.49534   Matrix Entry: _PHI_[5:3]
            CF4F5       29     0.01090   Matrix Entry: _PHI_[5:4]
            CF1F6       30    -0.29393   Matrix Entry: _PHI_[6:1]
            CF2F6       31    -0.39368   Matrix Entry: _PHI_[6:2]
            CF3F6       32    -0.43535   Matrix Entry: _PHI_[6:3]
            CF4F6       33     0.26818   Matrix Entry: _PHI_[6:4]
            CF5F6       34    -0.32008   Matrix Entry: _PHI_[6:5]
            VARE1       35     1.01085   Matrix Entry: _PHI_[7:7]
            VARE2       36     3.28232   Matrix Entry: _PHI_[8:8]
            VARE3       37     1.13545   Matrix Entry: _PHI_[9:9]
            VARE4       38     2.35994   Matrix Entry: _PHI_[10:10]
            VARE5       39     1.01348   Matrix Entry: _PHI_[11:11]
            VARE6       40     1.35461   Matrix Entry: _PHI_[12:12]
            VARE7       41     0.93305   Matrix Entry: _PHI_[13:13]
            VARE8       42     0.85291   Matrix Entry: _PHI_[14:14]
            VARE9       43     1.29834   Matrix Entry: _PHI_[15:15]
            VARE10      44     1.49501   Matrix Entry: _PHI_[16:16]
            VARE11      45     1.71303   Matrix Entry: _PHI_[17:17]
            VARE12      46     2.47972   Matrix Entry: _PHI_[18:18]
            VARE13      47     4.25951   Matrix Entry: _PHI_[19:19]
            VARE14      48     1.21680   Matrix Entry: _PHI_[20:20]
            VARE15      49     2.51170   Matrix Entry: _PHI_[21:21]
            VARE16      50     2.42260   Matrix Entry: _PHI_[22:22]
            VARE17      51     1.49770   Matrix Entry: _PHI_[23:23]
```

```
           Covariance Structure Analysis: Maximum Likelihood Estimation        7
                VARE18        52     2.24526   Matrix Entry: _PHI_[24:24]
                VARE19        53     2.37047   Matrix Entry: _PHI_[25:25]
```

```
           Covariance Structure Analysis: Maximum Likelihood Estimation        8
                         (Dual) Quasi-Newton Minimization:
                  Broyden - Fletcher - Goldfarb - Shanno Method (BFGS)
                             Maximum Iterations= 200
                            Maximum Function Calls= 500
                       Maximum Absolute Gradient Criterion= 0.001
                Number of Estimates= 53 Lower Bounds= 0 Upper Bounds= 0
                    Line Search Method 2: start alpha=1 sigma=0.4
 Minimization Start: Active Constraints= 0 Criterion= 1.232
 Maximum Gradient Element= 0.202
 Iter   res nfun act    mincrit   maxgrad   difcrit      alpha    descent
    0     1    1    0    1.2317    0.2018         0     1.0000    -0.1866
    1     1    4    0    1.2009    0.2760    0.0308     0.3477    -0.2101
    2     1    5    0    1.1424    0.2475    0.0585     0.6954    -0.1946
    3     1    7    0    1.1106    0.0752    0.0318     0.2649    -0.0543
    4     1    8    0    1.0948    0.1065    0.0158     0.5298    -0.0617
    5     1   10    0    1.0829    0.1101    0.0119     0.3959    -0.0420
    6     1   11    0    1.0714    0.1723    0.0115     0.7917    -0.0323
    7     1   13    0    1.0612    0.1052    0.0102     0.5737    -0.0255
    8     1   15    0    1.0538    0.0696   0.00739     0.5523    -0.0150
    9     1   16    0    1.0500    0.1084   0.00382     1.1045    -0.0165
   10     1   18    0    1.0441    0.0808   0.00585     0.6799   -0.00902
   11     1   20    0    1.0416    0.0499   0.00255     0.5530   -0.00332
   12     1   21    0    1.0405    0.0418   0.00107     1.1059   -0.00218
   13     1   23    0    1.0397    0.0298  0.000777     0.7279   -0.00171
   14     1   25    0    1.0392   0.00999  0.000523     0.6141   -0.00055
   15     1   26    0    1.0389    0.0170  0.000283     1.2281   -0.00065
   16     1   28    0    1.0385   0.00777  0.000387     1.1979   -0.00041
   17     1   29    0    1.0383    0.0157  0.000233     2.3958   -0.00064
   18     1   31    0    1.0377   0.00826   0.00058     1.8024   -0.00032
   19     1   32    0    1.0376    0.0185  0.000157     3.6049   -0.00076
   20     1   34    0    1.0372   0.00477  0.000361     0.9230   -0.00025
   21     1   35    0    1.0371   0.00777  0.000102     1.8460   -0.00029
   22     1   37    0    1.0368   0.00435  0.000277     1.8965   -0.00016
   23     1   38    0    1.0366   0.00435  0.000243     3.7931   -0.00014
   24     1   40    0    1.0365   0.00280  0.000135     1.9195    -0.0001
   25     1   42    0    1.0364   0.00257  0.000097     1.9795   -0.00005
   26     1   44    0    1.0363   0.00162   0.00003     1.2180   -0.00002
   27     1   45    0    1.0363   0.00207  0.000017     2.4361   -0.00002
   28     1   47    0    1.0363  0.000914   0.00001     1.1646   -0.00002
 Minimization Results: Iterations= 28 Function Calls= 47 Derivative Calls= 30
 Active Constraints= 0 Criterion= 1.04 Maximum Gradient Element= 0.000914
 Descent= -0.0000174
 NOTE: Convergence criterion satisfied.
```

Output 5.1: PROC CALIS Output Pages 1-8, Analysis of Initial Measurement Model, Investment Model Study

Page 1 of Output 5.1 identifies the endogenous and exogenous variables of the analysis. Pages 2–3 reproduce the structural equations being analyzed, and these should be reviewed to verify that PROC CALIS did in fact analyze the intended model. Page 3 also indicates which variances and covariances will be estimated, and which will be fixed. The table at the top of page 4 provides information concerning the number of observations, variables, "informations" (data points), and parameters included in the analysis, and these should be reviewed as well. Finally, page 8 provides the iteration history, and indicates whether the convergence criterion was satisfied. If the information presented on these pages appears to be in order, you may proceed to assess the fit between model and data, along with other results of the analysis.

Assessing the Fit between Model and Data

When conducting a confirmatory factor analysis, you begin with a model that predicts the existence of a specific number of latent factors, and predicts which indicator variables load on each factor.

You then test the model by measuring these variables in a sample of subjects drawn from the population of interest. If the model provides a reasonably good approximation to reality, it should do a good job of accounting for the observed relationships in this data set. In other words, the model should provide a good fit to the data.

The procedures for determining whether a path-analytic model fits the data were presented in the last chapter, and each of these procedures can also be used to assess the fit of a confirmatory factor model. A few modifications will be necessary (since CFA models tend to be somewhat more complex than path analytic models) but the basic strategy for assessing fit remains the same. The process begins by reviewing overall goodness of fit indices (such as the chi-square test, the CFI, and the NNFI), and then proceeds to indices that provide a more detailed assessment of fit (significance tests for factor loadings, R^2 values, normalized residuals, and modification indices).

Step 1: Reviewing the chi-square test. When the proper assumptions are met (e.g., large sample, multivariate normal distribution), the chi-square test provides a statistical test of the null hypothesis that the model fits the data. If the model provides a good fit, the chi-square value will be relatively small, and the corresponding *p* value will be relatively large (above .05 and preferably closer to 1.00). As was mentioned in the last chapter, however, with large samples and real-world data, the chi-square statistic will very frequently be significant even if the model provides a good fit (James et al., 1982). This is particularly true with CFA models, which tend to be more complex than simple path analysis models. For these reasons, it is frequently appropriate to conclude that a CFA model fits the data even if *p* is significant.

In real-world applications, therefore, it has become common practice to seek a model with a *relatively small* chi-square value, rather than necessarily seek a model with a *nonsignificant* chi-square. Just how small the chi-square must be depends on the degrees of freedom (*df*) associated with the analysis. If the model analyzed is the true model, then chi-square has an expected value equal to the *df* (Marsh, Balla, & McDonald, 1988). That is, if *df* = 100 for the chi-square test, then chi-square is also expected to equal 100 if the model analyzed is the true model. This means that researchers typically attempt to develop a model with a chi-square value that is small in relation to the *df*. Many use the informal criterion that the model may be acceptable if the chi-square value is less than twice the size of the degrees of freedom. Therefore, if *df* = 100, chi-square should be less than 200 for the model to be considered acceptable.

It should be noted however, that this criterion for model fit has a number of weaknesses, and should be used only with caution. First, the requirement that the chi-square/*df* ratio should be less than 2 is obviously somewhat arbitrary. Why not demand a ratio of 1.5? Or 2.5? Second, it has been shown that the chi-square/*df* ratio is affected by sample size, and that the same model may display significantly different ratios with small samples than with large samples (Marsh et al., 1988). For these reasons, we advise that this criterion be used only as a very rough rule of thumb, if at all, and be supplemented with other criteria that are not affected by sample size.

The chi-square goodness of fit statistic appears on page 14 of the current output in the table of goodness of fit statistics. This table is reproduced here as Output 5.2.

```
                          The SAS System                              14

         Covariance Structure Analysis: Maximum Likelihood Estimation

         Fit criterion . . . . . . . . . . . . . . . . . . .     1.0363
         Goodness of Fit Index (GFI) . . . . . . . . . . .       0.9064
         GFI Adjusted for Degrees of Freedom (AGFI)  . . .       0.8702
         Root Mean Square Residual (RMR) . . . . . . . . .       0.2373
         Chi-square = 247.6750      df = 137     Prob>chi**2 = 0.0001
         Null Model Chi-square:     df = 171                  2459.6733
         RMSEA Estimate  . . . . . .  0.0583  90%C.I.[0.0464, 0.0696]
         Bentler's Comparative Fit Index . . . . . . . . .       0.9516
         Normal Theory Reweighted LS Chi-square  . . . . .     234.4078
         Akaike's Information Criterion  . . . . . . . . .     -26.3250
         Consistent Information Criterion  . . . . . . . .    -640.1725
         Schwarz's Bayesian Criterion  . . . . . . . . . .    -503.1725
         McDonald's (1989) Centrality. . . . . . . . . . .       0.7941
         Bentler & Bonett's (1980) Non-normed Index. . . .       0.9396
         Bentler & Bonett's (1980) Normed Index. . . . . .       0.8993
         James, Mulaik, & Brett (1982) Parsimonious Index.      0.7205
         Z-Test of Wilson & Hilferty (1931) . . . . . . .        5.4582
         Bollen (1986) Normed Index Rho1 . . . . . . . . .       0.8743
         Bollen (1988) Non-normed Index Delta2 . . . . . .       0.9524
         Hoelter's (1983) Critical N . . . . . . . . . . .          161
```

Output 5.2: Table of Goodness of Fit Indices, Analysis of Initial Measurement Model, Investment Model Study

The chi-square test for the present model is included as the fifth line of Output 5.2. This line is again reproduced here:

```
Chi-square = 247.6750        df = 137      Prob>chi**2 = 0.0001
```

The entry labelled "Prob>chi**2" gives the *p* value for the test, which at .0001 is highly significant. Technically, this indicates that the model does not fit. However, because of the problems with this significance test mentioned earlier, this finding by itself will not cause you to reject the model.

The chi-square/*df* ratio for this model is 1.81 (since 247.6750 / 137 = 1.81). Because the ratio is less than 2, the model may be acceptable according to the rule of thumb discussed previously. Still, you will check additional indices before making any decisions.

Step 2: Reviewing the non-normed fit index and the comparative fit index. The previous chapter recommended the non-normed fit index, or NNFI (Bentler and Bonett, 1980) and the comparative fit index, or CFI (Bentler, 1989) as overall goodness of fit indices. These indices are generally preferable to the normed fit index, or NFI (Bentler & Bonett, 1980) as they are less likely to produce biased estimates in small samples (Bentler, 1989; Marsh et al., 1988). Values over .9 on the NNFI and CFI indicate an acceptable fit.

Values for the CFI and NNFI for the current analysis appear as the 8th and 14th entries, respectively, in the fit index summary table of Output 5.2. These entries are reproduced here.

```
Bentler's Comparative Fit Index . . . . . . . .    0.9516

Bentler & Bonett's (1980) Non-normed Index. . . .   0.9396
```

Both indices are greater than .9, indicating that the present model may provide an acceptable fit. Again, however, you will not draw any conclusions until all results have been reviewed.

Step 3: Reviewing significance tests for factor loadings. Remember that, in this analysis, a factor loading is equivalent to a path coefficient from a latent factor to an indicator variable. Therefore, a nonsignificant factor loading means that the involved indicator variable is not doing a good job of measuring the underlying factor, and should possibly be reassigned or dropped. Pages 15–16 of the current output present the nonstandardized factor loadings for the 19 indicators, along with the corresponding standard errors and large-sample *t* values (this information appears under the heading "Manifest Variable Equations"). These are reproduced here in Output 5.3.

```
                          The SAS System                          15
         Covariance Structure Analysis: Maximum Likelihood Estimation
                      Manifest Variable Equations

          V1       =       2.1735*F1      +   1.0000 E1
          Std Err          0.1286 LV1F1
          t Value         16.9018


          V2       =       2.4393*F1      +   1.0000 E2
          Std Err          0.1543 LV2F1
          t Value         15.8086


          V3       =       2.5449*F1      +   1.0000 E3
          Std Err          0.1347 LV3F1
          t Value         18.8927


          V4       =       2.3734*F1      +   1.0000 E4
          Std Err          0.1580 LV4F1
          t Value         15.0242


          V5       =       1.5951*F2      +   1.0000 E5
          Std Err          0.1051 LV5F2
          t Value         15.1797


          V6       =       1.8279*F2      +   1.0000 E6
          Std Err          0.1125 LV6F2
          t Value         16.2482


          V7       =       1.8024*F2      +   1.0000 E7
          Std Err          0.1086 LV7F2
          t Value         16.5890


          V8       =       0.9415*F3      +   1.0000 E8
          Std Err          0.0944 LV8F3
          t Value          9.9730


          V9       =       0.8964*F3      +   1.0000 E9
          Std Err          0.0943 LV9F3
          t Value          9.5089


          V10      =       1.2921*F3      +   1.0000 E10
          Std Err          0.1135 LV10F3
          t Value         11.3794


          V11      =       2.1428*F4      +   1.0000 E11
          Std Err          0.1709 LV11F4
          t Value         12.5358


          V12      =       2.3272*F4      +   1.0000 E12
          Std Err          0.1775 LV12F4
          t Value         13.1099
```

```
V13      =      1.0931*F4      +   1.0000 E13
Std Err         0.1565 LV13F4
t Value         6.9856

V14      =      1.7714*F5      +   1.0000 E14
Std Err         0.1235 LV14F5
```

Covariance Structure Analysis: Maximum Likelihood Estimation   16

```
t Value         14.3410

V15      =      1.5724*F5      +   1.0000 E15
Std Err         0.1359 LV15F5
t Value         11.5696

V16      =      1.0300*F5      +   1.0000 E16
Std Err         0.1215 LV16F5
t Value         8.4786

V17      =      1.3614*F6      +   1.0000 E17
Std Err         0.1302 LV17F6
t Value         10.4573

V18      =      1.4985*F6      +   1.0000 E18
Std Err         0.1266 LV18F6
t Value         11.8382

V19      =      1.5921*F6      +   1.0000 E19
Std Err         0.1412 LV19F6
t Value         11.2759
```

Variances of Exogenous Variables

Variable	Parameter	Estimate	Standard Error	t Value
F1		1.000000	0	0.000
F2		1.000000	0	0.000
F3		1.000000	0	0.000
F4		1.000000	0	0.000
F5		1.000000	0	0.000
F6		1.000000	0	0.000
E1	VARE1	1.453251	0.172294	8.435
E2	VARE2	2.512149	0.274058	9.166
E3	VARE3	0.942263	0.161894	5.820
E4	VARE4	2.926274	0.307410	9.519
E5	VARE5	1.176567	0.141685	8.304
E6	VARE6	1.123778	0.155095	7.246
E7	VARE7	0.977237	0.143165	6.826
E8	VARE8	1.122631	0.137736	8.151

E9	VARE9	1.179031	0.137753	8.559
E10	VARE10	1.301960	0.201908	6.448
E11	VARE11	2.142124	0.485055	4.416
E12	VARE12	1.826149	0.548805	3.327
E13	VARE13	4.381591	0.421740	10.389
E14	VARE14	1.280446	0.236968	5.403
E15	VARE15	2.451907	0.284243	8.626
E16	VARE16	2.451876	0.244926	10.011
E17	VARE17	2.151958	0.262947	8.184
E18	VARE18	1.619494	0.251286	6.445
E19	VARE19	2.238328	0.309491	7.232

```
      Covariance Structure Analysis: Maximum Likelihood Estimation      17

                   Covariances among Exogenous Variables
     ---------------------------------------------------------------------
                                             Standard
           Parameter          Estimate         Error        t Value
     ---------------------------------------------------------------------
     F2   F1   CF1F2           0.619199       0.046089       13.435
     F3   F1   CF1F3           0.438536       0.066112        6.633
     F3   F2   CF2F3           0.533795       0.062352        8.561
     F4   F1   CF1F4          -0.025940       0.072645       -0.357
     F4   F2   CF2F4          -0.224532       0.071155       -3.156
     F4   F3   CF3F4          -0.092181       0.081553       -1.130
     F5   F1   CF1F5           0.712953       0.044027       16.194
     F5   F2   CF2F5           0.635077       0.051991       12.215
     F5   F3   CF3F5           0.516257       0.068866        7.497
     F5   F4   CF4F5           0.007703       0.078803        0.098
     F6   F1   CF1F6          -0.259091       0.071774       -3.610
     F6   F2   CF2F6          -0.374124       0.068976       -5.424
     F6   F3   CF3F6          -0.423776       0.075095       -5.643
     F6   F4   CF4F6           0.253639       0.076059        3.335
     F6   F5   CF5F6          -0.300077       0.076943       -3.900
```

Output 5.3: Nonstandardized Factor Loadings, Exogenous Variable Variances, and Covariances among Exogenous Variables for Initial Measurement Model, Invesment Model Study

Before reviewing the results of the *t* tests, first verify that there are no near-zero standard errors (such as .0003). Standard errors this low may indicate an estimation problem. For the factor loadings, these standard errors appear in the section headed "Manifest Variable Equations" to the right of the heading "Std Err ." No problematic standard errors of this nature appear in the current output.

The *t* values presented on these pages represent large-sample *t* tests of the null hypothesis that the factor loading is equal to zero in the population. *t* values greater than 1.960 are significant at *p* < .05; those greater than 2.576 are significant at *p* < .01; and those greater than 3.291 are significant at *p* < .001. The obtained *t* values in the output show that all factor loadings were significant at *p* < .001.

Standardized loadings appear on pages 18 and 19 of the output, in the section headed "Equations with Standardized Coefficients." They are reproduced here in Output 5.4. A given factor loading appears just above its name (i.e., the standardized factor loading for V1 on F1 is .8745, and this appears just above the name "LV1F1" on output page 18). This output shows that the standardized loadings range in size from .46 to .93, and that only two are under .60. It can therefore be said that all loadings were at least moderately large.

```
                          The SAS System                          18

            Covariance Structure Analysis: Maximum Likelihood Estimation

               Equations with Standardized Coefficients

          V1        =        0.8745*F1       + 0.4850 E1
                             LV1F1

          V2        =        0.8385*F1       + 0.5448 E2
                             LV2F1

          V3        =        0.9343*F1       + 0.3564 E3
                             LV3F1

          V4        =        0.8112*F1       + 0.5847 E4
                             LV4F1

          V5        =        0.8269*F2       + 0.5623 E5
                             LV5F2

          V6        =        0.8650*F2       + 0.5017 E6
                             LV6F2

          V7        =        0.8768*F2       + 0.4809 E7
                             LV7F2

          V8        =        0.6643*F3       + 0.7475 E8
                             LV8F3

          V9        =        0.6366*F3       + 0.7712 E9
                             LV9F3

          V10       =        0.7496*F3       + 0.6619 E10
                             LV10F3

          V11       =        0.8258*F4       + 0.5640 E11
                             LV11F4
```

```
V12       =       0.8648*F4      + 0.5022 E12
                  LV12F4

V13       =       0.4629*F4      + 0.8864 E13
                  LV13F4

V14       =       0.8427*F5      + 0.5383 E14
                  LV14F5

V15       =       0.7086*F5      + 0.7056 E15
                  LV15F5

V16       =       0.5495*F5      + 0.8355 E16
                  LV16F5

V17       =       0.6802*F6      + 0.7330 E17
                  LV17F6

V18       =       0.7622*F6      + 0.6473 E18
                  LV18F6
```

Covariance Structure Analysis: Maximum Likelihood Estimation   19

```
V19       =       0.7287*F6      + 0.6848 E19
                  LV19F6
```

Variances of Endogenous Variables

	Variable	Estimate	R-squared
1	V1	6.177239	0.764741
2	V2	8.462500	0.703143
3	V3	7.418794	0.872990
4	V4	8.559396	0.658121
5	V5	3.720909	0.683796
6	V6	4.464852	0.748306
7	V7	4.225786	0.768744
8	V8	2.009106	0.441229
9	V9	1.982532	0.405290
10	V10	2.971406	0.561837
11	V11	6.733877	0.681888
12	V12	7.242161	0.747845
13	V13	5.576505	0.214276
14	V14	4.418458	0.710205
15	V15	4.924352	0.502085
16	V16	3.512720	0.302001
17	V17	4.005361	0.462731
18	V18	3.864986	0.580983
19	V19	4.773149	0.531058

```
          Correlations among Exogenous Variables
          ------------------------------------
                  Parameter            Estimate
          ------------------------------------
            F2    F1    CF1F2          0.619199
            F3    F1    CF1F3          0.438536
            F3    F2    CF2F3          0.533795
            F4    F1    CF1F4         -0.025940
            F4    F2    CF2F4         -0.224532
            F4    F3    CF3F4         -0.092181
            F5    F1    CF1F5          0.712953
            F5    F2    CF2F5          0.635077
            F5    F3    CF3F5          0.516257
            F5    F4    CF4F5          0.007703
            F6    F1    CF1F6         -0.259091
            F6    F2    CF2F6         -0.374124
            F6    F3    CF3F6         -0.423776
            F6    F4    CF4F6          0.253639
            F6    F5    CF5F6         -0.300077
```

Output 5.4: Standardized Factor Loadings, Exogenous Variable Variances, and Correlations among Exogenous Variables for Initial Measurement Model, Investment Model Study

Step 4: Reviewing the residual matrix and normalized residual matrix. The residual matrix, normalized residual matrix, and distribution of normalized residuals appeared on pages 10-13 of the output, and are reproduced in Output 5.5.

```
                        The SAS System                               10

                        Residual Matrix

          V1          V2          V3          V4          V5          V6          V7

V1    0.002957    0.006293    0.014877   -0.270414    0.318178    0.345099    0.242397
V2    0.006293   -0.000219    0.020487    0.441025   -0.377943   -0.634098   -0.658953
V3    0.014877    0.020487    0.001382   -0.054335    0.092719   -0.278719   -0.000707
V4   -0.270414    0.441025   -0.054335    0.002080    0.314254    0.108292    0.635854
V5    0.318178   -0.377943    0.092719    0.314254    0.000132   -0.009448   -0.019427
V6    0.345099   -0.634098   -0.278719    0.108292   -0.009448   -0.000083    0.024580
V7    0.242397   -0.658953   -0.000707    0.635854   -0.019427    0.024580    0.001350
V8    0.321426    0.200574    0.265452    0.238994    0.119484    0.204144   -0.075536
V9   -0.122824   -0.356800   -0.359882   -0.051344   -0.081506    0.035780    0.171054
V10   0.264233   -0.173532   -0.094190   -0.154343   -0.162322    0.017951   -0.165556
V11   0.449824    0.754598    0.176804   -0.156605   -0.038468   -0.030769    0.242953
V12  -0.136383    0.249026   -0.264193   -0.565367   -0.147588    0.102214   -0.109404
V13  -0.108511   -0.013214   -0.352127   -0.091523   -0.109267   -0.055024    0.039646
V14   0.176093   -0.463672    0.112617   -0.014556   -0.038781   -0.053220    0.003530
V15  -0.042436   -0.656078   -0.108730    0.092227    0.427514    0.134603    0.093498
```

```
V16    0.151006   -0.014080    0.331379   -0.037542   -0.128792   -0.181927   -0.312042
V17    0.065238    0.423848    0.107298   -0.772938   -0.345545   -0.261342   -0.336782
V18    0.184035   -0.105249    0.163322   -0.608696    0.166106    0.260376    0.185864
V19   -0.010567    0.103655    0.186746   -1.111572   -0.036162    0.123826   -0.054004
```

Covariance Structure Analysis: Maximum Likelihood Estimation   11

Residual Matrix

	V8	V9	V10	V11	V12	V13	V14
V1	0.321426	-0.122824	0.264233	0.449824	-0.136383	-0.108511	0.176093
V2	0.200574	-0.356800	-0.173532	0.754598	0.249026	-0.013214	-0.463672
V3	0.265452	-0.359882	-0.094190	0.176804	-0.264193	-0.352127	0.112617
V4	0.238994	-0.051344	-0.154343	-0.156605	-0.565367	-0.091523	-0.014556
V5	0.119484	-0.081506	-0.162322	-0.038468	-0.147588	-0.109267	-0.038781
V6	0.204144	0.035780	0.017951	-0.030769	0.102214	-0.055024	-0.053220
V7	-0.075536	0.171054	-0.165556	0.242953	-0.109404	0.039646	0.003530
V8	-0.001217	-0.065866	0.019591	-0.148639	-0.186959	-0.048924	0.047406
V9	-0.065866	-0.000068	0.036088	-0.023896	0.055896	0.273082	0.201308
V10	0.019591	0.036088	0.000770	-0.026628	0.193675	0.117989	0.025117
V11	-0.148639	-0.023896	-0.026628	0.000148	-0.000916	-0.021309	-0.012878
V12	-0.186959	0.055896	0.193675	-0.000916	-0.000680	0.015409	-0.060040
V13	-0.048924	0.273082	0.117989	-0.021309	0.015409	-0.006905	0.198394
V14	0.047406	0.201308	0.025117	-0.012878	-0.060040	0.198394	-0.000054
V15	-0.119712	-0.005925	-0.126896	-0.239013	0.013610	-0.337925	-0.010138
V16	-0.083733	-0.078207	-0.234725	0.435260	0.092479	0.230149	-0.024349
V17	-0.018219	-0.046334	0.096883	0.433598	0.359493	-0.136618	-0.201664
V18	0.157736	0.201066	-0.003121	-0.141009	-0.254956	0.080985	0.201471
V19	-0.256442	-0.312003	0.043032	-0.031822	-0.104849	-0.049524	0.024195

	V15	V16	V17	V18	V19
V1	-0.042436	0.151006	0.065238	0.184035	-0.010567
V2	-0.656078	-0.014080	0.423848	-0.105249	0.103655
V3	-0.108730	0.331379	0.107298	0.163322	0.186746
V4	0.092227	-0.037542	-0.772938	-0.608696	-1.111572
V5	0.427514	-0.128792	-0.345545	0.166106	-0.036162
V6	0.134603	-0.181927	-0.261342	0.260376	0.123826
V7	0.093498	-0.312042	-0.336782	0.185864	-0.054004
V8	-0.119712	-0.083733	-0.018219	0.157736	-0.256442
V9	-0.005925	-0.078207	-0.046334	0.201066	-0.312003

```
        V10  -0.126896  -0.234725   0.096883  -0.003121   0.043032
        V11  -0.239013   0.435260   0.433598  -0.141009  -0.031822
        V12   0.013610   0.092479   0.359493  -0.254956  -0.104849
        V13  -0.337925   0.230149  -0.136618   0.080985  -0.049524
        V14  -0.010138  -0.024349  -0.201664   0.201471   0.024195
        V15  -0.000391   0.085416  -0.494332   0.057033  -0.228175
        V16   0.085416  -0.000844   0.177027   0.190506   0.152217
        V17  -0.494332   0.177027  -0.001360   0.041019  -0.156292
        V18   0.057033   0.190506   0.041019   0.000170   0.071372
        V19  -0.228175   0.152217  -0.156292   0.071372   0.001076
               Average Absolute Residual = 0.1607
        Average Off-diagonal Absolute Residual = 0.1784
```

```
        Covariance Structure Analysis: Maximum Likelihood Estimation      12

                      Normalized Residual Matrix

              V1         V2         V3         V4         V5         V6         V7

    V1     0.0052     0.0109     0.0264    -0.4699     0.9384     0.9219     0.6640
    V2     0.0109    -0.0003     0.0315     0.6638    -0.9588    -1.4578    -1.5537
    V3     0.0264     0.0315     0.0020    -0.0842     0.2466    -0.6709    -0.0017
    V4    -0.4699     0.6638    -0.0842     0.0027     0.7967     0.2489     1.4990
    V5     0.9384    -0.9588     0.2466     0.7967     0.0004    -0.0292    -0.0614
    V6     0.9219    -1.4578    -0.6709     0.2489    -0.0292    -0.0002     0.0698
    V7     0.6640    -1.5537    -0.0017     1.4990    -0.0614     0.0698     0.0035
    V8     1.3697     0.7321     1.0278     0.8689     0.6497     1.0095    -0.3835
    V9    -0.5282    -1.3140    -1.4067    -0.1883    -0.4476     0.1787     0.8774
    V10    0.9183    -0.5168    -0.2971    -0.4581    -0.7180     0.0722    -0.6830
    V11    1.0803     1.5484     0.3874    -0.3195    -0.1177    -0.0858     0.6964
    V12   -0.3158     0.4927    -0.5583    -1.1123    -0.4349     0.2746    -0.3020
    V13   -0.2864    -0.0298    -0.8481    -0.2052    -0.3702    -0.1701     0.1260
    V14    0.4623    -1.0491     0.2657    -0.0330    -0.1355    -0.1684     0.0115
    V15   -0.1090    -1.4498    -0.2520     0.2036     1.4501     0.4144     0.2954
    V16    0.4751    -0.0380     0.9444    -0.1011    -0.5302    -0.6813    -1.1998
    V17    0.2008     1.1157     0.3009    -2.0245    -1.3569    -0.9350    -1.2377
    V18    0.5750    -0.2813     0.4647    -1.6189     0.6605     0.9428     0.6912
    V19   -0.0297     0.2496     0.4788    -2.6631    -0.1297     0.4044    -0.1812


              V8         V9         V10        V11        V12        V13        V14

    V1     1.3697    -0.5282     0.9183     1.0803    -0.3158    -0.2864     0.4623
    V2     0.7321    -1.3140    -0.5168     1.5484     0.4927    -0.0298    -1.0491
    V3     1.0278    -1.4067    -0.2971     0.3874    -0.5583    -0.8481     0.2657
    V4     0.8689    -0.1883    -0.4581    -0.3195    -1.1123    -0.2052    -0.0330
    V5     0.6497    -0.4476    -0.7180    -0.1177    -0.4349    -0.3702    -0.1355
    V6     1.0095     0.1787     0.0722    -0.0858     0.2746    -0.1701    -0.1684
    V7    -0.3835     0.8774    -0.6830     0.6964    -0.3020     0.1260     0.0115
    V8    -0.0066    -0.4709     0.1112    -0.6252    -0.7582    -0.2263     0.2368
    V9    -0.4709    -0.0004     0.2079    -0.1012     0.2282     1.2719     1.0155
```

V10	0.1112	0.2079	0.0028	-0.0921	0.6456	0.4488	0.1021
V11	-0.6252	-0.1012	-0.0921	0.0002	-0.0017	-0.0503	-0.0366
V12	-0.7582	0.2282	0.6456	-0.0017	-0.0010	0.0349	-0.1644
V13	-0.2263	1.2719	0.4488	-0.0503	0.0349	-0.0136	0.6192
V14	0.2368	1.0155	0.1021	-0.0366	-0.1644	0.6192	-0.0001
V15	-0.5729	-0.0286	-0.4956	-0.6430	0.0353	-0.9990	-0.0289
V16	-0.4798	-0.4518	-1.1009	1.3864	0.2840	0.8056	-0.0869
V17	-0.0977	-0.2505	0.4252	1.2805	1.0227	-0.4464	-0.7319
V18	0.8574	1.1022	-0.0139	-0.4228	-0.7363	0.2692	0.7416
V19	-1.2567	-1.5418	0.1725	-0.0860	-0.2728	-0.1482	0.0803

Covariance Structure Analysis: Maximum Likelihood Estimation   13

Normalized Residual Matrix

	V15	V16	V17	V18	V19
V1	-0.1090	0.4751	0.2008	0.5750	-0.0297
V2	-1.4498	-0.0380	1.1157	-0.2813	0.2496
V3	-0.2520	0.9444	0.3009	0.4647	0.4788
V4	0.2036	-0.1011	-2.0245	-1.6189	-2.6631
V5	1.4501	-0.5302	-1.3569	0.6605	-0.1297
V6	0.4144	-0.6813	-0.9350	0.9428	0.4044
V7	0.2954	-1.1998	-1.2377	0.6912	-0.1812
V8	-0.5729	-0.4798	-0.0977	0.8574	-1.2567
V9	-0.0286	-0.4518	-0.2505	1.1022	-1.5418
V10	-0.4956	-1.1009	0.4252	-0.0139	0.1725
V11	-0.6430	1.3864	1.2805	-0.4228	-0.0860
V12	0.0353	0.2840	1.0227	-0.7363	-0.2728
V13	-0.9990	0.8056	-0.4464	0.2692	-0.1482
V14	-0.0289	-0.0869	-0.7319	0.7416	0.0803
V15	-0.0009	0.2965	-1.7066	0.1999	-0.7205
V16	0.2965	-0.0026	0.7266	0.7947	0.5718
V17	-1.7066	0.7266	-0.0037	0.1434	-0.4961
V18	0.1999	0.7947	0.1434	0.0005	0.2250
V19	-0.7205	0.5718	-0.4961	0.2250	0.0025

Average Normalized Residual = 0.5019
Average Off-diagonal Normalized Residual = 0.5574

Rank Order of 10 Largest Normalized Residuals

V19,V4	V17,V4	V17,V15	V18,V4	V7,V2	V11,V2	V19,V9
-2.6631	-2.0245	-1.7066	-1.6189	-1.5537	1.5484	-1.5418

	V7,V4	V6,V2	V15,V5
	1.4990	-1.4578	1.4501

```
                 Distribution of Normalized Residuals
                    (Each * represents 2 residuals)

 -2.75000 -   -2.50000    1    0.53% |
 -2.50000 -   -2.25000    0    0.00% |
 -2.25000 -   -2.00000    1    0.53% |
 -2.00000 -   -1.75000    0    0.00% |
 -1.75000 -   -1.50000    4    2.11% | **
 -1.50000 -   -1.25000    6    3.16% | ***
 -1.25000 -   -1.00000    5    2.63% | **
 -1.00000 -   -0.75000    5    2.63% | **
 -0.75000 -   -0.50000   14    7.37% | *******
 -0.50000 -   -0.25000   22   11.58% | ***********
 -0.25000 -          0   43   22.63% | *********************
        0 -    0.25000   34   17.89% | *****************
  0.25000 -    0.50000   17    8.95% | ********
  0.50000 -    0.75000   13    6.84% | ******
  0.75000 -    1.00000   11    5.79% | *****
  1.00000 -    1.25000    7    3.68% | ***
  1.25000 -    1.50000    6    3.16% | ***
  1.50000 -    1.75000    1    0.53% |
```

Output 5.5: Residual Matrix, Normalized Residual Matrix, and Distribution of Normalized Residuals for Initial Measurement Model, Investment Model Study

If the model provides a good fit to the data, entries in the residual matrix are expected to be zero or near-zero. Unfortunately, the residuals in this matrix are difficult to interpret if the unit of measurement varies much from indicator variable to indicator variable (which will often be the case). Therefore, it is usually easier to instead review the normalized residual matrix, which appears on pages 12-13 of Output 5.5. Normalized residuals over 2.00 are generally considered large and therefore problematic.

Before inspecting the individual normalized residuals, it is good practice to first review the "Distribution of Normalized Residuals" table, which appears on page 13 of the output. This table summarizes the number of residuals that fall into the interval from 0.00 to 0.25, the number that fall in the interval from 0.25 to 0.50, and so forth. When a model provides a good fit to the data, you expect to see that the distribution

- is centered on zero

- is symmetrical

- contains no or few large residuals.

The residual summary table on page 13 of this output is centered on zero, but is not perfectly symmetrical. The bars on the lower half of the table (which represent the number of positive residuals) are somewhat longer than the bars on the upper half of the table.

The normalized residuals themselves appear on pages12-13, under the heading "Normalized Residual Matrix." The rank order of the ten largest normalized residuals appears at the bottom of the table on page 13 (in most cases, only this portion of the table will have to be reviewed). This rank order table shows that there is a large residual for V19 and V4, and another large residual for V17 and V4. Notice that both large residuals involve the manifest variable V4, which was predicted to load on F1 (see Figure 5.8); and the other two variables (V19 and V17) both load on F6.

To determine why these residuals are large, you must compare corresponding entries from the actual covariance matrix and the predicted covariance matrix. Output 5.6 reproduces the matrix of actual covariances, as well as the matrix of predicted covariances from the analysis.

```
                          The SAS System                                    4

                             Covariances

              V1          V2          V3          V4          V5          V6          V7

    V1     6.180196    5.308122    5.546157    4.888152    2.464884    2.805058    2.668055
    V2     5.308122    8.462281    6.228355    6.230589    2.031349    2.126764    2.063412
    V3     5.546157    6.228355    7.420176    5.985788    2.606280    2.601627    2.839476
    V4     4.888152    6.230589    5.985788    8.561476    2.658444    2.794552    3.284657
    V5     2.464884    2.031349    2.606280    2.658444    3.721041    2.906172    2.855537
    V6     2.805058    2.126764    2.601627    2.794552    2.906172    4.464769    3.319067
    V7     2.668055    2.063412    2.839476    3.284657    2.855537    3.319067    4.227136
    V8     1.218841    1.207762    1.316229    1.218966    0.921153    1.122795    0.830305
    V9     0.731560    0.602093    0.640510    0.881639    0.681724    0.910382    1.033461
    V10    1.495767    1.208643    1.347803    1.190484    0.937818    1.278627    1.077541
    V11    0.329010    0.619006    0.035344   -0.288533   -0.805927   -0.910217   -0.624232
    V12   -0.267593    0.101766   -0.417826   -0.708648   -0.981087   -0.852912   -1.051212
    V13   -0.170142   -0.082383   -0.424290   -0.158823   -0.500768   -0.503655   -0.402729
    V14    2.921095    2.617099    3.326718    2.982969    1.755710    2.003128    2.031205
    V15    2.394132    2.078533    2.744228    2.752945    2.020373    1.959896    1.893340
    V16    1.747037    1.777178    2.200158    1.705314    0.914581    1.013699    0.866912
    V17   -0.701403   -0.436568   -0.790355   -1.610105   -1.157979   -1.192328   -1.254787
    V18   -0.659809   -1.052313   -0.824729   -1.530169   -0.728143   -0.764365   -0.824588
    V19   -0.907129   -0.902575   -0.863031   -2.090612   -0.986278   -0.964933   -1.127582
```

```
        Covariance Structure Analysis: Maximum Likelihood Estimation        5

                             Covariances

              V8          V9          V10         V11         V12         V13         V14

    V1     1.218841    0.731560    1.495767    0.329010   -0.267593   -0.170142    2.921095
    V2     1.207762    0.602093    1.208643    0.619006    0.101766   -0.082383    2.617099
    V3     1.316229    0.640510    1.347803    0.035344   -0.417826   -0.424290    3.326718
    V4     1.218966    0.881639    1.190484   -0.288533   -0.708648   -0.158823    2.982969
    V5     0.921153    0.681724    0.937818   -0.805927   -0.981087   -0.500768    1.755710
    V6     1.122795    0.910382    1.278627   -0.910217   -0.852912   -0.503655    2.003128
    V7     0.830305    1.033461    1.077541   -0.624232   -1.051212   -0.402729    2.031205
```

V8	2.007889	0.778103	1.236111	-0.334617	-0.388941	-0.143797	0.908453
V9	0.778103	1.982464	1.194277	-0.200957	-0.136401	0.182758	1.021068
V10	1.236111	1.194277	2.972176	-0.281848	-0.083507	-0.012206	1.206741
V11	-0.334617	-0.200957	-0.281848	6.734025	4.985966	2.321072	0.016364
V12	-0.388941	-0.136401	-0.083507	4.985966	7.241481	2.559356	-0.028282
V13	-0.143797	0.182758	-0.012206	2.321072	2.559356	5.569600	0.213311
V14	0.908453	1.021068	1.206741	0.016364	-0.028282	0.213311	4.418404
V15	0.644586	0.721725	0.921959	-0.213057	0.041799	-0.324684	2.775281
V16	0.416907	0.398427	0.452309	0.452262	0.110945	0.238823	1.800191
V17	-0.561413	-0.563482	-0.648548	1.173526	1.163093	0.240840	-0.925342
V18	-0.440160	-0.368161	-0.823620	0.673434	0.629570	0.496454	-0.595085
V19	-0.891690	-0.916791	-0.828727	0.833501	0.834937	0.391902	-0.822124

	V15	V16	V17	V18	V19
V1	2.394132	1.747037	-0.701403	-0.659809	-0.907129
V2	2.078533	1.777178	-0.436568	-1.052313	-0.902575
V3	2.744228	2.200158	-0.790355	-0.824729	-0.863031
V4	2.752945	1.705314	-1.610105	-1.530169	-2.090612
V5	2.020373	0.914581	-1.157979	-0.728143	-0.986278
V6	1.959896	1.013699	-1.192328	-0.764365	-0.964933
V7	1.893340	0.866912	-1.254787	-0.824588	-1.127582
V8	0.644586	0.416907	-0.561413	-0.440160	-0.891690
V9	0.721725	0.398427	-0.563482	-0.368161	-0.916791
V10	0.921959	0.452309	-0.648548	-0.823620	-0.828727
V11	-0.213057	0.452262	1.173526	0.673434	0.833501
V12	0.041799	0.110945	1.163093	0.629570	0.834937
V13	-0.324684	0.238823	0.240840	0.496454	0.391902
V14	2.775281	1.800191	-0.925342	-0.595085	-0.822124
V15	4.923961	1.704946	-1.136696	-0.650021	-0.979400
V16	1.704946	3.511876	-0.243742	-0.272637	-0.339859
V17	-1.136696	-0.243742	4.004001	2.081068	2.011205
V18	-0.650021	-0.272637	2.081068	3.865156	2.457146
V19	-0.979400	-0.339859	2.011205	2.457146	4.774225

Determinant = 143562421 (Ln = 18.782)

Covariance Structure Analysis: Maximum Likelihood Estimation   9

Predicted Model Matrix

	V1	V2	V3	V4	V5	V6	V7
V1	6.177239	5.301829	5.531280	5.158566	2.146706	2.459959	2.425658
V2	5.301829	8.462500	6.207868	5.789564	2.409292	2.760862	2.722365
V3	5.531280	6.207868	7.418794	6.040123	2.513561	2.880346	2.840183
V4	5.158566	5.789564	6.040123	8.559396	2.344189	2.686260	2.648803
V5	2.146706	2.409292	2.513561	2.344189	3.720909	2.915619	2.874964
V6	2.459959	2.760862	2.880346	2.686260	2.915619	4.464852	3.294487
V7	2.425658	2.722365	2.840183	2.648803	2.874964	3.294487	4.225786
V8	0.897415	1.007187	1.050776	0.979972	0.801669	0.918651	0.905842
V9	0.854385	0.958893	1.000392	0.932983	0.763230	0.874602	0.862407
V10	1.231534	1.382175	1.441993	1.344827	1.100140	1.260676	1.243097

V11	-0.120814	-0.135592	-0.141460	-0.131928	-0.767458	-0.879448	-0.867185
V12	-0.131210	-0.147260	-0.153633	-0.143281	-0.833500	-0.955126	-0.941808
V13	-0.061631	-0.069169	-0.072163	-0.067300	-0.391502	-0.448631	-0.442375
V14	2.745002	3.080772	3.214101	2.997525	1.794491	2.056348	2.027675
V15	2.436568	2.734611	2.852958	2.660718	1.592859	1.825293	1.799842
V16	1.596031	1.791258	1.868779	1.742855	1.043374	1.195626	1.178954
V17	-0.766640	-0.860416	-0.897653	-0.837166	-0.812434	-0.930986	-0.918005
V18	-0.843844	-0.947064	-0.988051	-0.921473	-0.894249	-1.024741	-1.010452
V19	-0.896562	-1.006230	-1.049777	-0.979040	-0.950116	-1.088760	-1.073578

	V8	V9	V10	V11	V12	V13	V14
V1	0.897415	0.854385	1.231534	-0.120814	-0.131210	-0.061631	2.745002
V2	1.007187	0.958893	1.382175	-0.135592	-0.147260	-0.069169	3.080772
V3	1.050776	1.000392	1.441993	-0.141460	-0.153633	-0.072163	3.214101
V4	0.979972	0.932983	1.344827	-0.131928	-0.143281	-0.067300	2.997525
V5	0.801669	0.763230	1.100140	-0.767458	-0.833500	-0.391502	1.794491
V6	0.918651	0.874602	1.260676	-0.879448	-0.955126	-0.448631	2.056348
V7	0.905842	0.862407	1.243097	-0.867185	-0.941808	-0.442375	2.027675
V8	2.009106	0.843969	1.216520	-0.185978	-0.201982	-0.094873	0.861047
V9	0.843969	1.982532	1.158189	-0.177061	-0.192297	-0.090324	0.819760
V10	1.216520	1.158189	2.971406	-0.255220	-0.277182	-0.130195	1.181625
V11	-0.185978	-0.177061	-0.255220	6.733877	4.986881	2.342381	0.029242
V12	-0.201982	-0.192297	-0.277182	4.986881	7.242161	2.543947	0.031758
V13	-0.094873	-0.090324	-0.130195	2.342381	2.543947	5.576505	0.014917
V14	0.861047	0.819760	1.181625	0.029242	0.031758	0.014917	4.418458
V15	0.764298	0.727650	1.048855	0.025956	0.028190	0.013241	2.785419
V16	0.500640	0.476634	0.687034	0.017002	0.018465	0.008673	1.824539
V17	-0.543194	-0.517148	-0.745431	0.739928	0.803600	0.377458	-0.723678
V18	-0.597896	-0.569227	-0.820499	0.814442	0.884527	0.415470	-0.796556
V19	-0.635248	-0.604788	-0.871759	0.865323	0.939786	0.441425	-0.846319

Covariance Structure Analysis: Maximum Likelihood Estimation 9

Predicted Model Matrix

	V15	V16	V17	V18	V19
V1	2.436568	1.596031	-0.766640	-0.843844	-0.896562
V2	2.734611	1.791258	-0.860416	-0.947064	-1.006230
V3	2.852958	1.868779	-0.897653	-0.988051	-1.049777
V4	2.660718	1.742855	-0.837166	-0.921473	-0.979040
V5	1.592859	1.043374	-0.812434	-0.894249	-0.950116
V6	1.825293	1.195626	-0.930986	-1.024741	-1.088760
V7	1.799842	1.178954	-0.918005	-1.010452	-1.073578
V8	0.764298	0.500640	-0.543194	-0.597896	-0.635248
V9	0.727650	0.476634	-0.517148	-0.569227	-0.604788
V10	1.048855	0.687034	-0.745431	-0.820499	-0.871759
V11	0.025956	0.017002	0.739928	0.814442	0.865323
V12	0.028190	0.018465	0.803600	0.884527	0.939786
V13	0.013241	0.008673	0.377458	0.415470	0.441425

```
V14    2.785419    1.824539   -0.723678   -0.796556   -0.846319
V15    4.924352    1.619531   -0.642364   -0.707053   -0.751225
V16    1.619531    3.512720   -0.420769   -0.463143   -0.492077
V17   -0.642364   -0.420769    4.005361    2.040049    2.167497
V18   -0.707053   -0.463143    2.040049    3.864986    2.385774
V19   -0.751225   -0.492077    2.167497    2.385774    4.773149
           Determinant = 406137621 (Ln = 19.822)
```

Output 5.6: Actual Covariances and Predicted Covariances for Initial Measurement Model, Investment Model Study

The original covariance matrix is headed "Covariances", and appears on pages 4–5 of Output 5.6. You can find the actual covariance between V4 and V19 where the column headed V4 intersects with the row headed V19. You can see that the actual covariance between these two variables is –2.091.

The predicted covariance matrix is headed "Predicted Model Matrix", and appears on pages 9–10 of the output. The predicted covariance between V4 and V19 appears where the column headed V4 intersects with the row headed V19. You can see that the predicted covariance between these two variables was only –0.979.

The fact that this predicted covariance was so much smaller than the actual covariance means that your confirmatory factor model *underpredicts* the strength of the relationship between V4 and V19. Reviewing the corresponding covariance and predicted covariance for V4 and V17 reveals that your model also underpredicts the strength of the relationship between those two variables. This suggests the possibility that V4 is influenced by the latent factor F6 (which affects V17 and V19). Is it possible that V4 should be reassigned, so that you drop the path from F1 to V4, and add a new path from F6 to V4?

This would probably not be a satisfactory solution because V4 does seem to be doing a good job of measuring F1. The standardized coefficient for the path from F1 to V4 was fairly large at .81 (see Output 5.4, page 18), and was statistically significant (see Output 5.3, page 15).

A more likely interpretation is that V4 is a complex variable, that it is affected by both F1 and F6. It is possible that the most effective way of improving the model's fit is to modify it so that there is a path from F6 to V4 as well as a path from F1 to V4. But this is generally undesirable when performing path analysis with latent variables, as you usually desire to use only **factorially simple indicators**: indicators that measure only one factor. Rather than reassign V4, a better choice in this case would be to drop it from the analysis entirely. (But it is still premature to do this at the present time; you will first review the modification indices before making final decisions.)

This section describes only one way that residuals can be used to identify specification errors in a multiple-indicator measurement model. Additional procedures are discussed by Anderson and Gerbing (1988). For example, consider a situation in which a measurement model includes multiple factors, each with multiple indicators, and with no reverse-coded items. Anderson and Gerbing describe how the residuals may be reviewed to identify the following misspecifications:

- **An indicator variable is assigned to the wrong factor.** If, for example, the indicator variable was incorrectly assigned to factor F1, it will demonstrate large negative normalized residuals with the other indicator variables that were (correctly) assigned to F1. If the model contains another factor to which the indicator should have been assigned (say, F6), the indicator will display large positive residuals for the variables that were (correctly) assigned to this factor.

- **An indicator variable is multidimensional.** If an indicator is actually influenced by more than one factor, it will often display large normalized residuals for the indicator variables of more than one factor. Typically, these will be the only large normalized residuals that are displayed by the other indicators.

Note that the normalized residuals for the current analysis did not display a pattern of residuals that fits neatly into either of the above scenarios. This is because indicator V4 was reverse-coded, relative to V17 and V19, and although V4 appears to be multidimensional, it also appears to be influenced by only one other factor in addition to the one to which it was (correctly) assigned. The point being made here is that the pattern of large residuals obtained from PROC CALIS, when combined with other information, can often be helpful in identifying specification errors in the measurement model.

Modifying the Measurement Model

Strictly speaking, researchers do not normally review modification indices to assess a model's fit. In general, the indices discussed earlier in this section are consulted first, and if they provide evidence of a poor or questionable fit, you may then turn to the modification indices to determine which specific modifications might best improve the fit.

The results of the current analysis have provided mixed evidence concerning the fit of your measurement model. First, the chi-square test was significant, suggesting rejection of the model, but it has already been said that this is an excessively sensitive test in most applied situations, and that a significant chi-square alone should not be used as evidence of a poor fit. In support of the model, the chi-square/*df* ratio, the CFI, and the NNFI were all within an acceptable range. In addition, significance tests revealed that all factor loadings were significantly different from zero.

The only real problem involved the residuals and normalized residuals. The summary table for the normalized residuals was asymmetrical. In addition, the nature of the two large normalized residuals suggested that one indicator, V4, may be multifactorial. These findings indicate that it may be possible to substantially improve the fit of the measurement model, and therefore justify reviewing the modification indices.

The MODIFICATION option included in the PROC CALIS statement requests that two modification indices be computed (see the earlier section titled "Overview of the PROC CALIS program"). The first modification index is the Wald test, which identifies parameters that should possibly be dropped from the model. The second modification index is the Lagrange multiplier, which identifies parameters that should possibly be added.

The Wald test. The last chapter noted that when researchers modify a model based on modification indices they risk creating a model that will not generalize to other samples or to the population. With this in mind, it was also stated that it is generally safer to drop existing parameters than it is to add new ones. For this reason, The Wald test will be consulted first.

The results of the Wald test for the present analysis appear on page 37 of the output, and are reproduced as Output 5.7.

```
                           The SAS System                          37


                    Stepwise Multivariate Wald Test
        ---------------------------------------------------------------
                    Cumulative Statistics         Univariate Increment
        Parameter  Chi-Square    D.F.    Prob     Chi-Square    Prob
        ---------------------------------------------------------------
        CF4F5       0.009556      1      0.9221    0.009556     0.9221
        CF1F4       0.289501      2      0.8652    0.279945     0.5967
        CF3F4       1.732103      3      0.6298    1.442602     0.2297
```

Output 5.7: Wald Test Results for Initial Measurement Model, Investment Model Study

The Wald test estimates the change in model chi-square that would result from fixing a given parameter at zero. In other words, this test estimates the change in chi-square that would result from eliminating, say, a specific path or covariance from the model. The first parameter listed is the one that would result in the least change in chi-square if deleted, the second parameter would result in the second-least change, and so forth.

In one sense, dropping one of the parameters listed in the Wald test does not really improve the fit of the model (dropping any parameter will almost always increase chi-square, at least to a small degree). But fixing one of these parameters at zero does free up one degree of freedom for the analysis (e.g., the degrees of freedom for the current analysis would increase from 137 to 138 if one parameter were dropped). Freeing this degree of freedom decreases the ratio of chi-square to *df*, and this index is one of your informal measures of model fit. In addition, if the ratio can be decreased enough, it is even possible that a significant chi-square test will become nonsignificant (another indication of model fit).

The Wald test table of Output 5.7 suggests that three different parameters could be dropped without causing a significant increase in chi-square. These parameters are the covariances between F4 and F5, between F1 and F4, and between F3 and F4. This should come as no surprise, as the *t* test reported on page 17 of Output 5.3 has already shown that the covariances between these pairs of variables were all nonsignificant (and the results on page 19 of Output 5.4 showed that the correlations between these pairs of variables were near-zero).

Despite this, these parameters will not be dropped from the model. With confirmatory factor analysis, all factors are normally allowed to covary during the analysis, and therefore you will not drop these factor covariances simply because they were nonsignificant.

What you were really concerned about finding with the Wald test was some indication that a factor loading could be dropped from the model without hurting the model chi-square. Such a finding would indicate that an indicator variable was not doing a good job of measuring the factor to which it was assigned. However, it comes as no surprise that no factor loadings were reported in the Wald test; remember that the *t* tests for all of the factor loadings were significant when they were reviewed earlier.

The Lagrange multiplier test. The Lagrange multiplier test estimates the reduction in model chi-square that would result from freeing a fixed parameter and allowing it to be estimated. In other words, for a CFA the Lagrange multiplier estimates the degree to which chi-square would improve if a new factor loading or covariance were added to your model. Lagrange multipliers appear in the current output on pages 20-37.

Two matrices of Lagrange multiplier tests are provided. First, the **phi matrix** is presented over pages 20-33. It contains indices for every possible combination of latent factors (the F variables) and residual terms (the E terms).

You, on the other hand, are most interested in Lagrange multipliers that tell you whether you should add a new path from some latent factor (an F variable) to some indicator (a V variable). These tests appear in the second matrix of Lagrange multipliers, the **gamma matrix**, which appears on pages 34-37 of the output. The gamma matrix for the current analysis is reproduced in Output 5.8.

```
                         The SAS System                              34
          Covariance Structure Analysis: Maximum Likelihood Estimation

          Lagrange Multiplier and Wald Test Indices _GAMMA_[19:6]
                           General Matrix
               Univariate Tests for Constant Constraints
          -----------------------------------------------
          |  Lagrange Multiplier  or  Wald Index      |
          -----------------------------------------------
          |  Probability  | Approx Change of Value  |
          -----------------------------------------------

                       F1                    F2                    F3

     V1        285.671[LV1F1 ]          8.963                 5.378
                                        0.003    0.383        0.020    0.276

     V2        249.911[LV2F1 ]         18.910                 2.757
                                        0.000   -0.698        0.097   -0.250

     V3        356.932[LV3F1 ]          1.037                 1.535
                                        0.309   -0.130        0.215   -0.146
```

V4	225.726[LV4F1]		6.434		0.585	
			0.011	0.429	0.444	0.121
V5	0.837		230.423[LV5F2]		0.342	
	0.360	0.107			0.559	-0.071
V6	1.578		264.004[LV6F2]		1.483	
	0.209	-0.156			0.223	0.158
V7	0.157		275.196[LV7F2]		0.438	
	0.692	0.047			0.508	-0.083
V8	4.569		1.173		99.461[LV8F3]	
	0.033	0.220	0.279	0.129		
V9	3.465		0.233		90.419[LV9F3]	
	0.063	-0.189	0.629	0.056		
V10	0.101		2.172		129.491[LV10F3]	
	0.750	-0.042	0.141	-0.230		
V11	2.326		0.358		0.290	
	0.127	0.192	0.550	0.083	0.590	-0.077
V12	1.565		0.277		0.077	
	0.211	-0.169	0.598	-0.078	0.781	0.042
V13	0.315		0.017		0.321	
	0.574	-0.080	0.895	-0.020	0.571	0.090
V14	0.087		0.166		2.812	
	0.768	0.080	0.683	-0.087	0.094	0.317
V15	1.573		3.088		0.415	
	0.210	-0.306	0.079	0.352	0.519	-0.116

```
        Covariance Structure Analysis: Maximum Likelihood Estimation          35

        Lagrange Multiplier and Wald Test Indices _GAMMA_[19:6]
                           General Matrix
        Univariate Tests for Constant Constraints (Contd.)


                     F1                    F2                    F3

       V16       1.293                 2.913                 2.472
                 0.256    0.218        0.088   -0.285        0.116   -0.250

       V17       0.000                 5.094                 0.042
                 0.984   -0.002        0.024   -0.297        0.838   -0.030

       V18       0.087                 4.218                 3.379
                 0.769    0.035        0.040    0.271        0.066    0.279

       V19       0.082                 0.000                 2.902
                 0.775   -0.038        0.997   -0.001        0.088   -0.284
```

```
        Covariance Structure Analysis: Maximum Likelihood Estimation          36

        Lagrange Multiplier and Wald Test Indices _GAMMA_[19:6]
                           General Matrix
        Univariate Tests for Constant Constraints (Contd.)


                     F4                    F5                    F6

       V1        0.289                 3.243                 0.248
                 0.591    0.052        0.072    0.294        0.618    0.052

       V2        4.162                14.889                 1.185
                 0.041    0.248        0.000   -0.787        0.276    0.145

       V3        0.607                 1.259                 4.118
                 0.436   -0.073        0.262    0.185        0.042    0.209

       V4        3.024                 0.201                22.675
                 0.082   -0.224        0.654    0.096        0.000   -0.670

       V5        0.509                 0.666                 0.171
                 0.476   -0.065        0.415    0.108        0.679   -0.043

       V6        0.075                 0.226                 0.550
                 0.784    0.026        0.634   -0.066        0.458    0.080

       V7        0.137                 0.080                 0.126
                 0.712    0.034        0.777   -0.040        0.723   -0.037

       V8        1.416                 0.110                 0.130
                 0.234   -0.106        0.740    0.041        0.719   -0.040
```

V9	0.080		0.351		0.103	
	0.777	0.025	0.554	0.071	0.748	-0.035
V10	0.697		0.730		0.379	
	0.404	0.092	0.393	-0.136	0.538	0.087
V11	157.147[LV11F4]		0.117		0.214	
			0.732	0.045	0.644	0.071
V12	171.871[LV12F4]		0.147		0.152	
			0.701	-0.055	0.697	-0.065
V13	48.799[LV13F4]		0.013		0.019	
			0.911	0.017	0.889	-0.023
V14	0.014		205.665[LV14F5]		0.115	
	0.906	-0.016			0.735	0.050
V15	0.512		133.856[LV15F5]		1.742	
	0.474	-0.091			0.187	-0.191
V16	1.402		71.886[LV16F5]		1.721	
	0.236	0.138			0.190	0.171
V17	3.708		1.851		109.355[LV17F6]	
	0.054	0.243	0.174	-0.178		

```
          Covariance Structure Analysis: Maximum Likelihood Estimation          37

             Lagrange Multiplier and Wald Test Indices _GAMMA_[19:6]
                               General Matrix
                  Univariate Tests for Constant Constraints (Contd.)


                      F4                      F5                      F6

     V18           1.702                   2.254                  140.144[LV18F6]
                   0.192   -0.162          0.133    0.195

     V19           0.183                   0.082                  127.145[LV19F6]
                   0.669   -0.059          0.775   -0.041


            Rank order of 10 largest Lagrange multipliers in _GAMMA_

                V4 : F6                 V2 : F2                 V2 : F5
           22.6750 : 0.000        18.9100 : 0.000        14.8888 : 0.000

                V1 : F2                 V4 : F2                 V1 : F3
            8.9629 : 0.003         6.4338 : 0.011         5.3776 : 0.020

               V17 : F2                 V8 : F1                V18 : F2
            5.0940 : 0.024         4.5686 : 0.033         4.2183 : 0.040

                                        V2 : F4
                                   4.1620 : 0.041
```

Output 5.8: Lagrange Multiplier Indices for the Gamma Matrix, Initial Model, Investment Model Study

The rank order of the 10 largest Lagrange multipliers appears at the bottom of the matrix, on page 37. Predictably, the largest index was for the V4:F6 relationship. The value of the Lagrange multiplier is 22.68, which means that the model chi-square is expected to decrease by a value of 22.68 if you add a path from F6 (alternatives) to V4.

This finding is very consistent with the pattern of large residuals observed earlier. Those residuals showed that your initial measurement model underpredicted the relationship between V4 and V17 and the relationship between V4 and V19. This suggested the possibility that V4 may be influenced by the same factor that affects V17 and V19, namely the F6 factor.

Modifying the model. In a situation such as this, you have a number of options. One alternative is simply to add the path from F6 to V4, and leave the remainder of the model untouched. This means that V4 would become a complex variable, affected by more than one factor.

This alternative, however, is generally undesirable when developing a measurement model. This is because the theoretical model (to be assessed later) will be more easily interpreted if all of the indicators are unifactorial (if each indicator loads on only one factor). In most cases, it is preferable to reassign or completely drop an indicator from a model rather than assign it to two factors simultaneously.

Should you reassign V4, so that it loads on F6 but not on F1 (the factor to which it was originally assigned)? This option is also unsatisfactory, as the output has already shown that V4 displays a large and statistically significant loading for F1.

In this situation, the best alternative is to drop V4 from the analysis entirely. This should not create any identification problems for F1, because this factor will still be measured by three indicators even after V4 is deleted (this is why it is important to have at least four or five indicators for each factor in the initial measurement model).

The revised measurement model appears in Figure 5.9. You can see that this model is identical to that presented in Figure 5.8, except that the path from F1 (Commitment) to V4 has been deleted. In fact, for the sake of clarity, the variable V4 itself has been deleted from the model.

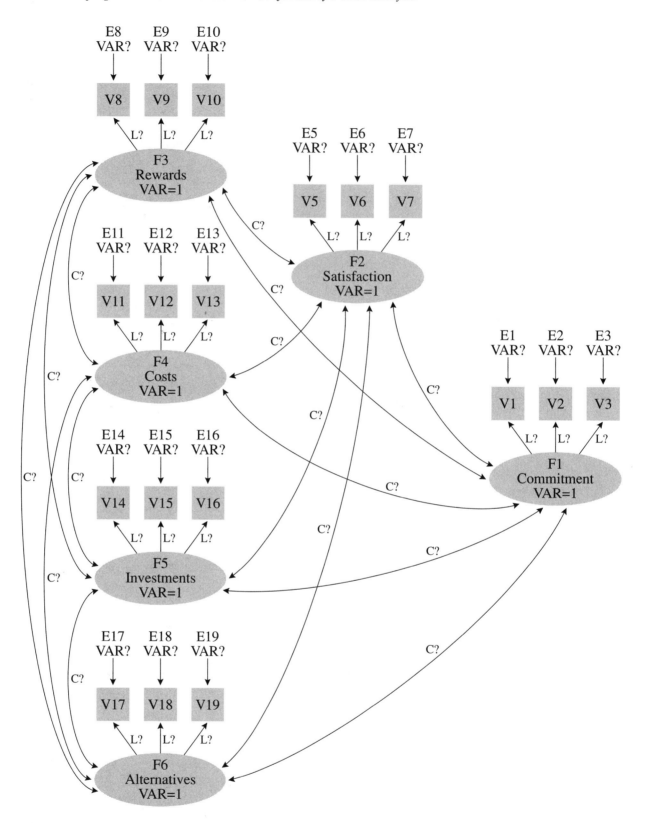

Figure 5.9: The Revised (and Final) Measurement Model

Estimating the Revised Measurement Model

The revised model must now be estimated to see if it provides an acceptable fit to the data. Here is a portion of the revised PROC CALIS program that will estimate the new model:

```
 1      PROC CALIS  COVARIANCE  CORR  RESIDUAL  MODIFICATION ;
 2         LINEQS
 3            V1  = LV1F1   F1 + E1,
 4            V2  = LV2F1   F1 + E2,
 5            V3  = LV3F1   F1 + E3,
 6
 7            V5  = LV5F2   F2 + E5,
 8            V6  = LV6F2   F2 + E6,
 9            V7  = LV7F2   F2 + E7,
10            V8  = LV8F3   F3 + E8,
11            V9  = LV9F3   F3 + E9,
12            V10 = LV10F3  F3 + E10,
13            V11 = LV11F4  F4 + E11,
14            V12 = LV12F4  F4 + E12,
15            V13 = LV13F4  F4 + E13,
16            V14 = LV14F5  F5 + E14,
17            V15 = LV15F5  F5 + E15,
18            V16 = LV16F5  F5 + E16,
19            V17 = LV17F6  F6 + E17,
20            V18 = LV18F6  F6 + E18,
21            V19 = LV19F6  F6 + E19;
22         STD
23            F1 = 1,
24            F2 = 1,
25            F3 = 1,
26            F4 = 1,
27            F5 = 1,
28            F6 = 1,
29            E1-E3  =  VARE1-VARE3,
30            E5-E19 = VARE5-VARE19;
                  .
                  .
                  .
```

Only two changes had to be made to the initial program so that it would estimate the revised model. First, because the path from F1 to V4 had been dropped from the initial model, the equation for V4 was deleted from the PROC CALIS program. In the initial program, this equation appeared on line 6 of the program; you can see that this line has been blanked out in the revised program. The second change involved the STD statement. In the initial program, the variance for E4 was estimated in the STD section of the program. The revised program does not estimate the variance for this residual term; instead, it estimates the variances for E1 through E3 (line 29), and the variances for E5 through E19 (line 30).

It was not necessary to revise the COV statement in the program because only the covariances between latent factors were specified in this statement. In the revised program, the factors are allowed to covary in the same way.

After reviewing the SAS log and output files to verify that the revised program ran properly, the same fit indices discussed earlier should then be consulted to see if the revised model provides a better fit to the data.

Page 13 of the output for this program provides the overall goodness of fit indices for the revised measurement model (these indices appear in Output 5.9). You can see that the model chi-square value for the revised model is 180.87, with 120 degrees of freedom. This chi-square value is still statistically significant ($p < .001$). However, this chi-square for the revised model represents a substantial drop from that observed with the initial measurement model, where chi-square was 247.68, with 137 degrees of freedom. By eliminating the indicator V4 from the analysis, model chi-square decreased by a quantity of 66.81, while the degrees of freedom decreased by only 17. Model modifications are generally more desirable if they can bring about a decrease in chi-square that is relatively large, compared to the change in degrees of freedom. Therefore, it so far appears that dropping V4 was a good move.

```
                          The SAS System                              13
            Covariance Structure Analysis: Maximum Likelihood Estimation

      Fit criterion . . . . . . . . . . . . . . . . .        0.7568
      Goodness of Fit Index (GFI) . . . . . . . . . .        0.9251
      GFI Adjusted for Degrees of Freedom (AGFI) . . .       0.8933
      Root Mean Square Residual (RMR) . . . . . . . .        0.1970
      Chi-square = 180.8733      df = 120     Prob>chi**2 = 0.0003
      Null Model Chi-square:     df = 153               2167.7711
      RMSEA Estimate  . . . . . .  0.0463  90%C.I.[0.0316, 0.0594]
      Bentler's Comparative Fit Index . . . . . . . .        0.9698
      Normal Theory Reweighted LS Chi-square  . . . . .    174.0149
      Akaike's Information Criterion  . . . . . . . . .    -59.1267
      Consistent Information Criterion  . . . . . . .     -596.8034
      Schwarz's Bayesian Criterion  . . . . . . . . .     -476.8034
      McDonald's (1989) Centrality. . . . . . . . . .        0.8809
      Bentler & Bonett's (1980) Non-normed Index. . . .      0.9615
      Bentler & Bonett's (1980) Normed Index. . . . . .      0.9166
      James, Mulaik, & Brett (1982) Parsimonious Index.     0.7189
      Z-Test of Wilson & Hilferty (1931) . . . . . . .       3.4488
      Bollen (1986) Normed Index Rho1 . . . . . . . . .      0.8936
      Bollen (1988) Non-normed Index Delta2 . . . . . .      0.9703
      Hoelter's (1983) Critical N . . . . . . . . . . .        195
```

Output 5.9: Goodness of Fit Indices for Revised Measurement Model, Investment Model Study

In addition, the chi-square/*df* ratio for the revised model is 1.51. This is down from the ratio of 1.81 observed for the initial model.

Bentler's Comparative Fit Index (CFI) appears on the 8th line of the table, and Bentler and Bonett's non-normed index (NNFI) appears on the 14th line. The table shows that the CFI and NNFI for the revised model are .970 and .962, respectively. These indices are not only acceptable, they are also somewhat higher than those observed with the initial model.

The *t* tests that test the significance of the factor loadings appear on pages 14-15 of Output 5.10. The *t* test for a specific loading appears below that loading, and to the right of the heading "t Value". For example, you can see that the *t* test for the first loading (LV1F1) is 17.1002. Remember that these tests are significant if the observed *t* value is greater than 1.96. With this in mind, you can see that the *t* tests on pages 14-15 of Output 5.10 show that the factor loadings for all 18 indicator variables were significantly different from zero.

The standardized factor loadings appear on page 17 of Output 5.10, in the section headed "Equations with Standardized Coefficients". Remember that a standardized coefficient appears above the name for that coefficient. For example, you can see that the first standardized coefficient on page 17 (again, for LV1F1) is 0.8848. The results presented on page 17 reveal that all standardized loadings except one are greater than .60.

```
                        The SAS System                              14

            Covariance Structure Analysis: Maximum Likelihood Estimation

                    Manifest Variable Equations

        V1        =        2.1995*F1      +    1.0000 E1
        Std Err            0.1286 LV1F1
        t Value           17.1002

        V2        =        2.3976*F1      +    1.0000 E2
        Std Err            0.1565 LV2F1
        t Value           15.3233

        V3        =        2.5515*F1      +    1.0000 E3
        Std Err            0.1358 LV3F1
        t Value           18.7822

        V5        =        1.5960*F2      +    1.0000 E5
        Std Err            0.1051 LV5F2
        t Value           15.1909

        V6        =        1.8291*F2      +    1.0000 E6
        Std Err            0.1124 LV6F2
        t Value           16.2677

        V7        =        1.7999*F2      +    1.0000 E7
        Std Err            0.1088 LV7F2
        t Value           16.5471
```

```
V8        =       0.9436*F3     +    1.0000 E8
Std Err           0.0943 LV8F3
t Value           10.0074

V9        =       0.8934*F3     +    1.0000 E9
Std Err           0.0943 LV9F3
t Value           9.4780

V10       =       1.2930*F3     +    1.0000 E10
Std Err           0.1135 LV10F3
t Value           11.3945

V11       =       2.1434*F4     +    1.0000 E11
Std Err           0.1709 LV11F4
t Value           12.5437

V12       =       2.3264*F4     +    1.0000 E12
Std Err           0.1774 LV12F4
t Value           13.1102

V13       =       1.0935*F4     +    1.0000 E13
Std Err           0.1565 LV13F4
t Value           6.9865

V14       =       1.7725*F5     +    1.0000 E14
Std Err           0.1234 LV14F5
t Value           14.3622

V15       =       1.5689*F5     +    1.0000 E15
Std Err           0.1359 LV15F5
```

```
Covariance Structure Analysis: Maximum Likelihood Estimation       15
          t Value           11.5440

V16       =       1.0320*F5     +    1.0000 E16
Std Err           0.1214 LV16F5
t Value           8.5015

V17       =       1.3670*F6     +    1.0000 E17
Std Err           0.1301 LV17F6
t Value           10.5088

V18       =       1.4937*F6     +    1.0000 E18
Std Err           0.1266 LV18F6
t Value           11.8010

V19       =       1.5917*F6     +    1.0000 E19
Std Err           0.1412 LV19F6
t Value           11.2756
```

```
                     Variances of Exogenous Variables
-------------------------------------------------------------------------
                                          Standard
Variable    Parameter     Estimate         Error           t Value
-------------------------------------------------------------------------
F1                        1.000000            0              0.000
F2                        1.000000            0              0.000
F3                        1.000000            0              0.000
F4                        1.000000            0              0.000
F5                        1.000000            0              0.000
F6                        1.000000            0              0.000
E1          VARE1         1.341222         0.178768          7.503
E2          VARE2         2.714458         0.298148          9.104
E3          VARE3         0.907155         0.189215          4.794
E5          VARE5         1.173210         0.141620          8.284
E6          VARE6         1.116879         0.154983          7.206
E7          VARE7         0.987059         0.143843          6.862
E8          VARE8         1.117200         0.137411          8.130
E9          VARE9         1.184284         0.137807          8.594
E10         VARE10        1.300405         0.201671          6.448
E11         VARE11        2.138657         0.484792          4.411
E12         VARE12        1.829435         0.548111          3.338
E13         VARE13        4.383078         0.421888         10.389
E14         VARE14        1.275822         0.236444          5.396
E15         VARE15        2.461919         0.284382          8.657
E16         VARE16        2.447250         0.244514         10.009
E17         VARE17        2.134937         0.262436          8.135
E18         VARE18        1.633744         0.250969          6.510
E19         VARE19        2.239552         0.309334          7.240
```

```
   Covariance Structure Analysis: Maximum Likelihood Estimation       16

                Covariances among Exogenous Variables
-------------------------------------------------------------------------
                                          Standard
        Parameter            Estimate      Error          t Value
-------------------------------------------------------------------------
F2    F1    CF1F2            0.608607     0.047237         12.884
F3    F1    CF1F3            0.440008     0.066338          6.633
F3    F2    CF2F3            0.534118     0.062324          8.570
F4    F1    CF1F4           -0.016384     0.072951         -0.225
F4    F2    CF2F4           -0.224614     0.071161         -3.156
F4    F3    CF3F4           -0.092439     0.081541         -1.134
F5    F1    CF1F5            0.714356     0.044263         16.139
F5    F2    CF2F5            0.634714     0.052018         12.202
F5    F3    CF3F5            0.516184     0.068847          7.498
F5    F4    CF4F5            0.007920     0.078794          0.101
F6    F1    CF1F6           -0.223309     0.073170         -3.052
F6    F2    CF2F6           -0.375187     0.068953         -5.441
F6    F3    CF3F6           -0.424229     0.075081         -5.650
F6    F4    CF4F6            0.255036     0.076039          3.354
F6    F5    CF5F6           -0.300531     0.076938         -3.906
```

```
      Covariance Structure Analysis: Maximum Likelihood Estimation      17
               Equations with Standardized Coefficients
               V1       =     0.8848*F1     +  0.4659 E1
                                 LV1F1


               V2       =     0.8242*F1     +  0.5663 E2
                                 LV2F1


               V3       =     0.9369*F1     +  0.3497 E3
                                 LV3F1


               V5       =     0.8274*F2     +  0.5616 E5
                                 LV5F2


               V6       =     0.8659*F2     +  0.5003 E6
                                 LV6F2


               V7       =     0.8755*F2     +  0.4832 E7
                                 LV7F2


               V8       =     0.6660*F3     +  0.7460 E8
                                 LV8F3


               V9       =     0.6345*F3     +  0.7729 E9
                                 LV9F3


               V10      =     0.7500*F3     +  0.6615 E10
                                 LV10F3


               V11      =     0.8260*F4     +  0.5636 E11
                                 LV11F4


               V12      =     0.8645*F4     +  0.5026 E12
                                 LV12F4


               V13      =     0.4630*F4     +  0.8864 E13
                                 LV13F4


               V14      =     0.8433*F5     +  0.5374 E14
                                 LV14F5


               V15      =     0.7071*F5     +  0.7071 E15
                                 LV15F5


               V16      =     0.5506*F5     +  0.8347 E16
                                 LV16F5


               V17      =     0.6832*F6     +  0.7302 E17
                                 LV17F6


               V18      =     0.7598*F6     +  0.6502 E18
                                 LV18F6


               V19      =     0.7286*F6     +  0.6850 E19
                                 LV19F6
```

```
         Covariance Structure Analysis: Maximum Likelihood Estimation          18

                    Variances of Endogenous Variables
           ------------------------------------------------
              Variable        Estimate         R-squared
           ------------------------------------------------
                 1    V1       6.178870         0.782934
                 2    V2       8.462812         0.679249
                 3    V3       7.417150         0.877695
                 4    V5       3.720272         0.684644
                 5    V6       4.462308         0.749708
                 6    V7       4.226846         0.766479
                 7    V8       2.007629         0.443523
                 8    V9       1.982431         0.402610
                 9    V10      2.972152         0.562470
                10    V11      6.732711         0.682348
                11    V12      7.241358         0.747363
                12    V13      5.578734         0.214324
                13    V14      4.417416         0.711184
                14    V15      4.923295         0.499945
                15    V16      3.512193         0.303213
                16    V17      4.003619         0.466748
                17    V18      3.864850         0.577281
                18    V19      4.773116         0.530799


                   Correlations among Exogenous Variables
                 --------------------------------------------
                            Parameter           Estimate
                 --------------------------------------------
                 F2   F1     CF1F2              0.608607
                 F3   F1     CF1F3              0.440008
                 F3   F2     CF2F3              0.534118
                 F4   F1     CF1F4             -0.016384
                 F4   F2     CF2F4             -0.224614
                 F4   F3     CF3F4             -0.092439
                 F5   F1     CF1F5              0.714356
                 F5   F2     CF2F5              0.634714
                 F5   F3     CF3F5              0.516184
                 F5   F4     CF4F5              0.007920
                 F6   F1     CF1F6             -0.223309
                 F6   F2     CF2F6             -0.375187
                 F6   F3     CF3F6             -0.424229
                 F6   F4     CF4F6              0.255036
                 F6   F5     CF5F6             -0.300531
```

Output 5.10: PROC CALIS Ouput Pages 14-18 for Revised Measurement Model, Investment Model Study

The normalized residuals appear in Output 5.11. Page 12 of this output presents a table that summarizes the frequency of normalized residuals that appear within specific intervals (this table is headed "Distribution of Normalized Residuals"). In the bars that constitute this table, each asterisk represents one residual, so the longer bars identify the intervals in which many residuals appear. You can see that the distribution of normalized residuals in this table is fairly symmetrical, and is centered around zero.

Just above the "Distribution of Normalized Residuals" table is a section headed "Rank order of 10 Largest Normalized Residuals". This section shows that none of the normalized residuals exceeded 2.00 in absolute value (the largest residual was only −1.6995 for the V17:V15 variable pair). This is the pattern of results that you would expect when there is an acceptable fit between model and data.

```
                                 The SAS System                                11

                  Covariance Structure Analysis: Maximum Likelihood Estimation

                             Normalized Residual Matrix

               V1          V2          V3          V5          V6          V7

    V1       0.0024      0.0601     -0.1158      0.9696      0.9537      0.7092
    V2       0.0601     -0.0007      0.1718     -0.7585     -1.2536     -1.3353
    V3      -0.1158      0.1718      0.0045      0.3415     -0.5761      0.1100
    V5       0.9696     -0.7585      0.3415      0.0023     -0.0399     -0.0540
    V6       0.9537     -1.2536     -0.5761     -0.0399      0.0060      0.0764
    V7       0.7092     -1.3353      0.1100     -0.0540      0.0764      0.0008
    V8       1.3012      0.7756      0.9944      0.6351      0.9940     -0.3904
    V9      -0.5717     -1.2547     -1.4167     -0.4385      0.1879      0.8957
    V10      0.8483     -0.4631     -0.3271     -0.7267      0.0623     -0.6826
    V11      0.9757      1.4431      0.2739     -0.1150     -0.0828      0.6946
    V12     -0.4256      0.3822     -0.6776     -0.4337      0.2763     -0.3056
    V13     -0.3450     -0.0889     -0.9117     -0.3686     -0.1683      0.1250
    V14      0.3564     -0.9500      0.2266     -0.1389     -0.1727      0.0203
    V15     -0.1818     -1.3483     -0.2672      1.4631      0.4268      0.3192
    V16      0.3945      0.0263      0.9092     -0.5384     -0.6904     -1.1998
    V17     -0.0926      0.7799     -0.0323     -1.3329     -0.9097     -1.2186
    V18      0.2315     -0.6776      0.0752      0.6611      0.9440      0.6847
    V19     -0.3539     -0.1217      0.1129     -0.1190      0.4160     -0.1767
```

	V8	V9	V10	V11	V12	V13
V1	1.3012	-0.5717	0.8483	0.9757	-0.4256	-0.3450
V2	0.7756	-1.2547	-0.4631	1.4431	0.3822	-0.0889
V3	0.9944	-1.4167	-0.3271	0.2739	-0.6776	-0.9117
V5	0.6351	-0.4385	-0.7267	-0.1150	-0.4337	-0.3686
V6	0.9940	0.1879	0.0623	-0.0828	0.2763	-0.1683
V7	-0.3904	0.8957	-0.6826	0.6946	-0.3056	0.1250
V8	0.0014	-0.4644	0.0910	-0.6214	-0.7547	-0.2240
V9	-0.4644	0.0002	0.2257	-0.1014	0.2275	1.2716
V10	0.0910	0.2257	0.0001	-0.0888	0.6485	0.4505
V11	-0.6214	-0.1014	-0.0888	0.0021	-0.0005	-0.0534
V12	-0.7547	0.2275	0.6485	-0.0005	0.0002	0.0353
V13	-0.2240	1.2716	0.4505	-0.0534	0.0353	-0.0179
V14	0.2254	1.0279	0.0967	-0.0390	-0.1669	0.6178
V15	-0.5726	-0.0085	-0.4887	-0.6450	0.0335	-0.9999
V16	-0.4915	-0.4476	-1.1087	1.3850	0.2824	0.8046
V17	-0.0761	-0.2455	0.4444	1.2590	1.0016	-0.4587
V18	0.8580	1.0854	-0.0192	-0.4291	-0.7412	0.2655
V19	-1.2475	-1.5495	0.1777	-0.0988	-0.2847	-0.1555

Covariance Structure Analysis: Maximum Likelihood Estimation   12

Normalized Residual Matrix

	V14	V15	V16	V17	V18	V19
V1	0.3564	-0.1818	0.3945	-0.0926	0.2315	-0.3539
V2	-0.9500	-1.3483	0.0263	0.7799	-0.6776	-0.1217
V3	0.2266	-0.2672	0.9092	-0.0323	0.0752	0.1129
V5	-0.1389	1.4631	-0.5384	-1.3329	0.6611	-0.1190
V6	-0.1727	0.4268	-0.6904	-0.9097	0.9440	0.4160
V7	0.0203	0.3192	-1.1998	-1.2186	0.6847	-0.1767
V8	0.2254	-0.5726	-0.4915	-0.0761	0.8580	-1.2475
V9	1.0279	-0.0085	-0.4476	-0.2455	1.0854	-1.5495
V10	0.0967	-0.4887	-1.1087	0.4444	-0.0192	0.1777
V11	-0.0390	-0.6450	1.3850	1.2590	-0.4291	-0.0988
V12	-0.1669	0.0335	0.2824	1.0016	-0.7412	-0.2847
V13	0.6178	-0.9999	0.8046	-0.4587	0.2655	-0.1555
V14	0.0024	-0.0156	-0.1031	-0.7157	0.7384	0.0854
V15	-0.0156	0.0015	0.2983	-1.6995	0.1902	-0.7229
V16	-0.1031	0.2983	-0.0010	0.7398	0.7952	0.5777
V17	-0.7157	-1.6995	0.7398	0.0010	0.1370	-0.5224
V18	0.7384	0.1902	0.7952	0.1370	0.0009	0.2512
V19	0.0854	-0.7229	0.5777	-0.5224	0.2512	0.0025

Average Normalized Residual = 0.4672
Average Off-diagonal Normalized Residual = 0.5219

```
              Rank Order of 10 Largest Normalized Residuals

     V17,V15    V19,V9    V15,V5    V11,V2    V9,V3    V16,V11   V15,V2
     -1.6995   -1.5495    1.4631    1.4431   -1.4167    1.3850   -1.3483

                          V7,V2    V17,V5    V8,V1
                         -1.3353   -1.3329    1.3012


                 Distribution of Normalized Residuals
                   (Each * represents 1 residuals)

     -1.75000 -   -1.50000    2    1.17% | **
     -1.50000 -   -1.25000    6    3.51% | ******
     -1.25000 -   -1.00000    4    2.34% | ****
     -1.00000 -   -0.75000    6    3.51% | ******
     -0.75000 -   -0.50000   15    8.77% | ***************
     -0.50000 -   -0.25000   18   10.53% | ******************
     -0.25000 -          0   33   19.30% | *********************************
            0 -    0.25000   39   22.81% | ***************************************
      0.25000 -    0.50000   15    8.77% | ***************
      0.50000 -    0.75000   10    5.85% | **********
      0.75000 -    1.00000   14    8.19% | **************
      1.00000 -    1.25000    3    1.75% | ***
      1.25000 -    1.50000    6    3.51% | ******
```

Output 5.11: Normalized Residuals and Distribution of Normalized Residuals for Revised Measurement Model, Investment Model Study

In summary, the preceding output suggests that the revised measurement model

- provides an acceptable fit to the data (as indicated by the CFI and NNFI)

- displays no nonsignificant factor loadings

- displays no large normalized residuals.

Combined, these findings provide support for the revised model. However, before you accept this revised model as your final measurement model, you will first perform a few additional tests to assess the reliability and validity of the factors and indicators.

Assessing Reliability and Validity of Constructs and Indicators

One of the most important advantages offered by latent-variable analyses is the opportunity that they provide to assess the reliability and validity of the study's variables. Broadly speaking, **reliability** refers to consistency of measurement. A test is reliable if, for example, it provides essentially the same set of scores for a group of subjects upon repeated testing. Reliability may be assessed in a variety of ways (e.g., you may assess test-retest reliability, alternate-forms reliability, and so forth).

Validity, on the other hand, refers to the extent to which an instrument measures what it is intended to measure. If, for example, you develop a scale designed to measure locus of control, and scores on the scale do in fact reflect subjects' underlying levels of locus of control, then the scale is valid. As with reliability, there are a number of different ways that validity can be studied (e.g., content validity, criterion-related validity).

Reliability is not an all-or-nothing phenomenon; rather, it is assessed along a continuum. An instrument may reflect a relatively high level of reliability, a relatively low level, or any amount in between. The same is true for validity.

This section shows how the results of a CFA (confirmatory factor analysis) using PROC CALIS can be used to assess item reliability, composite reliability, variance extracted estimates, convergent validity, and discriminant validity. Combined, these procedures provide evidence concerning the extent to which the indicators used in the study are producing reliable data and are measuring what they are intended to measure.

Indicator reliability. The reliability of an indicator variable is defined as the square of the correlation between a latent factor and that indicator. In other words, the reliability indicates the percent of variation in the indicator that is explained by the factor that it is supposed to measure (Long, 1983a).

The reliability of an indicator can be computed in a very straightforward way by simply squaring the standardized factor loadings obtained in the analysis. For example, the loading that represents the path from F1 to V1 was given the name LV1F1. The standardized loadings for all indicators are provided on page 17 of Output 5.10, and you can see that the standardized loading for LV1F1 is .8848. The square of this loading is .783, meaning that the reliability for V1 is .783.

Fortunately, it is not necessary to actually perform this calculation, as the squares of these loadings are provided on page 18 of Output 5.10. The first table on page 18 is titled "Variances of the Endogenous Variables," and the last column of the table (titled "R-squared") indicates the percent of variance in each indicator that is accounted for by the common factor to which it was assigned. These R^2 values are the indicator reliabilities.

You can see that the indicator reliabilities vary from a low of .21 for V13, to a high of .88 for V3. Some factors are measured by indicators which all display relatively high reliabilities. For example, F1 (Commitment) is measured by V1, V2, and V3, and the reliabilities for these indicators are .78, .68, and .88, respectively. Other factors are assessed by indicators with relatively low reliabilities. For example, F3 (Rewards) is assessed by V8, V9, and V10, and the reliabilities for these indicators are only .44, .40, and .56, respectively. It will be interesting to see whether the composite reliability for F3 is unacceptably low (this topic is covered next).

Composite reliability. When conducting research with a multiple-item scale, researchers often compute the coefficient alpha reliability estimate for that scale (Cronbach, 1951). Coefficient alpha is an index of internal consistency reliability; with other factors equal, alpha will be high if the various items that constitute the scale are strongly correlated with one another. Along with other interpretations, coefficient alpha may be interpreted conceptually as an estimate of the correlation between a given scale and an alternate form of the scale that includes the same number of items (Carmines & Zeller, 1988). Coefficient alpha reliability was covered in Chapter 3 of this text.

Similarly, when performing confirmatory factor analysis, it is possible to compute a **composite reliability** index for each latent factor included in the model. This index is analogous to coefficient alpha, and reflects the internal consistency of the indicators measuring a given factor. The formula for this composite reliability index (adapted from Fornell and Larcker [1981]) is presented here:

$$\texttt{Composite reliability} = \frac{(\Sigma\ \texttt{L}_i)^2}{(\Sigma\ \texttt{L}_i)^2 + \Sigma\ \texttt{Var}(\texttt{E}_i)}$$

where $\texttt{L}_i$ = the standardized factor loadings for that factor

$\texttt{Var}(\texttt{E}_i)$ = the error variance associated with the individual indicator variables.

Computing the composite reliability for each scale will be easier if you first prepare a table that summarizes all of the necessary information. Table 5.2 provides the information needed to compute composite reliabilities for the present measurement model. You can see that Table 5.2 includes the standardized loading for each indicator, along with each indicator's reliability (defined earlier as the square of the standardized loading). The last column in the table provides the error variance associated with each indicator.

Table 5.2

Information Needed to Compute Composite Reliability and Variance Extracted Estimates

Construct and Indicators	Standardized Loading	Indicator Reliability [a]	Error Variance [b]
Commitment (F1)			
V1	.885	.783	.217
V2	.824	.679	.321
V3	.937	.878	.122
Satisfaction (F2)			
V5	.827	.685	.315
V6	.866	.750	.250
V7	.876	.766	.234
Rewards (F3)			
V8	.666	.445	.555
V9	.635	.403	.597
V10	.750	.562	.438
Costs (F4)			
V11	.826	.682	.318
V12	.865	.747	.253
V13	.463	.214	.786
Investment size (F5)			
V14	.843	.711	.289
V15	.707	.500	.500
V16	.551	.303	.697
Alternative value (F6)			
V17	.683	.467	.533
V18	.760	.577	.423
V19	.729	.531	.469

[a] Calculated as the square of the standardized factor loading.

[b] Calculated as 1 minus the indicator reliability.

The error variance is calculated as $1 - L_i^2$, or 1 minus the square of the standardized factor loading. Because the reliability is the square of the factor loading, you can calculate error variances by simply subtracting the reliability estimates from 1. Thus, for indicator V1, $1 - .783 = .217$; for indicator V2, $1 - .679 = .321$; and so forth. You are encouraged to always prepare a table with the columns similar to those in Table 5.2 before computing composite reliabilities.

With the error variances computed, you may now insert the values of Table 5.2 into the appropriate parts of the formula for composite reliability. The numerator of the formula appears here:

$$(\Sigma\ L_i)^2$$

Remember that operations that appear within parentheses are performed prior to performing operations outside of the parentheses. The "Σ" symbol in this portion of the equation means that you are to sum the factor loadings (which are symbolized by L_i). After this is done, you will square the resulting sum (as is indicated by the superscript ^{2} that appears outside of the parentheses).

Here, these operations are performed on the loadings for V1, V2, and V3 on F1:

$$\Sigma\ L_i\ =\ .885 + .824 + .937$$
$$=\ 2.646$$

Now, the square of this sum is calculated:

$$(\Sigma\ L_i)^2\ =\ (2.646)^2$$
$$=\ 7.001$$

This quantity is now inserted in the appropriate sections of the equation:

$$\frac{(\Sigma L_i)^2}{(\Sigma L_i)^2 + \Sigma\ Var(E_i)}\ =\ \frac{7.001}{7.001 + \Sigma\ Var(E_i)}$$

To calculate $\Sigma\ Var(E_i)$, you will simply sum the error variances associated with V1, V2, and V3 (from Table 5.2):

$$\Sigma\ Var(E_i)\ =\ .217 + .321 + .122$$
$$=\ .660$$

This sum is inserted in the appropriate location in the formula, and you may now calculate the composite reliability for F1:

$$\frac{(\Sigma L_i)^2}{(\Sigma SL_i)^2 + \Sigma\ Var(E_i)}\ =\ \frac{7.001}{7.001 + .660}$$
$$=\ \frac{7.001}{7.661}$$
$$=\ .914$$

So the composite reliability for F1 (the commitment construct) is .918. You should generally think of .60 or .70 as being the minimally acceptable level of reliability for instruments used in research (.70 is preferable). Clearly, the composite reliability for the commitment construct exceeds this requirement.

For purposes of contrast, the composite reliability for the F3 construct (Rewards) is computed here:

$$\frac{(\Sigma L_i)^2}{(\Sigma L_i)^2 + \Sigma \, Var(E_i)} = \frac{(.666 + .635 + .750)^2}{(.666 + .635 + .750)^2 + (.555 + .597 + .438)}$$

$$= \frac{(2.051)^2}{(2.051)^2 + (1.590)}$$

$$= \frac{4.207}{4.207 + 1.590}$$

$$= \frac{4.207}{5.797}$$

$$= .726$$

And so the composite reliability for Rewards is .726, not as high as the reliability for Commitment, but above the minimally acceptable level.

Table 5.3 provides the reliabilities for all variables included in the final measurement model. The third column of figures provides reliability data; composite reliabilities for the latent factors are flagged with a superscript [b] symbol, and reliabilities for the individual indicators are indented two spaces. You may choose to use a format similar to this when summarizing properties of a measurement model in a published research article.

Table 5.3

Properties of the Revised Measurement Model

Construct and Indicators	Standardized Loading	t [a]	Reliability	Variance Extracted Estimate
Commitment (F1)			.914 [b]	.780
V1	.885	17.10	.783	
V2	.824	15.32	.679	
V3	.937	18.78	.878	
Satisfaction (F2)			.892 [b]	.734
V5	.827	15.19	.685	
V6	.866	16.27	.750	
V7	.876	16.55	.766	
Rewards (F3)			.726 [b]	.470
V8	.666	10.01	.445	
V9	.635	9.48	.403	
V10	.750	11.39	.562	
Costs (F4)			.774 [b]	.548
V11	.826	12.54	.682	
V12	.865	13.11	.747	
V13	.463	6.99	.214	
Investment size (F5)			.748 [b]	.505
V14	.843	18.36	.711	
V15	.707	11.54	.500	
V16	.551	8.50	.303	
Alternative value (F6)			.768 [b]	.525
V17	.683	10.51	.467	
V18	.760	11.80	.577	
V19	.729	11.28	.531	

[a] All t tests were significant at $p < .001$.

[b] Denotes composite reliability.

Variance extracted estimates.   Fornell and Larcker (1981) discuss an index called the **variance extracted estimate**, which assesses the amount of variance that is captured by an underlying factor in relation to the amount of variance due to measurement error. The formula appears here:

$$\text{Variance extracted} = \frac{\Sigma L_i^2}{\Sigma L_i^2 + \Sigma \text{ var}(E_i)}$$

Notice that the preceding differs from the formula for composite reliability in that the ΣL_i term is no longer within parentheses. This means that each factor loading is squared first, and then these squared factor loadings are summed. Because a squared factor loading for an indicator is equivalent to that indicator's reliability, this is equivalent to simply summing the reliabilities for a given factor's indicators. To illustrate, the variance extracted estimate for F1 is calculated below by summing the reliabilities and error variance terms from Table 5.2:

$$\frac{\Sigma L_i^2}{\Sigma L_i^2 + \Sigma \text{ Var}(E_i)} = \frac{(.783 + .679 + .878)}{(.783 + .679 + .878) + (.217 + .321 + .122)}$$

$$= \frac{2.340}{2.340 + .660}$$

$$= \frac{2.340}{3.000}$$

$$= .780$$

So the variance extracted estimate for the commitment factor was .78, meaning that 78% of the variance is captured by your commitment construct, and only 22% $(1 - .78 = .22)$ is due to measurement error. Fornell and Larcker (1981) suggest that it is desirable that constructs exhibit estimates of .50 or larger, because estimates less than .50 indicate that variance due to measurement error is larger than the variance captured by the factor. This may call into question the validity of the latent construct as well as its indicators. You are cautioned, however, that this test is quite conservative; very often variance extracted estimates will be below .50, even when reliabilities are acceptable.

The last column of Table 5.3 provides variance extracted estimates for the six constructs of your study. Note that all indices exceed the .50 criteria recommended by Fornell and Larcker (1981), except for rewards (F3), for which the variance extracted estimate was .47. Taken as a group, however, the constructs in the model performed fairly well.

Convergent validity. Convergent validity and discriminant validity are usually associated with use of the multitrait, multimethod (MTMM) approach to validation (Campbell & Fiske, 1959) in which multiple constructs are each assessed using more than one assessment method. It has been asserted that the MTMM approach provides a stronger test of convergent (and discriminant) validity than is afforded by the procedures to be discussed here (e.g., Netemeyer, Johnston, & Burton, 1990; Schmitt & Stults, 1986; Widaman, 1985). Nonetheless, the procedures discussed here do shed some light on the convergent and discriminant validity of the measures to be used in a path analysis with latent variables, and are useful for situations in which it is not possible to follow the MTMM approach.

Stated simply, **convergent validity** is demonstrated when different instruments are used to measure the same construct, and scores from these different instruments are strongly correlated. For example, imagine that tests using different methods (say, a written test and an oral test) are both used to measure some technical skill in a sample of subjects. Further imagine that the correlation between these two tests is quite high at .70. Assuming some additional requirements are met, this finding would be viewed as evidence of convergent validity; the strong correlation suggests that both instruments are measuring what they were intended to measure, even though they used different methods. In short, the correlation shows that both instruments were measuring the same construct.

In the present study, on the other hand, convergent validity is assessed instead by reviewing the *t* tests for the factor loadings. If all factor loadings for the indicators measuring the same construct are statistically significant (greater than twice their standard errors) this is viewed as evidence supporting the convergent validity of those indicators. The fact that all *t* tests are significant shows that all indicators are effectively measuring the same construct (Anderson & Gerbing, 1988).

For example, the standardized factor loadings from the current analysis and the *t* tests for these loadings are presented in Table 5.3. Consider the convergent validity of the three indicators measuring commitment: V1, V2, and V3. The results show that the *t* values for these three indicators range from 15.32 to 18.78. These *t* values are all significantly different from zero at $p < .001$ (because all three *t* values exceed the critical *t* of 3.29 for $p = .001$). These results support the convergent validity of V1, V2, and V3 as measures of commitment. A quick review of the indicators for the remaining constructs shows that all demonstrate significant *t* values, supporting their convergent validity as well.

Discriminant validity. **Discriminant validity**, on the other hand, is demonstrated when different instruments are used to measure different constructs, and the correlations between the measures of these different constructs are relatively weak. A test displays discriminant validity when it is demonstrated that the test does *not* measure a construct that it was not designed to measure.

As with convergent validity, discriminant validity is often studied using the multitrait, multimethod procedure. For example, assume that you are studying psychological needs, and have developed an instrument to measure the need for power. You not only want to demonstrate that your scale successfully measures the need for power, but you also want to demonstrate that it does not measure a similar psychological need, the need for achievement. You can obtain the evidence you need by using the multitrait, multimethod approach. Assume that you administer the following four instruments to a sample of subjects:

Test A: Your new measure of the need for power (a self-report scale)

Test B: A previously validated test of the need for power (a projective test)

Test C: A previously validated test of the need for achievement (a self-report scale)

Test D: A second previously validated test of the need for achievement (a projective test)

Note that you are assessing multiple traits (need for power versus need for achievement) and are measuring each with multiple methods (a self-report scale versus a projective test).

When the data are analyzed, you will hope for a number of results. First, you hope that Test A will show a relatively strong correlation with Test B. This would mean that your new need for power scale is strongly related to another measure of the need for power. This outcome would demonstrate convergent validity for the new scale.

To support the discriminant validity of the scale, you will also hope that Test A (a test of the need for power) will show relatively weak correlations with Tests C and D (two tests of the need for achievement). At the very least, you will hope that Test A will show a weaker correlation with Tests C and D than it shows with Test B. This outcome will provide evidence supporting the discriminant validity of Test A. It will show that Test A is apparently not measuring the need for achievement.

The multitrait, multimethod approach provides a relatively strong test of discriminant validity. Unfortunately, these tests cannot be repeated for the present investment model study, as multiple methods were not used to assess the different constructs. Nonetheless, some evidence regarding discriminant validity may still be obtained from the present analysis through use of three procedures:

- The chi-square difference test

- The confidence interval test

- The variance extracted test

Remember that the procedures discussed here do not necessarily have to be performed as a matter of course each time a latent-variable model is analyzed, but are recommended in cases where discriminant validity is in doubt.

With the **chi-square difference test**, you assess the discriminant validity of two constructs by

- estimating the standard measurement model in which all factors are allowed to covary

- creating a new measurement model identical to the previous one, except that the correlation between the two factors of interest is fixed at 1

- computing the chi-square difference statistic for the two models.

Discriminant validity is demonstrated if chi-square is significantly lower for the first model, as this suggests that the better model was the one in which the two constructs were viewed as distinct (but correlated) factors (Anderson & Gerbing, 1988; Bagozzi & Phillips, 1982).

To illustrate, this procedure will be used to assess the discriminant validity of the constructs commitment (F1) and investment size (F5). Output 5.12 presents "Correlations Among the Exogenous Variables" from your analysis of the revised measurement model in which all factors were allowed to covary (the program that produced this output appeared in the earlier section headed "Estimating the Revised Measurement Model"). Notice that this table shows that the correlation between F1 and F5 is .714 (this is the 7th entry down in the table). In one respect, this is encouraging, as the investment model predicts that investments have a positive causal effect on commitment, and this correlation is consistent with that prediction. What is disturbing, on the other hand, is the size of the correlation. These two constructs are so strongly correlated that it is reasonable to question whether you are really measuring two different constructs at all. It is possible that items V1-V3 and items V14-V16 are really measuring the same underlying construct. If true, then these items lack discriminant validity.

```
                        The SAS System                          18
              Correlations among Exogenous Variables
              -----------------------------------------
                    Parameter            Estimate
              -----------------------------------------
              F2    F1    CF1F2           0.608607
              F3    F1    CF1F3           0.440008
              F3    F2    CF2F3           0.534118
              F4    F1    CF1F4          -0.016384
              F4    F2    CF2F4          -0.224614
              F4    F3    CF3F4          -0.092439
              F5    F1    CF1F5           0.714356
              F5    F2    CF2F5           0.634714
              F5    F3    CF3F5           0.516184
              F5    F4    CF4F5           0.007920
              F6    F1    CF1F6          -0.223309
              F6    F2    CF2F6          -0.375187
              F6    F3    CF3F6          -0.424229
              F6    F4    CF4F6           0.255036
              F6    F5    CF5F6          -0.300531
```

Output 5.12: Correlations among the Exogenous Variables for Revised Measurement Model, Investment Model Study

To assess the discriminant validity of F1 and F5, you will modify your PROC CALIS program so that the covariance between the two factors is fixed at 1. This will require that you change only one equation in the COV statement of your program (the COV statement had appeared in the earlier section of this chapter titlted "Overview of the PROC CALIS program"). Here is the revised COV statement:

```
30          COV
31               F1  F2  =  CF1F2,
32               F1  F3  =  CF1F3,
33               F1  F4  =  CF1F4,
34               F1  F5  =  1,
35               F1  F6  =  CF1F6,
36               F2  F3  =  CF2F3,
37               F2  F4  =  CF2F4,
38               F2  F5  =  CF2F5,
39               F2  F6  =  CF2F6,
40               F3  F4  =  CF3F4,
41               F3  F5  =  CF3F5,
42               F3  F6  =  CF3F6,
43               F4  F5  =  CF4F5,
44               F4  F6  =  CF4F6,
45               F5  F6  =  CF5F6;
```

Notice that this statement is identical to the original COV statement with one exception; Line 34 of the new statement fixes the covariance between F1 and F5 at 1. The model created as a result of this modification will be referred to as the **unidimensional model**, and the model in which the correlation between F1 and F5 is a free parameter to be estimated will be referred to as the **standard measurement model.**

When estimated, the unidimensional model displayed a model chi-square of 251.14 with 121 *df* (the output from this analysis is not reproduced here). The summary table for your standard measurement model (from Output 5.9) has already shown that the chi-square for that model was 180.87 with 120 *df*. You may now calculate the difference in chi-square between the two models:

$$
\begin{array}{r}
251.14 \\
-\ 180.87 \\
\hline
70.27
\end{array}
$$

So the difference in chi-squares was 70.27. To determine whether this value is statistically significant, you must find the critical value of chi-square for the degrees of freedom associated with the test. The *df* for the test is found by subtracting the *df* for the two models:

$$
\begin{array}{r}
121 \\
-\ 120 \\
\hline
1
\end{array}
$$

Since there is 1 *df* associated with this chi-square difference test, you turn to a table of chi-square (provided in Appendix C of this text) and find that, with 1 *df*, the critical values of chi-square are 3.841 at $p = .05$, 6.635 at $p = .01$, and 10.827 at $p = .001$. Since your observed chi-square difference value was 70.27, the difference between the two models was clearly significant at $p < .001$. In other words, the standard measurement model in which the factors were viewed as distinct but correlated constructs provided a fit that was significantly better than the fit provided by the unidimensional model. In short, this test supports the discriminant validity of F1 and F5.

In some cases, you may want to test the discriminant validity for every possible pair of F factors. This would require a series of tests, in which the covariance between just two factors is fixed at 1, the model is estimated, and you compute the difference between the resulting chi-square and the chi-square for the standard measurement model. In the present case, this would result in 15 different models (and consequently 15 difference tests), as there are 15 separate covariances between the six factors in the model.

Performing such a large number of tests, however, creates problems involving the overall significance level for the family of tests. If you perform just one difference test and use the critical value of chi-square associated with $p = .05$, it is clear that the significance level for that test is .05. However, if you perform a series of tests, the overall significance level for that series of tests will be larger. The overall significance level for a family of tests can be computed with this formula (Anderson & Gerbing, 1988; Finn, 1974):

$$
a_0 = 1 - (1 - a_i)^t
$$

where

a_0 = the overall significance level for the family of tests

a_i = the significance level used for each individual difference test

t = the number of tests performed.

For example, imagine that you perform just two tests, and use the significance level of .05 for both individual tests (this means that you used a critical value of 3.841 for both tests). What is the actual overall significance level for the family of tests? That is, what is the probability that you will incorrectly reject a true null hypothesis in at least one of the tests? You may find this by inserting the appropriate figures in the preceding formula:

$$a_0 = 1 - (1 - a_i)^t$$

$$= 1 - (1 - .05)^2$$

$$= 1 - (.95)^2$$

$$= 1 - .9025$$

$$= .0975$$

So the actual significance level for the series of tests was actually .0975. This means that there was over 9% chance that you would incorrectly reject a true null hypothesis for at least one of the two tests. This actual significance level is higher than the standard level of .05, making this an unacceptable state of affairs.

The formula shows that the overall significance level quickly reaches an unacceptable level when many individual tests are performed. For example, if you set p at .001 for individual tests and perform just eight comparisons, the overall significance level is actually .12 for the series of tests.

You can do two things to deal with this problem. First, you should normally use a relatively small p value for the individual tests. This means using the critical chi-square value for $p = .001$, rather than $p = .05$. Second, you should perform as few individual tests as are necessary, conducting only those tests that are of substantive importance. For example, if factor F1 is being measured by new, unvalidated indicators, and factors F2–F6 are measured by older tests and scales with proven validity, then you may want to perform just those tests that specifically assess the discriminant validity of F1.

In addition to the chi-square difference test, you may also perform a **confidence interval test** to assess the discriminant validity of two factors. This test involves calculating a confidence interval of plus or minus 2 standard errors around the correlation between the factors, and determining whether this interval includes 1.0. If it does not include 1.0, discriminant validity is demonstrated (Anderson & Gerbing, 1988).

For the final measurement model, the information necessary to perform this test is again reproduced in Output 5.13. This output presents the covariances (correlations, in this case) between all latent factors, along with the standard errors for these correlations.

```
                        The SAS System                              16

           Covariance Structure Analysis: Maximum Likelihood Estimation

                 Covariances among Exogenous Variables
       --------------------------------------------------------------------
                                              Standard
              Parameter        Estimate        Error         t Value
       --------------------------------------------------------------------
       F2   F1   CF1F2          0.608607       0.047237        12.884
       F3   F1   CF1F3          0.440008       0.066338         6.633
       F3   F2   CF2F3          0.534118       0.062324         8.570
       F4   F1   CF1F4         -0.016384       0.072951        -0.225
       F4   F2   CF2F4         -0.224614       0.071161        -3.156
       F4   F3   CF3F4         -0.092439       0.081541        -1.134
       F5   F1   CF1F5          0.714356       0.044263        16.139
       F5   F2   CF2F5          0.634714       0.052018        12.202
       F5   F3   CF3F5          0.516184       0.068847         7.498
       F5   F4   CF4F5          0.007920       0.078794         0.101
       F6   F1   CF1F6         -0.223309       0.073170        -3.052
       F6   F2   CF2F6         -0.375187       0.068953        -5.441
       F6   F3   CF3F6         -0.424229       0.075081        -5.650
       F6   F4   CF4F6          0.255036       0.076039         3.354
       F6   F5   CF5F6         -0.300531       0.076938        -3.906
```

Output 5.13: Covariances among Exogenous Variables, Investment Model Study

Once again, you can see that the correlation between F1 and F5 is .714, and the standard error for this estimate is .044. To compute the confidence interval for this correlation, you first multiply this standard error by 2:

```
    2 X .044 = .088
```

The lower boundary for the confidence interval will be two standard errors below the correlation:

```
    .714 - .088 = .626
```

The upper boundary for the confidence interval will be two standard errors above the correlation:

```
    .714 + .088 = .802
```

So the confidence interval for the relationship between F1 and F5 ranges from .626 to .802. This confidence interval does not include the value of 1.0, meaning that it is very unlikely that the actual population correlation between F1 and F5 is 1.0. This finding supports the discriminant validity of the measures.

Finally, discriminant validity may also be assessed with a **variance extracted test** (Fornell & Larcker, 1981; Netemeyer et al., 1990). With the test, you review the variance extracted estimates (as described previously) for the two factors of interest, and compare these estimates to the square of the correlation between the two factors. Discriminant validity is demonstrated if both variance extracted estimates are greater than this squared correlation.

In the present study, the correlation between factors F1 and F5 was .714, and the square of this correlation is .510. Variance extracted estimates were calculated earlier and appear in Table 5.3. You can see that the variance extracted estimate was .780 for F1, and was .505 for investment size. Because the variance extracted estimate for investment size was less than the square of the interfactor correlation, this test fails to support the discriminant validity of the two factors.

In summary, your analyses provided mixed support for the discriminant validity of the commitment and investment size measures. The chi-square difference test and the confidence interval test suggested that indicators V1-V3 and indicators V14-V16 were measuring two distinct constructs, while the variance extracted test did not.

Characteristics of an "Ideal Fit" for the Measurement Model

A measurement model provides an ideal fit to the data when it displays the following characteristics:

- The *p* value for the model chi-square test should be nonsignificant (should be greater than .05); the closer to 1.00, the better.

- The chi-square/*df* ratio should be less than 2.

- The comparative fit index (CFI) and the non-normed fit index (NNFI) should both exceed .9; the closer to 1.00, the better.

- The absolute value of the *t* statistics for each factor loading should exceed 1.96, and the standardized factor loadings should be nontrivial in size.

- The distribution of normalized residuals should be symmetrical and centered on zero, and relatively few (or no) normalized residuals should exceed 2.0 in absolute value.

- Composite reliabilities for the latent factors should exceed .70 (.60 at the very least).

- Variance extracted estimates for the latent factors should exceed .50.

- Discriminant validity for questionable pairs of factors should be demonstrated through the chi-square difference test, the confidence interval test, or the variance extracted test.

Remember that these characteristics represent an ideal that very often will not be attained with real-world data even when the measurement model is quite good. A model's fit need not meet all of these criteria in order to be deemed acceptable. In particular, requiring that the model chi-square be nonsignificant is an excessively strict requirement in most applied situations.

For example, the final measurement model for the investment model study demonstrated all of the preceding characteristics, with the exception that the model chi-square was significant, and one test failed to support the discriminant validity of factors F1 and F5. Overall, however, the results revealed reasonable levels of reliability and validity for the final measurement model.

Having established that you have developed an acceptable measurement model, you may now turn your attention to the analysis that you are most interested in: the test of the theoretical model that specifies causal relationships between the latent factors (as illustrated in Figure 5.5). The following chapter shows how to test this theoretical model.

Conclusion: On to Path Analysis with Latent Variables?

This chapter has shown how to use the CALIS procedure to perform confirmatory factor analysis. In some cases, your only purpose in performing a confirmatory factor analysis will be to test the factor structure underlying a set of data. This will often be the case when some theory describes the factor structure that should underly a data set, and you want to test this prediction empirically. In these situations, your analysis will essentially begin and end with the confirmatory factor analysis.

In other cases, however, your use of confirmatory factor analysis will merely be the first step in a two-step process of theory testing. This will be the case when you actually wish to test a path model that specifies causal relationships between a number of latent variables. To test such models, you will typically (a) use confirmatory factor analysis to develop an acceptable measurement model, and (b) then modify this measurement model so that you can perform a path analysis with latent variables. The current chapter has shown how to use PROC CALIS to develop measurement models, and this material constitutes the necessary foundation for the following chapter which will show how to use PROC CALIS to perform path analysis with latent variables.

References

Anderson, J.C. & Gerbing, D.W. (1988). Structural equation modeling in practice: A review and recommended two-step approach. *Psychological Bulletin, 103,* 411-423.

Bagozzi, R.P. & Phillips, L.W. (1982). Representing and testing organizational theories: A holistic construal. *Administrative Science Quarterly, 27,* 459-489.

Bentler, P. M. (1989). *EQS structural equations program*. Los Angeles: BMDP Statistical Software.

Bentler, P.M. & Bonett, D.G. (1980). Significance tests and goodness-of-fit in the analysis of covariance structures. *Psychological Bulletin, 88,* 588-606.

Bentler, P.M. & Chou, C. (1987). Practical issues in structural modeling. *Sociological Methods & Research, 16,* 78-117.

Bollen, K. A. (1989). *Structural equations with latent variables.* New York: John Wiley & Sons.

Campbell, D.T. & Fiske, D.W. (1959). Convergent and discriminant validation by the multitrait-multimethod matrix. *Psychological Bulletin, 56,* 81-105.

Carmines, E.G. & Zeller, R.A. (1988). *Reliability and validity assessment.* Beverly Hills: Sage.

Cronbach, L.J. (1951). Coefficient alpha and the internal structure of tests. *Psychometrika, 16,* 297-334.

Finn, J.D. (1974). *A general model for multivariate analysis.* New York: Holt, Rinehart, & Winston.

Fornell, C. & Larcker, D.F. (1981). Evaluating structural equation models with unobservable variables and measurement error. *Journal of Marketing Research, 18,* 39-50.

James, L.R., Mulaik, S.A., & Brett, J.M. (1982). *Causal analysis.* Beverly Hills: Sage.

Joreskog, K. G. & Sorbom, D. (1989). *LISREL 7: A guide to the program and applications, 2nd edition.* Chicago: SPSS Inc.

Lomax, R.G. (1982). A guide to LISREL-type structural equation modeling. *Behavior Research Methods & Instrumentation, 14,* 1-8.

Long, J.S. (1983a). *Confirmatory factor analysis: A preface to LISREL.* Sage University Paper Series on Quantitative Application in the Social Sciences, 07-033. Beverly Hills: Sage.

Long, J.S. (1983b). *Covariance structure models: An introduction to LISREL.* Sage University Paper Series on Quantitative Application in the Social Sciences, 07-034. Beverly Hills: Sage.

MacCallum, R.C., Roznowski, M., & Necowitz, L. B. (1992). Model modifications in covariance structure analysis: The problem of capitalization on chance. *Psychological Bulletin, 111,* 490-504.

Marsh, H.W., Balla, J.R., & McDonald, R.P. (1988). Goodness-of-fit indexes in confirmatory factor analysis: The effect of sample size. *Psychological Bulletin, 103,* 391-410.

Netemeyer, R.G., Johnston, M.W., & Burton, S. (1990). Analysis of Role conflict and role ambiguity in a structural equations framework. *Journal of Applied Psychology, 75,* 148-157.

Rusbult, C. E. (1980). Commitment and satisfaction in romantic associations: A test of the investment model. *Journal of Experimental Social Psychology, 16,* 172-186.

SAS Institute Inc. (1989). *SAS/STAT users guide, version 6, fourth edition, volume 1.* Cary, NC: SAS Institute Inc.

Schmitt, N. & Stults, D.M. (1986). Methodology review: Analysis of multitrait-multimethod matrices. *Applied Psychological Measurement, 10,* 1-22.

Smith, P.C., Kendall, L.M., & Hulin, C.L. (1969). *The measurement of satisfaction in work and retirement; A strategy for the study of attitudes.* Chicago: Rand McNally.

Widaman, K.F. (1985). Hierarchically nested covariance structure models for multitrait-multimethod data. *Applied Psychological Measurement, 9,* 1-26.

Chapter 6

PATH ANALYSIS WITH LATENT VARIABLES

Overview. This chapter continues the discussion of path analysis with latent variables that began in the preceding chapter. This chapter focuses on the second step of Anderson and Gerbing's (1988) recommended two-step procedure for testing causal models with latent variables. It shows how to use PROC CALIS to perform path analysis with latent variables: how to prepare the necessary program figure and PROC CALIS program, how to determine whether the model provides an acceptable fit to the data, and how to use modification indices to develop a better-fitting model when necessary. In addition to the indices introduced in the preceding chapter, this chapter shows how to compute and use indices that reflect the goodness of fit and parsimony that is displayed by just the structural (theoretical) portion of the model. This chapter shows how the results of the analysis can be summarized in figures, tables, and text for a scholarly journal. Finally, it presents (in abbreviated form) two additional examples in which the two-step approach is used to analyze causal models with latent variables. The first of these involves the analysis of a nonstandard model: a model in which some constructs in the structural portion of the model are measured with single indicators, rather than multiple indicators.

Recapitulation: Basic Concepts in Path Analysis with Latent Variables

The concept of path analysis with latent variables was actually introduced in Chapter 5, "Developing Measurement Models with Confirmatory Factor Analysis." Chapter 5 indicated that causal models with latent (unobserved) variables are often referred to as structural equation models, covariance structure models, latent variable models, and causal models with unmeasured variables. The first widely available software that was capable of estimating these models was the LISREL program (Joreskog & Sorbom, 1989), and as a consequence these models are often called LISREL-type models.

Path Analysis with Manifest Variables versus Latent Variables

In some ways, performing path analysis with latent variables is similar to performing path analysis with manifest variables, which was covered in Chapter 4, "Path Analysis with Manifest Variables." For example, in Chapter 4, you learned how to use path analysis with manifest variables to test a causal model derived from the investment model (Rusbult, 1980). That model predicted that relationship commitment was causally determined by satisfaction, investment size, and alternative value, while relationship satisfaction was determined by the rewards and costs associated with the relationship. In this chapter, on the other hand, you will learn how to test the same causal model using path analysis with latent variables.

One of the differences between the two procedures involves the number of indicator variables that are used to represent the underlying constructs in the causal model. In path analysis with manifest variables, each construct of interest is measured by just *one* indicator variable. For example, in Chapter 4, the variable Commitment was measured by just one observed variable (the subjects' scores on a single commitment scale). With path analysis with latent variables, on the other hand, each construct of interest is measured by *multiple* indicator variables. For example, in this chapter you will learn how to perform a path analysis in which the latent construct commitment is measured by three observed variables, the latent construct satisfaction is measured by three different observed variables, and so forth.

Chapter 5 pointed out that performing path analysis with latent variables has a number of important advantages over performing path analysis with manifest variables. For example, when you perform path analysis with latent variables, you have the opportunity to assess the convergent and discriminant validity of your measures. You also have the opportunity to work with perfectly reliable causes and effects within your structural model. These are powerful advantages, and have no doubt contributed to the increasing popularity of path analysis with latent variables.

A Two-step Approach to Path Analysis with Latent Variables

Chapter 5 indicated that this text will follow a two-step approach for performing path analysis with latent variables, as recommended by Anderson and Gerbing (1988). The first step of this process involves using confirmatory factor analysis to develop an acceptable measurement model. A **measurement model** is a factor-analytic model in which you identify the latent constructs of interest and indicate which observed variables will be used to measure each latent construct. In a measurement model, you do not specify any causal relationships between the latent constructs themselves. Instead, you allow each latent construct to covary (correlate) with every other latent construct. Chapter 5 discussed a number of procedures that you can use to verify that your measurement model displays an acceptable fit to the data, and also showed how to modify the model to achieve a better fit.

Once you have developed a measurement model that dispays an acceptable fit, you are free to move on to the second step of the two-step procedure recommended by Anderson and Gerbing. In this phase, you modify the measurement model so that it now specifies causal relationships between some of the latent variables. You make these modifications so that the model comes to represent the theoretical causal model that you want to test. The resulting theoretical model is a **combined model** that actually consists of two components:

- A **measurement model** (that specifies relationships between the latent constructs that their indicator variables)

- A **structural model** (that specifies causal relationships between the latent constructs themselves)

When you perform path analysis with latent variables, you perform a simultaneous test that determines whether this combined model, as a whole, provides an acceptable fit to the data. If it does, then your theoretical model has survived an attempt at disconfirmation, and you obtain some support for its predictions.

This chapter focuses on the second step of Anderson and Gerbing's two-step procedure. It shows you how to modify a measurement model so that it specifies causal relationships between the latent constructs. It reviews a number of procedures and indices that can be used to determine whether the resulting theoretical model provides an acceptable and parsimonious fit to the data. It also shows how to use modification indices to achieve a better model fit, when necessary.

The Importance of Reading Chapter 4 and Chapter 5 First

If you are like most readers, it is necessary that you read Chapter 4 and Chapter 5 of this text before reading this chapter. Chapter 4 discusses a number of basic issues in path analysis, and provides an introduction to PROC CALIS (the SAS System procedure used to perform path analysis with latent variables). Unless you are already familiar with path analysis and the CALIS procedure, you should read Chapter 4 before beginning this one.

Chapter 5 discusses not only confirmatory factor analysis, but also introduces basic issues in path analysis with latent variables. In fact, Chapter 5 and the current chapter were designed to be used together as a two-part introduction to causal modeling with latent variables. The current chapter assumes that you are already familiar with the Anderson-Gerbing (1988) two-step approach to causal modeling, along with all of the other topics introduced in Chapter 5. Even if you are only interested in the topic of path analysis with latent variables, you will still need to understand the concepts taught in that chapter before beginning this one.

Testing the Fit of the Theoretical Model from the Investment Model Study

This chapter shows how to perform path analysis with latent variables in order to test a theoretical causal model derived from Rusbult's (1980) investment model. The theoretical model to be tested here is similar to the one first presented in Chapter 4, and it specifies the relationship between six constructs:

- **Commitment**: the subject's intention to maintain a current romantic relationship

- **Satisfaction**: the subject's emotional response to the current relationship

- **Investment size**: the amount of time and effort that the subject has put into the current relationship

- **Alternative value**: perceived attractiveness of the subject's alternatives to the current relationship

- **Rewards**: the subject's perceptions of the number of good things associated with the current relationship

- **Costs**: the subject's perceptions of the number of bad things associated with the current relationship

The theoretical model to be tested here predicts that commitment is causally determined by satisfaction, investment size, and alternative value, while satisfaction is causally determined by rewards and costs.

This analysis actually began in the preceding chapter. That chapter presented an initial measurment model (illustrated in that chapter as Figure 5.6) in which the latent construct commitment was measured by four indicator variables, while the remaining five latent constructs were measured by three indicator variables each. Eventually, one of the indicator variables was dropped so that the measurement model would achieve a better fit to the data. The resulting revised measurement model displayed a generally acceptable fit to the data, and is reproduced here as Figure 6.1.

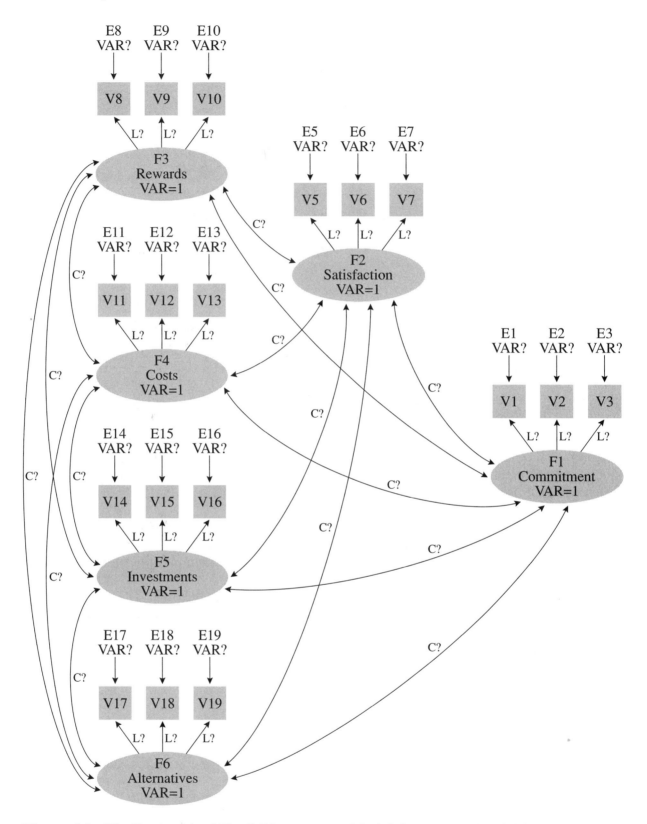

Figure 6.1: The Revised (and Final) Measurement Model, Investment Model Study (from Chapter 5)

Notice that Figure 6.1 is a standard confirmatory factor analytic model, in that no unidirectional causal relationships are predicted to exist between any of the latent contructs (i.e., the F variables). Instead, each latent construct is allowed to freely covary (correlate) with every other latent construct. This is represented by the curved, double-headed arrows that connect the various F variables in Figure 6.1.

In this chapter, you will learn how to convert this measurement model into a theoretical model that predicts causal relationships between some of these F variables. A measurement model is converted into a theoretical model by replacing some of the curved, double-headed arrows (that predict simple correlations) with straight, single-headed arrows (that predict unidirectional causal effects).

The theoretical model to be tested in this phase of the study is reproduced here as Figure 6.2. Notice how Commitment (F1) and Satisfaction (F2) are no longer connected to the other F variables by curved, double-headed arrows. Instead, straight, single-headed arrows now point from Satisfaction (F2), Investments (F5), and Alternatives (F6) to Commitment (F1). This represents the prediction that these three latent constructs will have direct causal effects on Commitment. Similarly, straight, single-headed arrows now point from Rewards (F3) and Costs (F4) to Satisfaction (F2), consistent with the prediction that these two constructs will have direct effects on Satisfaction.

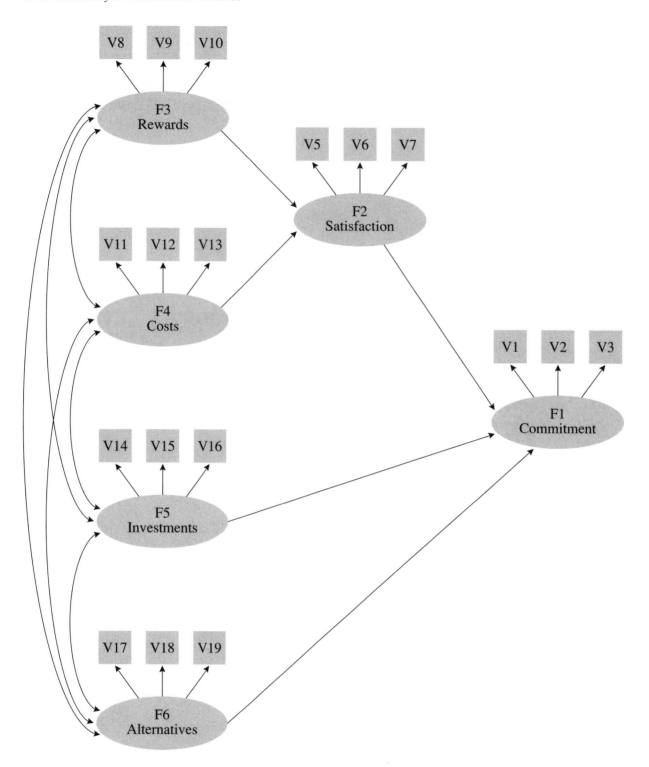

Figure 6.2: The Initial Theoretical Model

The steps followed in testing the theoretical model of Figure 6.2 are essentially the same as those followed in step 1 (from Chapter 5); you will first prepare a program figure, and use this figure to modify your PROC CALIS program. Because the steps in doing this should by now have become familiar to you (if not tedious), this will be done in an abbreviated form. In reviewing the results of the analysis, you will use the same fit indices discussed earlier to assess overall model fit, and will learn some new indices to assess fit in just the structural portion of the model.

The Rules for Structural Equation Modeling

Chapters 4 and 5 have provided a number of rules to follow when performing either path analysis with manifest variables or confirmatory factor analysis. These rules are again listed here:

Rule 1: In general, only exogenous variables are allowed to have covariances.

Rule 2: A residual term must be identified for each endogenous variable in the model.

Rule 3: Exogenous variables do not have residual terms.

Rule 4: Variances should be estimated for every exogenous variable in the model, including residual terms.

Rule 5: In most cases, covariances should be estimated for every possible pair of manifest exogenous variables; covariances are not estimated for endogenous variables.

Rule 6: For simple recursive models, covariances should not be estimated for residual terms.

Rule 7: One equation should be created for each endogenous variable, with that variable's name to the left of the equals sign.

Rule 8: Variables that have a direct effect on that endogenous variable are listed to the right of the equals sign.

Rule 9: Exogenous variables, including residual terms, are never listed to the left of the equals sign.

Rule 10: To estimate a path coefficient for a given independent variable, a unique path coefficient name should be created for the path coefficient associated with that independent variable.

Rule 11: The last term in each equation should be the residual (disturbance) term for that endogenous variable; this E (or D) term will have no name for its path coefficient.

Rule 12: To estimate a parameter, create a name for that parameter.

Rule 13: To fix a parameter at a given numerical value, insert that value in the place of the parameter's name.

Rule 14: To constrain two or more parameters to be equal, use the same name for those parameters.

Rule 15: In confirmatory factor analysis, the variances of the latent F variables are usually fixed at 1.

Many of these rules will also apply when performing path analysis with latent variables; specific rules will be mentioned when they become relevant in the sections to follow. This chapter will also introduce three additional rules that are pertinent to the types of analyses to be discussed here. The new rules are presented here for purposes of future reference, so all of the rules relevant to structural equation modeling will be grouped together in one location:

Rule 16: In path analysis with latent variables, the variances of the exogenous F variables are free parameters to be estimated.

Rule 17: In path analysis with latent variables, one factor loading for each F variable should be fixed at 1.

Rule 18: In a confirmatory factor analysis of a nonstandard model, the variance of a manifest structural variable should be a free parameter to be estimated.

You can tell by reading the preceding rules that Rules 16 and 17 are relevant to path analysis with latent variables, and Rule 18 is relevant to confirmatory factor analysis of nonstandard models. The meaning of these new rules will be explained in greater detail in later sections of this chapter.

Preparing the Program Figure

Remember that this chapter deals only with causal models in which the structural portion of a model is recursive (unidirectional). This means the model will contain no reciprocal relationships or feedback loops. The "Conclusion" section at the end of this chapter will list some additional references that you can use to learn about nonrecursive models.

When preparing the program figure for a theoretical model, you must first verify that the structural portion of the model is not **saturated** (i.e., is not just-identified). The structural portion of a model is saturated if every structural variable is related to every other structural variable by either a curved arrow or causal path.

For example, consider the model in Figure 6.2. The structural variables in this model are the variables that constitute the structural portion of the system, and in this case the structural variables are the latent F variables displayed in ovals: Commitment, Satisfaction, Rewards, Costs, Investments, and Alternatives. The structural portion of this model would be saturated if every oval were directly connected to every other oval by either a curved or straight arrow.

Fortunately, you can see that this is not the case. You know that this model is not saturated because

- the latent variable Rewards is not directly connected to Commitment in any way

- Costs is similarly not directly connected to Commitment

- Investments is not directly connected to Satisfaction

- Alternatives is not connected to Satisfaction.

If the structural portion of the model had been saturated, it would have been possible to estimate the model and test the overall model for goodness of fit, but it would not have been possible to test just the structural portion of the model for goodness of fit. Because the four paths just described have been eliminated, however, the structural portion of the model is not saturated and may be tested. You may therefore proceed with the development of the program figure.

Step 1: Identifying disturbance terms for endogenous variables. The program figure for a theoretical model is prepared by following the same steps used with the measurement model. First, the disturbance terms for all endogenous variables are identified. Earlier, it was said that a disturbance term represents causal effects on a dependent variable due to such factors as random shocks, misspecifications, and omitted independent variables.

There are generally two types of disturbance terms in a path-analytic model with latent variables: the E terms and the D terms. These are illustrated in Figure 6.3.

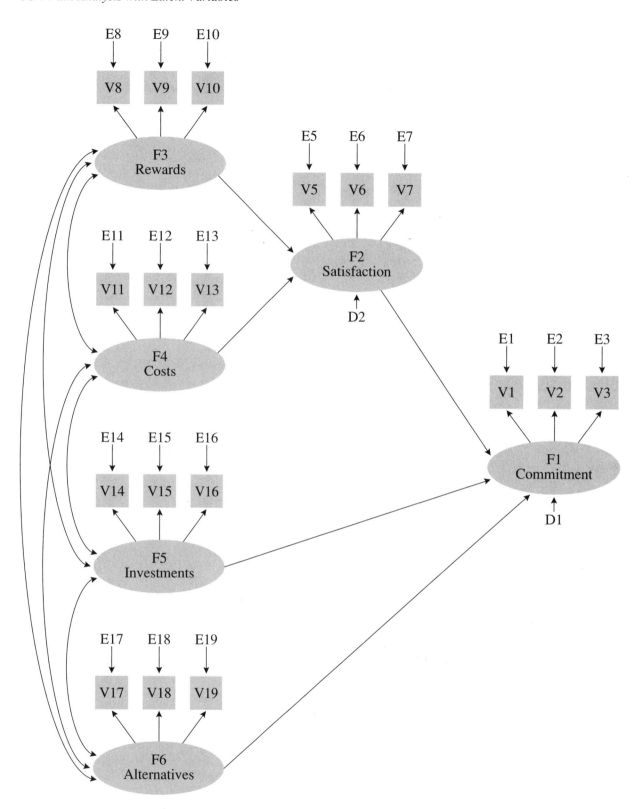

Figure 6.3: The Initial Theoretical Model, Including Disturbance (Residual) Terms for Endogenous Variables

First, the "E" terms are the disturbance terms for manifest endogenous variables. These were earlier referred to as residual terms or error terms, and the three names are in fact interchangeable in this analysis. In Figure 6.3, the E terms for all manifest endogenous variables have been identified. This was done in exactly the same way as when the program figure for the measurement model was prepared.

Second, the "D" terms are the Disturbance terms for latent endogenous variables (the F variables). Figure 6.3 shows that two of the latent factors are endogenous variables: Commitment (which is affected by three independent variables) and Satisfaction (affected by two). Therefore, there is a causal arrow drawn from the disturbance term D1 to F1 (Commitment), as well as a causal arrow drawn from D2 to F2 (Satisfaction). Notice that there are no disturbance terms for F3, F4, F5, or F6. This is consistent with Rule 3, which stated that exogenous variables do not have residual (disturbance) terms.

Step 2: Identifying all parameters to be estimated. Figure 6.4 identifies all the parameters to be estimated (or fixed) for the current analysis. Most of these parameters are the same ones estimated for the measurement model.

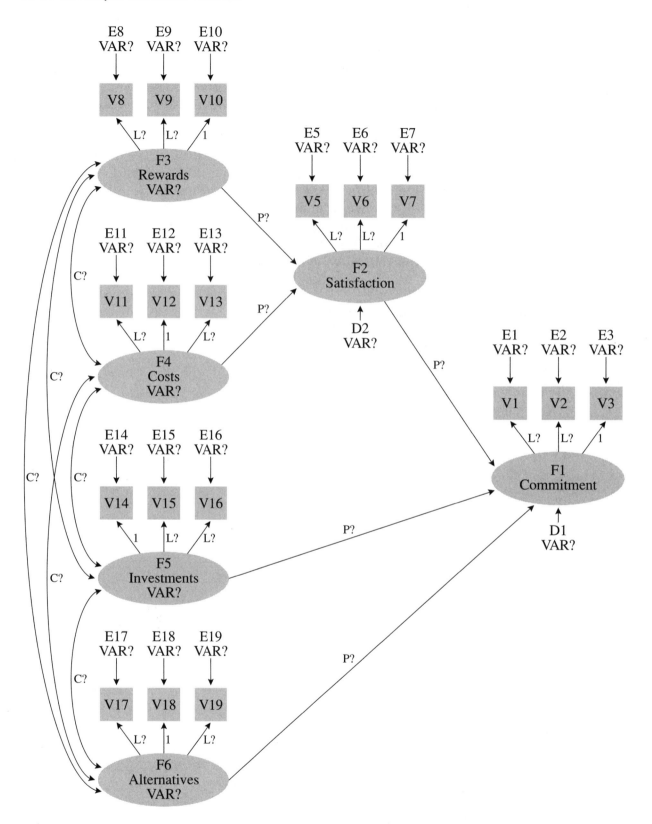

Figure 6.4: The Initial Theoretical Model, after Identifying All Parameters to Be Estimated (Completed Program Figure for Path Analysis with Latent Variables)

First, you must identify covariances to be estimated with the C? symbol. Rule 5 tells you to estimate covariances for every possible pair of exogenous variables, but Rule 6 says that you generally will not estimate covariances involving residual (disturbance) terms. In practice, this will generally mean estimating covariances for every possible pair of variables that are part of the structural model, and are exogenous variables within that structural model. Figure 6.4 shows that you will therefore estimate every possible covariance between variables F3, F4, F5, and F6. All of the remaining exogenous variables in this model are disturbance terms (E or D terms), so no covariances involving these terms will be estimated.

Next, place the VAR? symbol just beneath the name of each variable whose variance is to be estimated. As usual, variances will be estimated for all disturbance terms (notice the VAR? below all E and D terms).

However, it is at this point that you will note one of the differences between models for path analysis with latent variables versus models for confirmatory factor analysis (CFA). When performing CFA, it was noted that the variances of the latent exogenous F factors are generally fixed at 1 in order to solve the problem of scale indeterminancy. However, when performing a path analysis with latent variables, the variances of these F variables should be free parameters to be estimated. This is reflected in Figure 6.4: Notice that the VAR? symbol appears below the name of each of the exogenous F variables. This point is important enough to be summarized in a rule:

Rule 16:   In path analysis with latent variables, the variances of the exogenous F variables are free parameters to be estimated.

But if you estimate the variance for the exogenous F variables, what about the problem of scale indeterminancy? In this type of analysis, the indeterminancy problem is typically solved by adhering to Rule 17:

Rule 17:   In path analysis with latent variables, one factor loading for each F variable should be fixed at 1.

Remember that the scale indeterminacy problem (as explained in Chapter 5) involves the fact that an F variable is an unobserved variable that has no established unit of measurement. However, by fixing at 1 the path from the F variable to one of its manifest indicators, the unit of measurement for the F variable becomes equal to the unit of measurement for that indicator variable (minus its error term). For this reason, you should fix at 1 the factor loading for the indicator variable that, in some way, best represents that latent construct (Joreskog & Sorbom, 1989).

But which indicator best represents the F variable? One way to make this decision is to review the results of the confirmatory factor analysis (CFA) of the measurement model, and identify the indicator that had the largest standardized loading for that factor. In the subsequent path analysis with latent variables, the factor loading for this indicator may be fixed at one.

For example, consider the latent variable F1 in Figure 6.4. According to the program figure, the paths from F1 to V1 and V2 are free parameters to be estimated in this analysis (this is signified by the L? next to their causal arrows). However, a "1" appears next to the path from F1 to V3.

This means that this path will be fixed at 1 in order to solve the scale determinacy problem. This particular path has been fixed at 1 because in the CFA of the measurement model (reported in Chapter 5) the indicator V3 displayed the largest standardized coefficient out of all of the variables that were predicted to load on F1.

To verify this, review Output 5.10 from Chapter 5. The loadings of interest are the standardized factor loadings that appear on page 17 of the output, under the heading "Equations with Standardized Coefficients". You can see that three variables (V1, V2, and V3) were predicted to load on factor 1 (F1). The standardized factor loadings for these three variables were .8848 (for V1), .8242 (for V2), and .9369 (for V3). Because V3 displayed the largest factor loading for F1 in the confirmatory factor analysis, you conclude that V3 is the indicator that best represents F1. For that reason, you fix at 1 the path that goes from F1 to V3 when you perform the path analysis with latent variables.

This process was then repeated for each of the remaining F variables. That is, the results of the CFA (from Output 5.10 in Chapter 5) were reviewed to identify the indicator that displayed the largest standardized loading for each latent factor. These are the factor loadings that have been fixed at 1 in Figure 6.4.

With this done, all that remains is to identify the path coefficients to be estimated. This is done by placing the symbol "P?" on each of the causal paths that constitute the structural model in the figure. In Figure 6.4, this meant placing the symbol on the path from F2 to F1, from F3 to F2, and so forth.

As a final step before preparing the SAS program, you are well advised to verify that the model is identified. In fact, this should be done each time the model is modified, as any modification has the potential of resulting in an unidentified model. The preceding chapters provided a number of references on procedures for assessing model identification; in particular, see the section titled "Step 8: Verifying that the Model is Overidentified" from Chapter 4, and the section titled "Step 4: Verifying that the model is overidentified" from Chapter 5.

Preparing the SAS Program

In presenting programs, this chapter uses a system of notation based in part on on the system developed by Bentler (1989) for the EQS structural equations program (e.g., latent factors are represented by the letter F, and so forth). The modified system used by this text was introduced in Chapter 4, and further developed in Chapter 5.

Here is the entire SAS program (minus the DATA step) that was used to analyze the model portrayed in Figure 6.4:

```
1       PROC CALIS   COVARIANCE   CORR   RESIDUAL   MODIFICATION ;
2          LINEQS
3             V1  = LV1F1  F1 + E1,
4             V2  = LV2F1  F1 + E2,
5             V3  =        F1 + E3,
```

```
 6          V5   = LV5F2   F2 + E5,
 7          V6   = LV6F2   F2 + E6,
 8          V7   =         F2 + E7,
 9          V8   = LV8F3   F3 + E8,
10          V9   = LV9F3   F3 + E9,
11          V10  =         F3 + E10,
12          V11  = LV11F4  F4 + E11,
13          V12  =         F4 + E12,
14          V13  = LV13F4  F4 + E13,
15          V14  =         F5 + E14,
16          V15  = LV15F5  F5 + E15,
17          V16  = LV16F5  F5 + E16,
18          V17  = LV17F6  F6 + E17,
19          V18  =         F6 + E18,
20          V19  = LV19F6  F6 + E19,
21          F1   = PF1F2 F2 + PF1F5 F5 + PF1F6 F6 + D1,
22          F2   = PF2F3 F3 + PF2F4 F4             + D2;
23      STD
24          E1-E3   = VARE1-VARE3,
25          E5-E19 = VARE5-VARE19,
26          F3-F6   = VARF3-VARF6,
27          D1-D2   = VARD1-VARD2;
28      COV
29          F3 F4 = CF3F4,
30          F3 F5 = CF3F5,
31          F3 F6 = CF3F6,
32          F4 F5 = CF4F5,
33          F4 F6 = CF4F6,
34          F5 F6 = CF5F6;
35      VAR   V1 V2 V3 V5-V19 ;
36      RUN;
```

The easiest way to create the SAS program that will perform the path analysis with latent variables is to simply modify the program that had performed the confirmatory factor analysis of the corresponding measurement model (or, better still, to modify a copy of that program). Many aspects of the program that tests the theoretical model are identical to the program used with the measurement model (e.g., the PROC CALIS statement). Therefore, to save time, those aspects will not be reviewed again here. Instead, this section will merely discuss how the program for the measurement model must be changed so that it can perform the path analysis with latent variables.

For purposes of reference, the section titled "Preparing the SAS Program" in Chapter 5 discussed the DATA step and the various statements that constitute a PROC CALIS program. The initial measurement model and the revised measurement model for the investment model study appear in Chapter 5 in the sections titled "Overview of the PROC CALIS Program" and "Estimating the Revised Measurement Model," respectively. It is the revised measurement model that will be modified here to perform path analysis with latent variables.

The LINEQS statement. The LINEQS statement serves two functions in a path analysis with latent variables: it indicates which factor loadings are to be estimated or fixed, and it specifies the causal relations between variables in the structural model. Each will be discussed in turn.

Lines 3-20 of the preceding program indicate which manifest variables load on which latent factors in the theoretical model. Notice that this portion of the program is identical to that used to estimate the revised measurement model, with one important difference: the coefficient names for some of the factor loadings have been blanked out. It is in this way that the factor loadings are fixed at 1.

For example, consider the following equation from the program:

```
3          V1   = LV1F1   F1 + E1,
```

This line estimates the factor loading for the path from F1 to V1. You know that this parameter is estimated, because the name for the parameter (LV1F1) appears just before the name of the variable where the path originates (F1).

In contrast, notice how the following line is different:

```
5          V3   =          F1 + E3,
```

In the preceding equation, the name for the factor loading (LV3F1) has been omitted from the equation; it does not appear just before the F1. The earlier chapter on path analysis with manifest variables (Chapter 4) indicated that this has the effect of fixing that parameter at 1. Therefore, you know from reviewing this equation that the factor loading LV3F1 is fixed at 1.

Reviewing the SAS program shows that the factor loadings for all of the following indicators have been fixed at 1: V3, V7, V10, V12, V14, and V18. Note how this is consistent with the program figure appearing in Figure 6.4.

Lines 21-22 of the preceding program contain the equations that specify the causal relations between the F variables that constitute the structural portion of the theoretical model in Figure 6.4. These equations are reproduced here:

```
21     F1 = PF1F2 F2 + PF1F5 F5 + PF1F6 F6 + D1,
22     F2 = PF2F3 F3 + PF2F4 F4              + D2;
```

In preparing these equations, you should adhere to the recommendations that appear in the section titled "The LINEQS Statement" from Chapter 4 of this text, "Path Analysis with Manifest Variables" (pp. 170-177). In the present case, however, remember that you are dealing with latent variables (F variables) rather than manifest variables (V variables).

In each equation, the name of the dependent variable appears to the left of the equal sign, and the names of the independent variables having direct effects on that dependent variable appear to the right of the equal sign. For example, you can see that in line 21, F1 is the dependent variable and F2, F5, and F6 are independent variables (along with the disturbance term, D1).

In most analyses, you will estimate a path coefficient for each independent F variable in the equation but not for the disturbance D term. To estimate a path coefficient, simply create a name for that path coefficient, following the same conventions discussed in Chapter 4: (a) begin the name with the letter "P" (for **P**ath coefficient); (b) continue with the name of the relevant dependent variable being affected (such as "F1"), and end with the name of the relevant independent variable where the path originates (such as "F2"). Using these conventions, when you see a path coefficient named "PF1F2", you will know that this is the name for the path coefficient that represents the effect of F2 on F1.

In the LINEQS equation, place the name for a given path coefficient to the immediate left of the name of the independent variable where the path originates. For example, in line 21 (shown earlier), you can see that the path coefficient named "PF1F2" appears to the immediate left of the independent variable "F2." At this time, review the equations that appear in lines 21 and 22 on the previous page, and note how they comply with these recommendations.

The STD statement. Rule 4 (presented earlier) stated that variances are to be estimated for every exogenous variable in the model, including residual terms, and Figure 6.4 adheres to that rule. In that figure, the VAR? symbol is used to identify variables whose variances are to be estimated, and this included all of the residual, or disturbance, terms (all E and D variables). In addition, the figure shows that you are also to estimate variances for the exogenous F variables: F3, F4, F5, and F6. The STD statement that reflects this aspect of the figure is presented here:

```
23          STD
24              E1-E3   = VARE1-VARE3,
25              E5-E19  = VARE5-VARE19,
26              F3-F6   = VARF3-VARF6,
27              D1-D2   = VARD1-VARD2;
```

In some ways, the STD statement is identical to that used with the CFA of the measurement model. Specifically, it still contains equations to estimate variances of the E terms (lines 24 and 25).

However, it also differs from the previous STD statement in three important ways. First, it no longer contains any equations for F1 or F2. This is because F1 and F2 are now endogenous variables, and you do not estimate variances for endogenous variables. Second, the variances for F3, F4, F5, and F6 are now estimated rather than fixed at 1 (line 26). As was discussed before, it is now possible to estimate these variances because you establish their scale by fixing one factor loading at 1 for each F variable. Finally, you now estimate variances for the disturbance terms D1 and D2 (line 27). With this done, your STD statement now estimates all variances indicated by the program figure.

The COV statement. One of the ways that the theoretical model of Figure 6.4 differs from the measurement model of Figure 6.1 involves the covariances. With the theoretical model, there are no longer any covariances between F1 or F2 and any of the other F variables. This is consistent with Rule 1, which stated that only exogenous variables are allowed to have covariances. Because F1 and F2 are now endogenous variables, the SAS program must be modified so that it no longer estimates covariances between F1 or F2 and any other variable.

This is done in the following COV statement:

```
28          COV
29              F3 F4 = CF3F4,
30              F3 F5 = CF3F5,
31              F3 F6 = CF3F6,
32              F4 F5 = CF4F5,
33              F4 F6 = CF4F6,
34              F5 F6 = CF5F6;
```

Notice that none of the equations above include either F1 or F2. However, the statement does estimate every possible covariance between F3, F4, F5, and F6. This is because these latent variables are still exogenous variables in the theoretical model.

Interpreting the Results of the Analysis

If you specify LINESIZE=80 and PAGESIZE=60 in the OPTIONS statement, the preceding program would produce 43 pages of output, and this output would be similar to that created in the analysis of the measurement model (from Chapter 5). The following table indicates the pages on which various results are printed.

- Page 1 includes a list of the endogenous variables and exogenous variables specified in the LINEQS statement.

- Pages 2-3 include the general form of the structural equations specified in the LINEQS statement.

- The top of page 4 provides general information about the estimation problem, including

  - the number of observations
  - the number of variables
  - the amount of independent information in the data matrix (number of "data points")
  - the number of parameters to be estimated.

  Below this information, the output includes some univariate statistics for the manifest variables, and this is followed by the first part of the covariance matrix to be analyzed.

- Page 5 presents the remainder of the covariance matrix to be analyzed.

- Page 6 includes a vector of initial parameter estimates.

- Page 7 includes the iteration history.

- Pages 8-12 include the predicted model covariance matrix, the residual matrix, and the normalized residual matrix.

- Page 13 includes a bar chart displaying the distribution of normalized residuals.

- Page 14 includes a variety of goodness of fit indices, to be discussed later.

- Pages 15-17 include equations containing parameter estimates (factor loadings and path coefficients), their approximate standard errors and *t* values. These equations correspond to those constructed in the LINEQS statement. These are followed by estimates of variances and covariances, along with corresponding approximate standard errors and *t* values.

- Page 18 includes equations containing the standardized parameter estimates (factor loadings and path coefficients).

- Page 19 includes estimates of the endogenous variable variances and the R^2 values for each endogenous variable. Finally, the correlations among exogenous variables are presented.

- Pages 20-43 include modification indices (Lagrange multiplier and Wald test results).

Once the SAS program for estimating the theoretical model has been executed, the SAS log and SAS output files should be reviewed to verify that the program ran correctly. This should be done in the usual way, as described in the section titled "Making Sure that the SAS log and Output Files Look Right" from Chapter 5. The information shown on output pages 2–7 are particularly important for this purpose, and so these pages are reproduced here as Output 6.1. (Remember that the analyses reported in this chapter are fictitious and should not be viewed as legitimate tests of the investment model.)

```
          Covariance Structure Analysis: Pattern and Initial Values        2
                       Manifest Variable Equations
                            Initial Estimates

V1       =      .      *F1      + 1.0000 E1
                       LV1F1

V2       =      .      *F1      + 1.0000 E2
                       LV2F1

V3       =   1.0000 F1          + 1.0000 E3

V5       =      .      *F2      + 1.0000 E5
                       LV5F2

V6       =      .      *F2      + 1.0000 E6
                       LV6F2
```

```
V7      =      1.0000 F2     + 1.0000 E7

V8      =         .    *F3    + 1.0000 E8
                      LV8F3

V9      =         .    *F3    + 1.0000 E9
                      LV9F3

V10     =      1.0000 F3      + 1.0000 E10

V11     =         .    *F4    + 1.0000 E11
                      LV11F4

V12     =      1.0000 F4      + 1.0000 E12

V13     =         .    *F4    + 1.0000 E13
                      LV13F4

V14     =      1.0000 F5      + 1.0000 E14

V15     =         .    *F5    + 1.0000 E15
                      LV15F5

V16     =         .    *F5    + 1.0000 E16
                      LV16F5

V17     =         .    *F6    + 1.0000 E17
                      LV17F6

V18     =      1.0000 F6      + 1.0000 E18

V19     =         .    *F6    + 1.0000 E19
                      LV19F6
```

```
                    Latent Variable Equations
                        Initial Estimates

F1      =      .    *F2    + .    *F5    + .    *F6    + 1.0000 D1
              PF1F2        PF1F5        PF1F6
```

```
         Covariance Structure Analysis: Pattern and Initial Values     3

F2      =      .    *F3    + .    *F4    + 1.0000 D2
              PF2F3        PF2F4
```

```
                 Variances of Exogenous Variables
            ------------------------------------------
            Variable     Parameter     Estimate
            ------------------------------------------
            F3           VARF3                    .
            F4           VARF4                    .
```

```
        F5         VARF5                    .
        F6         VARF6                    .
        E1         VARE1                    .
        E2         VARE2                    .
        E3         VARE3                    .
        E5         VARE5                    .
        E6         VARE6                    .
        E7         VARE7                    .
        E8         VARE8                    .
        E9         VARE9                    .
        E10        VARE10                   .
        E11        VARE11                   .
        E12        VARE12                   .
        E13        VARE13                   .
        E14        VARE14                   .
        E15        VARE15                   .
        E16        VARE16                   .
        E17        VARE17                   .
        E18        VARE18                   .
        E19        VARE19                   .
        D1         VARD1                    .
        D2         VARD2                    .


        Covariances among Exogenous Variables
        -------------------------------------
              Parameter          Estimate
        -------------------------------------

        F4    F3   CF3F4               .
        F5    F3   CF3F5               .
        F5    F4   CF4F5               .
        F6    F3   CF3F6               .
        F6    F4   CF4F6               .
        F6    F5   CF5F6               .
```

```
    Covariance Structure Analysis: Maximum Likelihood Estimation        4

        240 Observations       Model Terms        1
         18 Variables          Model Matrices     4
        171 Informations       Parameters        47

        VARIABLE              Mean          Std Dev

        V1                      0         2.486000000
        V2                      0         2.909000000
        V3                      0         2.724000000
        V5                      0         1.929000000
        V6                      0         2.113000000
        V7                      0         2.056000000
        V8                      0         1.417000000
        V9                      0         1.408000000
        V10                     0         1.724000000
        V11                     0         2.595000000
```

V12	0	2.691000000
V13	0	2.360000000
V14	0	2.102000000
V15	0	2.219000000
V16	0	1.874000000
V17	0	2.001000000
V18	0	1.966000000
V19	0	2.185000000

Covariances

	V1	V2	V3	V5	V6	V7
V1	6.1801960	5.3081221	5.5461566	2.4648839	2.8050582	2.6680548
V2	5.3081221	8.4622810	6.2283552	2.0313489	2.1267641	2.0634119
V3	5.5461566	6.2283552	7.4201760	2.6062796	2.6016270	2.8394758
V5	2.4648839	2.0313489	2.6062796	3.7210410	2.9061716	2.8555373
V6	2.8050582	2.1267641	2.6016270	2.9061716	4.4647690	3.3190666
V7	2.6680548	2.0634119	2.8394758	2.8555373	3.3190666	4.2271360
V8	1.2188411	1.2077615	1.3162286	0.9211534	1.1227954	0.8303053
V9	0.7315602	0.6020932	0.6405105	0.6817240	0.9103818	1.0334607
V10	1.4957665	1.2086430	1.3478025	0.9378181	1.2786270	1.0775414
V11	0.3290097	0.6190061	0.0353439	-0.8059266	-0.9102170	-0.6242324
V12	-0.2675930	0.1017655	-0.4178262	-0.9810875	-0.8529125	-1.0512122
V13	-0.1701418	-0.0823829	-0.4242902	-0.5007684	-0.5036547	-0.4027293
V14	2.9210947	2.6170993	3.3267177	1.7557102	2.0031282	2.0312046
V15	2.3941324	2.0785329	2.7442284	2.0203729	1.9598962	1.8933396
V16	1.7470365	1.7771779	2.2001585	0.9145813	1.0136991	0.8669124
V17	-0.7014025	-0.4365682	-0.7903550	-1.1579787	-1.1923279	-1.2547871
V18	-0.6598093	-1.0523133	-0.8247291	-0.7281435	-0.7643651	-0.8245876
V19	-0.9071290	-0.9025754	-0.8630313	-0.9862784	-0.9649331	-1.1275824

Covariance Structure Analysis: Maximum Likelihood Estimation   5

Covariances

	V8	V9	V10	V11	V12	V13
V1	1.2188411	0.7315602	1.4957665	0.3290097	-0.2675930	-0.1701418
V2	1.2077615	0.6020932	1.2086430	0.6190061	0.1017655	-0.0823829
V3	1.3162286	0.6405105	1.3478025	0.0353439	-0.4178262	-0.4242902
V5	0.9211534	0.6817240	0.9378181	-0.8059266	-0.9810875	-0.5007684
V6	1.1227954	0.9103818	1.2786270	-0.9102170	-0.8529125	-0.5036547
V7	0.8303053	1.0334607	1.0775414	-0.6242324	-1.0512122	-0.4027293
V8	2.0078890	0.7781030	1.2361114	-0.3346175	-0.3889410	-0.1437972
V9	0.7781030	1.9824640	1.1942769	-0.2009568	-0.1364014	0.1827584
V10	1.2361114	1.1942769	2.9721760	-0.2818481	-0.0835071	-0.0122059
V11	-0.3346175	-0.2009568	-0.2818481	6.7340250	4.9859655	2.3210718
V12	-0.3889410	-0.1364014	-0.0835071	4.9859655	7.2414810	2.5593563
V13	-0.1437972	0.1827584	-0.0122059	2.3210718	2.5593563	5.5696000
V14	0.9084529	1.0210675	1.2067414	0.0163641	-0.0282824	0.2133110
V15	0.6445862	0.7217253	0.9219590	-0.2130573	0.0417993	-0.3246841

V16	0.4169069	0.3984274	0.4523086	0.4522618	0.1109445	0.2388226
V17	-0.5614126	-0.5634816	-0.6485481	1.1735265	1.1630933	0.2408404
V18	-0.4401599	-0.3681610	-0.8236203	0.6734336	0.6295702	0.4964543
V19	-0.8916898	-0.9167910	-0.8287268	0.8335010	0.8349366	0.3919016

	V14	V15	V16	V17	V18	V19
V1	2.9210947	2.3941324	1.7470365	-0.7014025	-0.6598093	-0.9071290
V2	2.6170993	2.0785329	1.7771779	-0.4365682	-1.0523133	-0.9025754
V3	3.3267177	2.7442284	2.2001585	-0.7903550	-0.8247291	-0.8630313
V5	1.7557102	2.0203729	0.9145813	-1.1579787	-0.7281435	-0.9862784
V6	2.0031282	1.9598962	1.0136991	-1.1923279	-0.7643651	-0.9649331
V7	2.0312046	1.8933396	0.8669124	-1.2547871	-0.8245876	-1.1275824
V8	0.9084529	0.6445862	0.4169069	-0.5614126	-0.4401599	-0.8916898
V9	1.0210675	0.7217253	0.3984274	-0.5634816	-0.3681610	-0.9167910
V10	1.2067414	0.9219590	0.4523086	-0.6485481	-0.8236203	-0.8287268
V11	0.0163641	-0.2130573	0.4522618	1.1735265	0.6734336	0.8335010
V12	-0.0282824	0.0417993	0.1109445	1.1630933	0.6295702	0.8349366
V13	0.2133110	-0.3246841	0.2388226	0.2408404	0.4964543	0.3919016
V14	4.4184040	2.7752811	1.8001906	-0.9253424	-0.5950846	-0.8221237
V15	2.7752811	4.9239610	1.7049465	-1.1366961	-0.6500205	-0.9794000
V16	1.8001906	1.7049465	3.5118760	-0.2437418	-0.2726370	-0.3398593
V17	-0.9253424	-1.1366961	-0.2437418	4.0040010	2.0810680	2.0112051
V18	-0.5950846	-0.6500205	-0.2726370	2.0810680	3.8651560	2.4571461
V19	-0.8221237	-0.9794000	-0.3398593	2.0112051	2.4571461	4.7742250

Determinant = 56874416 (Ln = 17.856)

Some initial estimates computed by instrumental variable method.
Some initial estimates computed by two-stage LS method.

Covariance Structure Analysis: Maximum Likelihood Estimation   6

Vector of Initial Estimates

LV1F1	1	0.89069	Matrix Entry:	_BETA_[1:19]
LV2F1	2	0.89461	Matrix Entry:	_BETA_[2:19]
LV5F2	3	0.92545	Matrix Entry:	_BETA_[4:20]
LV6F2	4	1.04052	Matrix Entry:	_BETA_[5:20]
PF1F2	5	0.52397	Matrix Entry:	_BETA_[19:20]
LV8F3	6	0.84001	Matrix Entry:	_GAMMA_[7:1]
LV9F3	7	0.63254	Matrix Entry:	_GAMMA_[8:1]
LV11F4	8	0.86916	Matrix Entry:	_GAMMA_[10:2]
LV13F4	9	0.44411	Matrix Entry:	_GAMMA_[12:2]
LV15F5	10	0.89697	Matrix Entry:	_GAMMA_[14:3]
LV16F5	11	0.57209	Matrix Entry:	_GAMMA_[15:3]
LV17F6	12	1.04325	Matrix Entry:	_GAMMA_[16:4]
LV19F6	13	1.16728	Matrix Entry:	_GAMMA_[18:4]
PF1F5	14	0.69577	Matrix Entry:	_GAMMA_[19:3]
PF1F6	15	0.12538	Matrix Entry:	_GAMMA_[19:4]
PF2F3	16	0.72184	Matrix Entry:	_GAMMA_[20:1]
PF2F4	17	-0.13164	Matrix Entry:	_GAMMA_[20:2]
VARF3	18	1.59016	Matrix Entry:	_PHI_[1:1]

CF3F4	19	-0.26849	Matrix Entry:	_PHI_[2:1]
VARF4	20	5.77864	Matrix Entry:	_PHI_[2:2]
CF3F5	21	1.10070	Matrix Entry:	_PHI_[3:1]
CF4F5	22	0.04126	Matrix Entry:	_PHI_[3:2]
VARF5	23	3.14953	Matrix Entry:	_PHI_[3:3]
CF3F6	24	-0.75485	Matrix Entry:	_PHI_[4:1]
CF4F6	25	0.86782	Matrix Entry:	_PHI_[4:2]
CF5F6	26	-0.77760	Matrix Entry:	_PHI_[4:3]
VARF6	27	1.90358	Matrix Entry:	_PHI_[4:4]
VARE1	28	0.93251	Matrix Entry:	_PHI_[5:5]
VARE2	29	3.16834	Matrix Entry:	_PHI_[6:6]
VARE3	30	0.80545	Matrix Entry:	_PHI_[7:7]
VARE5	31	1.06336	Matrix Entry:	_PHI_[8:8]
VARE6	32	1.10513	Matrix Entry:	_PHI_[9:9]
VARE7	33	1.12406	Matrix Entry:	_PHI_[10:10]
VARE8	34	0.88586	Matrix Entry:	_PHI_[11:11]
VARE9	35	1.34622	Matrix Entry:	_PHI_[12:12]
VARE10	36	1.38202	Matrix Entry:	_PHI_[13:13]
VARE11	37	2.36858	Matrix Entry:	_PHI_[14:14]
VARE12	38	1.46284	Matrix Entry:	_PHI_[15:15]
VARE13	39	4.42987	Matrix Entry:	_PHI_[16:16]
VARE14	40	1.26888	Matrix Entry:	_PHI_[17:17]
VARE15	41	2.38998	Matrix Entry:	_PHI_[18:18]
VARE16	42	2.48109	Matrix Entry:	_PHI_[19:19]
VARE17	43	1.93220	Matrix Entry:	_PHI_[20:20]
VARE18	44	1.96158	Matrix Entry:	_PHI_[21:21]
VARE19	45	2.18051	Matrix Entry:	_PHI_[22:22]
VARD1	46	3.23324	Matrix Entry:	_PHI_[23:23]
VARD2	47	2.12335	Matrix Entry:	_PHI_[24:24]

Covariance Structure Analysis: Maximum Likelihood Estimation   7

(Dual) Quasi-Newton Minimization:
Broyden - Fletcher - Goldfarb - Shanno Method (BFGS)
Maximum Iterations= 200
Maximum Function Calls= 500
Maximum Absolute Gradient Criterion= 0.001
Number of Estimates= 47 Lower Bounds= 0 Upper Bounds= 0
Line Search Method 2: start alpha=1 sigma=0.4
Minimization Start: Active Constraints= 0 Criterion= 1.048
Maximum Gradient Element= 0.226

Iter	res	nfun	act	mincrit	maxgrad	difcrit	alpha	descent
0	1	1	0	1.0477	0.2262	0	1.0000	-0.1909
1	1	3	0	1.0104	0.4643	0.0374	0.4086	-0.2937
2	1	5	0	0.9666	0.1600	0.0438	0.3521	-0.1024
3	1	6	0	0.9448	0.1819	0.0218	0.7042	-0.0387
4	1	8	0	0.9349	0.0816	0.00988	0.5119	-0.0241
5	1	10	0	0.9291	0.0585	0.00586	0.4822	-0.0141
6	1	11	0	0.9224	0.0562	0.00664	0.9645	-0.00791
7	1	12	0	0.9200	0.1811	0.00241	1.9289	-0.0182
8	1	14	0	0.9137	0.0317	0.00630	0.6745	-0.00406
9	1	15	0	0.9117	0.0321	0.00203	1.3490	-0.00291

```
10  1  17  0  0.9100   0.0244   0.00164   1.1346 -0.00294
11  1  19  0  0.9089   0.0118   0.00115   0.7650 -0.00108
12  1  20  0  0.9086   0.0280  0.000247   1.5299 -0.00108
13  1  22  0  0.9081   0.0106  0.000533   0.9865 -0.00045
14  1  24  0  0.9079  0.00605   0.00017   0.7542  -0.0003
15  1  25  0  0.9077  0.00615  0.000251   1.5085 -0.00028
16  1  27  0  0.9075  0.00460  0.000155   1.0931  -0.0001
17  1  28  0  0.9075   0.0110  0.000056   2.1863 -0.00016
18  1  30  0  0.9074  0.00241  0.000116   1.4604 -0.00005
19  1  31  0  0.9073  0.00415  0.000069   2.9207 -0.00006
20  1  33  0  0.9072  0.00274  0.000052   1.8722 -0.00005
21  1  35  0  0.9072  0.00376  0.000049   1.9415 -0.00005
22  1  37  0  0.9071  0.00400  0.000042   1.6464 -0.00004
23  1  39  0  0.9071  0.00242  0.000029   1.3572 -0.00002
24  1  40  0  0.9071  0.00231  0.000026   2.7144 -0.00003
25  1  41  0  0.9071  0.00349  0.000037   4.0000 -0.00004
26  1  43  0  0.9070  0.00201  0.000041   1.9345 -0.00003
27  1  44  0  0.9070  0.00458  0.000012   3.8689 -0.00006
28  1  46  0  0.9070  0.00111  0.000034   1.0850 -9.57E-6
29  1  47  0  0.9070  0.00114  0.000014   2.1700 -0.00002
30  1  49  0  0.9069  0.00152   0.00002   2.0252 -0.00001
31  1  50  0  0.9069  0.00107  0.000015   4.0000  -6.9E-6
32  1  52  0  0.9069 0.000573  5.324E-6   1.5443  -6.9E-6
Minimization Results: Iterations= 32 Function Calls= 52 Derivative Calls= 33
Active Constraints= 0 Criterion= 0.907 Maximum Gradient Element= 0.000573
Descent= -69E-7

NOTE: Convergence criterion satisfied.
```

Output 6.1: CALIS Output Pages 2-7 for Analysis of Initial Theoretical Model, Investment Model Study

You may begin your assessment of the fit of the theoretical model by following the same procedures used with the measurement model. Once this is done, however, this chapter will introduce some additional indices that are particularly useful for evaluating the fit of theoretical models.

Step 1: Reviewing the chi-square test. Output 6.2 presents the goodness of fit indices for the current model. These indices appeared on page 14 of the output. This table shows that the model chi-square value for the theoretical model was 216.75 with 124 *df* (*p* = .0001). Although a nonsignificant chi-square would have shown support for your model, this significant chi-square does not necessarily indicate a bad fit. The chi-square/*df* ratio for this model is 1.75, which meets the informal rule-of-thumb criteria that the ratio should be below 2.0.

```
                        The SAS System                          14

   Covariance Structure Analysis: Maximum Likelihood Estimation

   Fit criterion . . . . . . . . . . . . . . . . . .      0.9069
   Goodness of Fit Index (GFI) . . . . . . . . . . .      0.9094
   GFI Adjusted for Degrees of Freedom (AGFI)  . . .      0.8750
   Root Mean Square Residual (RMR) . . . . . . . . .      0.2680
   Chi-square = 216.7521      df = 124    Prob>chi**2 = 0.0001
   Null Model Chi-square:     df = 153                2167.7711
   RMSEA Estimate  . . . . . . 0.0561   90%C.I.[0.0434, 0.0682]
   Bentler's Comparative Fit Index . . . . . . . . .      0.9540
   Normal Theory Reweighted LS Chi-square  . . . . .    214.2723
   Akaike's Information Criterion  . . . . . . . . .    -31.2479
   Consistent Information Criterion  . . . . . . . .   -586.8471
   Schwarz's Bayesian Criterion  . . . . . . . . . .   -462.8471
   McDonald's (1989) Centrality. . . . . . . . . . .      0.8243
   Bentler & Bonett's (1980) Non-normed Index. . . .      0.9432
   Bentler & Bonett's (1980) Normed Index. . . . . .      0.9000
   James, Mulaik, & Brett (1982) Parsimonious Index.      0.7294
   Z-Test of Wilson & Hilferty (1931) . . . . . . .       4.8757
   Bollen (1986) Normed Index Rho1 . . . . . . . . .      0.8766
   Bollen (1988) Non-normed Index Delta2 . . . . . .      0.9546
   Hoelter's (1983) Critical N . . . . . . . . . . .         167
```

Output 6.2: Goodness of Fit Indices for Initial Theoretical Model, Investment Model Study

Step 2: Reviewing the non-normed index and comparative fit index. Output 6.2 shows that the comparative fit index (CFI) for the theoretical model was .954, a bit lower than the CFI of .970 observed for the measurement model, but still in the acceptable range. The non-normed fit index (NNFI) for the theoretical model is .943, whereas the NNFI for the measurement model was .962.

Step 3: Reviewing the significance tests for factor loadings and path coefficients. As before, it is good practice to review the standard errors for the factor loadings and path coefficients, as near-zero standard errors may indicate estimation problems. These parameters appear on pages 15-16 of Output 6.3, and none of the standard errors appear to be unacceptably small.

```
                          The SAS System                          15

             Covariance Structure Analysis: Maximum Likelihood Estimation

                        Manifest Variable Equations

V1       =       0.8606*F1      +   1.0000 E1
Std Err          0.0424 LV1F1
t Value         20.3190

V2       =       0.9387*F1      +   1.0000 E2
Std Err          0.0533 LV2F1
t Value         17.6237

V3       =       1.0000 F1      +   1.0000 E3

V5       =       0.8818*F2      +   1.0000 E5
Std Err          0.0564 LV5F2
t Value         15.6455

V6       =       1.0215*F2      +   1.0000 E6
Std Err          0.0606 LV6F2
t Value         16.8603

V7       =       1.0000 F2      +   1.0000 E7

V8       =       0.7656*F3      +   1.0000 E8
Std Err          0.0972 LV8F3
t Value          7.8770

V9       =       0.7346*F3      +   1.0000 E9
Std Err          0.0957 LV9F3
t Value          7.6730

V10      =       1.0000 F3      +   1.0000 E10

V11      =       0.9069*F4      +   1.0000 E11
Std Err          0.1010 LV11F4
t Value          8.9830

V12      =       1.0000 F4      +   1.0000 E12

V13      =       0.4662*F4      +   1.0000 E13
Std Err          0.0712 LV13F4
t Value          6.5509

V14      =       1.0000 F5      +   1.0000 E14
```

```
V15       =       0.8458*F5      +   1.0000 E15
Std Err           0.0842 LV15F5
t Value          10.0429


V16       =       0.5700*F5      +   1.0000 E16
Std Err           0.0709 LV16F5
t Value           8.0376


V17       =       0.9151*F6      +   1.0000 E17
Std Err           0.1058 LV17F6
t Value           8.6509
```

```
          Covariance Structure Analysis: Maximum Likelihood Estimation          16

V18       =.      1.0000 F6      +   1.0000 E18


V19       =       1.0729*F6      +   1.0000 E19
Std Err           0.1203 LV19F6
t Value           8.9214


                        Latent Variable Equations

F1        =       0.4610*F2      +   0.7568*F5      +   0.0997*F6      +   1.0000 D1
Std Err           0.0909 PF1F2       0.1033 PF1F5       0.1093 PF1F6
t Value           5.0731             7.3273             0.9128


F2        =       0.9732*F3      -   0.1209*F4      +   1.0000 D2
Std Err           0.1317 PF2F3       0.0508 PF2F4
t Value           7.3887            -2.3799



                        Variances of Exogenous Variables
```

| | | | Standard | |
Variable	Parameter	Estimate	Error	t Value
F3	VARF3	1.400268	0.264566	5.293
F4	VARF4	5.508715	0.840503	6.554
F5	VARF5	3.275707	0.450937	7.264
F6	VARF6	2.221639	0.377512	5.885
E1	VARE1	1.347998	0.179564	7.507
E2	VARE2	2.710495	0.297981	9.096
E3	VARE3	0.896566	0.190027	4.718
E5	VARE5	1.201289	0.144662	8.304
E6	VARE6	1.082749	0.156619	6.913
E7	VARE7	0.986775	0.146893	6.718
E8	VARE8	1.187937	0.135535	8.765
E9	VARE9	1.227853	0.136283	9.010
E10	VARE10	1.574893	0.193830	8.125
E11	VARE11	2.209074	0.489654	4.511
E12	VARE12	1.744981	0.564560	3.091
E13	VARE13	4.373650	0.421043	10.388

E14	VARE14	1.151034	0.249249	4.618
E15	VARE15	2.583846	0.295751	8.737
E16	VARE16	2.451646	0.245741	9.977
E17	VARE17	2.144295	0.262714	8.162
E18	VARE18	1.645024	0.251047	6.553
E19	VARE19	2.217876	0.309455	7.167
D1	VARD1	2.932420	0.389995	7.519
D2	VARD2	1.772546	0.268442	6.603

Covariance Structure Analysis: Maximum Likelihood Estimation   17

Covariances among Exogenous Variables

Parameter			Estimate	Standard Error	t Value
F4	F3	CF3F4	-0.272922	0.236811	-1.152
F5	F3	CF3F5	1.379742	0.225244	6.126
F5	F4	CF4F5	0.026116	0.328619	0.079
F6	F3	CF3F6	-0.826094	0.177110	-4.664
F6	F4	CF4F6	0.899434	0.293128	3.068
F6	F5	CF5F6	-0.801279	0.232593	-3.445

Covariance Structure Analysis: Maximum Likelihood Estimation   18

Equations with Standardized Coefficients

V1 = 0.8800*F1 + 0.4749 E1
        LV1F1

V2 = 0.8186*F1 + 0.5743 E2
        LV2F1

V3 = 0.9352 F1 + 0.3542 E3

V5 = 0.8230*F2 + 0.5680 E5
        LV5F2

V6 = 0.8704*F2 + 0.4923 E6
        LV6F2

V7 = 0.8756 F2 + 0.4830 E7

V8 = 0.6392*F3 + 0.7690 E8
        LV8F3

V9 = 0.6172*F3 + 0.7868 E9
        LV9F3

V10 = 0.6860 F3 + 0.7276 E10

```
V11      =      0.8199*F4      + 0.5725 E11
                      LV11F4

V12      =      0.8715 F4      + 0.4905 E12

V13      =      0.4636*F4      + 0.8860 E13
                      LV13F4

V14      =      0.8602 F5      + 0.5099 E14

V15      =      0.6896*F5      + 0.7241 E15
                      LV15F5

V16      =      0.5502*F5      + 0.8351 E16
                      LV16F5

V17      =      0.6816*F6      + 0.7317 E17
                      LV17F6

V18      =      0.7580 F6      + 0.6523 E18

V19      =      0.7318*F6      + 0.6815 E19
                      LV19F6

F1       =      0.3321*F2      + 0.5479*F5      + 0.0595*F6      + 0.6850 D1
                      PF1F2              PF1F5              PF1F6

F2       =      0.6394*F3      - 0.1576*F4      + 0.7392 D2
                      PF2F3              PF2F4
```

Covariance Structure Analysis: Maximum Likelihood Estimation   19

Variances of Endogenous Variables

	Variable	Estimate	R-squared
1	V1	5.976361	0.774445
2	V2	8.217024	0.670137
3	V3	7.146201	0.874539
4	V5	3.723366	0.677365
5	V6	4.467301	0.757628
6	V7	4.230339	0.766739
7	V8	2.008726	0.408612
8	V9	1.983389	0.380932
9	V10	2.975161	0.470653
10	V11	6.740177	0.672253
11	V12	7.253696	0.759436
12	V13	5.571147	0.214946
13	V14	4.426741	0.739982
14	V15	4.927348	0.475611
15	V16	3.515790	0.302676
16	V17	4.004684	0.464553

```
         17    V18        3.866663          0.574562
         18    V19        4.775096          0.535533
         19    F1         6.249635          0.530785
         20    F2         3.243564          0.453519


               Correlations among Exogenous Variables
               ------------------------------------
                    Parameter              Estimate
               ------------------------------------
               F4    F3    CF3F4          -0.098267
               F5    F3    CF3F5           0.644228
               F5    F4    CF4F5           0.006148
               F6    F3    CF3F6          -0.468368
               F6    F4    CF4F6           0.257103
               F6    F5    CF5F6          -0.297026
```

Output 6.3: PROC CALIS Output Pages 15-19 from Analysis of Initial Theoretical Model, Investment Model Study

The factor loadings on pages 15-16 of Output 6.3 are represented with coefficient names that begin with the "L" prefix (such as "LV1F1"). The results show that all factor loadings that were tested have *t* value greater than 1.96 and are therefore significantly different from zero.

Of greater interest in this analysis, however, are the path coefficients for the causal paths that constitute the structural portion of the model. These path coefficients are represented with coefficient names that begin with the "P" prefix (such as "PF1F2"), and appear under the heading "Latent Variable Equations" on page 16 of Output 6.3.

These results show that all of the path coefficients were significant except for the path from F6 (Alternative value) to F1 (Commitment), which displayed a nonsignificant *t* value of 0.91. Consistent with this, the standardized path coefficients for the latent variable equations presented on page 18 of the output show that the standardized path coefficient for the path from F6 to F1 was quite small .0595. This is an important finding because if later you decide to modify the model, deleting the path from F6 to F1 may be the place to start.

Step 4: Reviewing R^2 values for latent endogenous variables. The R^2 values for the study's endogenous variables are presented on page 19 of the output. Of particular interest are the R^2 values for the structural model's latent endogenous variables F1 (Commitment) and F2 (Satisfaction). The results on page 19 show that the independent F variables accounted for 53% of the variance in Commitment and 45% of the variance in Satisfaction.

Step 5: Reviewing the residual matrix and normalized residual matrix. Output 6.4 presents the actual covariances, predicted covariances, residual matrix, normalized residual matrix, and distribution of normalized residuals from the analysis. If the model provides a good fit, you expect the distribution of normalized residuals to be symmetrical and centered around zero. Unfortunately, the table of normalized residuals on page 13 of the output does not show this pattern. The residuals are centered around zero, but the distribution is somewhat asymmetrical due to one outlying residual at the bottom of the table (in the interval from 3.5 to 3.75).

```
                          The SAS System                              4

                             Covariances

           V1          V2          V3          V5          V6          V7

  V1     6.1801960   5.3081221   5.5461566   2.4648839   2.8050582   2.6680548
  V2     5.3081221   8.4622810   6.2283552   2.0313489   2.1267641   2.0634119
  V3     5.5461566   6.2283552   7.4201760   2.6062796   2.6016270   2.8394758
  V5     2.4648839   2.0313489   2.6062796   3.7210410   2.9061716   2.8555373
  V6     2.8050582   2.1267641   2.6016270   2.9061716   4.4647690   3.3190666
  V7     2.6680548   2.0634119   2.8394758   2.8555373   3.3190666   4.2271360
  V8     1.2188411   1.2077615   1.3162286   0.9211534   1.1227954   0.8303053
  V9     0.7315602   0.6020932   0.6405105   0.6817240   0.9103818   1.0334607
  V10    1.4957665   1.2086430   1.3478025   0.9378181   1.2786270   1.0775414
  V11    0.3290097   0.6190061   0.0353439  -0.8059266  -0.9102170  -0.6242324
  V12   -0.2675930   0.1017655  -0.4178262  -0.9810875  -0.8529125  -1.0512122
  V13   -0.1701418  -0.0823829  -0.4242902  -0.5007684  -0.5036547  -0.4027293
  V14    2.9210947   2.6170993   3.3267177   1.7557102   2.0031282   2.0312046
  V15    2.3941324   2.0785329   2.7442284   2.0203729   1.9598962   1.8933396
  V16    1.7470365   1.7771779   2.2001585   0.9145813   1.0136991   0.8669124
  V17   -0.7014025  -0.4365682  -0.7903550  -1.1579787  -1.1923279  -1.2547871
  V18   -0.6598093  -1.0523133  -0.8247291  -0.7281435  -0.7643651  -0.8245876
  V19   -0.9071290  -0.9025754  -0.8630313  -0.9862784  -0.9649331  -1.1275824
```

```
       Covariance Structure Analysis: Maximum Likelihood Estimation      5

                             Covariances

           V8          V9          V10         V11         V12         V13

  V1     1.2188411   0.7315602   1.4957665   0.3290097  -0.2675930  -0.1701418
  V2     1.2077615   0.6020932   1.2086430   0.6190061   0.1017655  -0.0823829
  V3     1.3162286   0.6405105   1.3478025   0.0353439  -0.4178262  -0.4242902
  V5     0.9211534   0.6817240   0.9378181  -0.8059266  -0.9810875  -0.5007684
  V6     1.1227954   0.9103818   1.2786270  -0.9102170  -0.8529125  -0.5036547
  V7     0.8303053   1.0334607   1.0775414  -0.6242324  -1.0512122  -0.4027293
  V8     2.0078890   0.7781030   1.2361114  -0.3346175  -0.3889410  -0.1437972
  V9     0.7781030   1.9824640   1.1942769  -0.2009568  -0.1364014   0.1827584
  V10    1.2361114   1.1942769   2.9721760  -0.2818481  -0.0835071  -0.0122059
  V11   -0.3346175  -0.2009568  -0.2818481   6.7340250   4.9859655   2.3210718
  V12   -0.3889410  -0.1364014  -0.0835071   4.9859655   7.2414810   2.5593563
  V13   -0.1437972   0.1827584  -0.0122059   2.3210718   2.5593563   5.5696000
  V14    0.9084529   1.0210675   1.2067414   0.0163641  -0.0282824   0.2133110
  V15    0.6445862   0.7217253   0.9219590  -0.2130573   0.0417993  -0.3246841
  V16    0.4169069   0.3984274   0.4523086   0.4522618   0.1109445   0.2388226
```

V17	-0.5614126	-0.5634816	-0.6485481	1.1735265	1.1630933	0.2408404
V18	-0.4401599	-0.3681610	-0.8236203	0.6734336	0.6295702	0.4964543
V19	-0.8916898	-0.9167910	-0.8287268	0.8335010	0.8349366	0.3919016

	V14	V15	V16	V17	V18	V19
V1	2.9210947	2.3941324	1.7470365	-0.7014025	-0.6598093	-0.9071290
V2	2.6170993	2.0785329	1.7771779	-0.4365682	-1.0523133	-0.9025754
V3	3.3267177	2.7442284	2.2001585	-0.7903550	-0.8247291	-0.8630313
V5	1.7557102	2.0203729	0.9145813	-1.1579787	-0.7281435	-0.9862784
V6	2.0031282	1.9598962	1.0136991	-1.1923279	-0.7643651	-0.9649331
V7	2.0312046	1.8933396	0.8669124	-1.2547871	-0.8245876	-1.1275824
V8	0.9084529	0.6445862	0.4169069	-0.5614126	-0.4401599	-0.8916898
V9	1.0210675	0.7217253	0.3984274	-0.5634816	-0.3681610	-0.9167910
V10	1.2067414	0.9219590	0.4523086	-0.6485481	-0.8236203	-0.8287268
V11	0.0163641	-0.2130573	0.4522618	1.1735265	0.6734336	0.8335010
V12	-0.0282824	0.0417993	0.1109445	1.1630933	0.6295702	0.8349366
V13	0.2133110	-0.3246841	0.2388226	0.2408404	0.4964543	0.3919016
V14	4.4184040	2.7752811	1.8001906	-0.9253424	-0.5950846	-0.8221237
V15	2.7752811	4.9239610	1.7049465	-1.1366961	-0.6500205	-0.9794000
V16	1.8001906	1.7049465	3.5118760	-0.2437418	-0.2726370	-0.3398593
V17	-0.9253424	-1.1366961	-0.2437418	4.0040010	2.0810680	2.0112051
V18	-0.5950846	-0.6500205	-0.2726370	2.0810680	3.8651560	2.4571461
V19	-0.8221237	-0.9794000	-0.3398593	2.0112051	2.4571461	4.7742250

Determinant = 56874416 (Ln = 17.856)

Covariance Structure Analysis: Maximum Likelihood Estimation   8

Predicted Model Matrix

	V1	V2	V3	V5	V6	V7
V1	5.9763615	5.0483873	5.3782507	1.8348716	2.1255784	2.0808355
V2	5.0483873	8.2170236	5.8663268	2.0013862	2.3184746	2.2696713
V3	5.3782507	5.8663268	7.1462007	2.1321574	2.4699646	2.4179724
V5	1.8348716	2.0013862	2.1321574	3.7233656	2.9216605	2.8601603
V6	2.1255784	2.3184746	2.4699646	2.9216605	4.4673006	3.3133081
V7	2.0808355	2.2696713	2.4179724	2.8601603	3.3133081	4.2303390
V8	1.0575987	1.1535758	1.2289508	0.9422924	1.0915839	1.0686063
V9	1.0146880	1.1067709	1.1790877	0.9040601	1.0472943	1.0252490
V10	1.3813722	1.5067317	1.6051821	1.2307660	1.4257616	1.3957497
V11	-0.2497824	-0.2724501	-0.2902521	-0.7451392	-0.8631948	-0.8450248
V12	-0.2754134	-0.3004071	-0.3200358	-0.8216003	-0.9517700	-0.9317355
V13	-0.1284095	-0.1400627	-0.1492144	-0.3830652	-0.4437559	-0.4344150
V14	2.5960119	2.8315999	3.0166176	1.1812646	1.3684176	1.3396127
V15	2.1957697	2.3950358	2.5515283	0.9991422	1.1574407	1.1330769
V16	1.4796338	1.6139106	1.7193640	0.6732785	0.7799490	0.7635313
V17	-0.6343754	-0.6919450	-0.7371569	-0.7364964	-0.8531827	-0.8352234
V18	-0.6932362	-0.7561474	-0.8055543	-0.8048325	-0.9323456	-0.9127200
V19	-0.7437527	-0.8112482	-0.8642555	-0.8634810	-1.0002861	-0.9792303

	V8	V9	V10	V11	V12	V13
V1	1.0575987	1.0146880	1.3813722	-0.2497824	-0.2754134	-0.1284095
V2	1.1535758	1.1067709	1.5067317	-0.2724501	-0.3004071	-0.1400627
V3	1.2289508	1.1790877	1.6051821	-0.2902521	-0.3200358	-0.1492144
V5	0.9422924	0.9040601	1.2307660	-0.7451392	-0.8216003	-0.3830652
V6	1.0915839	1.0472943	1.4257616	-0.8631948	-0.9517700	-0.4437559
V7	1.0686063	1.0252490	1.3957497	-0.8450248	-0.9317355	-0.4344150
V8	2.0087260	0.7874867	1.0720658	-0.1895074	-0.2089534	-0.0974230
V9	0.7874867	1.9833890	1.0285681	-0.1818184	-0.2004754	-0.0934702
V10	1.0720658	1.0285681	2.9751609	-0.2475233	-0.2729225	-0.1272481
V11	-0.1895074	-0.1818184	-0.2475233	6.7401770	4.9960542	2.3293743
V12	-0.2089534	-0.2004754	-0.2729225	4.9960542	7.2536962	2.5683989
V13	-0.0974230	-0.0934702	-0.1272481	2.3293743	2.5683989	5.5711474
V14	1.0563503	1.0134902	1.3797416	0.0236853	0.0261157	0.0121763
V15	0.8934866	0.8572346	1.1670188	0.0200336	0.0220893	0.0102990
V16	0.6020818	0.5776531	0.7864033	0.0134998	0.0148850	0.0069400
V17	-0.5787684	-0.5552856	-0.7559527	0.7464677	0.8230652	0.3837482
V18	-0.6324697	-0.6068080	-0.8260941	0.8157290	0.8994336	0.4193545
V19	-0.6785581	-0.6510264	-0.8862920	0.8751716	0.9649758	0.4499130

Covariance Structure Analysis: Maximum Likelihood Estimation   9

Predicted Model Matrix

	V14	V15	V16	V17	V18	V19
V1	2.5960119	2.1957697	1.4796338	-0.6343754	-0.6932362	-0.7437527
V2	2.8315999	2.3950358	1.6139106	-0.6919450	-0.7561474	-0.8112482
V3	3.0166176	2.5515283	1.7193640	-0.7371569	-0.8055543	-0.8642555
V5	1.1812646	0.9991422	0.6732785	-0.7364964	-0.8048325	-0.8634810
V6	1.3684176	1.1574407	0.7799490	-0.8531827	-0.9323456	-1.0002861
V7	1.3396127	1.1330769	0.7635313	-0.8352234	-0.9127200	-0.9792303
V8	1.0563503	0.8934866	0.6020818	-0.5787684	-0.6324697	-0.6785581
V9	1.0134902	0.8572346	0.5776531	-0.5552856	-0.6068080	-0.6510264
V10	1.3797416	1.1670188	0.7864033	-0.7559527	-0.8260941	-0.8862920
V11	0.0236853	0.0200336	0.0134998	0.7464677	0.8157290	0.8751716
V12	0.0261157	0.0220893	0.0148850	0.8230652	0.8994336	0.9649758
V13	0.0121763	0.0102990	0.0069400	0.3837482	0.4193545	0.4499130
V14	4.4267414	2.7706724	1.8670357	-0.7332442	-0.8012785	-0.8596681
V15	2.7706724	4.9273475	1.5791840	-0.6201957	-0.6777408	-0.7271281
V16	1.8670357	1.5791840	3.5157895	-0.4179229	-0.4567001	-0.4899800
V17	-0.7332442	-0.6201957	-0.4179229	4.0046838	2.0330056	2.1811518
V18	-0.8012785	-0.6777408	-0.4567001	2.0330056	3.8666630	2.3835308
V19	-0.8596681	-0.7271281	-0.4899800	2.1811518	2.3835308	4.7750955

Determinant = 141385485 (Ln = 18.767)

Residual Matrix

	V1	V2	V3	V5	V6	V7
V1	0.2038345	0.2597348	0.1679059	0.6300123	0.6794798	0.5872192
V2	0.2597348	0.2452574	0.3620284	0.0299627	-0.1917105	-0.2062594
V3	0.1679059	0.3620284	0.2739753	0.4741222	0.1316625	0.4215034
V5	0.6300123	0.0299627	0.4741222	-0.0023246	-0.0154889	-0.0046230
V6	0.6794798	-0.1917105	0.1316625	-0.0154889	-0.0025316	0.0057585
V7	0.5872192	-0.2062594	0.4215034	-0.0046230	0.0057585	-0.0032030
V8	0.1612423	0.0541857	0.0872778	-0.0211389	0.0312115	-0.2383010
V9	-0.2831278	-0.5046778	-0.5385773	-0.2223360	-0.1369125	0.0082117
V10	0.1143943	-0.2980888	-0.2573796	-0.2929479	-0.1471346	-0.3182083
V11	0.5787921	0.8914562	0.3255960	-0.0607874	-0.0470222	0.2207924
V12	0.0078203	0.4021727	-0.0977904	-0.1594872	0.0988576	-0.1194767
V13	-0.0417323	0.0576798	-0.2750758	-0.1177032	-0.0598987	0.0316857
V14	0.3250828	-0.2145006	0.3101000	0.5744456	0.6347106	0.6915919
V15	0.1983626	-0.3165029	0.1927002	1.0212307	0.8024555	0.7602627
V16	0.2674027	0.1632674	0.4807944	0.2413028	0.2337501	0.1033811
V17	-0.0670271	0.2553768	-0.0531981	-0.4214823	-0.3391451	-0.4195636
V18	0.0334270	-0.2961659	-0.0191748	0.0766890	0.1679805	0.0881324
V19	-0.1633763	-0.0913272	0.0012242	-0.1227974	0.0353530	-0.1483520

Covariance Structure Analysis: Maximum Likelihood Estimation   10

Residual Matrix

	V8	V9	V10	V11	V12	V13
V1	0.1612423	-0.2831278	0.1143943	0.5787921	0.0078203	-0.0417323
V2	0.0541857	-0.5046778	-0.2980888	0.8914562	0.4021727	0.0576798
V3	0.0872778	-0.5385773	-0.2573796	0.3255960	-0.0977904	-0.2750758
V5	-0.0211389	-0.2223360	-0.2929479	-0.0607874	-0.1594872	-0.1177032
V6	0.0312115	-0.1369125	-0.1471346	-0.0470222	0.0988576	-0.0598987
V7	-0.2383010	0.0082117	-0.3182083	0.2207924	-0.1194767	0.0316857
V8	-0.0008370	-0.0093837	0.1640457	-0.1451100	-0.1799876	-0.0463741
V9	-0.0093837	-0.0009250	0.1657088	-0.0191384	0.0640740	0.2762286
V10	0.1640457	0.1657088	-0.0029849	-0.0343248	0.1894154	0.1150422
V11	-0.1451100	-0.0191384	-0.0343248	-0.0061520	-0.0100887	-0.0083025
V12	-0.1799876	0.0640740	0.1894154	-0.0100887	-0.0122152	-0.0090427
V13	-0.0463741	0.2762286	0.1150422	-0.0083025	-0.0090427	-0.0015474
V14	-0.1478974	0.0075773	-0.1730002	-0.0073212	-0.0543981	0.2011347
V15	-0.2489004	-0.1355093	-0.2450598	-0.2330909	0.0197100	-0.3349831
V16	-0.1851749	-0.1792258	-0.3340947	0.4387620	0.0960595	0.2318825
V17	0.0173559	-0.0081960	0.1074046	0.4270588	0.3400281	-0.1429079
V18	0.1923098	0.2386470	0.0024738	-0.1422954	-0.2698634	0.0770999
V19	-0.2131317	-0.2657646	0.0575652	-0.0416706	-0.1300392	-0.0580114

	V14	V15	V16	V17	V18	V19
V1	0.3250828	0.1983626	0.2674027	-0.0670271	0.0334270	-0.1633763
V2	-0.2145006	-0.3165029	0.1632674	0.2553768	-0.2961659	-0.0913272
V3	0.3101000	0.1927002	0.4807944	-0.0531981	-0.0191748	0.0012242
V5	0.5744456	1.0212307	0.2413028	-0.4214823	0.0766890	-0.1227974
V6	0.6347106	0.8024555	0.2337501	-0.3391451	0.1679805	0.0353530
V7	0.6915919	0.7602627	0.1033811	-0.4195636	0.0881324	-0.1483520
V8	-0.1478974	-0.2489004	-0.1851749	0.0173559	0.1923098	-0.2131317
V9	0.0075773	-0.1355093	-0.1792258	-0.0081960	0.2386470	-0.2657646
V10	-0.1730002	-0.2450598	-0.3340947	0.1074046	0.0024738	0.0575652
V11	-0.0073212	-0.2330909	0.4387620	0.4270588	-0.1422954	-0.0416706
V12	-0.0543981	0.0197100	0.0960595	0.3400281	-0.2698634	-0.1300392
V13	0.2011347	-0.3349831	0.2318825	-0.1429079	0.0770999	-0.0580114
V14	-0.0083374	0.0046087	-0.0668451	-0.1920983	0.2061939	0.0375444
V15	0.0046087	-0.0033865	0.1257624	-0.5165004	0.0277202	-0.2522719
V16	-0.0668451	0.1257624	-0.0039135	0.1741811	0.1840631	0.1501208
V17	-0.1920983	-0.5165004	0.1741811	-0.0006828	0.0480624	-0.1699467
V18	0.2061939	0.0277202	0.1840631	0.0480624	-0.0015070	0.0736153
V19	0.0375444	-0.2522719	0.1501208	-0.1699467	0.0736153	-0.0008705

Average Absolute Residual = 0.1886
Average Off-diagonal Absolute Residual = 0.2057

Covariance Structure Analysis: Maximum Likelihood Estimation   11

Normalized Residual Matrix

	V1	V2	V3	V5	V6	V7
V1	0.3736	0.4659	0.3073	1.9283	1.8840	1.6718
V2	0.4659	0.3270	0.5812	0.0789	-0.4578	-0.5058
V3	0.3073	0.5812	0.4200	1.3159	0.3308	1.0871
V5	1.9283	0.0789	1.3159	-0.0068	-0.0478	-0.0146
V6	1.8840	-0.4578	0.3308	-0.0478	-0.0062	0.0163
V7	1.6718	-0.5058	1.0871	-0.0146	0.0163	-0.0083
V8	0.6895	0.1988	0.3395	-0.1132	0.1517	-1.1891
V9	-1.2220	-1.8678	-2.1149	-1.2027	-0.6722	0.0414
V10	0.3994	-0.8934	-0.8167	-1.2789	-0.5823	-1.2931
V11	1.4117	1.8545	0.7262	-0.1859	-0.1311	0.6327
V12	0.0184	0.8064	-0.2102	-0.4696	0.2654	-0.3295
V13	-0.1120	0.1320	-0.6752	-0.3990	-0.1853	0.1007
V14	0.8741	-0.4987	0.7527	2.1047	2.1133	2.3651
V15	0.5249	-0.7212	0.4622	3.5971	2.5726	2.5038
V16	0.8600	0.4507	1.4056	1.0158	0.8965	0.4074
V17	-0.2105	0.6847	-0.1526	-1.6610	-1.2177	-1.5476
V18	0.1066	-0.8068	-0.0559	0.3063	0.6110	0.3293
V19	-0.4693	-0.2240	0.0032	-0.4420	0.1159	-0.4996

	V8	V9	V10	V11	V12	V13
V1	0.6895	-1.2220	0.3994	1.4117	0.0184	-0.1120
V2	0.1988	-1.8678	-0.8934	1.8545	0.8064	0.1320
V3	0.3395	-2.1149	-0.8167	0.7262	-0.2102	-0.6752
V5	-0.1132	-1.2027	-1.2789	-0.1859	-0.4696	-0.3990
V6	0.1517	-0.6722	-0.5823	-0.1311	0.2654	-0.1853
V7	-1.1891	0.0414	-1.2931	0.6327	-0.3295	0.1007
V8	-0.0046	-0.0677	0.9520	-0.6101	-0.7294	-0.2147
V9	-0.0677	-0.0051	0.9732	-0.0810	0.2613	1.2868
V10	0.9520	0.9732	-0.0110	-0.1186	0.6306	0.4375
V11	-0.6101	-0.0810	-0.1186	-0.0100	-0.0182	-0.0196
V12	-0.7294	0.2613	0.6306	-0.0182	-0.0184	-0.0204
V13	-0.2147	1.2868	0.4375	-0.0196	-0.0204	-0.0030
V14	-0.7243	0.0375	-0.6903	-0.0208	-0.1487	0.6274
V15	-1.1790	-0.6476	-0.9485	-0.6266	0.0511	-0.9905
V16	-1.0528	-1.0272	-1.5550	1.3963	0.2947	0.8117
V17	0.0929	-0.0442	0.4709	1.2605	0.9662	-0.4672
V18	1.0425	1.3041	0.0110	-0.4264	-0.7783	0.2563
V19	-1.0414	-1.3089	0.2303	-0.1125	-0.3378	-0.1736

Covariance Structure Analysis: Maximum Likelihood Estimation   12
Normalized Residual Matrix

	V14	V15	V16	V17	V18	V19
V1	0.8741	0.5249	0.8600	-0.2105	0.1066	-0.4693
V2	-0.4987	-0.7212	0.4507	0.6847	-0.8068	-0.2240
V3	0.7527	0.4622	1.4056	-0.1526	-0.0559	0.0032
V5	2.1047	3.5971	1.0158	-1.6610	0.3063	-0.4420
V6	2.1133	2.5726	0.8965	-1.2177	0.6110	0.1159
V7	2.3651	2.5038	0.4074	-1.5476	0.3293	-0.4996
V8	-0.7243	-1.1790	-1.0528	0.0929	1.0425	-1.0414
V9	0.0375	-0.6476	-1.0272	-0.0442	1.3041	-1.3089
V10	-0.6903	-0.9485	-1.5550	0.4709	0.0110	0.2303
V11	-0.0208	-0.6266	1.3963	1.2605	-0.4264	-0.1125
V12	-0.1487	0.0511	0.2947	0.9662	-0.7783	-0.3378
V13	0.6274	-0.9905	0.8117	-0.4672	0.2563	-0.1736
V14	-0.0206	0.0131	-0.2373	-0.6963	0.7580	0.1244
V15	0.0131	-0.0075	0.4377	-1.7840	0.0972	-0.7968
V16	-0.2373	0.4377	-0.0122	0.7147	0.7675	0.5636
V17	-0.6963	-1.7840	0.7147	-0.0019	0.1681	-0.5388
V18	0.7580	0.0972	0.7675	0.1681	-0.0043	0.2321
V19	0.1244	-0.7968	0.5636	-0.5388	0.2321	-0.0020

Average Normalized Residual = 0.6241
Average Off-diagonal Normalized Residual = 0.6894
Rank Order of 10 Largest Normalized Residuals

V15,V5	V15,V6	V15,V7	V14,V7	V9,V3	V14,V6	V14,V5
3.5971	2.5726	2.5038	2.3651	-2.1149	2.1133	2.1047

V5,V1	V6,V1	V9,V2
1.9283	1.8840	-1.8678

```
           Covariance Structure Analysis: Maximum Likelihood Estimation        13

                        Distribution of Normalized Residuals
                           (Each * represents 1 residuals)

   -2.25000 -    -2.00000    1    0.58%  | *
   -2.00000 -    -1.75000    2    1.17%  | **
   -1.75000 -    -1.50000    3    1.75%  | ***
   -1.50000 -    -1.25000    3    1.75%  | ***
   -1.25000 -    -1.00000    8    4.68%  | *******
   -1.00000 -    -0.75000    7    4.09%  | *******
   -0.75000 -    -0.50000   13    7.60%  | *************
   -0.50000 -    -0.25000   11    6.43%  | ***********
   -0.25000 -           0   40   23.39%  | ****************************************
          0 -     0.25000   21   12.28%  | ********************
    0.25000 -     0.50000   20   11.70%  | *******************
    0.50000 -     0.75000   11    6.43%  | ***********
    0.75000 -     1.00000   11    6.43%  | ***********
    1.00000 -     1.25000    3    1.75%  | ***
    1.25000 -     1.50000    7    4.09%  | *******
    1.50000 -     1.75000    1    0.58%  | *
    1.75000 -     2.00000    3    1.75%  | ***
    2.00000 -     2.25000    2    1.17%  | **
    2.25000 -     2.50000    1    0.58%  | *
    2.50000 -     2.75000    2    1.17%  | **
    2.75000 -     3.00000    0    0.00%  |
    3.00000 -     3.25000    0    0.00%  |
    3.25000 -     3.50000    0    0.00%  |
    3.50000 -     3.75000    1    0.58%  | *
```

Output 6.4: Actual Covariances, Predicted Model Matrix, Residual Matrix, Normalized Residual Matrix, and Distribution of Normalized Residuals from Analysis of Initial Theoretical Model, Investment Model Study

The bottom of page 12 provides the rank order of the 10 largest normalized residuals, and it is interesting to see that the three largest residuals involve the relationship between V15 (an indicator for F5: Investments) and V5, V6, and V7 (all indicators for F2: Satisfaction). Could it be that your theoretical model would be substantially improved if you were to add a path from Investments to Satisfaction? This is possible, but you will make no decisions until you have reviewed the modification indices (in a later section).

Step 6: Reviewing the parsimony ratio and the parsimonious normed fit index. With other factors held constant, the most desirable theoretical model is the most parsimonious model. In the broadest sense, the **parsimony** of a model refers to its simplicity. The principal of parsimony states that, when several theoretical explanations are equally satisfactory in accounting for some phenomenon, the preferred explanation is the one that is least complicated; the one that makes the fewest assumptions.

It is easy to see how this relates to causal modeling. The theoretical model presented in Figure 6.4 attempts to explain variability in satisfaction and commitment in romantic associations. It identifies four exogenous variables (Rewards, Costs, Investments, and Alternatives) that causally determine levels of satisfaction and commitment. It posits that the relationships between these constructs can be accounted for with just five causal paths: the two paths from Rewards and Costs to Satisfaction, and the three paths from Satisfaction, Investments and Alternatives to Commitment. And this model did a fairly good job of accounting for the observed covariances in the data (as is indicated by the relatively high CFI and NNFI values).

However, the model would have done an even better job of accounting for the observed covariances if you had made it more complicated. For example, the fit indices would have been somewhat higher if you had seen to it that each of the six latent factors was connected to every other latent factor by either a covariance (a curved arrow) or a causal path (a straight arrow).

The problem is that this hypothetical second version of the model is less parsimonious than the first. It might have resulted in a slight increase in some fit indices, but at the price of the model's parsimony. Because of the importance of this concept, it is necessary that you make use of some indices that reflect a model's level of parsimony.

One such index is the parsimony ratio (James et al., 1982). The parsimony ratio (PR) is easily calculated with the following formula:

$$PR = \frac{df_J}{df_0}$$

where

> df_J = the degrees of freedom for the model being studied

> df_0 = the degrees of freedom for the null model.

You are already familiar with the df_J: it is simply the degrees of freedom for the chi-square test for your model. Output 6.2 has already shown that the chi-square for the current model had 124 degrees of freedom.

The new term in the equation is df_0, the degrees of freedom for the null model. The **null model** is a model that predicts no relationships between any of the study's variables. In a null model, all paths and covariances between all variables have been deleted. PROC CALIS automatically computes the chi-square and degrees of freedom for this null model each time any model is analyzed. This information also appears among the goodness of fit indices of Output 6.2, just below the model chi-square. The null model chi-square for the current analysis is:

```
Null Model Chi-square:     df = 153          2167.7711
```

So the null model chi-square is 2,167.77 with 153 degrees of freedom. You may now insert the degrees of freedom for the theoretical model and for the null model into the preceding equation to calculate the parsimony ratio for the theoretical model:

$$PR = \frac{df_J}{df_0} = \frac{124}{153} = .810$$

The parsimony ratio for the theoretical model is therefore .810. You can see that the lowest possible value for the parsimony ratio is zero, and this value will be obtained for a fully-saturated model: one in which every V variable is connected to every other V variable by either a covariance or a causal path. A fully saturated model would be the least parsimonious system possible. The upper limit on the PR is 1.0, and this value will be obtained for the null model itself: the model that hypothesizes no relationships between any variables, and thus, the most parsimonious system possible.

The parsimony ratio can help you choose the best model when more than one demonstrates a good fit to the data. When choosing between two nested models that display an acceptable and similar fit, the more desirable model will be the one with the higher parsimony ratio.

The PR is also helpful because it can be used to compute a second index that also reflects the parsimony of a model: the parsimonious fit index (James et al., 1982). The **parsimonious fit index** (referred to here as the parsimonious normed-fit index, or PNFI) is calculated using the following formula:

$$PNFI = (PR)(NFI)$$

where

$$PR = \text{parsimony ratio}$$

$$NFI = \text{normed fit index.}$$

The NFI is the same normed fit index (Bentler & Bonett, 1980) discussed earlier. The NFI is a measure of the overall fit of a model that may range from zero to 1, with higher values reflecting a better fit. One problem with the NFI is that it does not take into account the parsimony of a model. In fact, it is possible to get higher values on the NFI simply by adding more and more paths and covariances to an initial model; that is, by making it less parsimonious. It would be helpful to have a version of the NFI that penalizes models for not being parsimonious.

The PNFI is just such an index. It is a single index that simultaneously reflects both the fit and the parsimony of the model. As the preceding equation shows, the PNFI is determined by simply multiplying the NFI for a model by the parsimony ratio.

We can obtain the NFI for the present model from the goodness of fit summary table from Output 6.2, on the line labeled "Bentler & Bonett's (1980) normed index." The model's NFI is .900. You insert this in the preceding formula to obtain the PNFI for the model:

$$PNFI = (PR) (NFI) = (.810) (.900) = .729$$

Fortunately, you do not actually have to perform this computation by hand, as it is also included in the summary table in Output 6.2, next to the heading "James, Mulaik, & Brett (1982) Parsimonious Index." Output 6.2 shows that PROC CALIS computes the PNFI to be .7294.

The PNFI is similar to the NFI in that higher values indicate a more desirable model. Like the PR, the PNFI can help in selecting a best model when more than one provides an acceptable fit to the data. There is no firm criterion as to how large the PNFI must be for the model to be acceptable, although .60 has been suggested as one ad-hoc criterion (Netemeyer et al., 1990; Williams & Hazur, 1986). In addition, Mulaik et al. (1989) indicates that it is possible to have acceptable models with parsimonious normed fit indices in the .50s.

Step 7: Reviewing the relative normed-fit index. Earlier, it was said that a theoretical model consists of two components: a measurement model that describes the relationships between the latent variables and their indicators, and a structural model that describes the causal relationships between the latent variables themselves. All of the goodness of fit indices discussed so far have reflected the overall fit of the measurement model and structural model combined.

This, unfortunately, creates a problem. Most theoretical models are similar to the one being tested here in that its measurement model consists of a relatively small number of latent variables, and a relatively large number of indicator variables. This means that the measurement portion of the model will usually contain many more parameters to be estimated compared to the structural portion of the model. For example, notice how, with the theoretical model tested here, a large number of factor loadings (from the measurement model) are estimated, but only a few path coefficients (from the structural model) are estimated.

As a consequence, indices of overall model fit (such as the NNFI and CFI) are often much more influenced by the fit of the measurement portion of the model than by the fit of the structural portion. In fact, if the measurement model provides a good fit, the overall model is likely to produce very high NNFI and CFI values even if there are serious misspecifications in the structural portion of the model (Mulaik et al., 1989). This presents a serious problem, as researchers are generally more interested in the fit of the structual model than the fit of the measurement model.

Fortunately, this problem can be solved by using the results of the analysis to calculate a relative normed-fit index (Mulaik et al., 1989). The relative normed-fit index (RNFI) reflects the fit in just the structural portion of the model, and is not influenced by the fit of the measurement model. Its interpretation is similar to that of the NFI, in that higher values (nearer 1) indicate that the hypothesized causal relations between the structural variables provide a good fit to the data.

The formula for the RNFI is as follows:

$$\text{RNFI} = \frac{F_u - F_j}{F_u - F_m - (df_j - df_m)}$$

where

F_u = model chi-square for the uncorrelated variables model

F_j = model chi-square for the model of interest

F_m = model chi-square for the measurement model

df_j = degrees of freedom for the model of interest

df_m = degrees of freedom for the measurement model.

Two of the models described here have already been discussed in this chapter, and one model is new. The model of interest is simply whatever model you are testing, in this case that version of the investment model presented in Figure 6.4. The measurement model is the confirmatory factor analysis model in which all F variables are allowed to covary; i.e., the model in which each latent F variable is connected to every other latent F variable by a curved, double-headed arrow (as presented in Figure 6.1).

The new model in the preceding equation is the uncorrelated factors model. This model is identical to measurement model, except that none of the F variables are allowed to covary. In other words, in this model none of the latent F variables is connected to any of the other latent F variables by either a curved or straight arrow. This model predicts that all of the F variables are mutually uncorrelated.

By now, you have been introduced to so many different models and so many different indices that it is necessary to begin organizing them in a table. This is done in Table 6.1. Table 6.1 presents goodness of fit and parsimony indices for the various models that either have been or will be covered in this chapter.

Table 6.1

Goodness of Fit and Parsimony Indices for the Investment Model Study (Standard Model)

Model	Chi-square	df	NFI	NNFI	CFI	PR	PNFI	RNFI	RPR	RPFI
				Combined model					Structural model	
M_O Null model	2167.77	153	.000	--	--	--	--	--	--	--
M_u Uncorrelated factors	519.98	135	.760	.783	.809	.882	.671	.000	1.000	0.000
M_t Theoretical model	216.75	124	.900	.943	.954	.810	.729	.905	.267	.242
M_{r1} Revised model 1	217.58	125	.900	.944	.954	.817	.735	.905	.333	.301
M_{r2} Revised model 2	183.20	124	.916	.964	.971	.810	.742	1.005	.267	.268
M_m Measurement model	180.87	120	.917	.962	.970	.784	.719	1.000	0.000	0.000

Note: $\underline{N}$ = 240. NFI = normed-fit index; NNFI = non-normed-fit index; CFI = comparative fit index; PR = parsimony ratio; PNFI = parsimonious normed-fit index; RNFI = relative normed-fit index; RPR = relative parsimony ratio; RPFI = relative parsimonious-fit index.

The first line of the table presents information about the null model: the model in which all variables are completely unrelated to all other variables. The chi-square for the null model is 2,167.77 with 153 *df*. Earlier, it was pointed out that this null-model chi-square is automatically printed by PROC CALIS in the table of fit indices.

The second line in Table 6.1 presents results for the uncorrelated factors model. Unfortunately, this information is not automatically printed by PROC CALIS; to obtain it, it is necessary to actually estimate a model in which the covariances between all latent variables have been fixed at zero. This, it turns out, is quite easy to do. All that is required is that you make a copy of the PROC CALIS program that estimated the (final) measurement model, and change the COV statement so that all covariances between F variables are fixed at zero.

Here is a copy of the PROC CALIS program that estimates the uncorrelated factors model for the present study:

```
1       PROC CALIS   COVARIANCE   CORR   RESIDUAL   MODIFICATION ;
2            LINEQS
3               V1   = LV1F1  F1 + E1,
4               V2   = LV2F1  F1 + E2,
5               V3   = LV3F1  F1 + E3,
6
7               V5   = LV5F2  F2 + E5,
8               V6   = LV6F2  F2 + E6,
9               V7   = LV7F2  F2 + E7,
```

```
10              V8   =  LV8F3   F3 + E8,
11              V9   =  LV9F3   F3 + E9,
12              V10  =  LV10F3  F3 + E10,
13              V11  =  LV11F4  F4 + E11,
14              V12  =  LV12F4  F4 + E12,
15              V13  =  LV13F4  F4 + E13,
16              V14  =  LV14F5  F5 + E14,
17              V15  =  LV15F5  F5 + E15,
18              V16  =  LV16F5  F5 + E16,
19              V17  =  LV17F6  F6 + E17,
20              V18  =  LV18F6  F6 + E18,
21              V19  =  LV19F6  F6 + E19;
22         STD
23              F1 = 1,
24              F2 = 1,
25              F3 = 1,
26              F4 = 1,
27              F5 = 1,
28              F6 = 1,
29              E1-E19 = VARE1-VARE19;
30         COV
31              F1 F2 = 0,
32              F1 F3 = 0,
33              F1 F4 = 0,
34              F1 F5 = 0,
35              F1 F6 = 0,
36              F2 F3 = 0,
37              F2 F4 = 0,
38              F2 F5 = 0,
39              F2 F6 = 0,
40              F3 F4 = 0,
41              F3 F5 = 0,
42              F3 F6 = 0,
43              F4 F5 = 0,
44              F4 F6 = 0,
45              F5 F6 = 0;
46         VAR  V1-V19 ;
47         RUN;
```

Notice that the CALIS, LINEQS, and STD statements in the above program are identical to those used when the (final) measurement model was estimated. However, the equations in the COV statement have been revised. With the estimation of the measurement model, the names for parameter estimates (such "CF1F2") had appeared to the right of the equals sign. This allowed all F variables to covary. In the revised program, however, only zeros appear to the right of the equals sign in the COV statement. This fixes all covariances between the F variables at 0, and therefore causes the program to estimate the uncorrelated factors model.

Table 6.1 shows that the chi-square for the uncorrelated factors model is 519.98, with 135 *df.* The CFI of .809 shows that this model does not provide an especially good fit to the data.

The third line of Table 6.1 summarizes results from the estimation of the theoretical model, and the last line of the table summarizes results for the (final) measurement model.

With all three models estimated, it is now possible to substitute the appropriate values in the formula to compute the relative normed-fit index for the theoretical model (note how each term in this equation may be found in Table 6.1):

$$RNFI = \frac{F_u - F_j}{F_u - F_m - (df_j - df_m)}$$

$$= \frac{519.98 - 216.75}{519.98 - 180.87 - (124 - 120)}$$

$$= \frac{303.23}{339.11 - (4)} = \frac{303.23}{335.11} = .905$$

And so, the RNFI for the theoretical model is .905. Remember that this indicates the fit demonstrated by just the structural portion of the theoretical model, irrespective of how well the latent variables were measured by their indicators. The RNFI of .905 indicates a minimally acceptable, although not outstanding, fit. The RNFI for the various models is presented in the column headed "RNFI" in Table 6.1.

Step 8: Reviewing the relative parsimony ratio and the relative parsimonious fit index.
Now that you have learned something about the fit of the structural portion of the model, you will investigate the parsimony of this part of the model. Remember that the structural portion of a model is saturated if every F variable is connected to every other F variable by either a covariance or a causal path. If the structural model is saturated, then it displays no parsimony, and its explanatory power cannot be tested. You make a model parsimonious by fixing some of the covariances and paths at zero, or by placing other constraints on its parameters.

A relative parsimony ratio can be computed to determine the parsimony of the structural portion of the model (Mulaik et al., 1989). The formula for the relative parsimony ratio (RPR) is:

$$RPR = \frac{df_j - df_m}{df_u - df_m}$$

390 *Path Analysis with Latent Variables*

where

df_j = the degrees of freedom for the model of interest

df_m = the degrees of freedom for the measurement model

df_u = the degrees of freedom for the uncorrelated factors model.

The model of interest in this case is the theoretical model. To compute the relative parsimony ratio for this model, you insert the appropriate terms from Table 6.1 into the equation:

$$RPR = \frac{124 - 120}{135 - 120} = \frac{4}{15} = .267$$

The RPR for the structural portion of a model may range from 0.0 (for the measurement model, in which every F variable is related to every other F variable) to 1.0 (for the uncorrelated factors model). The RPR for the preceding model was .267, which, by itself, does not tell you whether to accept or reject this model. However, if you arrive at a number of models that are equally acceptable according to your other criteria, the model with the higher RPR may be preferred.

The RNFI may now be multiplied by the RPR to produce the relative parsimonious-fit index, or RPFI (Mulaik et al., 1989). Remember that the RNFI provides information about the fit in just the structural portion of the model, while the RPR provides information about the parsimony of that part of the model. By multiplying them together, you produce a single index that simultaneously reflects both the fit and the parsimony in just the structural portion of the model. Here, the relative parsimonious-fit index is computed for the current model:

$$RPFI = (RNFI) \times (RPR) = (.905) \times (.267) = .242$$

When comparing models that all involve the same latent variables and all provide an acceptable fit to the data, the RPFI can be helpful in choosing the model that simultaneously maximizes both fit and parsimony in the structural portion of the model. With the RPFI, higher values are more desirable.

By now, you are likely confused by the proliferation of indices that have been described in this chapter (along with Chapters 4 and 5). Understanding these indices is made more difficult by the fact that some indices reflect the *fit* of a model, while others reflect its *parsimony*, and some indices refer to the *combined* measurement and structural portions of a model, while others refer only to the *structural* portion. To simplify matters, consider the following framework for categorizing these indices:

Indices for the Combined Model:

- **Measures of fit:**

  - Model chi-square test
  - NFI: Normed-fit index
  - NNFI: Non-normed-fit index
  - CFI: Comparative fit index

- **Measure of parsimony:**

  - PR: Parsimony ratio

- **Measure that reflects both fit and parsimony:**

  - PNFI: Parsimonious normed-fit index

Indices for the Structural Portion of the Model:

- **Measure of fit:**

  - RNFI: Relative normed-fit index

- **Measure of parsimony:**

  - RPR: Relative parsimony ratio

- **Measure that reflects both fit and parsimony:**

  - RPFI: Relative parsimonious-fit index

As a mnemonic device, remember that an index that begins with the letter "R" is a *relative* index, and therefore pertains only to the structural portion of the model. The remaining indices are for the combined model.

Step 9: Performing a chi-square difference test comparing the theoretical model to the measurement model. As a final test (before moving to the next stage of the analysis), you should perform a chi-square difference test to determine whether there is a significant difference between the fit provided by the theoretical model and the fit provided by the measurement model. A finding of no significant differences provides support for the nomological validity of the theoretical model (Anderson & Gerbing, 1988). If the theoretical model is successful in accounting for the observed relations between the F variables, there will not be a significant difference between the chi-square for the theoretical model and the chi-square for the measurement model. Reviewing how the measurement model differs conceptually from the theoretical model will help make this clear.

Earlier, it was said that the measurement model is basically a confirmatory factor analysis model in which the relations between all F variables are saturated, or just-identified. This is to say that each F variable is connected to every other F variable by a curved arrow. Because each F is connected to every other F, this measurement model does a perfect job of accounting for the covariances between the F variables. This is reflected in Table 6.1, which shows that the RNFI for the measurement model is a perfect 1.000.

In most cases, you hope to develop a theoretical model that is more parsimonious than the measurement model, but at the same time does nearly as good a job of accounting for the covariances between the latent F variables. A theoretical model is more parsimonious than a measurement model because the theoretical model is really a *constrained* version of the measurement model. This means that a theoretical model is basically a measurement model in which some of the covariances between F variables have been either replaced with unidirectional causal paths or fixed at zero. You can see this by comparing Figure 6.1 (the final measurement model) with Figure 6.4 (the theoretical model). In the measurement model, a covariance is estimated for F1 (Commitment) and F3 (Rewards). In the theoretical model, this covariance has been eliminated (fixed at zero). In the new model, there is no direct relationship between F1 and F3. The same is true for several other covariances in the measurement model.

If your theoretical model is correct, then fixing the covariance between F1 and F3 at zero should not dramatically hurt the theoretical model's fit to the data. This is because, if the theoretical model is correct, there should be essentially no direct causal relationship between F1 and F3. In other words, if the theoretical model is correct, it should provide a fit to the data that is nearly as good as the fit provided by the measurement model, even after eliminating these nonessential covariances.

The adequacy of the theoretical model can be determined by performing a chi-square difference test that compares the theoretical model (symbolized as M_t) to the measurement model (symbolized M_m). This is done by simply subtracting the chi-square values for the two models, as they appear in Table 6.1. This is done below:

$$M_t - M_m = 216.75 - 180.87 = 35.88$$

The resulting chi-square difference value (35.88 in this case) is distributed as chi-square, and the degrees of freedom for the test may be found by subtracting the corresponding degrees of freedom for the two models. These degrees of freedom may also be obtained from Table 6.1:

$$df_t - df_m = 124 - 120 = 4$$

With 4 degrees of freedom, the critical value of chi-square is 18.465 at $p < .001$. Your obtained chi-square difference value of 35.88 is clearly greater than this critical value, meaning that there is a significant difference between the fit provided by the theoretical model and the measurement model. In other words, the theoretical model provides a fit to the data that is significantly worse than the fit provided by the measurement model. This finding fails to support the theoretical model's predictions concerning the relationships between the F variables in the structural portion of the model. Apparently, your theoretical model contains some serious misspecifications, and will have to be modified if it is to fit the data.

Characteristics of an "Ideal Fit" for the Theoretical Model

Before moving on to the section on model modification, it will be helpful to first briefly summarize the results that you should expect to see if your model provides an ideal fit to the data. A theoretical model provides an ideal fit when it displays the following characteristics:

- The *p* value for the model chi-square test should be nonsignificant (should be greater than .05); the closer to 1.00, the better.

- The chi-square/*df* ratio should be less than 2.

- The CFI and the NNFI should both exceed .9; the closer to 1.00, the better.

- The absolute value of the *t* statistics for each factor loading and path coefficient should exceed 1.96, and the standardized factor loadings should be nontrivial in size (i.e., absolute values should exceed .05).

- The R^2 values for the latent endogenous variables should be relatively large, compared to what typically is obtained in research with these variables.

- The distribution of normalized residuals should be symmetrical and centered on zero, and relatively few (or no) normalized residuals should exceed 2.0 in absolute value.

- The combined model should demonstrate relatively high levels of parsimony and fit, as evidenced by the PR and the PNFI.

- The structural portion of the model should demonstrate relatively high levels of parsimony and fit, as evidenced by the RNFI, the RPR, and the RPFI.

- A chi-square difference test should reveal no significant difference between the theoretical model and the measurement model.

Remember that these represent an ideal that very often will not be attained with real-world data even with a theoretical model that is quite good. A model's fit need not meet all of these criteria in order to be deemed acceptable. In particular, requiring that the model chi-square be nonsignificant is unreasonably strict in most applied situations.

However, it *is* important that there be no significant difference between the fit of the (final) theoretical model and the measurement model. This is because a significant difference between the chi-square for the theoretical and measurement models shows that the theoretical model fails to successfully account for the observed covariances between the F variables in the structural portion of the model.

Given these criteria, your initial theoretical model clearly does not provide a fully acceptable fit to the data. Although some overall fit indices are in the acceptable range (such as the NNFI and CFI, as is summarized in Table 6.1), the model also demonstrates significant problems. Perhaps the most troubling is the finding that the theoretical model provides a fit to the data that is significantly worse than the fit provided by the measurement model. For this reason, it will be necessary to modify the model to attain a better fit.

A Decision-Tree Framework for Modifying the Theoretical Model

As complex as the concept of model modification has been so far in this chapter, it becomes even more complex when the model to be modified is a theoretical model. In many cases, this task can be made easier by following a structured approach to model modification described by Anderson and Gerbing (1988). Although a full discussion of this approach is beyond the scope of this chapter, some basic concepts are reviewed.

Essentially, the procedure involves performing a series of chi-square difference tests for four causal models:

- M_t, the initial theoretical model

- M_m, the measurement model

- M_c, a constrained model

- M_u, an unconstrained model

Two of these models have already been discussed: M_t, or the initial theoretical model, and M_m, the measurement model.

A third model is M_c, a **constrained model**. M_c is a constrained version of M_t, in the sense that one or more of the parameters (e.g., causal paths) in M_t is fixed at zero in M_c. Theoretical considerations determine which parameters should be fixed at zero, so that M_c represents a theoretically "next most likely" modification of M_t.

The fourth model is M_u, an **unconstrained model**. M_u is an unconstrained version of M_t, in that one or more parameters in M_t are freed to be estimated in M_u. As before, theoretical considerations determine which parameters should be freed, so that M_u constitutes a "next most likely" modification of M_t.

Anderson and Gerbing (1988) provide a decision-tree framework for a series of chi-square difference tests between pairs of these four models. The procedure results in the acceptance of a model that does not significantly differ from the measurement model, while at the same time is as parsimonious as possible. If none of the preceding models meets these criteria, the framework guides you through additional modifications (e.g., relaxing additional constraints) until an acceptable model is found. The authors point out that their approach shifts from being confirmatory to being increasingly exploratory as more modifications are made.

This framework is particularly well suited to situations in which the literature suggests competing explanations for some phenomenon of interest, with some of the explanatory models being more constrained than others. For those situations, you are encouraged to make use of the structured approach described in Anderson and Gerbing (1988).

In some situations, however, theory and prior research will not be adequate to allow the development of a series of a priori models, and you will be forced to rely more heavily on modification indices and other results to guide model modifications. The next section shows how this information can be used to arrive at a better-fitting theoretical model in those situations.

Using Modification Indices to Modify the Present Model

In Chapter 4, "Path Analysis with Manifest Variables," much was said concerning the dangers of data-driven model modifications. All of those warnings apply to the modification of latent-variable models as well. Whenever you modify models based on the results of an analysis, you run the risk of capitalizing on the chance characteristics of the sample data, and creating a new model that will not generalize to other samples.

Chapter 4 made six recommendations concerning things that you can do to minimize the dangers of data snooping:

- Use large samples.

- Make few modifications.

- Make only changes that can be meaningfully interpreted.

- Follow a parallel specification search procedure.

- Compare alternative a priori models.

- Carefully describe the limitations of your study.

Even a cursory review of these recommendations shows that the present fictitious study is poorly suited for a specification search. The sample is relatively small ($N = 240$), and it has been established that data-driven modifications are particularly risky in small samples (MacCallum, Roznowski, & Necowitz, 1992).

Nevertheless, the following sections will describe the steps to be followed in modifying the present model for purposes of illustration. Here, you will see how information from the Wald test and the Lagrange Multiplier test may be used to develop a better-fitting model.

The Wald test. Although models may be modified in any of a number of ways (e.g., by placing equality constraints on parameters), they are most frequently modified by either fixing causal paths at zero (e.g., eliminating a nonsignificant path from the model), or freeing causal paths to be estimated (i.e., adding new paths to the model). Of these alternatives, eliminating a nonsignificant path is less likely to capitalize on chance characteristics of the data, and is therefore less risky (Bentler & Chou, 1987). For this reason, you will first review the results of the analysis to identify any nonsignificant paths to be deleted.

This review should normally begin with the Wald test, as it identifies parameters that may be dropped without causing a significant decrease in model chi-square. The Wald test for your analysis of the theoretical model appears on page 43 of the Output 6.3, and is reproduced here as Output 6.5.

```
                    Stepwise Multivariate Wald Test                          43
------------------------------------------------------------------------------
                    Cumulative Statistics          Univariate Increment
   Parameter     Chi-Square     D.F.      Prob      Chi-Square      Prob
------------------------------------------------------------------------------
   CF4F5           0.006316      1       0.9367       0.006316      0.9367
   PF1F6           0.845457      2       0.6553       0.839141      0.3596
   CF3F4           2.738053      3       0.4338       1.892596      0.1689
```

Output 6.5: Wald Test Results for Initial Theoretical Model, Investment Model Study

Earlier, you learned that the Wald test estimates the change in model chi-square that would result from fixing a given parameter at zero. The first parameter listed in the preceding Wald test results is CF4F5, the covariance between F4 (Costs) and F5 (Investments), and the third entry in the table is CF3F4, the covariance between F3 (Rewards) and F4 (Costs). Covariances are generally estimated for all possible pairs of exogenous F variables in an analysis of this sort (unless there is theoretical reason that they be fixed at zero), so you will disregard the Wald test results for CF4F5 and CF3F4 for the moment.

In this case, you are more interested in finding unidirectional causal paths that may be eliminated without significantly hurting the model's fit, and the Wald test has identified just such a path. The parameter PF1F6 represents the path to F1 (Commitment) from F6 (Alternative value), and the univariate Wald test suggests that model chi-square will change only 0.84 (a nonsignificant amount) if this path were deleted. Remember that this finding is very consistent with what you learned when you reviewed the latent variable equations on page 16 of Output 6.3. There, you learned that the path coefficient for the path from F6 to F1 was nonsignificant.

The safest approach to modifying models is to change just one parameter at a time. You will therefore re-estimate your model with PF1F6 fixed at zero, and then review the results to see if any additional modifications are necessary.

Creating revised model 1. Figure 6.5 presents the program figure for revised model 1. You can see that this model is identical to the initial theoretical model (Figure 6.4) except that the path from Alternatives to Commitment has been deleted.

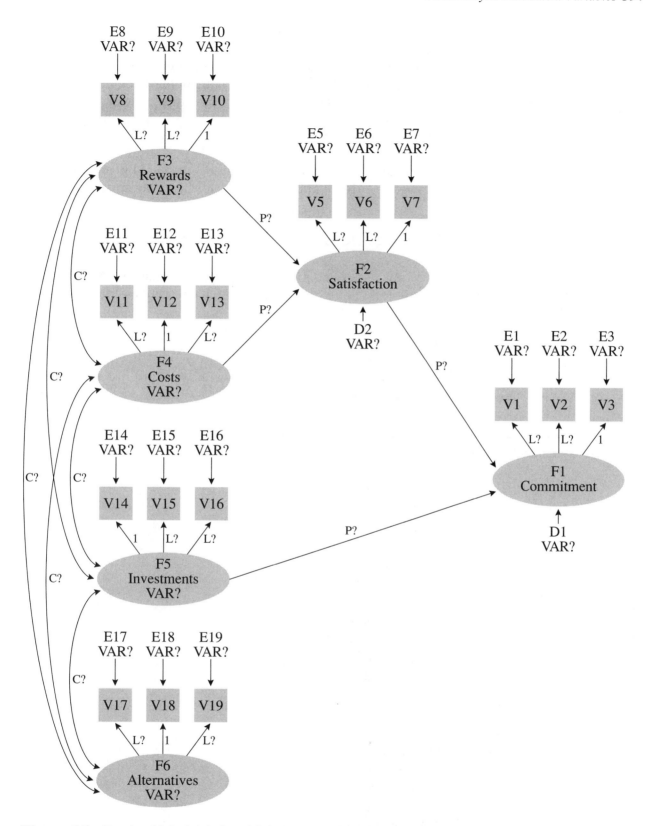

Figure 6.5: Revised Model 1, in which the Path from F6 (Alternatives) to F1 (Commitment) has Been Deleted

To create the PROC CALIS program that will estimate revised model 1, it is necessary to make just one small change in the program that had estimated the initial theoretical model. Specifically, the latent-variable equation for F1 in the LINEQS statement has to be modified so that F6 is no longer specified as an independent variable for F1. This can be done easily by making a copy of the original program, and then blanking out the path coefficient name (PF1F6) and the short name (F6) for the alternatives construct.

The PROC CALIS program that estimated the initial theoretical model appeared in an earlier section titled "Preparing the SAS Program." In that program, the latent variable equation for F1 had appeared in the following way:

```
21             F1  = PF1F2 F2 + PF1F5 F5 + PF1F6 F6 + D1,
```

In the PROC CALIS program that estimates revised model 1, the equation takes on the following form; notice that the path coefficient name and the short name for the alternatives construct has been blanked out:

```
21             F1  = PF1F2 F2 + PF1F5 F5                + D1,
```

No other changes to the program were necessary. The complete program (minus the DATA step) for estimating revised model 1 appears here:

```
1        PROC CALIS  COVARIANCE  CORR  RESIDUAL  MODIFICATION ;
2           LINEQS
3               V1  = LV1F1  F1 + E1,
4               V2  = LV2F1  F1 + E2,
5               V3  =        F1 + E3,
6               V5  = LV5F2  F2 + E5,
7               V6  = LV6F2  F2 + E6,
8               V7  =        F2 + E7,
9               V8  = LV8F3  F3 + E8,
10              V9  = LV9F3  F3 + E9,
11              V10 =        F3 + E10,
12              V11 = LV11F4 F4 + E11,
13              V12 =        F4 + E12,
14              V13 = LV13F4 F4 + E13,
15              V14 =        F5 + E14,
16              V15 = LV15F5 F5 + E15,
17              V16 = LV16F5 F5 + E16,
18              V17 = LV17F6 F6 + E17,
19              V18 =        F6 + E18,
20              V19 = LV19F6 F6 + E19,
21              F1  = PF1F2 F2 + PF1F5 F5                + D1,
22              F2  = PF2F3 F3 + PF2F4 F4                + D2;
23          STD
```

```
24              E1-E3   = VARE1-VARE3,
25              E5-E19  = VARE5=VARE19,
26              F3-F6   = VARF3-VARF6,
27              D1-D2   = VARD1-VARD2;
28          COV
29              F3  F4  = CF3F4,
30              F3  F5  = CF3F5,
31              F3  F6  = CF3F6,
32              F4  F5  = CF4F5,
33              F4  F6  = CF4F6,
34              F5  F6  = CF5F6;
35          VAR   V1 V2 V3 V5-V19 ;
36          RUN;
```

The preceding program produced 43 pages of output, but, to save space the output will not be reproduced here. Instead, the most important findings from the output have already been summarized in Table 6.1, to the right of "M_{r1} Revised model 1." You can see that the chi-square for the revised model is 217.58, which, with 125 degrees of freedom, provides a chi-square/*df* ratio of 1.74, almost identical to the ratio of 1.75 obtained with the initial theoretical model.

Before reviewing any additional fit indices, you will first perform a chi-square difference test comparing the theoretical model (M_t) to revised model 1 (M_{r1}). Finding a significant difference between the two models would indicate that the path from F6 to F1 had been an important path, and should not have been deleted. Model chi-square values are obtained from Table 6.1 to perform this test:

$$M_{r1} - M_t = 217.58 - 216.75 = 0.83$$

So the chi-square difference value is 0.83, which is quite close to the value of 0.839 that had been estimated by the Wald test. The degrees of freedom for the test are equal to the difference between the *df* for the two models:

$$df_{r1} - df_t = 125 - 124 = 1$$

A table of chi-square (in Appendix C) shows the critical value of chi-square with 1 degree of freedom is 3.841. Because your observed chi-square difference value of 0.83 is less than this, you conclude that there is not a significant difference between the fit provided by the theoretical model and that provided by revised model 1. Apparently, deleting the path from F6 to F1 did not hurt the model's fit.

Consistent with this, Table 6.1 reveals acceptable values for revised model 1 on the NNFI and CFI. Notice that the parsimony ratio (PR) for the revised model is .817, whereas the PR for the theoretical model was only .810. This slightly higher value for the revised model reflects the fact that this model posits fewer causal relationships than the theoretical model, and thus is more parsimonious. The superior parsimony of revised model 1 is similarly reflected in the other parsimony indices in Table 6.1: the PNFI, the RPR, and the RPFI.

Other results from the output (not reproduced here) provide a mixed picture regarding the fit of revised model 1. Although many results provided support for the model (e.g., all path coefficients are now significant), the distribution of normalized residuals was still somewhat asymmetrical, and the three largest normalized residuals involved the relationship between V15 (an indicator for F5, Investments) and V5, V6, and V7 (all indicators for F2, Satisfaction). You will remember that this was the same pattern of normalized residuals observed with the initial theoretical model, and led to speculation that it may be necessary to add a path from F5 to F2.

However, the critical test of the validity of revised model 1 is the chi-square difference test comparing the revised model to the measurement model. A significant difference between these two models suggests that revised model 1 is not successfully accounting for the relationships between the latent F variables that constitute the structural portion of the model.

Values of chi-square may be found in Table 6.1 and substituted in the following equation:

$$M_{r1} - M_m = 217.58 - 180.87 = 36.71$$

The degrees of freedom for the test are calculated in the usual way:

$$df_{r1} - df_m = 125 - 120 = 5$$

The critical value of chi-square with 5 degrees of freedom is 11.070 at $p < .05$ and 20.517 at $p < .001$. The obtained value of chi-square is greater than these, which indicates a significant difference between chi-squares for the two models. In other words, revised model 1 exhibits a fit to the data that is significantly worse than the fit displayed by the measurement model. Apparently, there is still a serious misspecification involving the relationships between the F variables in M_{r1}.

Although the pattern of normalized residuals provides clues as to how the model may be modified to obtain a better fit, it is generally much more efficient to go straight to the modification indices. As before, your first stop is the Wald test, as it is generally safer to drop parameters than to add them. But the Wald test reveals no path coefficient that may be dropped without hurting the model's fit, only covariances. This is consistent with the finding that all factor loadings and path coefficients are significant for revised model 1. Given that it apparently is not possible to drop nonsignificant paths from the model, you now turn your attention to identifying new paths that should be added. The Lagrange multiplier test is used for this purpose.

The Lagrange multiplier test. The MODIFICATION option requested in the PROC CALIS statement resulted in three matrices of Lagrange multiplier tests:

- A **phi matrix** whose rows and columns consisted of the model's exogenous variables (F3, F4, F5, F6, all E terms, and all D terms)

- A **gamma matrix**, whose columns consisted of the model's exogenous F variables (F3, F4, F5, and F6), and whose rows consisted of the model's endogenous variables (all V variables, F1, and F2)

- A **beta matrix**, whose rows and columns consisted of the study's endogenous variables (all V variables, F1, and F2)

The rank order of the 10 largest Lagrange multipliers for the phi matrix showed that these large values all involved adding paths or covariances connecting disturbance terms to disturbance terms, or connecting disturbance terms to F variables. Since it has already been explained that these types of relationships are normally not allowed in the relatively simple models discussed in this text, the results of the phi matrix will be disregarded.

Of the remaining Lagrange multiplier values, by far the largest was for the F2:F5 path. The test estimated that the model chi-square would decrease by 34.53 (a significant amount) if a path were added that went from F5 (Investments) to F2 (Satisfaction). The chapter on path analysis with manifest variables indicated that such a path could be defended on theoretical grounds, as it is consistent with some aspects of cognitive dissonance theory. You will therefore add this path and re-estimate the model, to see if it results in improved fit.

Creating revised model 2. Figure 6.6 displays the resulting model, revised model 2. This system is identical to revised model 1, except that a causal path has been added that leads from Investments to Satisfaction.

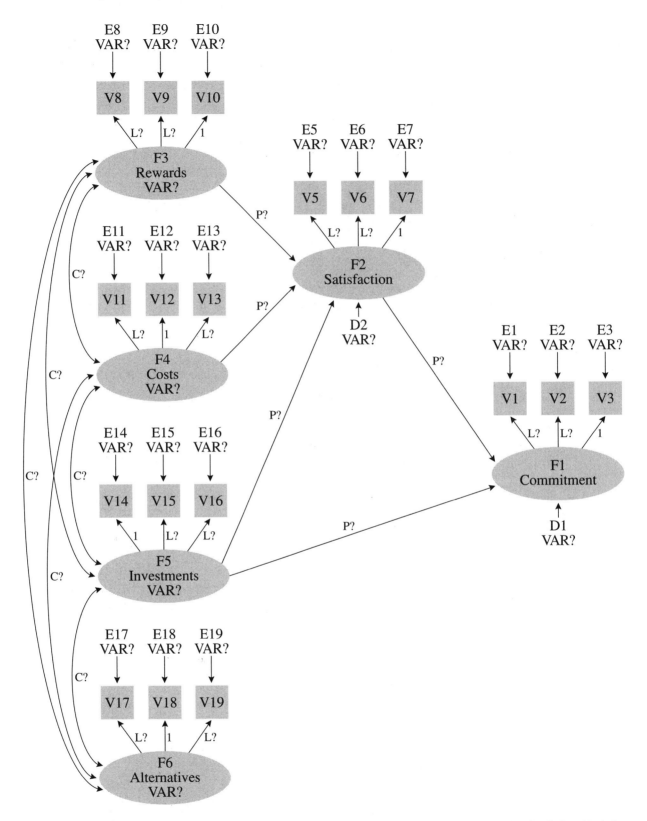

Figure 6.6: Revised Model 2, in which the Path from F5 (Investments) to F2 (Satisfaction) has Been Added.

The PROC CALIS program that estimates revised model 2 is identical to the one for revised model 1, except that the latent variable equation for F2 had been changed to reflect the new path from F5 to F2. The original version of this equation had been as follows:

```
22              F2   = PF2F3 F3 + PF2F4 F4                    + D2;
```

while the revised equation took this form:

```
22              F2   = PF2F3 F3 + PF2F4 F4 + PF2F5 F5    + D2;
```

Revised model 2 was therefore estimated and the goodness of fit and parsimony indices for this model are presented in Output 6.6, as well as in Table 6.1.

```
                            The SAS System                            13

          Covariance Structure Analysis: Maximum Likelihood Estimation

          Fit criterion . . . . . . . . . . . . . . . .      0.7665
          Goodness of Fit Index (GFI) . . . . . . . . . .    0.9243
          GFI Adjusted for Degrees of Freedom (AGFI) . . .   0.8957
          Root Mean Square Residual (RMR) . . . . . . . .    0.2082
          Chi-square = 183.1955      df = 124     Prob>chi**2 = 0.0004
          Null Model Chi-square:      df = 153              2167.7711
          RMSEA Estimate  . . . . . .  0.0449  90%C.I.[0.0301, 0.0579]
          Bentler's Comparative Fit Index . . . . . . . .    0.9706
          Normal Theory Reweighted LS Chi-square . . . . .   175.9803
          Akaike's Information Criterion . . . . . . . . .   -64.8045
          Consistent Information Criterion . . . . . . . .   -620.4037
          Schwarz's Bayesian Criterion . . . . . . . . . .   -496.4037
          McDonald's (1989) Centrality. . . . . . . . . .    0.8840
          Bentler & Bonett's (1980) Non-normed Index. . . .  0.9637
          Bentler & Bonett's (1980) Normed Index. . . . . .  0.9155
          James, Mulaik, & Brett (1982) Parsimonious Index.  0.7420
          Z-Test of Wilson & Hilferty (1931) . . . . . . .   3.3242
          Bollen (1986) Normed Index Rho1 . . . . . . . . .  0.8957
          Bollen (1988) Non-normed Index Delta2 . . . . . .  0.9710
          Hoelter's (1983) Critical N . . . . . . . . . . .  198
```

Output 6.6: Goodness of Fit Indices for Revised Model 2, Investment Model Study

Before proceeding with a detailed assessment of fit, it is first necessary to perform two chi-square difference tests. First, you will compare revised model 2 (M_{r2}) with revised model 1 (M_{r1}). Here, you hope to observe a significant chi-square difference value, as this will indicate that the model with the new path (M_{r2}) provides a fit to the data that is significantly better than the fit provided by the more constrained model (M_{r1}). As before, chi-square values for the test may be obtained from Table 6.1.

$$M_{r1} - M_{r2} = 217.58 - 183.20 = 34.38$$

The chi-square difference value for this comparison is 34.38, which is quite close to the value of 34.53 that was predicted by the Lagrange multiplier test. The degrees of freedom for the test are equal to the difference between the degrees of freedom for the two models, or $125 - 124 = 1$. The critical value of chi-square with 1 degree of freedom ($p < .001$) is 10.827, so this chi-square difference test is significant at $p < .001$. In other words, the test shows that revised model 2 (with the new path) provides a fit that is significantly superior to that of revised model 1.

So far so good. The second chi-square difference test involves comparing revised model 2 to the measurement model (M_m). In this case you hope for a *nonsignificant* chi-square difference value, as this will suggest that M_{r2} does a good job of accounting for the relationships between the F variables that constitute the structural portion of the model. The test is performed as follows:

$$M_{r2} - M_m = 183.20 - 180.87 = 2.33$$

The degrees of freedom for the test are $124 - 120 = 4$, and the critical value of chi-square ($p < .05$) for 4 degrees of freedom is 9.488. The observed chi-square difference value of 2.33 is less than this critical value, meaning that there is no significant difference in the fit provided by the two models. This finding supports the validity of revised model 2. Apparently, revised model 2 provides a fit to the data that is essentially as good as the fit provided by the measurement model.

Output 6.7 provides some additional results from the analysis of revised model 2. These results may be reviewed to determine whether revised model 2 provides an "ideal" fit to the data.

The SAS System   12

Average Normalized Residual = 0.4797
Average Off-diagonal Normalized Residual = 0.5349

Rank Order of 10 Largest Normalized Residuals

V17,V5	V11,V2	V17,V7	V17,V15	V19,V9	V8,V1	V17,V6
-1.8395	1.7835	-1.7535	-1.7012	-1.4587	1.4354	-1.4299

V15,V5	V15,V2	V11,V1
1.4267	-1.3810	1.3404

```
                    Distribution of Normalized Residuals
                      (Each * represents 2 residuals)

   -2.00000 -   -1.75000    2   1.17%  | *
   -1.75000 -   -1.50000    1   0.58%  |
   -1.50000 -   -1.25000    6   3.51%  | ***
   -1.25000 -   -1.00000    5   2.92%  | **
   -1.00000 -   -0.75000    5   2.92%  | **
   -0.75000 -   -0.50000   15   8.77%  | *******
   -0.50000 -   -0.25000   15   8.77%  | *******
   -0.25000 -          0   49  28.65%  | ***********************
          0 -    0.25000   28  16.37%  | **************
    0.25000 -    0.50000   12   7.02%  | ******
    0.50000 -    0.75000   10   5.85%  | *****
    0.75000 -    1.00000   14   8.19%  | *******
    1.00000 -    1.25000    2   1.17%  | *
    1.25000 -    1.50000    6   3.51%  | ***
    1.50000 -    1.75000    0   0.00%  |
    1.75000 -    2.00000    1   0.58%  |
```

```
        Covariance Structure Analysis: Maximum Likelihood Estimation        13

        Fit criterion . . . . . . . . . . . . . . . .      0.7665
        Goodness of Fit Index (GFI) . . . . . . . . .      0.9243
        GFI Adjusted for Degrees of Freedom (AGFI) . . .   0.8957
        Root Mean Square Residual (RMR) . . . . . . . .    0.2082
        Chi-square = 183.1955      df = 124    Prob>chi**2 = 0.0004
        Null Model Chi-square:     df = 153              2167.7711
        RMSEA Estimate . . . . . .  0.0449  90%C.I.[0.0301, 0.0579]
        Bentler's Comparative Fit Index . . . . . . . .    0.9706
        Normal Theory Reweighted LS Chi-square . . . . .   175.9803
        Akaike's Information Criterion . . . . . . . . .   -64.8045
        Consistent Information Criterion . . . . . . . .   -620.4037
        Schwarz's Bayesian Criterion . . . . . . . . . .   -496.4037
        McDonald's (1989) Centrality. . . . . . . . . .    0.8840
        Bentler & Bonett's (1980) Non-normed Index. . . .  0.9637
        Bentler & Bonett's (1980) Normed Index. . . . . .  0.9155
        James, Mulaik, & Brett (1982) Parsimonious Index. 0.7420
        Z-Test of Wilson & Hilferty (1931) . . . . . . .   3.3242
        Bollen (1986) Normed Index Rho1 . . . . . . . . .  0.8957
        Bollen (1988) Non-normed Index Delta2 . . . . . .  0.9710
        Hoelter's (1983) Critical N . . . . . . . . . .        198
```

```
        Covariance Structure Analysis: Maximum Likelihood Estimation        15

V18       =     1.0000 F6     +  1.0000 E18

V19       =     1.0618*F6     +  1.0000 E19
Std Err         0.1191 LV19F6
t Value         8.9190
```

Latent Variable Equations

```
F1      =      0.3589*F2    +   0.8039*F5     +   1.0000 D1
Std Err        0.1126 PF1F2     0.1296 PF1F5
t Value        3.1862           6.2045


F2      =      0.3633*F3    -   0.1614*F4     +   0.5148*F5     +   1.0000 D2
Std Err        0.1175 PF2F3     0.0472 PF2F4      0.0868 PF2F5
t Value        3.0910          -3.4197            5.9317
```

Variances of Exogenous Variables

Variable	Parameter	Estimate	Standard Error	t Value
F3	VARF3	1.659100	0.291847	5.685
F4	VARF4	5.482210	0.828133	6.620
F5	VARF5	3.130293	0.435356	7.190
F6	VARF6	2.256832	0.380421	5.932
E1	VARE1	1.354896	0.179438	7.551
E2	VARE2	2.705866	0.297378	9.099
E3	VARE3	0.892005	0.189055	4.718
E5	VARE5	1.176597	0.141947	8.289
E6	VARE6	1.107337	0.154862	7.150
E7	VARE7	0.991936	0.144302	6.874
E8	VARE8	1.125193	0.137324	8.194
E9	VARE9	1.173904	0.137271	8.552
E10	VARE10	1.316373	0.200691	6.559
E11	VARE11	2.188103	0.478392	4.574
E12	VARE12	1.779719	0.546928	3.254
E13	VARE13	4.370980	0.420766	10.388
E14	VARE14	1.293246	0.233121	5.548
E15	VARE15	2.467558	0.284073	8.686
E16	VARE16	2.447443	0.244432	10.013
E17	VARE17	2.166535	0.263409	8.225
E18	VARE18	1.606254	0.251713	6.381
E19	VARE19	2.229566	0.309892	7.195
D1	VARD1	2.928320	0.391701	7.476
D2	VARD2	1.582540	0.234447	6.750

Covariance Structure Analysis: Maximum Likelihood Estimation   16
Covariances among Exogenous Variables

		Parameter	Estimate	Standard Error	t Value
F4	F3	CF3F4	-0.278512	0.249006	-1.118
F5	F3	CF3F5	1.187097	0.220068	5.394
F5	F4	CF4F5	0.060595	0.322652	0.188
F6	F3	CF3F6	-0.837788	0.186891	-4.483
F6	F4	CF4F6	0.907508	0.294503	3.081
F6	F5	CF5F6	-0.803543	0.227429	-3.533

```
        Covariance Structure Analysis: Maximum Likelihood Estimation    17
                   Equations with Standardized Coefficients
V1      =     0.8838*F1      + 0.4679 E1
                  LV1F1

V2      =     0.8249*F1      + 0.5652 E2
                  LV2F1

V3      =     0.9381 F1      + 0.3464 E3

V5      =     0.8271*F2      + 0.5621 E5
                  LV5F2

V6      =     0.8673*F2      + 0.4978 E6
                  LV6F2

V7      =     0.8750 F2      + 0.4841 E7

V8      =     0.6632*F3      + 0.7484 E8
                  LV8F3

V9      =     0.6388*F3      + 0.7694 E9
                  LV9F3

V10     =     0.7467 F3      + 0.6651 E10

V11     =     0.8219*F4      + 0.5697 E11
                  LV11F4

V12     =     0.8689 F4      + 0.4951 E12

V13     =     0.4641*F4      + 0.8858 E13
                  LV13F4

V14     =     0.8412 F5      + 0.5407 E14

V15     =     0.7068*F5      + 0.7074 E15
                  LV15F5

V16     =     0.5511*F5      + 0.8344 E16
                  LV16F5

V17     =     0.6776*F6      + 0.7354 E17
                  LV17F6

V18     =     0.7643 F6      + 0.6448 E18

V19     =     0.7301*F6      + 0.6834 E19
                  LV19F6

F1      =     0.2526*F2      + 0.5561*F5      + 0.6691 D1
                  PF1F2              PF1F5

F2      =     0.2600*F3      - 0.2099*F4      + 0.5060*F5      + 0.6988 D2
                  PF2F3              PF2F4              PF2F5
```

```
Covariance Structure Analysis: Maximum Likelihood Estimation          18

                    Variances of Endogenous Variables
         ------------------------------------------------
              Variable         Estimate         R-squared
         ------------------------------------------------
               1    V1         6.188336          0.781056
               2    V2         8.469520          0.680517
               3    V3         7.433706          0.880005
               4    V5         3.724434          0.684087
               5    V6         4.469370          0.752239
               6    V7         4.232273          0.765626
               7    V8         2.008896          0.439894
               8    V9         1.982996          0.408015
               9    V10        2.975473          0.557592
              10    V11        6.742819          0.675491
              11    V12        7.261929          0.754925
              12    V13        5.571196          0.215432
              13    V14        4.423538          0.707645
              14    V15        4.931056          0.499588
              15    V16        3.515161          0.303747
              16    V17        4.005986          0.459176
              17    V18        3.863086          0.584205
              18    V19        4.774108          0.532988
              19    F1         6.541700          0.552361
              20    F2         3.240337          0.511613


                    Correlations among Exogenous Variables
             ------------------------------------------
                     Parameter           Estimate
             ------------------------------------------
              F4    F3    CF3F4         -0.092348
              F5    F3    CF3F5          0.520904
              F5    F4    CF4F5          0.014627
              F6    F3    CF3F6         -0.432960
              F6    F4    CF4F6          0.258002
              F6    F5    CF5F6         -0.302320


      Rank order of 10 largest Lagrange multipliers in _GAMMA_

          V2 : F5                V1 : F3                V17 : F4
      10.9899 : 0.001        6.1863 : 0.013         3.6918 : 0.055

          V18 : F3               V2 : F4                V1 : F5
      3.5986 : 0.058        3.0898 : 0.079         3.0605 : 0.080

          V19 : F3               V16 : F3               V18 : F5
      2.8862 : 0.089        2.7224 : 0.099         2.1320 : 0.144

                              V14 : F3
                          1.9742 : 0.160
```

```
        Rank order of 10 largest Lagrange multipliers in _BETA_

           V1 : V6              V3 : V1              V1 : V3
       16.7104 : 0.000      16.4797 : 0.000      15.9460 : 0.000


           F1 : V2              V2 : F2              V1 : F2
       15.6650 : 0.000      12.4099 : 0.000      12.3790 : 0.000


           V2 : V7              F1 : V1              V2 : V3
       11.8967 : 0.001       8.8174 : 0.003       8.2976 : 0.004


                                V3 : V2
                             8.1187 : 0.004
```

Output 6.7: Miscellaneous Results from Analysis of Revised Model 2, Investment Model Study

Some findings regarding revised model 2 are summarized here:

- Model chi-square was 183.20 with 124 *df* ($p = .0004$) (this appears on output page 13, to the right of the heading "Chi-square =").

- The chi-square/*df* ratio was 1.48 (lower than the informal rule of thumb of 2.00, and also lower than the ratio of 1.75 observed for the initial theoretical model).

- The CFI and NNFI were .971 and .964, respectively, well above the recommended level of .9 (these also appear on page 13).

- The *t* values for all factor loadings (from output pages 14-15) were statistically significant, and all standardized factor loadings and path coefficients (from output page 17) were nontrivial in absolute magnitude.

- R^2 values for the commitment and satisfaction constructs were quite large at .55 and .51, respectively (these are the last two entries in the "R-squared" column in the "Variances of Endogenous Variables" table from output page 18).

- The distribution of normalized residuals was fairly symmetrical and centered around zero, and no residuals exceeded 2.0 in absolute magnitude (this information appears on output page 12).

- Although the PR of .810 for revised model 2 was somewhat lower than the PR of .817 observed for revised model 1, the slight decrease in parsimony was offset by an increase in fit. This is reflected in the PNFI of .742 for M_{r2}, which was larger than the PNFI of .735 observed for M_{r1} (these indices appear in Table 6.1).

- In the same way, the structural portion of M_{r2} was less parsimonious than the structural portion of M_{r1}, in that the RPR and RPFI values were lower for M_{r2}. However, M_{r2}

displayed a better fit in the structural portion of the model, as can be seen by the larger RNFI value for M_{r2} (again, these indices appear in Table 6.1).

Technically, revised model 2 does not provide an "ideal fit," as described earlier, because the model chi-square was still statistically significant. However, as has been noted several times before, a good model will often demonstrate a significant chi-square with real-world data. The bulk of the remaining results, however, indicate that M_{r2} provides very acceptable levels of fit and parsimony. These results, coupled with the finding that revised model 2 provided a fit to the data that was not significantly worse than that of the measurement model, support M_{r2} as the study's "final" model.

Preparing a Formal Description of Results for a Paper

Figures and Tables

In most cases, it is best to use a figure to illustrate the causal model tested in your study. At a minimum, this will require one figure that illustrates your initial theoretical model. This figure should follow the standard conventions in which latent variables are portrayed as oval or circles and manifest variables are portrayed as squares or rectangles. Your predictions will be made even more clear if causal paths are labelled with "+" and "−" signs to indicate whether positive or negative relationships are anticipated.

Figure 6.7 illustrates one example of how the theoretical model for the present fictitious study could be presented. Notice that, in order to simplify the readers' task, the model in Figure 6.7 is much simpler than the program figures referred to throughout this chapter (e.g., it does not contain symbols for parameters to be estimated).

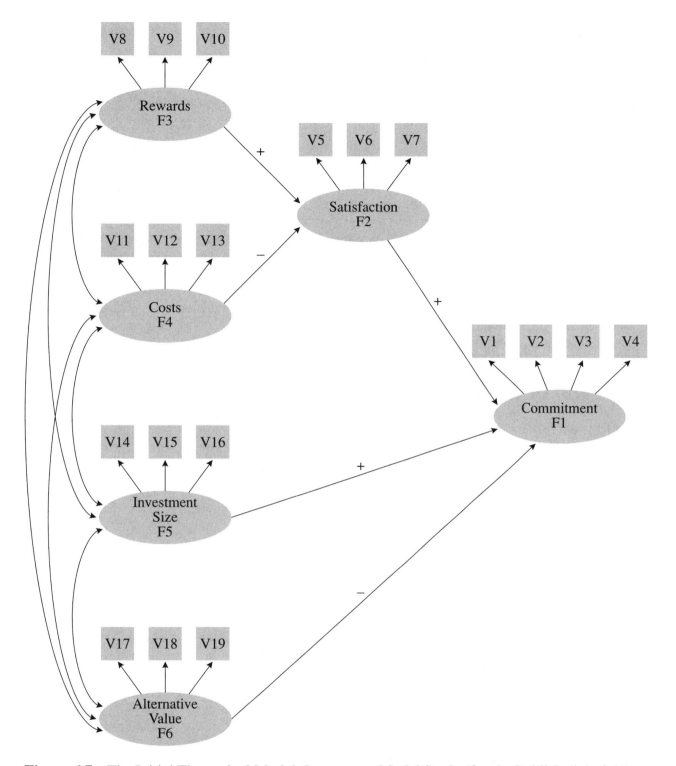

Figure 6.7: The Initial Theoretical Model, Investment Model Study (for the Published Article)

Later in the paper, you may want to present a second figure that illustrates the study's "final" model, after modifications (if any) have been made. For this second figure, you may choose to place the obtained standardized path coefficients on the causal paths that connect the study's F variables. This will help readers understand which independent variables exerted relatively strong effects, and which exerted relatively weak effects. An example of such a figure is presented later.

Plan on preparing at least four tables to summarize the results of the analysis. Most of these tables have already been presented in Chapter 5 and the current chapter. First, a table similar to Table 5.1 should be used to display the correlations or covariances between the manifest variables that were analyzed (this table appeared in Chapter 5, in the section titled "Preparing the SAS Program"). Many researchers who perform structural equation modeling omit this information, but it is important to include such a table because it allows other researchers to replicate your analyses (and even test competing causal models!). As always, if your matrix is a correlation matrix, be sure to include the standard deviations.

Next, a table similar to Table 5.3 should be used to display the properties of the measurement model (e.g., standardized factors loadings, reliabilities). This table also appeared in Chapter 5, in the section titled "Assessing the Reliability and Validity of Constructs and Indicators."

Third, a table similar to Table 6.1 (from the current chapter) should summarize parsimony indices and goodness of fit indices for the various models estimated.

Finally, if the initial theoretical model is revised and re-estimated several times, it may be best to present the standardized path coefficients for the various models in a single table. An example of such a table appears here as Table 6.2.

Table 6.2

Standardized Path Coefficients

Dependent Variable / Independent Variable	Theoretical Model	Revised Model 1	Revised Model 2
Commitment (F1)			
Satisfaction (F2)	.332***	.315***	.253**
Investment size (F5)	.548***	.537***	.556***
Alternative value (F6)	.060		
Satisfaction (F2)			
Rewards (F3)	.639***	.639***	.260**
Costs (F4)	-.158*	-.158*	-.210***
Investment size (F5)			.506***

Note. $\underline{N}$ = 240.

* $\underline{p}$ < .05; ** $\underline{p}$ < .01; *** $\underline{p}$ < .001.

Preparing Text for the Results Section of the Paper

There is a good deal of variability in the way that research papers report the results of a path analysis with latent variables. In part, this reflects the fact that latent variables analyses are somewhat new, relative to other statistical procedures, and researchers are still working out the details of just how they should be conducted and reported.

The way that you report the results of a path analysis with latent variables will depend, in part, on the purpose of your research. If your research was designed to test hypotheses related to the measurement model, then it is appropriate to discuss in some detail the tests that dealt specifically with the measurement model such as the procedures that assess the convergent and discriminant validity of measures and constructs. If measurement concerns are not a central focus of the research, less time can be spent on these matters.

The following is one approach to describing the analyses reported in this chapter. The report touches on several different aspects of the analysis; you should feel free to deal with any of these topics in greater or lesser detail, depending on the purpose of the research.

One word of warning: the following results section is very lengthy, and goes into a great deal of detail in describing the procedures that were performed and the results that were obtained. This was done for the sake of completeness, so that you would have a model to follow when reporting the results of your analyses (regardless of which aspects of the results you chose to emphasize). In its entirety, the following section is actually much longer and more detailed than would be allowed in many scholarly journals. When you write for a real-world journal, you should generally be more concise, and only explain in great detail those aspects of the analysis that are particularly relevant to your research questions.

Results
Overview of the Analysis

Data were analyzed using the SAS System's CALIS procedure (SAS Institute Inc., 1989), and the models tested were covariance structure models with multiple indicators for all latent constructs. Standard deviations and intercorrelations for the study's 19 manifest variables are presented in Table 5.1.

The present analysis followed a two-step procedure based in part on an approach recommended by Anderson and Gerbing (1988). In the first step, confirmatory factor analysis was used to develop a measurement model that demonstrated an acceptable fit to the data. In step two, the measurement model was modified so that it came to represent the theoretical (causal) model of interest. This theoretical model was then tested and revised until a theoretically meaningful and statistically acceptable model was found.

The Measurement Model

In path analysis with latent variables, a measurement model describes the nature of the relationship between (a) a number of latent variables, or factors, and (b) the manifest indicator variables that measure those latent variables. The model investigated in this study consisted of six latent variables corresponding to the six constructs of the investment model: commitment, satisfaction, rewards, costs, investment size, and alternative value. Each of the six latent variables was measured by at least three manifest indictor variables.

        <u>The initial measurement model</u>. This article follows
Bentler's (1989) convention of identifying latent variables
with the letter "F" (for Factor), and labelling manifest
variables with the letter "V" (for Variable). Figure 6.7
uses these conventions in identifying the six latent
constructs investigated in this study, as well as the
indicators that measure these constructs. The figure shows
that the commitment construct (F1) is measured by manifest
variables V1 through V4, the satisfaction construct (F2) is
measured by manifest variables V5 through V7, and so forth.

        The measurement model assessed in the first stages of
this analysis was not identical to the model in Figure 6.7,
because the model in that figure posits certain
unidirectional causal relationships between the latent
constructs. The measurement model, on the other hand,
posits no unidirectional paths between latent variables.
Instead, in a measurement model, a covariance is estimated
to connect each latent variable with every other latent
variable. In a figure, this would be indicated by a
curved, two-headed arrow connecting each F variable to
every other F variable. In other words, a measurement
model is equivalent to a confirmatory factor analysis model
in which each latent construct is allowed to covary with
every other latent construct.

        This measurement model was estimated using the maximum
likelihood method, and the chi-square value for the model
was statistically significant, $\chi^2(137, \underline{N} = 240) = 247.68$,
$\underline{p} < .001$. Technically, when the proper assumptions are
met, this chi-square statistic may be used to test the null
hypothesis that the model fits the data. In practice,
however, the statistic is very sensitive to sample size and
departures from multivariate normality, and will very often
result in the rejection of a well-fitting model. For this
reason, it has been recommended that the model chi-square
statistic be used as a goodness of fit index, with smaller
chi-square values (relative to the degrees of freedom)
indicative of a better model fit (James, Mulaik, & Brett,
1982; Joreskog & Sorbom, 1989).

        A number of other results, however, indicated that
there was in fact a problem with the model's fit. The
pattern of large normalized residuals, parameter
significance tests, and Lagrange multiplier tests showed
that one manifest indicator, V4, was apparently causally
affected by both the alternative value construct (F6), as
well as the construct that it was expected to be affected

by (commitment, F1). Because V4 was apparently a
multidimensional variable, it was eliminated from the
measurement model, and the measurement model was re-
estimated.

   The revised measurement model. Goodness of fit
indices for the respecified measurement model (M_m) are
presented in Table 6.1. This table shows that the revised
measurement model displayed values greater than .9 on the
non-normed-fit index (NNFI) and the comparative fit index
(CFI), indicative of an acceptable fit (Bentler & Bonett,
1980; Bentler, 1989). Therefore, model M_m was tentatively
accepted as the study's "final" measurement model, and a
number of tests were conducted to assess its reliability
and validity.
   Standardized factor loadings for the indicator
variables are presented in Table 5.3. The SAS System's
CALIS procedure provides approximate standard errors for
these coefficients which allow large-sample t tests of the
null hypothesis that the coefficients are equal to zero in
the population. The t scores obtained for the coefficients
in Table 5.3 range from 6.99 through 18.78, indicating that
all factor loadings were significant (p < .001). This
finding provides evidence supporting the convergent
validity of the indicators (Anderson & Gerbing, 1988).
   Table 5.3 also provides the reliabilities of the
indicators (the square of the factor loadings), along with
the composite reliability for each construct. Composite
reliability is a measure of internal consistency comparable
to coefficient alpha (Fornell & Larcker, 1981). All six
scales demonstrated acceptable levels of reliability, with
coefficients in excess of .70.
   The final column of Table 5.3 provides the variance
extracted estimate for each scale. This is a measure of
the amount of variance captured by a construct, relative to
the variance due to random measurement error (Fornell &
Larcker, 1981). Five of the six constructs demonstrated
variance extracted estimates in excess of .50, the level
recommended by Fornell and Larcker (1981).
   Combined, these findings generally support the
reliability and validity of the constructs and their
indicators. The revised measurement model (M_m) was
therefore retained as the study's final measurement model
against which other models would be compared.

The Structural Model

   The initial theoretical model. The theoretical model tested in the present study is identical to the one presented in Figure 6.7, with the exception that V4 was dropped as a measure of commitment (consistent with our findings in analyzing the measurement model). The analysis of this model may be described as a path analysis with latent variables.

   Goodness of fit indices for the model appear in Table 6.1, in the row headed "M_t Theoretical model." Values on the NNFI and CFI were acceptable for M_t (in excess of .9).

   However, a review of the model's residuals revealed that the distribution of normalized residuals was asymmetrical, and that seven of the normalized residuals were relatively large (in excess of 2.0, with one in excess of 3.0). In addition, one of the causal paths linking two latent constructs proved to be nonsignificant: Table 6.2 shows that the standardized path coefficient for the path from alternative value (F6) to commitment (F1) was only .060, ns.

   The nomological validity of a theoretical model can be tested by performing a chi-square difference test in which the theoretical model is compared to the measurement model. A finding of no significant difference indicates that the theoretical model is successful in accounting for the observed relationships between the latent constructs (Anderson & Gerbing, 1988).

   Therefore, the chi-square for the measurement model was subtracted from the chi-square for the theoretical model with the resulting chi-square difference value of 216.75 - 180.87 = 35.88. The degrees of freedom for the test are equal to the difference between the df for the two models, in this case 124 - 120 = 4. The critical chi-square value with 4 df is 18.465 (p < .001), and so this chi-square difference was significant. This finding shows that the theoretical model was unsuccessful in accounting for the relationships between the latent constructs.

   Combined, these results showed that the initial theoretical model did not provide an acceptable fit to the data. Therefore, a specification search was conducted to arrive at a better-fitting model.

Revised model 1. When conducting a specification search using data based on a relatively small sample (as in the present case, where N = 240), there is a danger that data-driven model modifications will capitalize on chance characteristics of the sample data and result in a final model that will not generalize to the population or to other samples (MacCallum, Roznowski, & Necowitz, 1992). We therefore began the search by attempting to identify parameters that could be dropped from the model without significantly hurting the model's fit, as it is generally safer to drop parameters than to add new parameters when modifying models (Bentler & Chou, 1987).

A Wald test (Bentler, 1989) suggested that it was possible to delete the path from alternative value to commitment without a significant increase in model chi-square. Therefore, this path was deleted, and the resulting model, revised model 1 (M_{r1}) was then estimated. Fit indices for this model appear in Table 6.1. Once again, overall goodness of fit indices for the model were acceptable, with values on the NNFI and CFI in excess of .9. Table 6.2 shows that the path coefficients in revised model 1 were statistically significant.

Dropping the alternatives-commitment path would be acceptable only if it did not result in a significant increase in model chi-square. A significant increase would indicate that M_{r1} provided a fit that was significantly worse than M_t. Therefore, a chi-square difference test was conducted, comparing M_t to M_{r1} (see Table 6.1 for model chi-square values). The chi-square difference for this comparison was equal to 217.58 - 216.75 = 0.83 which, with 1 df, was nonsignificant (p > .05).

Having passed this test, M_{r1} was next compared to M_m to determine whether it successfully accounted for the relationships between the latent constructs. The chi-square difference was calculated as 217.58 - 180.87 = 36.71, which, with 5 df, was statistically significant (p < .001). Once again, the model failed to provide an acceptable fit.

Revised model 2. Wald tests conducted in the course of analyzing M_{r1} did not reveal any additional causal paths between latent constructs that could be deleted without affecting the model's fit. We therefore reviewed results of Lagrange multiplier tests (Bentler, 1989) to identify new causal paths that should be added to the model.

A Lagrange multiplier test estimated that model chi-square for M_{r1} could be reduced by 34.53 if a causal path were added that went from investment size (F5) to satisfaction (F2). Adding such a path would be consistent with the prediction from cognitive dissonance theory (Festinger, 1957) that individuals often adjust their attitudes (i.e., become more satisfied) so that their attitudes will be consistent with their behaviors (i.e., the behavior of investing time and effort in a relationship). Because its addition could be justified on theoretical grounds, a path from investment size to satisfaction was added to M_{r1}. The resulting model, revised model 2 (M_{r2}), was then estimated.

Fit indices for revised model 2 are presented in Table 6.1. It can be seen that the fit indices (i.e., the NNFI and CFI) were not only above .9 but were also higher than those displayed by revised model 1 and the initial theoretical model. A chi-square difference test comparing M_{r2} to M_{r1} revealed a significant difference value of 217.58 – 183.20 = 34.38 (df = 1, p < .001). This finding shows that revised model 2 provided a fit to the data that was significantly better than the fit provided by revised model 1, thus justifying the addition of the new path.

Table 6.2 and Figure 6.8 display standardized path coefficients for revised model 2. It can be seen that all coefficients were significant and in the predicted direction. R^2 values showed that satisfaction and investment size accounted for 55% of the variance in commitment, while rewards, costs, and investment size accounted for 51% of the variance in satisfaction.

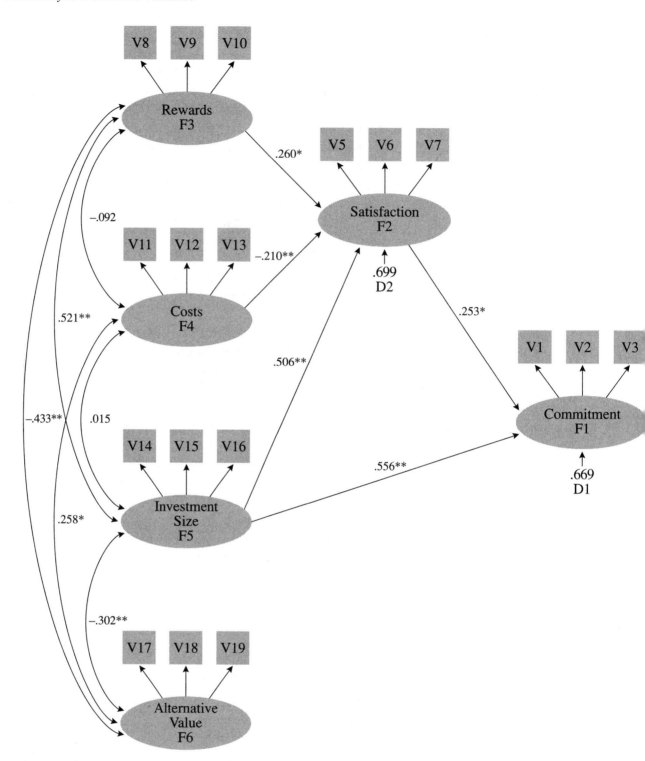

Figure 6.8: Revised Model 2, Investment Model Study: Standardized Path Coefficients Appear on Single-Headed Arrows; Correlations Appear on Curved Double-Headed Arrows, * *p* < .01; ** *p* < .001

The distribution of normalized residuals for revised model 2 was symmetrical and centered on zero. No normalized residuals were greater than 2.0 in absolute magnitude.

Table 6.1 also presents indices that reflect the parsimony of the models that were tested. The parsimony ratio, or PR (James et al., 1982) indicates the parsimony of the overall model, with higher values reflecting greater parsimony. The parsimonious normed-fit index (PNFI, James et al., 1982) is obtained by multiplying the parsimony ratio by the normed-fit index, resulting in a single index that reflects both the parsimony and the fit of the overall model.

These indices show that revised model 2 displayed a parsimony ratio of .810, which was somewhat lower than that of revised model 1, which displayed a PR of .817. However, this was more than offset by the superior fit achieved by M_{r2}, as is demonstrated by the fact that the parsimonious normed-fit index for M_{r2} was .742, while the PNFI for M_{r1} was only .735.

Table 6.1 also provides indices that represent the fit and parsimony in just the structural portion of a model; that is, the part of a model that describes just the relations between the latent variables (the F variables). For example, the relative normed-fit index, or RNFI (Mulaik et al., 1989) reveals the fit achieved in just the structural portion of the model, independent of the fit of the measurement model. In the same way, the relative parsimony ratio (RPR) reveals the parsimony of the structural portion of the model, regardless of the parsimony of the measurement model. Finally, the relative parsimonious fit index (RPFI) is obtained by multiplying the RNFI by the RPR. The RPFI indicates how well the model explains all possible relations among the F variables, from outside the data (Mulaik et al., 1989).

The RNFI indices in Table 6.1 show that revised model 2 demonstrated a fit to the data that was superior to that of revised model 1, due to the addition of the new path from investment size to satisfaction. On the other hand, the addition of this path did hurt the parsimony of revised model 2, as M_{r2} displayed values on both the RPR and RPFI that were slightly lower than those exhibited by the simpler (although poorer-fitting) M_{r1}.

As a final test, a chi-square difference test was used to compare the fit of M_{r2} with that of M_m. This comparison

```
resulted in a difference value of 183.20 - 180.87 = 2.33,
which, with 4 df was nonsignificant (p > .05).  The
nonsignificant chi-square indicated that M_r2 provided a fit
that was not significantly worse than that provided by a
measurement model in which all F varibles were free to
covary.  In other words, this finding showed that the
causal relationships described in revised model 2 were
successful in accounting for the observed relationships
between the latent constructs.
     Combined, these findings generally provide support for
revised model 2 over the other models tested.  M_r2 was
therefore retained as this study's "final" model, and is
displayed in Figure 6.8.  Standardized path coefficients
appear on the causal paths.
```

Finally, somewhere in the discussion section, you should caution that your final model was based on data-driven model modifications, and must therefore be regarded as tentative until the results are replicated in other samples. Providing a caution along these lines is particularly important when working with a sample of small or moderate size.

Additional Examples

Example 2: A Nonstandard Model

An introductory section in Chapter 5 made a distinction between standard and nonstandard latent-variable models. With **standard models**, all constructs that constitute the structural portion of the model (the "structural variables") are represented as F variables with multiple manifest indicators. For example, the causal model represented in Figure 6.8 is a standard model, because all of the structural variables (e.g., Commitment, Rewards) are latent variables with multiple indicators. To this point, Chapters 5 and 6 have focused almost exclusively on the analysis of standard models.

With a **nonstandard model**, on the other hand, at least one of the constructs that constitute the structural portion of the model is represented as a single manifest variable. In other words, this construct is measured by only one manifest variable; it is not measured by multiple manifest variables.

A nonstandard latent-variable model is presented in Figure 6.9. This figure is identical to previous figures portraying the investment model, except that the satisfaction construct is now measured by just one manifest variable (V5), and the rewards construct is measured by just one manifest variable (V8).

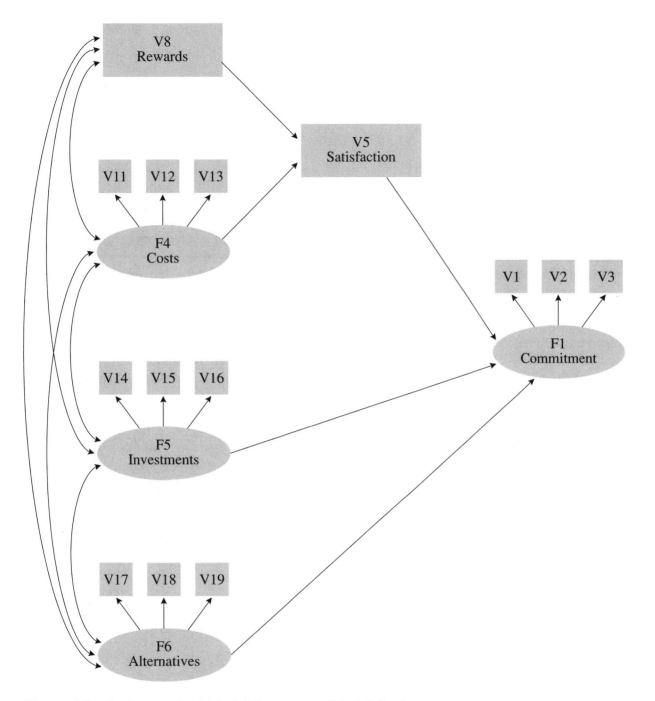

Figure 6.9: A Nonstandard Model, Investment Model Study

Ideally, a construct should be measured by just one manifest variable only when that manifest variable is a perfectly reliable measure of the construct. For example, consider the situation in which one construct included in a causal model is "subject age." It is probably possible to obtain just one measure of subject age that is perfectly reliable, and it is therefore not necessary to have multiple measures of this construct (multiple measures of age would most likely be perfectly correlated with each other). It is this type of situation that is ideally suited for the analysis of nonstandard models.

In contrast, the investment model study discussed in this chapter would probably not be well suited for the analysis of a nonstandard model. If you assume that each of the V variables in Figure 6.9 represent responses to individual items on a questionnaire, the figure shows that the satisfaction and rewards constructs were measured with just one item each, and you know that responses to single items tend to be unreliable. You can therefore expect that any analysis of a nonstandard model that includes V5 and V8 as measures of satisfaction and rewards, respectively, will produce biased parameter estimates. The reasons for this were discussed in Chapter 5 in the section headed "Advantages of Path Analysis with Latent Variables."

Nonetheless, this section will show how to analyze the nonstandard model portrayed in Figure 6.9, in order to illustrate the analysis of nonstandard model. Analyzing the model presented in Figure 6.9 (which should look fairly familar by now) will make it easy to demonstrate how the analysis of a nonstandard model differs from the analysis of a standard model.

The analysis of a nonstandard model should follow the same 2-phase procedure discussed earlier, in which a measurement model is first estimated and refined, and is then modified to represent the theoretical model of interest. Most of the "rules" discussed earlier will continue to apply to nonstandard models. There will be only a few minor changes in the way the program figures and PROC CALIS programs are prepared.

A nonstandard measurement model. Figure 6.10 presents the completed program figure for a confirmatory factor analysis of a nonstandard model. Notice that, in most respects, it is identical to the program figure for the CFA of the standard model that was presented in Figure 6.1. For example, the figure contains six structural variables: Commitment, Satisfaction, Rewards, Costs, Investments, and Alternatives. Each structural variable is connected to every other structural variable by a curved, double-headed arrow. For those structural variables that have multiple indicators, all factor loadings are free parameters to be estimated.

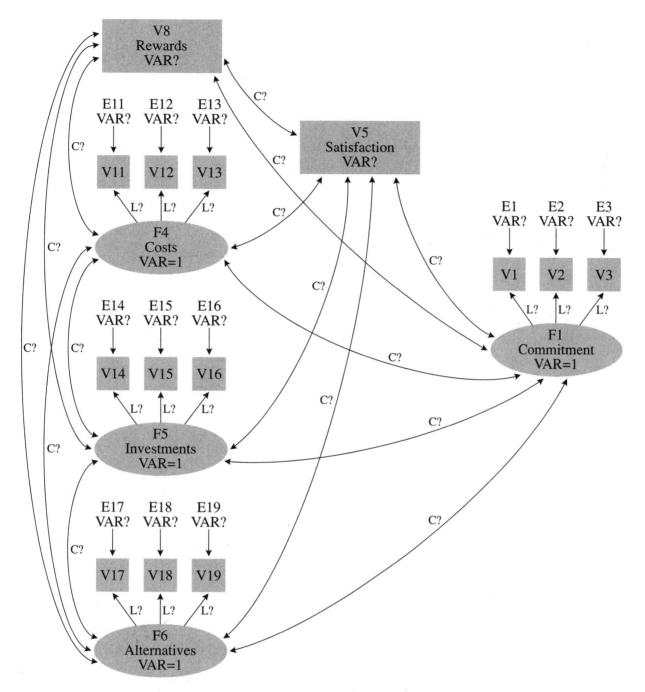

Figure 6.10: Completed Program Figure for Confirmatory Factor Analysis of Nonstandard Measurement Model, Investment Model Study

Finally, variances of all F variables are fixed at 1, consistent with Rule 15. It will be remembered that it was necessary to fix the variances of the F variables at 1 to solve the scale indeterminancy problem and assure that the model would be identified.

Now you come to a way in which the new program figure is different from the old. Notice that the VAR? symbol appears in the rectangles that represent the satisfaction construct and the rewards construct. This means that the variance for V5 (Satisfaction) should be a free parameter to be estimated, and that the variance for V8 (Rewards) should also be a free parameter to be estimated, consistent with the following rule:

Rule 18: In a confirmatory factor analysis of a nonstandard model, the variance of a *manifest structural* variable should be a free parameter to be estimated.

The words *manifest structural* have been placed in italics to avoid confusion. Remember that a structural variable is a variable that will later be part of the structural portion of the theoretical model. Figure 6.10 contains just six structural variables: F1, F4, F5, F6, V5, and V8. Of these, however, only V5 and V8 are *manifest* structural variables (the others are all latent structural variables). Therefore, only the variances V5 and V8 will be estimated.

That is essentially the only difference between the program figure of the CFA for a nonstandard versus a standard measurement model; in all other respects, the program figures are identical. The SAS program (minus the DATA step) to analyze this nonstandard measurement model appears here:

```
1      PROC CALIS  COVARIANCE  CORR  RESIDUAL  MODIFICATION ;
2         LINEQS
3             V1  = LV1F1   F1 + E1,
4             V2  = LV2F1   F1 + E2,
5             V3  = LV3F1   F1 + E3,
6             V11 = LV11F4  F4 + E11,
7             V12 = LV12F4  F4 + E12,
8             V13 = LV13F4  F4 + E13,
9             V14 = LV14F5  F5 + E14,
10            V15 = LV15F5  F5 + E15,
11            V16 = LV16F5  F5 + E16,
12            V17 = LV17F6  F6 + E17,
13            V18 = LV18F6  F6 + E18,
14            V19 = LV19F6  F6 + E19;
15        STD
16            V5  = VARV5,
17            V8  = VARV8,
18            F1  = 1,
19            F4  = 1,
20            F5  = 1,
21            F6  = 1,
22            E1-E3  = VARE1-VARE3,
23            E11-E19 = VARE11-VARE19;
24        COV
25            F1 V5  = CF1V5,
```

```
26                F1 V8 = CF1V8,
27                F1 F4 = CF1F4,
28                F1 F5 = CF1F5,
29                F1 F6 = CF1F6,
30                V5 V8 = CV5V8,
31                V5 F4 = CV5F4,
32                V5 F5 = CV5F5,
33                V5 F6 = CV5F6,
34                V8 F4 = CV8F4,
35                V8 F5 = CV8F5,
36                V8 F6 = CV8F6,
37                F4 F5 = CF4F5,
38                F4 F6 = CF4F6,
39                F5 F6 = CF5F6;
40          VAR  V1-V3 V11-V19;
41          RUN;
```

You can see that this program was written to analyze the nonstandard model of Figure 6.10 by following the same conventions used throughout Chapters 4, 5, and 6 of this text. In many respects, it is identical to the program presented in Chapter 5 to analyze the "standard" measurement model for the investment model study. There are some important differences, however.

For example, the LINEQS statement of the current program no longer contains lines to indicate which manifest varibles are indicators for the F2 latent variable. This is because F2 had represented the satisfaction construct in the earlier standard model, and satisfaction is now represented by V5; there simply *is* no F2 variable in the present nonstandard model. Similarly, there are no LINEQS statements for F3, the rewards construct of the standard model, because rewards are now measured with a single item, V8. LINEQS statements do, however, indicate which V variables serve as indicators for F1, F4, F5, and F6, as these latent variables have been retained in the nonstandard model.

Another important difference appears in lines 16 and 17, which request that the variances of V5 and V8 be estimated. As before, the variances for the F variables are fixed at 1.

In most other respects, the program is identical to the program for the CFA of the standard model, with the exception that "F2" from the earlier program has been replaced with "V5" in the above program. Similarly, "F3" from the earlier program has now been replaced with "V8." For example, notice that, in the COV statement, V5 and V8 are allowed to covary with the F variables in the same way that F2 and F3 had done in the earlier program.

The results created by the preceding program would be interpreted in the same way that you interpreted the results of the CFA of the standard measurement model. If necessary, these results could be used to modify the model in order to attain a better fit.

A nonstandard theoretical model. Figure 6.11 provides the completed program figure for the path analysis of the nonstandard model. Notice that this is a path analysis in which some of the structural variables are latent F variables, and others are manifest V variables.

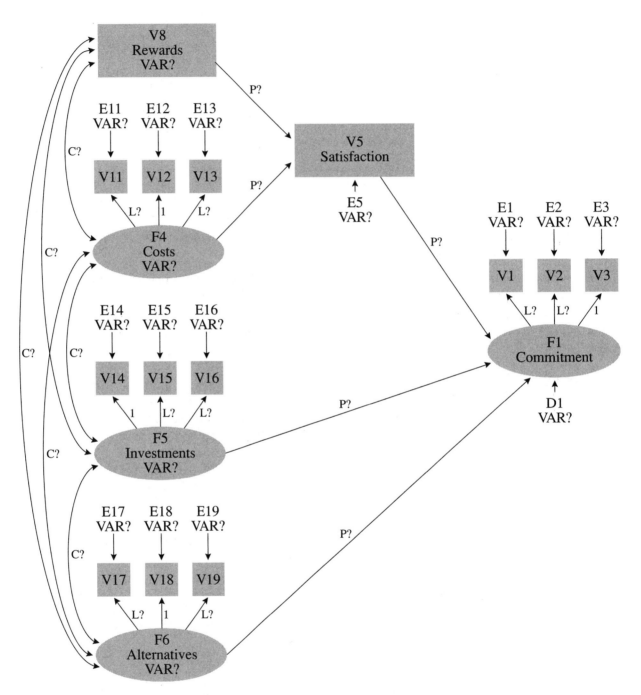

Figure 6.11: Completed Program Figure for Path Analysis of Nonstandard Model, Investment Model Study

No new rules are necessary to guide the preparation of this figure, as the figure was simply developed according to the rules already provided. For example:

- Variances are estimated for all of the exogenous structural variables (V8, F4, F5, and F6), consistent with Rule 4.

- V5 (Satisfaction) is an endogenous variable, so it has been given a disturbance term (E5), just as F1 (Commitment) also has been given a disturbance term (D1). This is consistent with Rule 2.

- V8 is the only V variable that does not have a disturbance term. This is because V8 is the only exogenous V variable (consistent with Rule 3).

- For each of the exogenous F variables (F4, F5, and F6), their variances are free parameters to be estimated, and one factor loading for each F variable is fixed at 1 (consistent with Rules 16 and 17).

The PROC CALIS program to analyze the nonstandard model of Figure 6.11 is presented here:

```
1      PROC CALIS   COVARIANCE   CORR   RESIDUAL   MODIFICATION ;
2         LINEQS
3            V1  = LV1F1   F1 + E1,
4            V2  = LV2F1   F1 + E2,
5            V3  =         F1 + E3,
6            V11 = LV11F4  F4 + E11,
7            V12 =         F4 + E12,
8            V13 = LV13F4  F4 + E13,
9            V14 =         F5 + E14,
10           V15 = LV15F5  F5 + E15,
11           V16 = LV16F5  F5 + E16,
12           V17 = LV17F6  F6 + E17,
13           V18 =         F6 + E18,
14           V19 = LV19F6  F6 + E19,
15           F1  = PF1V5 V5 + PF1F5 F5 + PF1F6 F6 + D1,
16           V5  = PV5V8 V8 + PV5F4 F4              + E5;
17        STD
18           V8      = VARV8,
19           E1-E3   = VARE1-VARE3,
20           E5      = VARE5,
21           E11-E19 = VARE11=VARE19,
22           F4-F6   = VARF4-VARF6,
23           D1      = VARD1;
24        COV
25           V8 F4 = CV8F4,
26           V8 F5 = CV8F5,
27           V8 F6 = CV8F6,
```

```
28              F4 F5 = CF4F5,
29              F4 F6 = CF4F6,
30              F5 F6 = CF5F6;
31          VAR  V1 V2 V3 V5 V8 V11-V19 ;
32          RUN;
```

The results from this program would be interpreted in the usual way. If necessary, the model could be modified to achieve a better fit.

Example 3: A Causal Model Predicting Victim Reactions to Sexual Harassment

This final section of the chapter will very briefly present a new theoretical model to be analyzed. It will also provide the program figures and PROC CALIS programs needed to conduct the analysis. All material needed to prepare the necessary programs has already been provided. The purpose of this section is to quickly apply these concepts to a model other than the investment model. As an informal exercise, you are encouraged to

- study the causal models as they appear in the text's figures

- actually prepare the PROC CALIS program that will analyze the model

- compare your programs against those presented in this text.

Chapter 4 described a fictitious role-playing study in which a large sample of women read a description of a woman who was being sexually harassed on the job. The scenarios that the women read varied with regard to the seriousness of the offense that the woman experienced. After reading the scenario, the subject was asked to imagine how she would feel if she had been the victim of the harassment, and to respond to a questionnaire that assessed the following constructs:

- The subject's **intention to report** the harassment to a higher level manager at the organization. With this variable, higher scores indicate greater intention to report.

- The **expected outcomes** of reporting the harassment, where higher scores indicate stronger belief that reporting the harassment will result in positive results for the victim.

- **Feminist ideology**, where higher scores indicate pro-feminist, egalitarian attitudes about sex roles.

- **Seriousness of the offense**, where higher scores reveal stronger belief that the woman in the scenario experienced a serious form of harassment.

- **Normative expectations**, where higher scores reflect stronger belief that the victim's family, friends, and coworkers would support her if she reported the harassment.

Figure 6.12 presents a theoretical model that describes predicted relationships between these five constructs. According to the model, intention to report is expected to be directly affected by expected outcomes, feminist ideology, and normative expectations. Expected outcomes, in turn, is expected to be affected by all three of the study's exogenous variables. (Remember that this model was constructed for purposes of illustration, and should not necessarily be regarded as a serious model of victim reactions to sexual harassment.)

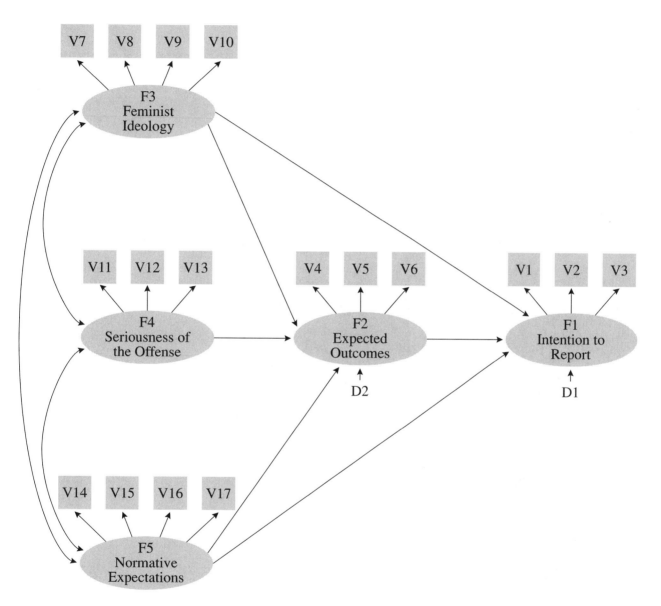

Figure 6.12: Initial Theoretical Model, Sexual Harassment Study

The model in Figure 6.12 is a standard latent variable model, in that all five constructs are represented as F variables with multiple indicators. Assume that the V variables are responses to individual items on a questionnaire.

In the first phase of testing this model, a confirmatory factor analysis would be performed to develop a satisfactory measurement model for the study. The program figure for this CFA is presented in Figure 6.13. Notice that all structural variables in the model (F variables, in this case) are allowed to freely covary with one another, in the usual way.

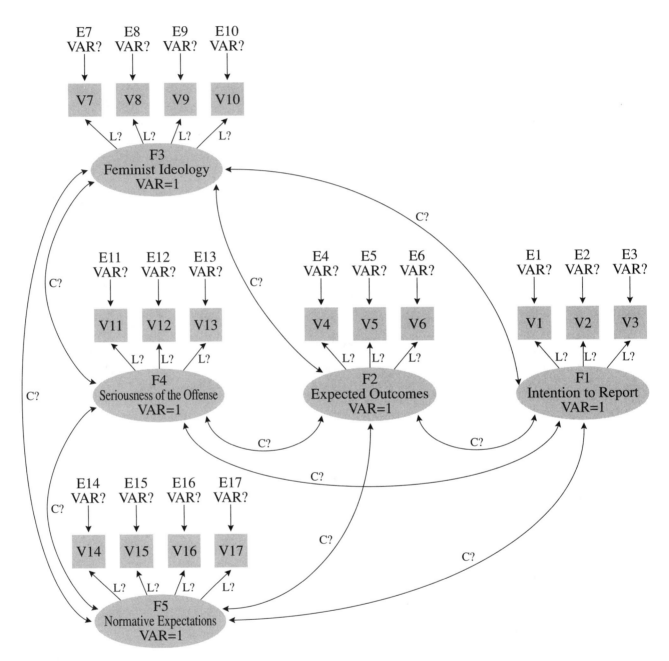

Figure 6.13: Program Figure for Confirmatory Factor Analysis of Measurement Model, Sexual Harassment Study

Here is the PROC CALIS program that would estimate the measurement model of Figure 6.13:

```
 1     PROC CALIS   COVARIANCE   CORR   RESIDUAL   MODIFICATION ;
 2        LINEQS
 3           V1  = LV1F1   F1 + E1,
 4           V2  = LV2F1   F1 + E2,
 5           V3  = LV3F1   F1 + E3,
 6           V4  = LV4F2   F2 + E4,
 7           V5  = LV5F2   F2 + E5,
 8           V6  = LV6F2   F2 + E6,
 9           V7  = LV7F3   F3 + E7,
10           V8  = LV8F3   F3 + E8,
11           V9  = LV9F3   F3 + E9,
12           V10 = LV10F3  F3 + E10,
13           V11 = LV11F4  F4 + E11,
14           V12 = LV12F4  F4 + E12,
15           V13 = LV13F4  F4 + E13,
16           V14 = LV14F5  F5 + E14,
17           V15 = LV15F5  F5 + E15,
18           V16 = LV16F5  F5 + E16,
19           V17 = LV17F5  F5 + E17;
20        STD
21           E1-E17 = VARE1-VARE17,
22           F1 = 1,
23           F2 = 1,
24           F3 = 1,
25           F4 = 1,
26           F5 = 1;
27        COV
28           F1 F2 = CF1F2,
29           F1 F3 = CF1F3,
30           F1 F4 = CF1F4,
31           F1 F5 = CF1F5,
32           F2 F3 = CF2F3,
33           F2 F4 = CF2F4,
34           F2 F5 = CF2F5,
35           F3 F4 = CF3F4,
36           F3 F5 = CF3F5,
37           F4 F5 = CF4F5;
38        VAR  V1-V17 ;
39        RUN;
```

Assume that the initial measurement model did not provide an acceptable fit, and that the indicator V17 had to be dropped from F5 (normative expectations), in order to achieve the desired fit. Having developed an adequate measurement model, the study now moves to step 2, and the estimation of the theoretical model. Figure 6.14 displays the completed program figure for the estimation of the theoretical model.

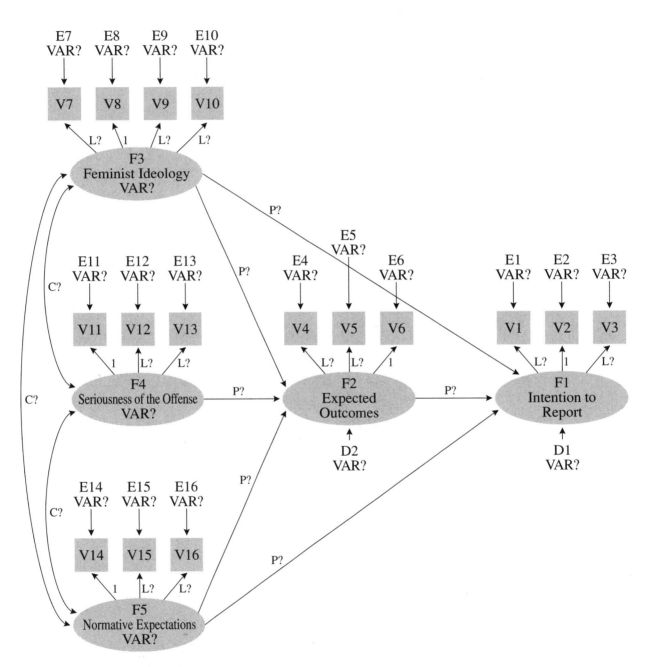

Figure 6.14: Program Figure for Path Analysis with Latent Variables, Sexual Harassment Study

The PROC CALIS program for estimating this model appears here:

```
1      PROC CALIS   COVARIANCE   CORR   RESIDUAL   MODIFICATION ;
2          LINEQS
3              V1  = LV1F1   F1 + E1,
4              V2  =         F1 + E2,
5              V3  = LV3F1   F1 + E3,
6              V4  = LV4F2   F2 + E4,
7              V5  = LV5F2   F2 + E5,
8              V6  =         F2 + E6,
9              V7  = LV7F3   F3 + E7,
10             V8  =         F3 + E8,
11             V9  = LV9F3   F3 + E9,
12             V10 = LV10F3 F3 + E10,
13             V11 =         F4 + E11,
14             V12 = LV12F4 F4 + E12,
15             V13 = LV13F4 F4 + E13,
16             V14 =         F5 + E14,
17             V15 = LV15F5 F5 + E15,
18             V16 = LV16F5 F5 + E16,
19             F1  = PF1F2 F2 + PF1F3 F3 + PF1F5 F5 + D1,
20             F2  = PF2F3 F3 + PF2F4 F4 + PF2F5 F5 + D2;
21         STD
22             E1-E16 = VARE1-VARE16,
23             F3 = VARF3,
24             F4 = VARF4,
25             F5 = VARF5,
26             D1 = VARD1,
27             D2 = VARD2;
28         COV
29             F3  F4 = CF3F4,
30             F3  F5 = CF3F5,
31             F4  F5 = CF4F5;
32         VAR   V1-V16 ;
33         RUN;
```

Notice that the preceding program adheres to the usual conventions for testing a standard theoretical model (e.g., the variances of the exogenous F variables are now free parameters to be estimated; one factor loading for each F variable has been fixed at 1). If a data set were actually analyzed using this program, the results would be interpreted according to the same guidelines presented earlier in this chapter.

Conclusion: Learning More about Latent Variable Models...

The chapter on path analysis with manifest variables stated that the best way to learn about that procedure was to practice performing the procedure. The same is true regarding path analysis with latent variables. Your next step should be to locate books or articles on latent variable models that provide both the correlation or covariance matrix that was analyzed, along with the results of the analysis. Reanalyze the data sets you find using PROC CALIS; if you obtain the same results reported by the authors, you have performed the analysis correctly (assuming the original researchers conducted their analyses correctly!).

This chapter has introduced some basic concepts in the analysis of latent-variable causal models, but has been somewhat narrow in focusing only on recursive models with cross-sectional data. You should now be ready to move on to more complex causal models such as those with reciprocal causation, feedback loops, and time-series designs. The following references should be useful in learning about these more complex topics: Bollen (1989); Fuller (1987); James et al. (1982); Kenny (1979); Loehlin (1987); Long (1983a, 1983b); Saris and Stronkhorst (1984); and Wiley (1973).

References

Anderson, J.C. & Gerbing, D.W. (1988). Structural equation modeling in practice: A review and recommended two-step approach. *Psychological Bulletin, 103,* 411-423.

Bentler, P.M. (1989). *EQS structural equations program manual.* Los Angeles: BMDP Statistical Software.

Bentler, P.M. & Bonett, D.G. (1980). Significance tests and goodness-of-fit in the analysis of covariance structures. *Psychological Bulletin, 88,* 588-606.

Bentler, P.M. & Chou, C. (1987). Practical issues in structural modeling. *Sociological Methods & Research, 16,* 78-117.

Bollen, K. A. (1989). *Structural equations with latent variables.* New York: John Wiley & Sons.

Festinger, L. (1957). *A theory of cognitive dissonance.* Stanford, CA: Stanford University Press.

Fornell, C. & Larcker, D.F. (1981). Evaluating structural equation models with unobservable variables and measurement error. *Journal of Marketing Research, 18,* 39-50.

Fuller, W.A. (1987). *Measurement error models.* New York: John Wiley & Sons.

James, L.R., Mulaik, S.A., & Brett, J.M. (1982). *Causal analysis.* Beverly Hills: Sage.

Joreskog, K.G. & Sorbom, D. (1989). *LISREL 7: A guide to the program and applications, 2nd edition.* Chicago: SPSS Inc.

Kenny, D.A. (1979). *Correlation and causality.* New York: John Wiley & Sons.

Loehlin, J.C. (1987). *Latent variable models.* Hillsdale, NJ: Lawrence Erlbaum Associates.

Long, J.S. (1983a). *Confirmatory factor analysis: A preface to LISREL.* Sage University Paper Series on Quantitative Application in the Social Sciences, 07-033. Beverly Hills: Sage.

Long, J.S. (1983b). *Covariance structure models: An introduction to LISREL.* Sage University Paper Series on Quantitative Application in the Social Sciences, 07-034. Beverly Hills: Sage.

MacCallum, R.C., Roznowski, M., & Necowitz, L. B. (1992). Model modifications in covariance structure analysis: The problem of capitalization on chance. *Psychological Bulletin, 111,* 490-504.

Mulaik, S.A., James, L.R., Van Alstine, J., Bennett, N., Lind., S., & Stilwell, C.D. (1989). Evaluation of goodness-of-fit indices for structural equation models. *Psychological Bulletin, 105,* 430-445.

Netemeyer, R.G., Johnston, M.W., & Burton, S. (1990). Analysis of role conflict and role ambiguity in a structural equations framework. *Journal of Applied Psychology, 75,* 148-157.

Rusbult, C.E. (1980). Commitment and satisfaction in romantic associations: A test of the investment model. *Journal of Experimental Social Psychology, 16,* 172-186.

Saris, W. & Stronkhorst, H. (1984). *Causal modeling in nonexperimental research.* Amsterdam, the Netherlands: Sociometric Research Foundation.

Wiley, D.E. (1973). "The identification problem for structural equation models with unmeasured variables," in Goldberger, A.S. and Duncan, O.D., eds., *Structural equation models in the social sciences.* New York: Academic Press.

Williams, L.J. & Hazer, J.T. (1986). Antecedents and consequences of satisfaction and commitment in turnover models: A reanalysis using latent variable structural equation methods. *Journal of Applied Psychology, 71,* 219-231.

Appendix A.1

INTRODUCTION TO SAS® PROGRAMS, SAS® LOGS, AND SAS® OUTPUT

> **Overview**. This appendix describes the three types of files that you will work with while using the SAS System: the SAS program, the SAS log, and the SAS output file. It presents a very simple SAS program, along with the log and output files produced by that program. This chapter provides the big picture regarding the steps you need to follow when performing data analyses with the SAS System.

Introduction: What is the SAS System?

The SAS System is a modular, integrated, and hardware-independent system of software. It is a particularly powerful tool for social scientists because it allows them to easily perform virtually *any* type of statistical analysis that may be required in the course of conducting social science research. The SAS System is comprehensive enough to perform the most sophisticated multivariate analyses, but is so easy to use that even undergraduates can perform simple analyses after only a short period of instruction.

In a sense, the SAS System may be viewed as a library of prewritten statistical algorithms. By submitting a short SAS program, you can access a prewritten procedure from the library and use it to analyze a set of data. For example, here are the SAS statements used to call up the algorithm that calculates Pearson correlation coefficients:

```
PROC CORR    DATA=D1;
   RUN;
```

The preceding statements will cause the SAS System to compute the Pearson correlation between all numeric variables in your data set. Being able to call up complex procedures with such a simple statement is what makes this system so powerful and so easy to use. By contrast, if you had to prepare your own programs to compute Pearson correlations by using a programming language such as FORTRAN or BASIC, it would require many statements, and there would be many opportunities for error. By using the SAS System instead, most of the work has already been completed, and you are able to focus on the *results* of the analysis rather than on the *mechanics* of obtaining those results.

Where is the SAS System installed? SAS Institute's computer software products are installed at over 25,000 sites in 112 countries. Approximately two-thirds of the installations are in business locations, 15% are government sites, and 18% are education sites. It is estimated that the Institute's software products are used by over three million people worldwide.

Three Types of SAS System Files

Subsequent appendices of this manual will provide details on how to write a SAS program: how to handle the DATA step, how to request specific statistical procedures, and so forth. This appendix however, simply presents a short SAS program and will discuss the output that is created by the program. Little elaboration will be offered. The purpose of this appendix is to provide a very general sense of what it is to submit a SAS program and interpret the results. You are encouraged to copy the program that appears in the following example, submit it for analysis, and verify that the resulting output matches the output reproduced here. This exercise will provide you with the big picture of what it is use the SAS System, and this perspective will facilitate learning the programming details presented later.

Briefly, you will work with three types of "files" when using the SAS System: One file contains the SAS program, one contains the SAS log, and one contains the SAS output. The following sections discuss the differences between these files.

The SAS Program

A SAS program consists of a set of statements written by the researcher or programmer. These statements provide the SAS System with the data to be analyzed, tell it about the nature of these data, and indicate which statistical analyses should be performed on the data.

With older computer systems, these statements were punched onto cards, usually with one statement per card. The cards were then read by a computer and the program was executed. Today, however, the use of *physical* cards has largely been phased out. With modern computer systems, the statements are almost always now typed as *card images* (or *data lines*) in a file in the computer's memory.

This section illustrates a simple SAS program by analyzing some fictitious data from a fictitious study. Assume that six high school students have taken the *Scholastic Assessment Test* (SAT). This test provides two scores for each student: a score on the SAT verbal test, and a score on the SAT math test. With both tests, scores may range from 200 to 800, with higher scores indicating higher levels of aptitude.

Assume that you now wish to obtain some simple descriptive statistics regarding the six students' scores on these two tests. For example, what is their *average* score on the SAT verbal test? On the SAT math test? What is the *standard deviation* of the scores on the two tests?

To perform these analyses, you prepare the following SAS program:

```
                    DATA D1;
                    INPUT    SUBJECT   SATV   SATM;
                    CARDS;
                    1 520 490
DATA step           2 610 590
                    3 470 450
                    4 410 390
                    5 510 460
                    6 580 350
                    ;
                    PROC MEANS    DATA=D1;
PROC step              VAR   SATV   SATM;
                       RUN;
```

The preceding shows that a SAS program consists of two parts: a **DATA step**, which is used to read data and create a SAS data set, and a **PROC step**, which is used to process or analyze the data. The differences between these steps are described in the next two sections.

The DATA step. In the DATA step, programming statements create and/or modify a SAS data set. Among other things, these statements may

- provide a name for the data set
- provide a name for the variables to be included in the data set
- provide the actual data to be analyzed.

In the preceding program, the DATA step begins with the DATA statement, and ends with the semicolon that immediately precedes the PROC MEANS statement.

The first statement of the preceding program begins with the word DATA, and specifies that the SAS System should create a data set to be called D1. The next line contains the INPUT statement, which indicates that three variables will be contained in this data set. The first variable will be called SUBJECT, and this variable will simply provide a subject number for each student. The second variable will be called SATV (for the SAT verbal test), and the third will be called SATM (for the SAT math test).

The CARDS statement indicates that card images containing your data will appear on the following lines. The first line after the CARDS statement contains the data (test scores) for subject 1. You can see that this first data line contains the numbers 520 and 490, meaning that subject 1 received a score of 520 on the SAT verbal test and a score of 490 on the SAT math test. The next data line shows that subject 2 received a score of 610 for the SAT verbal and a score of 590 for the SAT math, and so forth. The semicolon after the last data line signals the end of the data.

The PROC step. In contrast to the DATA step, the PROC step includes programming statements that request specific statistical analyses of the data. For example, the PROC step might request that correlations be performed between all quantitative variables, or might request that a *t* test be performed. In the preceding example, the PROC step consists of the last three lines of the program.

The first line after the DATA step is the PROC MEANS statement. This requests that the SAS System use a procedure called MEANS to analyze the data. The MEANS procedure computes means, standard deviations, and some other descriptive statistics for numeric variables in the data set. Immediately after the words PROC MEANS are the words DATA=D1. This tells the system that the data to be analyzed are in a data set named D1 (remember that D1 is the name of the data set just created).

Following the PROC MEANS statement is the VAR statement, which includes the names of two variables: SATV and SATM. This requests that the descriptive statistics be performed on just SATV (SAT verbal test scores) and SATM (SAT math test scores).

Finally, the last line of the program is the RUN statement that signals the end of the PROC step. If a SAS program requests multiple PROCs (procedures), you have two options for using the RUN statement:

- You may place a separate RUN statement following each PROC statement.

- You may place a single RUN statement following the last PROC statement.

What is the single most common programming error? For new users of the SAS System, the single most common error probably involves leaving off a required semicolon (;). Remember that every SAS statement must end with a semicolon (in the preceding program, notice that the DATA statement ends with a semicolon, as does the INPUT statement, the CARDS statement, the PROC MEANS statement, and the RUN statement). When you obtain an error in running a SAS program, one of your first things you should do is look over the program for missing semicolons.

The preceding program is a complete SAS program. However, with a few computer systems it may be necessary to add JCL (job control language) lines at the beginning and end of the SAS program. Different JCL is required by different systems, and the staff of your computer facility can show you what JCL you must use with your SAS programs. As an illustration, the preceding program appears here with JCL added:

```
//SMITH JOB IPSYC,SMITH
// EXEC SAS
DATA D1;
    INPUT  SUBJECT  SATV  SATM;
CARDS;
1 520 490
2 610 590
3 470 450
4 410 390
5 510 460
6 580 350
;
PROC MEANS   DATA=D1;
   VAR  SATV  SATM;
   RUN;
/*EOF
```

The first two lines and the last line of this program are examples of JCL. Remember that if JCL is required with your system, only the staff at your computer facility can show you the exact JCL you will need.

What editor will I use to write my SAS program? An **editor** is an application program that allows you to create lines of text such as the lines that constitute a SAS program. If you are working on a mainframe or midrange computer system, you may have a variety of editors that you can use to write your SAS programs; simply ask the appropriate staff at your computer facility.

For many users, it is wise to use the SAS Display Manager to edit SAS programs. The **SAS Display Manager** is an integrated system that allows users to create and edit SAS programs, submit them for interactive analysis, and view the results on their screens. This application is available at most locations where the SAS System is installed (including personal computers).

Once you have submitted the preceding program for analysis, the SAS System will create two types of files reporting the results of the analysis. One file is called the **SAS log**, or **log file** in this text. This file contains notes, warnings, error messages, and other information related to the execution of the SAS program. The other file is referred to as the **SAS output file**. The SAS output file contains the results of the statistical analyses requested in the SAS program.

The SAS log and output files are created in different ways on different computer systems. With personal computers, the log and output files usually appear in immediate memory, and may be printed or saved as permanent files on a hard drive or floppy disk if the user desires. With some mainframe and midrange systems, the log and output files are created by the system as separate computer files in the researcher's computer account. Users must then open these files to review the results of their analyses. With still other systems, the files are printed on paper as soon as the SAS program is executed.

The SAS Log

The SAS log is a listing of notes and messages that will help you verify that your SAS program was executed successfully. Specifically, the log provides:

- a reprinting of the SAS program that was submitted

- a listing of notes indicating how many variables and observations are contained in the data set

- a listing of any errors made in writing the SAS program.

Log A.1.1 provides a reproduction of the SAS log for the preceding program:

```
16:10 Monday, April 11, 1994

NOTE: Copyright(c) 1989 by SAS Institute Inc., Cary, NC USA.
NOTE: SAS (r) Proprietary Software Release 6.07   TS304
      Licensed to WINTHROP UNIVERSITY, Site 0008647001.

NOTE: Running on VAX Model 6000-510 Serial Number 12000003.


Welcome to the new SAS System, Release 6.07.

This message is seen by users when the NEWS option is specified.
You can replace this message with your own by editing the NEWS file.

Changes and enhancements available in SAS Release 6.07 are documented
in the online Host Help.

1             OPTIONS   LINESIZE=80   PAGESIZE=60    NODATE;
2
3             DATA D1;
4                INPUT   SUBJECT   SATV   SATM;
5             CARDS;

NOTE: The data set WORK.D1 has 6 observations and 3 variables.

12              ;
13            PROC MEANS    DATA=D1;
14               VAR SATV SATM;
15               RUN;

NOTE: The PROCEDURE MEANS printed page 1.

NOTE: SAS Institute Inc., SAS Campus Drive, Cary, NC USA 27513-2414
```

Log A.1.1: SAS Log for the Preceding Program

Notice that the statements constituting the SAS program are assigned line numbers, and are reproduced in the SAS log. The data lines are not normally reproduced as part of the SAS log unless they are specifically requested.

About halfway down the log, a note indicates that your data set contains 6 observations and 3 variables. You would check this note to verify that the data set contains all of the variables that you intended to input (in this case 3), and that it contains data from all of your subjects (in this case 6). So far, everything appears to be correct.

If you had made any errors in writing the SAS program, there also would have been ERROR messages in the SAS log. Often, these error messages provide you with some help in determining what was wrong with the program. For example, a message may indicate that SAS was expecting a program statement which was not included. Whenever you encounter an error message, read it carefully and review all of the program statements that preceded it. Often, the error appears in the program statements which immediately precede the error message, but in other cases the error may be hidden much earlier in the program.

If more than one error message is listed, do not panic; there still may be only one error. Sometimes a single error will cause a large number of error messages.

Once the error or errors have been identified, you must revise the original SAS program and resubmit it for analysis. After processing is complete, review the new SAS log to see if the errors have been eliminated. If the log indicates that the program ran correctly, you are free to review the results of the analyses in the SAS output file.

The SAS Output File

The SAS output file contains the results of the statistical analyses requested in the SAS program. Because the program in the previous example requested the MEANS procedure, the corresponding output file will contain means and other descriptive statistics for the variables analyzed. In this text, the SAS output file is sometimes referred to as the **lis file**. "Lis" is used as an abbreviation for "listing of results."

The following is a reproduction of the SAS output file that would be produced from the preceding SAS program:

```
    Variable  N        Mean       Std Dev      Minimum      Maximum
    -------------------------------------------------------------------
    SATV      6   516.6666667   72.5718035   410.0000000   610.0000000
    SATM      6   455.0000000   83.3666600   350.0000000   590.0000000
    -------------------------------------------------------------------
```

Output A.1.1: Results of the MEANS Procedure

Below the heading "Variable", the SAS System prints the names of each of the variables being analyzed. In this case, the variables are called SATV and SATM. To the right of the heading SATV, descriptive statistics for the SAT verbal test may be found. Figures for the SAT math test appear to the right of SATM.

Below the heading "N", the number of cases analyzed is reported. The average score on each variable is reproduced under "Mean", and standard deviations appear in the column headed "Std Dev". Minimum and maximum scores for the two variables appear in the remaining two columns. You can see that the mean score on the SAT verbal test was 516.67, and the standard deviation of these scores was 72.57. For the SAT math test, the mean was 455.00, and the standard deviation was 83.37.

The statistics included in the preceding output are printed by default. In Appendix A.4 you will learn that there are many additional statistics that you can request as options with PROC MEANS.

Conclusion

Regardless of the computing environment in which you work, the basics of using the SAS System remain the same: you prepare the SAS program, submit it for analysis, and review the resulting log and output files to gain insight into your data. This appendix has provided a quick overview of how the system is used; for more information on the fundamentals of creating SAS data sets, refer to Appendix A.2, "Data Input."

Appendix A.2

DATA INPUT

Overview. This appendix shows how to create a SAS data set. A SAS data set contains the information (independent variables, dependent variables, survey responses, etc.) that is analyzed using SAS procedures such as PROC MEANS or PROC GLM. The appendix begins with a simple illustrative example in which a SAS data set is created using the CARDS statement. In subsequent sections, additional guidelines show how to input the different types of data that are most frequently encountered in social science research.

Introduction: Inputting Questionnaire Data
versus Other Types of Data

This appendix shows how to create SAS data sets in a number of different ways, and it does this by illustrating how to input the type of data that are often obtained through **questionnaire research**. This type of research generally involves distributing structured questionnaires to a sample of subjects, and asking them to respond by circling or checking fixed responses. For example, subjects may be asked to indicate the extent to which they agree or disagree with a set of items by checking a 7-point scale in which 1 represents "strongly disagree", and 7 represents "strongly agree".

Because this appendix (and much of the entire text, for that matter) focuses on questionnaire research, some readers may be concerned that it will not be useful for analyzing data that are obtained using different methods. This concern is understandable, because the social sciences are so diverse, and so many different *types* of variables are investigated in social science research. These variables might be as different as "the number of aggressive acts performed by a child", "rated preferences for laundry detergents", or "levels of serotonin in the frontal lobes of chimpanzees".

However, because of the generality and flexibility of the basic principles of this discussion, you can expect, upon completing this appendix, to be prepared to input virtually any type of data obtained in social science research. The same may be said for the remaining appendices of this text: although this book emphasizes the analysis of questionnaire data, the concepts taught here can be readily applied to different types of data. This fact should become clear as the mechanics of using the SAS System are presented.

This text emphasizes the analysis of questionnaire data for two reasons. First, for better or for worse, many social scientists rely on questionnaire data in conducting their research. By focusing on this method, this text provides examples that will be meaningful to the single largest subgroup of readers. Second, questionnaire data often creates special input and analysis problems that are not generally encountered with other research methods (e.g., large numbers of variables, "check all that apply" variables). This text addresses some of the more common of these difficulties.

Keying Data: An Illustrative Example

Before data can be input and analyzed by the SAS System, they must be keyed in some systematic way. There are a number of different approaches to keying data, but to keep things simple, this appendix presents only the fixed format approach. With the **fixed format** method, each variable is assigned to a specific column (or set of columns) in the data set. The fixed format method has the advantage of being very general: You can use it for almost any type of research problem. An additional advantage is that researchers are probably less likely to make errors in inputting data if they stick to this format.

In the following example of how to key data, you will actually key some fictitious data from a fictitious study. Assume that you have developed a survey to measure attitudes toward *volunteerism*. A copy of the survey appears here:

Volunteerism Survey

Please indicate the extent to which you agree or disagree with each of the following statements. You will do this by circling the appropriate number to the left of that statement. The following format shows what each response number stands for:

        5 = Agree Strongly
        4 = Agree Somewhat
        3 = Neither Agree nor Disagree
        2 = Disagree Somewhat
        1 = Disagree Strongly

For example, if you "Disagree Strongly" with the first question, circle the "1" to the left of that statement. If you "Agree Somewhat," circle the "4," and so on.

```
-------------
 Circle Your
   Response
-------------
```

1 2 3 4 5   1. I feel a personal responsibility to help needy
                        people in my community.

1 2 3 4 5   2. I feel I am personally obligated to help homeless
                        families.

1 2 3 4 5   3. I feel no personal responsibility to work with poor
                        people in my town.

1 2 3 4 5   4. Most of the people in my community are willing to
                        help the needy.

1 2 3 4 5   5. A lot of people around here are willing to help
                        homeless families.

1 2 3 4 5   6. The people in my town feel no personal
                        responsibility to work with poor people.

1 2 3 4 5   7. Everyone should feel the responsibility to perform
                        volunteer work in their communities.

What is your age in years? _____

Further assume that you administer this survey to 10 subjects. For each of these individuals, you also obtain their IQ scores.

You then key your data as a file in a computer. All of the survey responses and information about subject 1 appear on the first line of this file. All of the survey responses and information about subject 2 appear on the second line of this file, and so forth. You keep the data lined up properly, so that responses to question 1 appear in column 1 for *all* subjects, responses to question 2 appear in column 2 for all subjects, and so forth. When you key in this way, your data set should look like this:

```
2234243 22  98   1
3424325 20 105   2
3242424 32  90   3
3242323  9 119   4
3232143  8 101   5
3242242 24 104   6
4343525 16 110   7
3232324 12  95   8
1322424 41  85   9
5433224 19 107  10
```

You can think of the preceding as a matrix of data consisting of 10 rows and 17 columns. The rows run horizontally (from left to right), and each row represents data for a different subject. The columns run vertically (up and down). For the most part, a given column represents a different variable that you measured or created (although, in some cases, a given variable is more than one column wide; more on this later).

For example, look at the last column in the matrix: the vertical column on the right side which goes from 1 (at the top) to 10 (at the bottom). This column codes the Subject Number variable. In other words, this variable simply tells us *which subject's* data are included on that line. For the top line, the value of Subject Number is 1, so you know that the top line includes data for subject 1; the second line down has the value 2 in the subject number column, so this second line down includes data for subject number 2, and so forth.

The *first* column of data includes subject responses to survey question 1. It can be seen that subject 1 circled a "2" in response to this item, while subject 2 circled a "3." The *second* column of data includes subjects responses to survey question 2, the *third* column codes question 3, and so forth. After keying responses to question 7, you left column 8 blank. Then, in columns 9 and 10, you keyed in each subjects' age. You can see that subject 1 was 22 years old, while subject 2 was 20 years old. You left column 11 blank, and then keyed the subjects' IQs in columns 12, 13, and 14 (IQ can be a 3-digit number, so it required three columns to key it). You left column 15 blank, and keyed subject numbers in columns 16 and 17.

The following table presents a brief coding guide to summarize how you keyed your data:

Column	Variable Name	Explanation
1	V1	Responses to survey question 1
2	V2	Responses to survey question 2
3	V3	Responses to survey question 3
4	V4	Responses to survey question 4
5	V5	Responses to survey question 5
6	V6	Responses to survey question 6
7	V7	Responses to survey question 7
8	(blank)	(blank)
9-10	AGE	Subject's age in years
11	(blank)	(blank)
12-14	IQ	Subject's IQ score
15	(blank)	(blank)
16-17	NUMBER	Subject's number

Guides similar to this will be used throughout this text to explain how data sets are arranged, so a few words of explanation are in order. This table simply identifies the specific columns in which variable values are keyed. For example, the first line of the preceding table indicates that in column 1 of the data set, the values of a variable called V1 are stored, and this variable includes responses to question 1. The next line shows that in column 2, the values of variable V2 are stored, and this variable includes responses to question 2. The remaining lines of the guide are interpreted in the same way. You can see, therefore, that it is necessary to read down the lines of this table to learn what is in each column of the data set.

A few important notes about how you should key data that you will analyze using SAS:

- **Make sure that you key the variables in the correct column**. For example, make sure that the data are lined up so that responses to question 6 always appear in column 6. If a subject happened to leave question 6 blank, then you should leave column 6 blank when you are keying your data (leave this column blank by simply striking the space bar on your keyboard). Then go on to type the subject's response to question 7 in column 7. Do *not* key a zero if the subject didn't answer a question; simply leave the space blank.

  It is also acceptable to key a period (.) instead of a blank space to represent missing data. When using this convention, if a subject has a missing value on a variable, key a single period in place of that missing value. If this variable happens to be more than one column wide, you should still key just one period. For example, if the variable occupies columns 12-14 (as does IQ in the table), key just one period in column 14; do not key three periods in columns 12, 13, and 14.

- **Right-justify numeric data**. You should shove over numeric variables to the right side of the columns in which they appear. For example, IQ is a 3-digit variable (it could assume values such as 112 or 150). However, the IQ score for many individuals is a 2-digit number (such as 99 or 87). Therefore, the 2-digit IQ scores should appear to the right side of this 3-digit column of values. A correct example of how to right-justify your data follows:

```
 99
109
100
 87
118
```

The following is *not* right-justified, and so is less preferable:

```
99
109
100
87
118
```

There are exceptions to this rule. For example, if numeric data contain decimal points, it is generally preferable to align the decimal points when keying the data, so that the decimal points appear in the same column. If there are no values to the right of the decimal point for a given subject, you may key zeros to the right of the decimal point. Here is an example of this approach:

```
  3.450
 12.000
  0.133
144.751
  0.000
```

The preceding data set includes scores for five subjects on just one variable. Assume that scores on this variable could range from 0.000 to 200.000. Subject 1 had a score of 3.45, subject 2 had a score of 12, and so forth. Notice that the scores have been keyed so that the decimal points are aligned in the same vertical column.

Notice also that, if a given subject's score does not include any digits to the right of the decimal point, zeros have been added. For example, subject 2 has a score of 12. However, this subject's score has been keyed as 12.000 so that it is easier to line it up with the other scores.

Technically, it is not always necessary to align subject data in this way in order to include it in a SAS data set. However, arranging data in such an orderly fashion generally decreases the likelihood of making an error when entering the data.

- **Left-justify character data**. Remember that character variables may include letters of the alphabet. In contrast to numeric variables, you should typically left-justify character variables. This means that you shoved over entries to the left, rather than the right.

  For example, imagine that you are going to key two character variables for each subject. The first variable will be called FIRST, and this variable will include each subect's first name. You will key this variable in columns 1–15. The second variable will be called LAST, and will include each subject's LAST name. You will key this variable in columns 16–25. Data for five subjects are reproduced here:

  ```
  Francis        Smith
  John           Wolf
  Jean           Adams
  Alice          OConnel
  ```

  The preceding shows that the first subject was named Francis Smith, the second was named John Wolf, and so forth. Notice that the value "Francis" has been shoved over to the left side of the columns that include the FIRST variable (columns 1–15) The same is true for "John", as well as the remaining first names. In the same way, "Smith" has been shoved over to the left side of the columns that include the LAST variable (columns 16–25). The same is true for the remaining last names.

- **The use of blank columns can be helpful but is not necessary**. Recall that, when you keyed your data, you left a blank column between V7 and the AGE variable, and another blank column between AGE and IQ. Leaving blank columns between variables can be helpful because it makes it easier for you to look at your data and see if something is keyed out of place. However, leaving blank columns is not necessary for the SAS System to accurately read your data, so this approach is optional.

Inputting Data Using the CARDS Statement

Now that you know how to key your data, you are ready to learn about the SAS statements that actually allow the computer to *read* the data and put them into a SAS data set. There are a variety of ways that you can input data, but this text focuses on only two: the use of the **CARDS statement** (which allows you to include the data lines within the SAS program itself), and the use of the **INFILE statement** (which allows you to include the data lines within an external file).

There are also a number of different *styles* of input that may be used when reading data. The "style of input" refers to the type of instructions that you provide concerning the location and format of your variables. Although the SAS System allows for list input, column input, and formatted input, this text presents only the formatted input style because of its ability to easily handle many different types of data sets. For additional information regarding styles of input, see *SAS language and procedures: usage, version 6, first edition* (particularly Chapter 3), as well as *SAS language: reference, version 6, first edition* (particularly Chapter 9).

Here is the general form for inputting data using the CARDS statement and the formatted input style:

```
DATA data-set-name;
   INPUT  #line-number    @column-number    (variable-name)   (column-width.)
                          @column-number    (variable-name)   (column-width.)
                          @column-number    (variable-name)   (column-width.) ;
CARDS;
keyed data are placed here
;
PROC name-of-desired-statistical-procedure      DATA=data-set-name ;
   RUN;
```

The following example shows a SAS program to analyze the preceding data set. In the following example, the numbers on the far left side are *not* actually part of the program. They are simply provided so that it will be easy to refer to specific lines of the program in explaining the meaning of the program in subsequent sections.

```
1          DATA D1;
2             INPUT    #1   @1    (V1)       (1.)
3                           @2    (V2)       (1.)
4                           @3    (V3)       (1.)
5                           @4    (V4)       (1.)
6                           @5    (V5)       (1.)
7                           @6    (V6)       (1.)
8                           @7    (V7)       (1.)
9                           @9    (AGE)      (2.)
10                          @12   (IQ)       (3.)
11                          @16   (NUMBER)   (2.)   ;
12         CARDS;
13         2234243 22   98  1
14         3424325 20  105  2
15         3242424 32   90  3
16         3242323  9  119  4
17         3232143  8  101  5
18         3242242 24  104  6
19         4343525 16  110  7
20         3232324 12   95  8
21         1322424 41   85  9
22         5433224 19  107 10
23         ;
24
25         PROC MEANS    DATA=D1;
26            RUN;
```

A few important notes about these data input statements:

- **The DATA statement**. Line 1 from the preceding program included the DATA statement, where the general form is:

      DATA data-set-name;

  In this case, you gave your data set the name D1, so the statement read

      DATA D1;

- **DATA set names and variable names**. The preceding paragraph stated that your data set was assigned the name D1 on line 1 of the program. In lines 2-11 of the program, the data set's variables are assigned names such as V1, V2, AGE, and IQ.

  You are free to assign a data set or a variable any name you like as long as it conforms to the following rules:

  - It must begin with a letter (rather than a number).
  - It may be no more than 8 characters long.
  - It may contain no special characters such as "*" or "#".
  - It may contain no blank spaces.

  Although the preceding data set was named D1, it could have been given any of an almost infinite number of other names. Below are examples of some other acceptable names for SAS data sets:

      SURVEY
      SURVEY1
      RESEARCH
      VOLUNT

- **The INPUT statement**. The INPUT statement has the following general form:

      INPUT   #line-number   @column-number   (variable-name)   (column-width.)
                             @column-number   (variable-name)   (column-width.)
                             @column-number   (variable-name)   (column-width.) ;

  Compare this general form to the actual INPUT statement that appears on lines 2-11 of the preceding SAS program, and note the values which were filled in to read your data. In the actual program, the word `INPUT` appears on line 2 and tells SAS that the INPUT statement has begun; the SAS System assumes that all of the instructions that follow are data input directions *until* it reads a semicolon (;). At that semicolon, the INPUT statement ends. In the actual program, the semicolon appears at the end of line 11.

- **Line number directions**. To the right of the word INPUT is the following:

    ```
    #line-number
    ```

    This tells SAS what line it should read from to find specific variables. This is important because in some cases there may be two or more lines of data for *each* subject. There will be more on this type of situation in a later section. For the present data, however, the situation is fairly simple: there was only one line of data for each subject, so your program includes the following line number direction (from line 2 of the actual program):

    ```
    INPUT    #1
    ```

    Technically, it is not really necessary to include line number directions when there is only one line of data for each subject (as in the present case). In this text, however, line number directions appear for the sake of consistency.

- **Column location, variable name, and column width directions**. To the right of the line number directions, you place the column location, variable name, and column width directions. The general form for this is as follows:

    ```
    @column-number    (variable-name)    (column-width.)
    ```

    Where `column-number` appears above, you key the number of the column in which a specific variable appears. If the variable occupies more than one column (such as IQ in columns 12, 13, and 14), you should key the number of the column in which it *begins* (e.g., column 12). Where `variable-name` appears, you will key the name that you have given to that variable. And where `column-width` appears, you will key how many columns are occupied by that variable. In the case of the preceding data, the first variable was V1, which appeared in column 1, and was only one column wide. The actual program, therefore, provides the following column location directions (from line 2):

    ```
    @1    (V1)    (1.)
    ```

    The preceding line tells SAS the following: "Go to column 1. In that column you will find a variable called V1. It is 1 column wide."

    IMPORTANT: Note that you must follow the column width with a period, so in this case the column width is (1.). It is important that you include this period; later you will learn how the period provides information about decimal places in a data set.

    Now that variable V1 has been read, you must give SAS the directions needed to read the remaining variables in the data set. The entire completed INPUT statement appears as follows. Note that the line number directions are given only once because all of these variables come from the same line (for a given subject). However, there are different column directions for the different variables. Note also how the column width is different for AGE, IQ, and NUMBER:

```
INPUT    #1    @1    (V1)        (1.)
               @2    (V2)        (1.)
               @3    (V3)        (1.)
               @4    (V4)        (1.)
               @5    (V5)        (1.)
               @6    (V6)        (1.)
               @7    (V7)        (1.)
               @9    (AGE)       (2.)
               @12   (IQ)        (3.)
               @16   (NUMBER)    (2.)   ;
```

IMPORTANT: Notice the semicolon which appears after the column width entry for the last variable (NUMBER). You must always end your input statement with a semicolon in this way. It is easy to make the mistake of leaving off this semicolon, so always check for this semicolon when you get an error message following the INPUT statement.

- **The CARDS or DATALINES statement**. The CARDS statement goes after the INPUT statement, and tells SAS that the raw data are to follow. Don't forget the semicolon after the word CARDS. In the preceding program, the CARDS statement appeared on line 12.

  It is also possible to use the newer DATALINES statement in place of the CARDS statement. The same rules concerning placement of the data and semicolons also apply with the DATALINES statement.

- **The data lines**. The data lines, of course, are the lines that contain the subjects' values on the numeric and/or character variables. In the preceding program, these appear on lines 13-22.

  The data lines should begin on the very next line after the CARDS (or DATALINES) statement; there should be no blank lines. These data lines begin on line 13 in the preceding program. On the very first line after the last of the data lines (line 23, in this case), you should add another semicolon to let SAS know that the data have ended. Do *not* place this semicolon at the end of the last line of data (that is, on the *same line* as the data), as this may cause an error.

  With respect to the data lines, the most important thing to remember is that you must key a given variable in the column that your INPUT statement says it is in. For example, if your input statement contains the following line:

```
@9    (AGE)    (2.)
```

then make sure that the variable AGE really is a 2-digit number keyed in columns 9 and 10.

- **PROC and RUN statements**. There is little to say about PROC and RUN statements at this time because most of the remainder of the text will be concerned with using SAS System procedures. Suffice it to say that a PROC statement asks SAS to perform some statistical analysis. To keep things simple, this section uses a procedure called PROC MEANS. PROC MEANS asks SAS to calculate means, standard deviations, and some other descriptive statistics for numeric variables. The preceding program includes the PROC MEANS statement on line 25.

In most cases, your program will end with a RUN statement. In the preceding program, the RUN statement appears on line 26. A RUN statement executes any previously entered SAS statements, and RUN statements are typically placed after every PROC. If your program includes a number of PROC statements in sequence, it is acceptable to place just one RUN statement after the final PROC.

If you submitted the preceding program for analysis, PROC MEANS would produce the results presented in Output A.2.1:

Variable	N	Mean	Std Dev	Minimum	Maximum
V1	10	3.0000000	1.0540926	1.0000000	5.0000000
V2	10	2.6000000	0.8432740	2.0000000	4.0000000
V3	10	3.2000000	0.7888106	2.0000000	4.0000000
V4	10	2.6000000	0.8432740	2.0000000	4.0000000
V5	10	2.9000000	1.1972190	1.0000000	5.0000000
V6	10	2.6000000	0.9660918	2.0000000	4.0000000
V7	10	3.7000000	0.9486833	2.0000000	5.0000000
AGE	10	20.3000000	10.2745641	8.0000000	41.0000000
IQ	10	101.4000000	9.9241568	85.0000000	119.0000000
NUMBER	10	5.5000000	3.0276504	1.0000000	10.0000000

Output A.2.1: Results of the MEANS Procedure

Additional Guidelines

Inputting string variables with the same prefix and different numeric suffixes. In this section, **prefix** refers to the first part of a variable's name, while a **suffix** refers to the last part. For example, think about our variables V1, V2, V3, V4, V5, V6, and V7. These are multiple variables with the same prefix (V) and different numeric suffixes (1, 2, 3, 4, 5, 6, and 7).

Variables such as this are sometimes referred to as **string variables**. Earlier, this chapter provided one way of inputting these variables, and this original INPUT statement is reproduced here:

```
INPUT   #1   @1    (V1)        (1.)
             @2    (V2)        (1.)
             @3    (V3)        (1.)
             @4    (V4)        (1.)
             @5    (V5)        (1.)
             @6    (V6)        (1.)
             @7    (V7)        (1.)
             @9    (AGE)       (2.)
             @12   (IQ)        (3.)
             @16   (NUMBER)    (2.)   ;
```

However, with string variables named in this way, there is a much easier way of writing the INPUT statement. You could have written it in this way:

```
INPUT   #1   @1    (V1-V7)     (1.)
             @9    (AGE)       (2.)
             @12   (IQ)        (3.)
             @16   (NUMBER)    (2.)   ;
```

The first line of this INPUT statement gives SAS the following directions: "Go to line #1. There, go to column 1. Beginning in column 1 you will find variables V1 through V7. Each variable is 1 column wide." With this second INPUT statement, SAS will read the data in exactly the same way that it would have using the original input statement.

As an additional example, imagine you had a 50-item survey instead of a 7-item survey. You called your variables Q1, Q2, Q3, and so forth instead of V1, V2, V3, and so forth. You keyed your data in the following way:

Column	Variable Name	Explanation
1-50	Q1-Q50	Responses to survey questions 1-50
51	(blank)	(blank)
52-53	AGE	Subject's age in years
54	(blank)	(blank)
55-57	IQ	Subject's IQ score
58	(blank)	(blank)
59-60	NUMBER	Subject's number

You could use the following INPUT to read these data:

```
INPUT    #1    @1     (Q1-Q50)    (1.)
               @52    (AGE)       (2.)
               @55    (IQ)        (3.)
               @59    (NUMBER)    (2.)   ;
```

Inputting character variables. This text deals with basically two types of variables: numeric variables and character variables. A **numeric variable** consists entirely of numbers–it contains no letters. For example, all of your variables from the preceding data set were numeric variables: V1 could assume only the values of 1, 2, 3, 4, or 5. Similarly, AGE could take on only numeric values. On the other hand, a **character variable** may consist of either numbers or alphabetic characters (letters), or both.

The following is the data set introduced earlier in this chapter. Remember that responses to the seven questions of the Volunteerism Survey are keyed in columns 1-7 of this data set, AGE is keyed in columns 9-10, IQ is keyed in columns 12-14, and subject number is in columns 16-17.

```
2234243 22   98   1
3424325 20  105   2
3242424 32   90   3
3242323  9  119   4
3232143  8  101   5
3242242 24  104   6
4343525 16  110   7
3232324 12   95   8
1322424 41   85   9
5433224 19  107  10
```

None of the preceding variables were character variables, but you could easily add a character variable to this data set. For example, you could determine the sex of each subject and create a new variable called SEX which codes the subjects' gender. If a subject were male, SEX would assume the value "M." If a subject were female, SEX would assume the value "F." In the following, the new SEX variable appears in column 19 (the last column):

```
2234243 22  98   1 M
3424325 20 105   2 M
3242424 32  90   3 F
3242323  9 119   4 F
3232143  8 101   5 F
3242242 24 104   6 M
4343525 16 110   7 F
3232324 12  95   8 M
1322424 41  85   9 M
5433224 19 107  10 F
```

You can see that subjects 1 and 2 were males, while subjects 3, 4, and 5 were females, and so forth.

IMPORTANT: You must use a special command within the INPUT statement to input a character variable. Specifically, in the column width region for the character variable, precede the column width with a dollar sign ("$"). For the preceding data set, you would use the following INPUT statement (note the dollar sign in the column width region for the SEX variable):

```
INPUT    #1   @1    (V1-V7)    (1.)
              @9    (AGE)      (2.)
              @12   (IQ)       (3.)
              @16   (NUMBER)   (2.)
              @19   (SEX)      ($1.)   ;
```

Using multiple lines of data for each subject. Very often a researcher obtains so much data from each subject that it is impractical to key all of the data on just one line. For example, imagine that you administer a 100-item questionnaire to a sample, and that you plan to key responses to question 1 in column 1, responses to question 2 in column 2, and so forth. Following this process, you are likely to run into difficulty, because you will need 100 columns to key all responses from a given subject, but most computer monitors allow only around 79 columns. If you continue keying data past column 79, your data are likely to wrap around or appear in some way that makes it difficult to verify that you are keying a given value in the correct column.

In situations in which you require a very large number of columns for your data, it is often best to divide each subject's data so that they appear on more than one line (in other words, it is often best to have multiple lines of data for each subject). To do this, it is necessary to modify your INPUT statement somewhat.

To illustrate, assume that you have obtained two additional variables for each subject in your study: their SAT verbal test scores and SAT math test scores. You decide to key your data so that there are two lines of data for each subject. On line 1 for a given subject, you key V1 through V7, AGE, IQ, NUMBER, and SEX (as above). On line 2 for that subject, you key SATV (the SAT verbal test score) in columns 1 through 3, and you key SATM (the SAT math test score) in columns 5 through 7:

```
2234243 22  98   1 M
520 490
3424325 20 105   2 M
440 410
3242424 32  90   3 F
390 420
3242323  9 119   4 F

3232143  8 101   5 F

3242242 24 104   6 M
330 340
4343525 16 110   7 F

3232324 12  95   8 M

1322424 41  85   9 M
380 410
5433224 19 107  10 F
640 590
```

It can be seen that the SATV score for subject 1 was 520, and the SATM score was 490.

IMPORTANT: When a given subject has no data for a variable which would normally appear on a given line, your data set must still include a line for that subject, even if it is blank. For example, subject 4 is only 9 years old, so she has not yet taken the SAT, and obviously does not have any SAT scores to key. Nonetheless, you still had to include a second line for subject 4, even though it was blank. Notice that blank lines also appear for subjects 5, 7, and 8, who were also too young to take the SAT.

Be warned that, with some text editors, it is necessary to create these blank lines by pressing the *return* key, thus creating a "hard" carriage return; with these editors, simply using the directional arrows on the keypad may not create the necessary hard return. This is especially true when keying data with PCs and ASCII files, but is less likely to be a problem when working on a mainframe computer. Problems in reading the data are also likely to occur if tabs have been used; it is generally best to avoid the use of tabs or other hidden codes when keying data.

The following coding guide tells us where each variable appears. Notice that this guide indicates the *line* on which a variable is located, as well as which *column* in which it is located.

Line	Column	Variable Name	Explanation
1	1–7	V1–V7	Survey questions 1–7
	8	(blank)	(blank)
	9–10	AGE	Subject's age in years
	11	(blank)	(blank)
	12–14	IQ	Subject's IQ score
	15	(blank)	(blank)
	16–17	NUMBER	Subject's number
	18	(blank)	(blank)
	19	SEX	Subject's sex
2	1–3	SATV	SAT-Verbal test score
	5–7	SATM	SAT-Math test score

IMPORTANT: When there are multiple lines of data for each subject, the INPUT statement must indicate which line a given variable is on. This is done with the line number command ("#") which was introduced earlier. You could use the following INPUT statement to read the preceding data set:

```
INPUT   #1   @1    (V1-V7)    (1.)
             @9    (AGE)      (2.)
             @12   (IQ)       (3.)
             @16   (NUMBER)   (2.)
             @19   (SEX)      ($1.)
        #2   @1    (SATV)     (3.)
             @5    (SATM)     (3.)   ;
```

This INPUT statement tells SAS to begin at line #1 for a given subject, to go to column 1 and find variables V1 through V7. It continues to tell SAS where it will find each of the other variables located on line #1, as before. After reading the SEX variable, SAS is told to move on to line #2. There, it is to go to column 1, and find the variable SATV, which is 3 columns wide. The variable SATM then begins in column 5, and is also 3 columns wide. Obviously, it is theoretically possible to have any number of lines of data for each subject, as long as you use these line number directions correctly.

Creating decimal places for numeric variables. Assume that you have determined the high school grade point ratios (GPRs) for a sample of 5 subjects. You could create a SAS data set containing these GPRs using the following program:

```
1           DATA D1;
2               INPUT    #1    @1    (GPR)    (4.);
3           CARDS;
4           3.56
5           2.20
6           2.11
7           3.25
8           4.00
9           ;
10          PROC MEANS    DATA=D1;
11              RUN;
```

The INPUT statement tells SAS to go to line 1, column 1, and find a variable called GPR which is 4 columns wide. Within the data set itself, values of GPR were keyed using a period as a decimal point, with two digits to the right of the decimal point.

This same data set could have been keyed and input in a slightly different way. For example, what if the data had been keyed without a decimal point, as follows?

```
356
220
211
325
400
```

It is still possible to have SAS insert a decimal point where it belongs, in front of the last two digits in each number. You do this in the column width command of the INPUT statement. With this column width command, you indicate how many columns the variable occupies, key a period, and then indicate how many columns of data should appear to the right of the decimal place. In the present case, the preceding GPR variable was only 3 columns wide, and two columns of data should have appeared to the right of the decimal place. So you would modify the SAS program in the following way (notice the column width command):

```
1          DATA D1;
2              INPUT    #1    @1    (GPR)    (3.2);
3          CARDS;
4          356
5          220
6          211
7          325
8          400
9          ;
10         PROC MEANS    DATA=D1;
11             RUN;
```

Inputting "check all that apply" questions as multiple variables. A "check all that apply" question is a special type of questionnaire item that is often used in social science research. These items generate data that must be input in a special way. The following is an example of a "check all that apply" item that could have appeared on your volunteerism survey:

```
Below is a list of activities.  Please place a check mark next to
any activity that you have engaged in in the past six months.

Check here
-----

_____  1. Did volunteer work at a shelter for the homeless.
_____  2. Did volunteer work at a shelter for battered women.
_____  3. Did volunteer work at a hospital or hospice.
_____  4. Did volunteer work for any other community agency or
           organization.
_____  5. Donated money to the United Way.
_____  6. Donated money to a church-sponsored charity.
_____  7. Donated money to any other charitable cause.
```

An inexperienced researcher might think of the preceding as a single question with seven possible responses, and try to key the data in a single column in the data set (say, in column 1). But this would lead to big problems: what would you key in column 1 if a subject checked more than one category?

One way out of this difficulty is to treat the seven possible responses above as seven different questions. When keyed, each of these questions will be treated as a separate variable and will appear in a separate column. For example, whether or not a subject checked activity 1 may be coded in column 1; whether the subject checked activity 2 may be coded in column 2, and so forth.

Researchers may code these variables by placing any values they like in these columns, but you should key a zero ("0") if the subject did not check that activitiy, and a one ("1") if the subject did check it. Why code the variables using zeros and ones? The reason is that this makes it easier to perform some types of analyses that you may later wish to perform. A variable that may assume only two values is called a **dichotomous variable**, and the process of coding dichotomous variables with ones and zeros is known as **dummy coding**.

Once a dichotomous variable has been dummy coded, it can be analyzed using a variety of SAS System procedures such as PROC CORR, which computes Pearson product-moment correlations between quantitative variables. Once a dichotomous variable has been dummy-coded, it is possible to compute the Pearson correlation between it and other numeric variables using PROC CORR. Similarly, you can use PROC REG to perform **multiple regression analysis**, a procedure that allows one to assess the nature of the relationship between a single criterion variable and multiple predictor variables. Once again, if a dichotomous variable has been appropriately dummy-coded, it can be used as a predictor variable in a multiple regression analysis. For these and other reasons, it is good practice to code dichotomous variables using zeros and ones.

The following coding guide summarizes how you could key responses to the preceding question:

Line	Column	Variable Name	Explanation
1	1-7	V1-V7	Responses regarding activities 1 through 7. For each activity, a 0 was recorded if the subject did not check the activity, and a 1 was recorded if the subject did check it.

When subjects have responded to a "check all that apply" item, it is often best to analyze the resulting data with the FREQ procedure. PROC FREQ indicates the raw number of people who appear in each category (in this case, PROC FREQ will indicate the number of people who did not check a given activity versus the number who did). It also provides percent of people who appear in each category, along with some additional information.

The following program inputs some fictitious data and requests frequency tables for each activity using PROC FREQ:

```
1        DATA D1;
2           INPUT    #1   @1    (V1-V7)    (1.);
3        CARDS;
4        0010000
5        1011111
6        0001001
7        0010000
8        1100000
9        ;
10       PROC FREQ     DATA=D1;
11          TABLES V1-V7;
12          RUN;
```

Data for the first subject appears on line 4 of the program. Notice that a one is keyed in column 3 for this subject, indicating that he or she did perform activity 3 ("did volunteer work at a hospital or hospice"), and that zeros are recorded for the remaining six activities, meaning that the subject did not perform those activities. The data keyed for subject 2 on line 5 shows that this subject performed all of the activities except for activity 2.

Inputting a Correlation or Covariance Matrix

There are times when, for reasons of either necessity or convenience, you may choose to analyze a correlation matrix or covariance matrix rather than raw data. The SAS System allows you to input such a matrix as data, and some (but not all) SAS System procedures may then be used to analyze the data set. For example, a correlation or covariance matrix can be analyzed using PROC REG, PROC FACTOR, or PROC CALIS, along with some other procedures.

Inputting a correlation matrix. This type of data input is sometimes necessary when a researcher obtains a correlation matrix from an earlier study (perhaps from an article published in a research journal) and wishes to perform further analyses on the data. You could input the published correlation matrix as a SAS System data set and analyze it in the same way you would analyze raw data.

For example, imagine that you have read an article that tested a social psychology theory called the *investment model* (Rusbult, 1980). The investment model identifies a number of variables that are believed to influence a person's satisfaction with and commitment to a romantic relationship. The following are short definitions for the variables that constitute the investment model:

Commitment: The person's intention to remain in the relationship

Satisfaction: The person's affective (emotional) response to the relationship

Rewards: The number of good things or benefits associated with the relationship

Costs: The number of bad things or hardships associated with the relationship

Investment size: The amount of time, energy, and personal resources put into the relationship

Alternative value: The attractiveness of one's alternatives to the relationship (e.g., attractiveness of alternative romantic partners)

One interpretation of the investment model predicts that commitment to the relationship is determined by satisfaction, investment size, and alternative value, while satisfaction with the relationship is determined by rewards and costs. The predicted relationships between these variables are portrayed in Figure A.2.1:

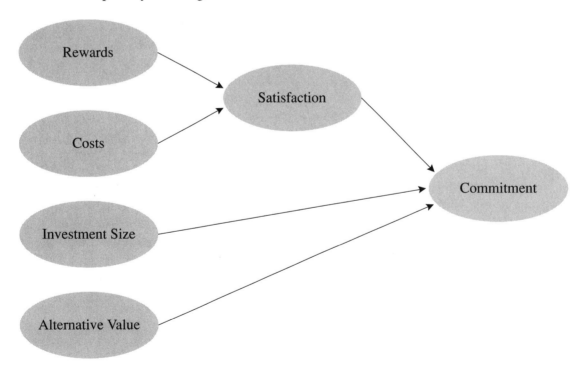

Figure A.2.1: Predicted Relationships between Investment Model Variables

Assume that you have read an article that reports an investigation of the investment model, and that the article included the following (fictitious) table:

Table A.2.1

Standard Deviations and Intercorrelations for All Variables

			Intercorrelations					
Variable	SD	1	2	3	4	5	6
1. Commitment	2.3192	1.0000					
2. Satisfaction	1.7744	.6742	1.0000				
3. Rewards	1.2525	.5501	.6721	1.0000			
4. Costs	1.4086	-.3499	-.5717	-.4405	1.0000		
5. Investments	1.5575	.6444	.5234	.5346	-.1854	1.0000	
6. Alternatives	1.8701	-.6929	-.4952	-.4061	.3525	-.3934	1.0000

Note: N = 240.

Armed with this information, you may now create a SAS data set that includes just these correlations and standard deviations. Here are the necessary data input statements:

```
1        DATA D1(TYPE=CORR) ;
2          INPUT _TYPE_ $ _NAME_ $ V1-V6 ;
3          LABEL
4              V1 ='COMMITMENT'
5              V2 ='SATISFACTION'
6              V3 ='REWARDS'
7              V4 ='COSTS'
8              V5 ='INVESTMENTS'
9              V6 ='ALTERNATIVES' ;
10       CARDS;
11       N    .   240     240     240     240     240     240
12       STD  .   2.3192  1.7744  1.2525  1.4086  1.5575  1.8701
13       CORR V1  1.0000  .       .       .       .       .
14       CORR V2  .6742   1.0000  .       .       .       .
15       CORR V3  .5501   .6721   1.0000  .       .       .
16       CORR V4  -.3499  -.5717  -.4405  1.0000  .       .
17       CORR V5  .6444   .5234   .5346   -.1854  1.0000  .
18       CORR V6  -.6929  -.4952  -.4061  .3525   -.3934  1.0000
19            ;
```

The following shows the general form for this DATA step; the statements assume that six variables are to be analyzed. The program would, of course, be modified if the analysis involved a different number of variables.

```
1          DATA data-set-name(TYPE=CORR) ;
2            INPUT _TYPE_ $ _NAME_ $ variable-list ;
3            LABEL
4               V1 ='long-name'
5               V2 ='long-name'
6               V3 ='long-name'
7               V4 ='long-name'
8               V5 ='long-name'
9               V6 ='long-name' ;
10         CARDS;
11         N       .    n       n       n       n       n       n
12         STD     .    std     std     std     std     std     std
13         CORR  V1  1.0000   .       .       .       .       .
14         CORR  V2    r     1.0000   .       .       .       .
15         CORR  V3    r       r     1.0000   .       .       .
16         CORR  V4    r       r       r     1.0000   .       .
17         CORR  V5    r       r       r       r     1.0000   .
18         CORR  V6    r       r       r       r       r     1.0000
19         ;
```

where

variable-list = List of variables (V1, V2, etc.).

long-name = Full name for the given variable (this will be used to label the variable when it appears in certain parts of the SAS output; if this is not desired, you can omit the entire LABEL statement).

n = Number of observations contributing to the correlation matrix. Each correlation in this matrix should be based on the same observations (and hence the same number of observations); this will automatically be the case if the matrix is created using the NOMISS option with PROC CORR, as discussed in Appendix A.5.

std = Standard deviation obtained for each variable. Technically, these standard deviations are not needed if you are performing an analysis on the *correlation* matrix. However, if you wish to instead perform an analysis on a variance-covariance matrix, then the standard deviations are required so that the SAS System can convert the correlation matrix into a variance-covariance matrix.

r = Correlations between the variables.

The observations that appear on lines 11-18 in the preceding program are easiest to understand if you think of the observations as a matrix with eight rows and eight columns. The first column in this matrix (running vertically) contains the _TYPE_ variable (notice that the INPUT statement tells SAS that the first variable it will read is a character variable named "_TYPE_"). If an "N" appears as a value in this _TYPE_ column, then the SAS System knows that sample sizes will appear on that line. If an "STD" appears as a value in the _TYPE_ column, then the system knows that standard deviations will appear on that line. Finally, if a "CORR" appears as a value in the _TYPE_ column, the system knows that correlation coefficients will appear on that line.

The second column in this matrix contains short names for the manifest variables. These names should appear only on the CORR lines; periods (for missing data) should appear where the N and STD lines intersect with this column.

Looking at the matrix from the other direction, you see eight rows running horizontally. The first row is the N row (or "line") which should contain

- the N symbol.
- a period for the missing variable name.
- the sample sizes for the variables, each separated by at least one blank space. The preceding program shows that the sample size was 240 for each variable.

The STD row (or line) should contain

- the STD symbol.
- the period for the missing variable name.
- the standard deviations for the variables, each separated by at least one blank space. If the STD line is omitted from a correlation matrix, the analysis can only be performed on correlations, not on covariances.

Finally, where rows 3-8 intersect with columns 3-8, the correlation coefficients should appear. The coefficients themselves appear below the diagonal, ones should appear on the diagonal, and periods appear above the diagonal (where the correlation coefficients would again appear if this were a full matrix). Be very careful in keying these correlations; one missing period can cause an error in reading the data.

You can see that the columns of data in this matrix are lined up in a fairly neat fashion. Technically, this neatness was not really required, as this INPUT statement is in free format. However, you should try to be equally neat when preparing your matrix, as this will minimize the chance of leaving out an entry and causing an error.

Inputting a covariance matrix. The procedure for inputting a covariance matrix is very similar to that used with a correlation matrix. An example is presented here:

```
 1          DATA D1(TYPE=COV) ;
 2            INPUT _TYPE_ $ _NAME_ $ V1-V6 ;
 3            LABEL
 4               V1 ='COMMITMENT'
 5               V2 ='SATISFACTION'
 6               V3 ='REWARDS'
 7               V4 ='COSTS'
 8               V5 ='INVESTMENTS'
 9               V6 ='ALTERNATIVES' ;
10          CARDS;
11          N      .    240     240     240     240     240     240
12          COV   V1 11.1284    .       .       .       .       .
13          COV   V2  5.6742  9.0054    .       .       .       .
14          COV   V3  4.5501  3.6721  6.8773    .       .       .
15          COV   V4 -3.3499 -5.5717 -2.4405 10.9936    .       .
16          COV   V5  7.6444  2.5234  3.5346 -4.1854  7.1185    .
17          COV   V6 -8.6329 -3.4952 -6.4061  4.3525 -5.3934  9.2144
18          ;
```

Notice that the DATA statement now specifies TYPE=COV rather than TYPE=CORR. The line providing standard deviations is no longer needed, and has been removed. The matrix itself now provides variances on the diagonal and covariances below the diagonal, and the beginning of each line now specifies COV to indicate that this is a covariance matrix. The remaining sections are identical to those used to input a correlation matrix.

Inputting Data Using the INFILE Statement Rather than the CARDS Statement

When working with a very large data set with many lines of data, it may be more convenient to input data using the INFILE statement rather than the CARDS statement. This involves

- adding an INFILE statement to your program
- placing your data lines in a second computer file, rather than in the computer file that contains your SAS program
- deleting the CARDS statement from your SAS program.

Your INFILE statement should appear *after* the DATA statement, but *before* the INPUT statement. The general form for a SAS program using the INFILE statement is as follows:

```
DATA data-set-name;
   INFILE  'name-of-data-file' ;
   INPUT  #line-number   @column-number   (variable-name)   (column-width.)
                         @column-number   (variable-name)   (column-width.)
                         @column-number   (variable-name)   (column-width.) ;

PROC name-of-desired-statistical-procedure    DATA=data-set-name;
   RUN;
```

Notice that the above is identical to the general form for a SAS program presented earlier, except an INFILE statement has been added, and the CARDS statement and data lines have been deleted.

To illustrate the use of the INFILE statement, consider Dr. Smith's volunteerism study. The data set itself is reproduced here:

```
2234243 22  98   1 M
3424325 20 105   2 M
3242424 32  90   3 F
3242323  9 119   4 F
3232143  8 101   5 F
3242242 24 104   6 M
4343525 16 110   7 F
3232324 12  95   8 M
1322424 41  85   9 M
5433224 19 107  10 F
```

If you were to input these data using the INFILE statement, you would key the data in a separate computer file, giving the file any name you like. Assume, in this case, that the preceding data file is named "VOLUNT."

Note: You must key these data lines beginning on line 1 of the computer file; do *not* leave any blank lines at the top of the file. Similarly, there should be no blank lines at the end of the file (unless a blank line is appropriate because of missing data for the last subject).

Once the data are keyed and saved in the file called VOLUNT, you could key the SAS program itself in a separate file. Perhaps you would give this file a name such as SURVEY.SAS. A SAS program which would input the preceding data and calculate means for the variables appears here:

```
1        DATA D1;
2           INFILE 'VOLUNT';
3           INPUT   #1   @1   (V1-V7)    (1.)
4                        @9   (AGE)      (2.)
5                        @12  (IQ)       (3.)
6                        @16  (NUMBER)   (2.)
7                        @19  (SEX)      ($1.)  ;
8
9        PROC MEANS   DATA=D1;
10          RUN;
```

Controlling the Size of the Ouput and Log Pages with the OPTIONS Statement

Although it is not really related to the topic of data input, the OPTIONS statement is introduced at this point in the text so that you can modify the size of your output pages and log pages, if necessary. For example, when printing your output on a 132-character printer, you may wish to modify your output so that each line may be up to 120-characters long. When working with a printer with a smaller platen width, however, you may wish to produce output that is less than 120 characters in length. The OPTIONS statement allows you to do this.

Here is the general form of the OPTIONS statement that will allow you to control the maximum number of characters and lines that appear on each page in output and log files:

```
OPTIONS   LINESIZE=x   PAGESIZE=y ;
```

With the preceding general form, "x" = the maximum number of characters that you wish to appear on each line, and "y" = the maximum number of lines that you wish to appear on each page. The values that you specify for LINESIZE= may range from 64 through 256. The values that you specify for PAGESIZE= may range from 15 through 32,767.

For example, to request output and log files in which each line may be up to 80 characters long and each page may contain up to 60 lines, use the following OPTIONS statement as the first line of your program:

```
OPTIONS   LINESIZE=80   PAGESIZE=60;
```

To request output and log files with lines that are up to 120 characters long, use the following:

```
OPTIONS   LINESIZE=120   PAGESIZE=60;
```

Conclusion

The material presented in this appendix has prepared you to input most of the types of data that are normally encountered in social science research. However, even when the data have been successfully input, they are not necessarily ready to be analyzed. Perhaps you have input raw data, and need to transform the data in some way before they can be analyzed. This is often the case with questionnaire data, as responses to multiple questions are often summed or averaged to create new variables to be analyzed. Or perhaps you have input data from a large and heterogenous sample, and wish to perform analyses on only a subgroup of that sample (such as the female, but not the male, respondents). In these situations, some form of *data manipulation* or *data subsetting* are called for, and the following appendix shows how to do this.

References

Rusbult, C. E. (1980). Commitment and satisfaction in romantic associations: A test of the investment model. *Journal of Experimental Social Psychology, 16*, 172-186.

SAS Institute Inc. (1989). *SAS language and procedures: usage, version 6, first edition*. Cary, NC: SAS Institute Inc.

SAS Institute Inc. (1990). *SAS language: reference, version 6, first edition*. Cary, NC: SAS Institute Inc.

Appendix A.3

WORKING WITH VARIABLES AND OBSERVATIONS IN SAS® DATA SETS

Overview. This appendix shows how to modify a data set so that existing variables are transformed or recoded or so that new variables are created. The appendix shows how to eliminate unwanted observations from a data set so you can perform analyses only on a specified subgroup or on subjects that have no missing data. This appendix also shows the correct use of arithmetic operators, IF-THEN control statements, and comparison operators. Finally, the appendix shows how to concatenate and merge existing data sets to create new data sets.

Introduction: Manipulating, Subsetting, Concatenating, and Merging Data

Very often, researchers obtain a data set in which the data are not yet in a form appropriate for analysis. For example, imagine that you are conducting research on job satisfaction. Perhaps you wish to compute the correlation between subject age and a single index of job satisfaction. You administer to 200 employees a 10-item questionnaire that assesses job satisfaction, and you key their responses to the 10 individual questionnaire items. You now need to add together each subject's response to those 10 items to arrive at a single composite score that reflects that subject's overall level of satisfaction. This computation is very easy to perform by including a number of data-manipulation statements in the SAS program. **Data-manipulation** statements are SAS statements that transform the data set in some way. They may be used to recode reversed variables, create new variables from existing variables, and perform a wide range of other tasks.

At the same time, your original data set may contain observations that you do not wish to include in your analyses. Perhaps you administered the questionnaire to hourly as well as nonhourly employees, and you wish to analyze only data from the hourly employees. In addition, you may wish to analyze data only from subjects who have usable data on all of the study's variables. In these situations, you may include data-subsetting statements to eliminate the unwanted people from the sample. **Data-subsetting** statements are SAS statements that eliminate unwanted observations from a sample, so that only a specified subgroup is included in the resulting data set.

In other situations, it may be necessary to concatenate or merge data sets before you can perform the analyses you desire. When you **concatenate** data sets, you combine two previously existing data sets that contain data on the same variables but from different subjects. The resulting concatenated data set contains aggregate data from all subjects. In contrast, when you **merge** data sets, you combine two data sets that involve the same subjects but contain different variables. For example, assume that data set D1 contains variables V1 and V2, while data set D2 contains variables V3 and V4. Assume further that both data sets have a variable called ID (identification number) that will be used to merge data from the same subjects. Once D1 and D2 have been merged, the resulting data set (D3) contains V1, V2, V3, and V4 as well as ID.

The SAS programming language is so comprehensive and flexible that it can perform virtually any type of manipulation, subsetting, concatenating, or merging task imaginable. A complete treatment of these capabilities would easily fill a book; therefore it is beyond the scope of this text. However, this chapter reviews some basic statements that can be used to solve a wide variety of problems that are commonly encountered in social science research (particularly in research that involves the analysis of questionnaire data). Readers needing additional help should consult *SAS language and procedures: usage, version 6, first edition* and *SAS language: reference, version 6, first edition.*

Placement of Data Manipulation and Data Subsetting Statements

The use of data manipulation and data subsetting statements are illustrated here with reference to the fictitious study described in the preceding chapter. In that chapter, you were asked to imagine that you had developed a 7-item questionnaire dealing with volunteerism, as shown in the following example.

Volunteerism Survey

Please indicate the extent to which you agree or disagree with each of the following statements. You will do this by circling the appropriate number to the left of that statement. The following format shows what each response number stands for:

```
5 = Agree Strongly
4 = Agree Somewhat
3 = Neither Agree nor Disagree
2 = Disagree Somewhat
1 = Disagree Strongly
```

For example, if you "Disagree Strongly" with the first question, circle the "1" to the left of that statement. If you "Agree Somewhat," circle the "4," and so on.

```
-------------
 Circle Your
  Response
-------------
```

1 2 3 4 5   1. I feel a personal responsibility to help
                         needy people in my community.

1 2 3 4 5   2. I feel I am personally obligated to help
                         homeless families.

1 2 3 4 5   3. I feel no personal responsibility to work
                         with poor people in my town.

1 2 3 4 5   4. Most of the people in my community are
                         willing to help the needy.

1 2 3 4 5   5. A lot of people around here are willing to
                         help homeless families.

1 2 3 4 5   6. The people in my town feel no personal
                         responsibility to work with poor people.

1 2 3 4 5   7. Everyone should feel the responsibility to
                         perform volunteer work in their communities.

What is your age in years? _____

Assume that you administer this survey to a number of subjects, and you also obtain information concerning sex, IQ scores, SAT verbal test scores, and SAT math test scores for each of the subjects. Once the data are keyed, you may wish to write a SAS program that includes some data-manipulation or data-subsetting statements to transform the raw data. But where within the SAS program should these statements appear?

In general, these statements should only appear somewhere within the *DATA step*. Remember that the DATA step begins with the DATA statement and ends as soon as the SAS System encounters a procedure. This means that if you prepare the DATA step, end the DATA step with a procedure, and then place some manipulation or subsetting statements immediately after the procedure, you will receive an error.

To avoid this error (and keep things simple), place your data-manipulation and data-subsetting statements in only two locations within a SAS program:

- immediately following the INPUT statement

- immediately following the creation of a new data set.

Immediately following the INPUT statement. The first of the two preceding guidelines indicates that the statements may be placed immediately following the INPUT statement. This guideline is illustrated by again referring to the study on volunteerism. Assume that you prepare the following SAS program to analyze data obtained in your study. In the following program, lines 11 and 12 indicate where you can place data-manipulation or data-subsetting statements in that program. (To conserve space, only some of the data lines are reproduced in the program.)

```
1       DATA D1;
2          INPUT    #1    @1    (V1-V7)    (1.)
3                         @9    (AGE)      (2.)
4                         @12   (IQ)       (3.)
5                         @16   (NUMBER)   (2.)
6                         @19   (SEX)      ($1.)
7                   #2    @1    (SATV)     (3.)
8                         @5    (SATM)     (3.)    ;
9
10
11       place data-manipulation statements and
12       data-subsetting statements here
13
14       CARDS;
15       2234243 22   98   1 M
16       520 490
17       3424325 20 105   2 M
18       440 410
19            .
```

```
20          .
21
22        5433224 19 107 10 F
23        640 590
24        ;
25
26        PROC MEANS  DATA=D1;
27           RUN;
```

Immediately after creating a new data set. The second guideline for placement provides another option regarding where you may place data-manipulation or data-subsetting statements: They may also be placed immediately following program statements that create a new data set. A new data set may be created at virtually any point in a SAS program (even after procedures have been requested).

At times, you may want to create a new data set so that, initially, it is identical to an existing data set (perhaps the one created with a preceding INPUT statement). If data-manipulation or data-subsetting statements follow the creation of this new data set, the new set displays the modifications requested by those statements.

To create a new data set which is identical to an existing data set, the general form is

```
DATA  new-data-set-name;
   SET  existing-data-set-name;
```

To create such a data set, use the following statements:

```
   DATA D2;
      SET D1;
```

These lines told the SAS System to create a new data set called D2 and to make this new data set identical to D1. Now that a new set has been created, you are free to write as many manipulation and subsetting statements as you like. However, once you write a procedure, that effectively ends the DATA step, and you cannot write any more manipulation or subsetting statements beyond that point (unless you create another data set later in the program).

The following is an example of how you might write your program so that the manipulation and subsetting statements follow the creation of the new data set:

```
1       DATA D1;
2          INPUT   #1    @1    (V1-V7)    (1.)
3                        @9    (AGE)      (2.)
4                        @12   (IQ)       (3.)
5                        @16   (NUMBER)   (2.)
6                        @19   (SEX)      ($1.)
```

```
 7                          #2    @1    (SATV)    (3.)
 8                                @5    (SATM)    (3.)    ;
 9
10        CARDS;
11        2234243 22   98   1 M
12        520 490
13        3424325 20 105   2 M
14        440 410
15           .
16           .
17
18        5433224 19 107 10 F
19        640 590
20        ;
21
22        DATA D2;
23           SET D1;
24
25        place data manipulation statements and
26        data subsetting statements here
27
28        PROC MEANS   DATA=D2;
29           RUN;
```

The SAS System creates two data sets according to the preceding program: D1 contains the original data, and D2 is identical to D1 except for any modifications requested by the data-manipulation and data-subsetting statements.

Notice that the MEANS procedure in line 28 requests the computation of some simple descriptive statistics. It is clear that these statistics are performed on the data from data set D2 because DATA=D2 appears in the PROC MEANS statement. If the statement instead specified DATA=D1, the analyses would have instead been performed on the original data set.

The INFILE statement versus the CARDS statement. The preceding program illustrates the use of the CARDS statement rather than the INFILE statement, but the guidelines regarding the placement of data-modifying statements are the same regardless of which approach is followed. The data-manipulation or data-subsetting statement should either immediately follow the INPUT statement *or* the creation of a new data set. When a program is written using the INFILE statement rather than the CARDS statement, data-manipulation and subsetting statements should appear after the INPUT statement but before the first procedure. For example, if your data are keyed into an external file called VOLUNTEER.DAT, you can write the following program. Notice where the manipulation and subsetting statements are placed.

```
1    DATA D1;
2       INFILE 'VOLUNTEER.DAT';
3       INPUT   #1   @1    (V1-V7)    (1.)
4                    @9    (AGE)      (2.)
5                    @12   (IQ)       (3.)
6                    @16   (NUMBER)   (2.)
7                    @19   (SEX)      ($1.)
8               #2   @1    (SATV)     (3.)
9                    @5    (SATM)     (3.)  ;
10
11       place data manipulation statements and
12       data subsetting statements here
13
14       PROC MEANS    DATA=D1;
15          RUN;
```

In the preceding program, the data-modifying statements again come immediately after the INPUT statement but before the first procedure, consistent with earlier recommendations.

Data Manipulation

Data manipulation involves performing some type of transformation on one or more variables in the DATA step. This section discusses several types of transformations that are frequently required in social science research: such as creating duplicate variables with new variable names, creating new variables from existing variables, recoding reversed items, and using IF-THEN/ELSE statements, along with some related procedures.

Creating duplicate variables with new variable names. Suppose you give a variable a certain name when it is input, but then you want the variable to have a different, perhaps more *meaningful* name when it appears later in the SAS program or in the SAS output. This can easily be accomplished with a statement written according to the following general form:

```
new-variable-name  =  existing-variable-name;
```

For example, in the preceding data set, the first 7 questions are given variables names of V1 through V7. Item 1 in the questionnaire says "I feel a personal responsibility to help needy people in my community." In the INPUT statement, this item was given a SAS variable name V1, which is not very meaningful. RESNEEDY, which stands for "responsible for the needy," is a more meaningful name. Similarly, RESHOME is more meaningful than V2, and NORES is more meaningful than V3.

One way to rename an existing variable is to create a new variable that is identical to the existing variable and assign a new, more meaningful name to this new variable. The following program renames V1, V2, and V3 in this way.

Note: This and later examples show only a portion of the entire program. However, enough of the program appears to illustrate where the remaining statements should be placed.

```
15        .
16        .
17
18        5433224  19  107  10  F
19        640  590
20        ;
21
22        DATA D2;
23           SET D1;
24
25        RESNEEDY  =  V1;
26        RESHOME   =  V2;
27        NORES     =  V3;
28
29        PROC MEANS    DATA=D2;
30           RUN;
```

Line 25 tells SAS to create a new variable called RESNEEDY and to set it identical to the existing variable, V1. Variables RESNEEDY and V1 now have identical data, but RESNEEDY has a more meaningful name, which will facilitate the reading of printouts when statistical analyses are performed later.

Note: When developing a new variable name, conform to the rules for naming SAS variables discussed in Appendix A.2 (e.g., begins with a letter, can be no longer than 8 characters, etc.). Also, note that each statement that creates a duplicate of an existing variable must end with a semicolon.

Duplicating variables versus renaming variables. Technically, the previous program did not really rename variables V1, V2, and V3. Rather, the program created duplicates of these variables and assigned new names to the duplicate variables. Therefore, the resulting data set contains both the original variables under their old names (V1, V2, and V3) as well as the duplicate variables under their new names (RESNEEDY, RESHOME, and NORES). If, for some reason, you literally need to rename the existing variables so that the old variable names no longer exist in the data set, you might consider using the RENAME statement. For details on the RENAME statement, see Chapter 9 of *SAS language: reference*.

Creating new variables from existing variables. It is often necessary to perform mathematical operations on existing variables and use the results to create a new variable. With SAS, the following symbols may be used in arithmetic operations:

> + (addition)
> - (subtraction)
> * (multiplication)
> / (division)
> = (equals)

When writing formulas, you should make heavy use of parentheses. Remember that operations enclosed within parentheses are performed first, and operations outside of the parentheses are performed later. To create a new variable by performing mathematical operations on existing variables, use the following general form:

```
new-variable-name  =  formula-including-existing-variables;
```

For example, two existing variables in your data set are SATV (SAT verbal test scores) and SATM (SAT math test scores). Suppose you wanted to create a new variable called SATCOMB. This variable includes each subject's combined SAT score. For a given subject, you need to add together that person's SATV score and SATM score; therefore, the SATCOMB value for that subject is the sum of the values for SATV and SATM. The program repeats this operation for each subject in the sample, using just one statement:

```
SATCOMB = (SATV + SATM);
```

The preceding statement tells SAS to create a new variable called SATCOMB and set it equal to the sum of SATV and SATM.

Suppose you wanted to calculate, for a given subject, the *average* of that subject's SATV and SATM scores. The new variable is called SATAVG. The program repeats this operation for each subject in the sample using the following statement:

```
SATAVG = (SATV + SATM) / 2;
```

The preceding statement tells SAS to create a new variable called SATAVG by adding together the values of SATV and SATM then dividing this sum by 2. The resulting quotient is that subject's score for SATAVG. You can also arrive at the same result by using two statements instead of one, as shown here:

```
SATCOMB = (SATV + SATM);
SATAVG  = SATCOMB/2;
```

Very often, researchers need to calculate the average of a subject's responses to several items on a questionnaire. For example, look at items 1 and 2 in the questionnaire shown previously. Both items seem to be measuring the subject's sense of personal responsibility to help the needy.

Rather than analyze responses to the items separately, it may be more useful to calculate (for a given subject) the average of his or her responses to those items. This average could then serve as that subject's score on some "personal responsibility" variable. For example, consider the following:

```
RESPONS = (V1 + V2) / 2;
```

The preceding statement tells SAS to create a new variable called RESPONS by adding together a given subject's scores for V1 and V2 then dividing the resulting sum by 2. The resulting quotient is that subject's score for RESPONS.

Note: When creating new variables in this fashion, be sure that all variables on the right side of the equals sign are *existing* variables. This means that they already exist in the data set, either because they are listed in the INPUT statement or because they were created with earlier data-manipulation statements.

Priority of operators in compound expressions. A SAS expression (for example, a formula) that contains just one operator is known as a **simple expression**. The following statement contains a simple expression. Notice that there is only one operator (+ sign) to the right of the = sign:

```
RESPONS = V1 + V2;
```

In contrast, a **compound expression** is one that contains more than one operator. A compound expression is illustrated in the following example; notice that several different operators appear to the right of the = sign:

```
RESPONS = V1 + V2 - V3 / V4 * V5;
```

When an expression contains more than one operator, the SAS System follows a set of rules that determine which operations are performed first, which are performed second, and so forth. The rules that pertain to mathematical operators (+, –, /, and *) are summarized here:

- Multiplication and division operators (* and /) have equal priority, and they are performed first.

- Addition and subtraction operators (+ and –) have equal priority, and they are performed second.

One point made in the preceding rules is that multiplication and division are performed prior to addition or subtraction. For example, consider the following statement:

```
RESPONS = V1 + V2 / V3;
```

Since division has priority over addition, the operations in the preceding statement would be executed in this sequence:

- V2 would first be divided by V3

- The resulting quotient would then be added to V1.

Notice that the division is performed first, even though the addition appears earlier in the formula (reading from left to right).

But what if multiple operators that have equal priority appear in the same statement? In this situation, the SAS System reads the formula from left to right, and performs the operations in that sequence. For example, consider the following:

```
RESPONSE = V1 + V2 - V3;
```

The preceding expression contains only addition and subtraction, operations that have equal priority. The SAS System therefore reads the statement from left to right: First, V1 is added to V2; then V3 is subtracted from the resulting sum.

Because different priority is given to different operators, it is unfortunately very easy to write a statement that results in operations being performed in some sequence other than the intended sequence. For example, imagine that you want to create a new variable called RESPONS. Each subject's score for RESPONS is created by adding the subject's values for V1, V2, and V3 and by dividing this sum by 3. Imagine further that you attempt to achieve this with the following statement:

```
RESPONS = V1 + V2 + V3 / 3;
```

The preceding statement will not create RESPONS in the manner that you intended. Because division has priority over addition, the SAS System performs the operations in the following order:

1. V3 is divided by 3
2. The resulting quotient is then added to V1 and V2.

Obviously, this is not what you intended.

To avoid mistakes such as this, it is important to use parentheses when writing formulas. Because operations that are included inside parentheses are performed first, the use of parentheses gives you control over the sequence in which operations are executed. For example, the following statement creates RESPONS in the way originally intended because the lower priority operations (adding together V1 plus V2 plus V3) are now included within parentheses:

```
RESPONSE = (V1 + V2 + V3) / 3;
```

This statement tells the SAS System to add together V1 plus V2 plus V3; whatever the sum is, divide it by 3.

This section has provided a brief introduction the priority of a few of the operators that can be used with the SAS System. To learn about the priority for other operators (such as exponentiation or comparison operators), see Chapter 10 of *SAS language: reference.*

Recoding reversed variables. Very often, a questionnaire contains a number of reversed items. A **reversed item** is a question stated so that its meaning is the opposite of the meaning of the other items in that group. For example, consider the meaning of the following items from the volunteerism survey:

```
1  2  3  4  5      1.   I feel a personal responsibility to help
                        needy people in my community.

1  2  3  4  5      2.   I feel I am personally obligated to help
                        homeless families.

1  2  3  4  5      3.   I feel no personal responsibility to work
                        with poor people in my town.
```

In a sense, all of these questions are measuring the same thing: whether the subject feels some sense of personal responsibility to help the needy. Items 1 and 2 are stated so that the more strongly you agree with the statement, the stronger is your sense of personal responsibility. This means that scores of 5 indicate a strong sense of responsibility and scores of 1 indicate a weak sense of responsibility. However, item 3 is a *reversed* item. It is stated so that the more strongly you agree with it, the weaker is your sense of personal responsibility. Here, scores of 1 indicate a strong sense of responsibility and scores of 5 indicate a weak sense of responsibility (which is just the reverse of items 1 and 2).

It would be nice if all three items were consistent so that scores of 5 always indicate a strong sense of responsibility and scores of 1 always indicate a weak sense of responsibility. This requires that you recode item 3 so that people who actually circle 5 are given, instead, a score of 1; people who actually circle 4, are given, instead, a score of 2; people who actually circle 2 are given a score of 4; and people who actually circle 1 are given a score of 5. This can be done very easily with the following statement:

```
V3 = 6 - V3;
```

The preceding statement tells SAS to create a new version of the variable V3, then take the number 6 and subtract from it the subject's existing (old) score for V3. The result is the subject's new score for V3. Notice that with this statement, if a person's old score for V3 was 5, his or her new score is, in fact, 1; if the old score was 1, the new score is 5, and so forth.

The general form for this recoding statement is as follows:

```
existing-variable  =  constant  -  existing-variable;
```

The constant is always equal to the number of response points on your survey plus 1. For example, the volunteerism survey included 5 response points: people could circle "1" for "Disagree Strongly" all the way through "5" for "Agree Strongly." It was a 5-point scale, so the constant was 5 + 1 = 6. What would the constant be if the following 7-point scale had been used instead?

> 7 = Agree Very Strongly
> 6 = Agree Strongly
> 5 = Agree Somewhat
> 4 = Neither Agree nor Disagree
> 3 = Disagree Somewhat
> 2 = Disagree Strongly
> 1 = Disagree Very Strongly

It would be 8 because 7 + 1 = 8, and the recoding statement would read

```
V3 = 8 - V3;
```

> *Where* **should the recoding statements go?** In most cases, reversed items should be recoded *before* other data manipulations are performed on them. For example, assume that you want to create a new variable called RESPONS, which stands for "personal responsibility." With this scale, higher scores indicate higher levels of perceived personal responsibility. For a given subject, his or her score on this scale is the average of his or her responses to items 1, 2, and 3 from the survey. Because item 3 is a reversed item, it is important that it be recoded before it is added in with items 1 and 2 in calculating this scale score. Therefore, the correct sequence of statements is as follows:
>
> ```
> V3 = 6 - V3;
> RESPONS = (V1 + V2 + V3) / 3;
> ```
>
> The following sequence is *not* correct:
>
> ```
> RESPONS = (V1 + V2 + V3) / 3;
> V3 = 6 - V3;
> ```

Using IF-THEN control statements. An IF-THEN control statement allows you to make sure that operations are performed on a given subject's data only if certain conditions are true regarding that subject. The following comparison operators may be used with IF-THEN statements:

```
        =     is equal to
       NE     is not equal to
  GT or >     is greater than
       GE     is greater than or equal to
  LT or <     is less than
       LE     is less than or equal to
```

The general form for an IF-THEN statement is as follows:

```
IF  expression  THEN  statement ;
```

The expression usually consists of some comparison involving existing variables. The statement usually involves some operation performed on existing variables or new variables. For example, assume that you want to create a new variable called SATVGRP for "SAT-verbal group." This variable will be created so that

- if you do not know what a subject's SAT verbal test score is, that subject will have a score of "." (for "missing data")

- if subject's score is under 500 on the SAT verbal test, the subject will have a score of 0 for SATVGRP

- if the subject's score is 500 or greater on the SAT verbal test, the subject will have a score of 1 for SATVGRP.

Assume that the variable SATV already exists in your data set and that it contains each subject's score for the SAT verbal test. You can use it to create the new variable SATVGRP by writing the following statements:

```
SATVGRP = .;
IF SATV LT 500 THEN SATVGRP = 0;
IF SATV GE 500 THEN SATVGRP = 1;
```

The preceding statements tell the SAS System to create a new variable called SATVGRP and begin by setting everyone's score as equal to "." (missing). If a subject's score for SATV is less than 500, then set his or her score for SATVGRP equal to 0. If a subject's score for SATV is greater than or equal to 500, then set his or her score for SATVGRP equal to 1.

Using ELSE statements. In reality, you can perform the preceding operations more efficiently by using the ELSE statement. The general form for using the ELSE statement, in conjunction with the IF-THEN statement, is presented as follows:

```
IF  expression  THEN  statement  ;
    ELSE  IF  expression  THEN  statement;
```

The ELSE statement provides alternative actions that the SAS System may take if the original IF expression is not true. For example, consider the following:

```
1     SATVGRP = .;
2     IF SATV LT 500 THEN SATVGRP = 0;
3     ELSE  IF SATV GE 500 THEN SATVGRP = 1;
```

The preceding tells the SAS System to create a new variable called SATVGRP and initially assign all subjects a value of "missing." If a given subject has an SATV score less than 500, the system assigns that subject a score of 0 for SATVGRP. Otherwise, if the subject has an SATV score greater than or equal to 500, then the system assigns that subject a score of 1 for SATVGRP.

Obviously, the preceding statements are identical to the earlier statements that created SATVGRP, except that the word ELSE has been added to the beginning of the third line. In fact, these two approaches actually result in assigning exactly the same values for SATVGRP to each subject. So what, then, is the advantage of including the ELSE statement? The answer has to do with efficiency. When an ELSE statement is included, the actions specified in that statement are executed only if the expression in the preceding IF statement is not true.

For example, consider the situation in which subject 1 has a score for SATV that is less than 500. Line 2 in the preceding statements assigns that subject a score of 0 for SATVGRP. The SAS System then ignores line 3 (because it contains the ELSE statement), thus saving computer time. If line 3 did not contain the word ELSE, the SAS System would have executed the line, checking to see whether the SATV score for subject 1 is greater than or equal to 500 (which is actually unnecessary, given what was learned in line 2).

A word of caution regarding missing data is relevant at this point. Notice that line 2 of the preceding program assigns subjects to group 0 (under SATVGRP) if their values for SATV are less than 0. Unfortunately, a value of "missing" (that is, a value of ".") for SATV is viewed as being less than 500 (actually, it is viewed as being less than 0) by the SAS System. This means that subjects with missing data for SATV are assigned to group 0 under SATVGRP by line 2 of the preceding program. This is not desirable.

To prevent this from happening, you may rewrite the program in the following way:

```
1    SATVGRP = .;
2    IF SATV GE 200 AND SATV LT 500 THEN SATVGRP = 0;
3    ELSE  IF SATV GE 500 THEN SATVGRP = 1;
```

Line 2 of the program now tells the SAS System to assign subjects to group 0 only if their values for SATV are both greater than or equal to 200 and less than 500. This modification to the line involves the use of the conditional AND statement, which is discussed in greater detail in the following section.

Finally, remember that the ELSE statement should only be used in conjunction with a preceding IF statement. In addition, remember to always place the ELSE statement *immediately* following the relevant IF statement.

Using the conditional statements AND and OR. As the preceding section suggests, you can also use the conditional statement AND within an IF-THEN statement or an ELSE statement. For example, consider the following:

```
SATVGRP = .;
IF SATV GT 400 AND SATV LT 500 THEN SATVGRP = 0;
ELSE IF SATV GE 500 THEN SATVGRP = 1;
```

The second statement in the preceding program tells SAS that if SATV is greater than 400 and less than 500, then give this person a score for SATVGRP of 0. This means that subjects are given a value of 0 *only* if they are both over 400 and under 500. What happens to people who have a score of 400 or less for SATV? They are given a value of "." for SATVGRP. That is, they are classified as having missing data for SATVGRP. This is because they (along with everyone else) were given a value of "." in the first statement, and neither of the later statements replaces that "." with 0 or 1. However, for people over 400, one of the later statements replaces the "." with either 0 or 1.

You can also use the conditional statement OR within an IF-THEN statement or an ELSE statement. For example, assume that you have a variable in your data set called ETHNIC. With this variable, subjects were assigned the value 5 if they were Caucasian, 6 if they were African-American, or 7 if they were Asian-American. Assume that you now wish to create a new variable called MAJORITY. Subjects will be assigned a value of 1 for this variable if they are in the majority group (i.e., if they are Caucasians), and they will be assigned a value of 0 for this variable if they are in a minority group (if they are either African-Americans or Asian-Americans). The creation of this variable can be achieved with the following statements:

```
MAJORITY=.;
IF ETHNIC = 5 THEN MAJORITY = 1;
ELSE IF ETHNIC = 6 OR ETHNIC = 7 THEN MAJORITY = 0;
```

In the preceding statements, all subjects are first assigned a value of "missing" for MAJORITY. If their value for ETHNIC is 5, their value for MAJORITY changes to 1 and the SAS System ignores the following ELSE statement. If their value for ETHNIC is not 5, then the SAS System moves on to the ELSE statement. There, if the subject's value for ETHNIC is either 6 or 7, the subject is assigned a value of 0 for MAJORITY.

Working with character variables. When working with character variables (variables in which the values consists of letters rather than numbers), you must enclose values within single quotation marks in the IF-THEN and ELSE statements. For example, suppose you want to create a new variable called SEXGRP. With this variable, males are given a score of 0, and females are given a score of 1. The variable SEX already exists in your data set, and it is a character variable in which males are coded with the letter M and females are coded with the letter F. You can create the new SEXGRP variable using the following statements:

```
SEXGRP = .;
IF SEX = 'M' THEN SEXGRP = 0;
ELSE IF SEX = 'F' THEN SEXGRP = 1;
```

Using the IN operator. The IN operator makes it easy to determine whether a given value is among a specified list of values. Because of this, a single IF statement including the IN operator can perform comparisons that could otherwise require a large number of IF statements. The general form for using the IN operator is as follows:

```
IF  variable  IN  (value-1,value-2, ...value-n)  THEN  statement;
```

Notice that each value in the preceding list of values must be separated by a comma.

For example, assume that you have a variable in your data set called MONTH. The values assumed by this variable are the numbers 1 through 12. With these values, 1 represents January, 2 represents February, 3 represents March, and so forth. Assume that these values for MONTH indicate the month in which a given subject was born, and that you have data for 100 subjects.

Imagine that you now wish to create a new variable called SEASON. This variable will indicate the season in which each subject was born. Subjects are assigned values for SEASON according to the following guidelines:

- Subjects are assigned a value of 1 for SEASON if they were born in January, February, or March (months 1, 2, or 3).

- Subjects are assigned a value of 2 for SEASON if they were born in April, May, or June (months 4, 5, or 6).

- Subjects are assigned a value of 3 for SEASON if they were born in July, August, or September (months 7, 8, or 9).

- Subjects are assigned a value of 4 for SEASON if they were born in October, November, or December (months 10, 11, or 12).

One way to create the new SEASON variable involves using four IF-THEN statements, as shown here:

```
SEASON = .;
IF   MONTH = 1  OR MONTH = 2  OR MONTH = 3   THEN   SEASON = 1;
IF   MONTH = 4  OR MONTH = 5  OR MONTH = 6   THEN   SEASON = 2;
IF   MONTH = 7  OR MONTH = 8  OR MONTH = 9   THEN   SEASON = 3;
IF   MONTH = 10 OR MONTH = 11 OR MONTH = 12 THEN   SEASON = 4;
```

However, the same results can be achieved somewhat more easily by using the IN operator within the IF-THEN statements, as shown here:

```
SEASON = .;
IF   MONTH IN (1,2,3)    THEN SEASON = 1;
IF   MONTH IN (4,5,6)    THEN SEASON = 2;
IF   MONTH IN (7,8,9)    THEN SEASON = 3;
IF   MONTH IN (10,11,12) THEN SEASON = 4;
```

In the preceding example, all variable values are numbers. However, the IN operator may also be used with character variables. As always, it is necessary to enclose all character variable values within single quotation marks. For example, assume that MONTH is actually a character variable that assumes values such as "Jan," "Feb," "Mar," and so forth. Assume further that SEASON assumes the values "Winter," Spring," Summer," and "Fall." Under these circumstances, the preceding statements would be modified in the following way:

```
SEASON = '.';
IF   MONTH IN ('Jan', 'Feb', 'Mar')  THEN SEASON = 'Winter';
IF   MONTH IN ('Apr', 'May', 'Jun')  THEN SEASON = 'Spring';
IF   MONTH IN ('Jul', 'Aug', 'Sep')  THEN SEASON = 'Summer';
IF   MONTH IN ('Oct', 'Nov', 'Dec')  THEN SEASON = 'Fall';
```

Data Subsetting

Using a simple subsetting statement. Often, it is necessary to perform an analysis on only a subset of the subjects who are included in the data set. For example, you may wish to review the mean survey responses provided by just the female subjects. A subsetting IF statement may be used to accomplish this, and the general form is presented here:

```
DATA  new-data-set-name;
   SET  existing-data-set-name;
IF  comparison;

PROC  name-of-desired-statistical-procedure    DATA=new-data-set-name;
   RUN;
```

The comparison described in the preceding statements generally includes some existing variable and at least one comparison operator. The following statements allow you to calculate the mean survey responses for only the female subjects.

```
15          .
16          .
17
18       5433224 19 107 10 F
19       640 590
20       ;
21
22       DATA D2;
23          SET D1;
24
25          IF SEX = 'F';
26
27       PROC MEANS    DATA=D2;
28          RUN;
```

The preceding statements tell SAS to create a new data set called D2 and to make it identical to D1. However, the program keeps a subject's data only if SEX has a value of F. Then the program executes the MEANS procedure for the data which are retained.

Using comparison operators. All of the comparison operators described previously can be used in a subsetting IF statement. For example, consider the following:

```
DATA D2;
   SET D1;

   IF SEX = 'F' AND AGE GE 65;

PROC MEANS    DATA=D2;
   RUN;
```

The preceding statements analyze only data from women who are 65 or older.

Eliminating observations with missing data for some variables. One of the most common difficulties encountered by researchers in the social sciences is the problem of missing data. Briefly, **missing data** involves not having scores for all variables for all subjects in a data set. This section discusses the problem of missing data, and shows how a subsetting IF statement may be used to deal with it.

Assume that you administer your volunteerism survey to 100 subjects, and you use their scores to calculate a single volunteerism score for each subject. You also obtain a number of additional variables regarding the subjects. The SAS names for the study's variables and their descriptions are as follows:

VOLUNT: Subject scores on the volunteerism questionnaire, where higher scores reveal greater intention to engage in unpaid prosocial activities

SATV: Subject scores on the verbal subtest of the Scholastic Assessment Test

SATM: Subject scores on the math subtest of the Scholastic Assessment Test

IQ: Subject scores on a standard intelligence test

Assume further that you obtained scores for VOLUNT, SATV, and SATM for all 100 of the subjects. However, due to a recordkeeping error, you were able to obtain IQ scores for only 75 of the subjects.

You now wish to analyze your data using a procedure called multiple regression (this procedure is covered in Hatcher & Stepanski [1994]; you do not need to understand multiple regression to understand the points to be made here, however). In Analysis #1, VOLUNT is the criterion variable, and SATV and SATM are the predictor variables. The multiple regression equation for Analysis #1 is represented in the following PROC REG statement:

Analysis #1:

```
PROC REG   DATA=D1;
   MODEL VOLUNT  =  SATV   SATM ;
   RUN;
```

When you review the results of the analysis, note that the analysis is based on 100 subjects. This makes sense because you had complete data on all of the variables included in this analysis.

In Analysis #2, VOLUNT is again the criterion variable, but this time the predictor variables will include SATV and SATM as well as IQ. The equation for Analysis #2 is as follows:

Analysis #2:

```
PROC REG   DATA=D1;
   MODEL VOLUNT  =  SATV   SATM   IQ;
   RUN;
```

When you review the results of analysis #2, however, you see that you have encountered a problem. The SAS output indicates that the analysis is based on only 75 subjects. At first you may not understand this because you know that there are 100 subjects in the data set. But then you remember that you did not have complete data for one of the variables. You had values for the IQ variable for only 75 subjects. The REG procedure (and many other SAS procedures) includes in the analysis only those subjects who have complete data for *all* of the variables being analyzed with that procedure. For Analysis #2, this means that any subject with missing data for IQ will be eliminated from the sample. Twenty-five subjects had missing data for IQ and were therefore eliminated.

Why were these 25 subjects not eliminated from Analysis #1? Because that analysis did not involve the IQ variable. It only involved VOLUNT, SATV, and SATM, and all 100 subjects had complete data for those three variables.

In a situation such as this, you have a number of options with respect to how you might perform these analyses and summarize the results for publication. One option is to retain the results described previously. You could report that you performed one analysis on all 100 subjects and a second analysis on just the 75 subjects who had complete data for the IQ variable.

This approach might leave you open to criticism, however. The beginning of your research paper probably reported demographic characteristics for all 100 subjects (how many were female, mean age, etc.). However, you may not have a section providing demographics for the subgroup of 75. This might lead readers to wonder if the subgroup differed in some important way from the aggregate group.

There are statistical reasons that this approach might cause problems as well. For example, you might wish to test the significance of the difference between the R^2 value obtained from Analysis #1 and the R^2 value obtained from Analysis #2. (This test is described in Chapter 13 of Hatcher & Stepanski [1994].) When performing this test, it is important that both R^2 values be based on exactly the same subjects in both analyses. This is obviously not the case in your study, as 25 of the subjects used in Analysis #1 were not used in Analysis #2.

In situations such as this, you are usually better-advised to ensure that every analysis you perform is performed on exactly the same sample of people. This means that, in general, any subject who has missing data for variables to be included in any (reported) analysis should be deleted from the sample before the analyses are performed. In your case, therefore, it is best to see to it that both Analysis #1 and Analysis #2 are performed on only those 75 subjects who had complete data for all four variables (VOLUNT, SATV, SATM, and IQ). Fortunately, this may easily be done using a subsetting IF statement.

With the SAS System, a missing value is represented in the computer's memory with a period (a "."). You can take advantage of this fact to eliminate any subject with missing data for any variable being analyzed. For example, consider the following subsetting IF statement:

```
DATA D2;
   SET D1;
IF VOLUNT NE . AND SATV NE . AND
   SATM   NE . AND IQ   NE . ;
```

The preceding statements tell the system to

1) create a new data set named D2, and make it an exact copy of D1;

2) retain a subject in this new data set *only* if (for that subject)

- VOLUNT is not equal to missing

- SATV is not equal to missing

- SATM is not equal to missing

- IQ is not equal to missing.

In other words, the system creates a new data set named D2. That new data set contains only the 75 subjects who have complete data for all four variables of interest. You may now specify DATA=D2 in all SAS procedures, with the result that all analyses will be performed on exactly the same 75 subjects.

The following SAS program shows where these statements should be placed:

```
14          .
15          .
16          .
17
18          5433224 19 107 10 F
19          640 590
20          ;
21
22          DATA D2;
23             SET D1;
24
25          IF VOLUNT NE . AND SATV NE . AND
26             SATM   NE . AND IQ   NE . ;
27
28          PROC REG   DATA=D2;
29             MODEL VOLUNT =  SATV  SATM ;
30             RUN;
31
32          PROC REG   DATA=D2;
33             MODEL VOLUNT =  SATV  SATM  IQ ;
34             RUN;
```

Obviously, the subsetting IF statement must appear in the program before the procedures that request the modified data set (data set D2, in this case).

How should I key missing data? If you are keying data and come to a subject with missing data for some variable, you do not need to record a "." to represent the missing data . As long as your data are being input using the CARDS statement and the conventions discussed here, it is acceptable to simply leave that column (or those columns) blank by hitting the space bar on your keyboard. The SAS System, however, will internally give that subject a missing data value (a ".") for the concerned variable. In some cases, however, it may be useful to key a "." for variables with missing data, as this may make it easier to keep your place when keying information.

When using a subsetting IF statement to eliminate subjects with missing data, exactly *which* variables should be included in that statement? In most cases, it should be those variables–and only those variables–that are ultimately discussed in the published article. This means that you may not know exactly which variables to include until you actually begin analyzing the data. For example, imagine that you conduct your study and obtain data for the following number of subjects for each of the following variables:

Variable	Number of Subjects with Valid Data for This Variable
VOLUNT	100
SATV	100
SATM	100
IQ	75
AGE	10

As before, you obtained complete data for all 100 subjects for VOLUNT, SATV, and SATM, and you obtained data for 75 subjects on IQ. But notice the last variable. You obtained information concerning the subjects' age for only 10 subjects. What would happen if you included the variable AGE in the subsetting IF statement, as shown here?

```
IF VOLUNT NE . AND SATV NE . AND
   SATM   NE . AND IQ   NE . AND AGE NE . ;
```

This IF statement causes the system to eliminate from the sample anyone who does not have complete data for all five variables. Since only 10 subjects have values for the AGE variable, you know that the resulting data set includes no more than 10 subjects. Obviously, this sample is too small for most statistical procedures. At this point, you have to decide whether to gather more data or forget about doing any analyses with the AGE variable.

In summary, one approach for identifying those variables to be included in the subsetting if statement is to

- perform some initial analyses

- decide which variables will be included in the final analyses (for the published paper)

- include all of those variables in the subsetting IF statement

- perform again all analyses on this reduced data set so that all analyses reported in the paper are performed on exactly the same sample.

Of course, there will sometimes be circumstances in which it is neither necessary nor desirable that all analyses be performed on exactly the same group of subjects. The purpose of the research, along with other considerations, should determine when this is appropriate.

A More Comprehensive Example

Often, a single SAS program will contain a large number of data-manipulation and subsetting statements. Consider the following example, which makes use of the INFILE statement rather than the CARDS statement:

```
1    DATA D1;
2        INFILE 'VOLUNTEER.DAT' ;
3        INPUT    #1    @1    (V1-V7)    (1.)
4                       @9    (AGE)      (2.)
5                       @12   (IQ)       (3.)
6                       @16   (NUMBER)   (2.)
7                       @19   (SEX)      ($1.)
8                 #2    @1    (SATV)     (3.)
9                       @5    (SATM)     (3.)   ;
10
11       DATA D2;
12          SET D1;
13
14       V3 = 6 - V3;
15       V6 = 6 - V6;
16       RESPONS = (V1 + V2 + V3) / 3;
17       TRUST   = (V4 + V5 + V6) / 3;
18       SHOULD = V7;
19
20       PROC MEANS   DATA=D2;
21          RUN;
22
23       DATA D3;
24          SET D2;
25          IF SEX = 'F';
26
27       PROC MEANS   DATA=D3;
28          RUN;
29
30       DATA D4;
31          SET D2;
32          IF SEX = 'M';
33
34       PROC MEANS   DATA=D4;
35          RUN;
```

In the preceding program, lines 11 and 12 create a new data set called D2 and set it identical to D1. All of the data-manipulation commands which appear between those lines and PROC MEANS on line 20 are performed on this data set called D2. Notice that a new variable called

TRUST is created on line 17. TRUST is the average of a subject's responses to items 4, 5, and 6. Look over these items on the volunteerism survey to see why the name TRUST makes sense. On line 18, variable V7 is duplicated, and the resulting new variable is called SHOULD. Why does this make sense? PROC MEANS appears on line 20, so the means and other descriptive statistics are calculated for all of the quantitative variables in the most recently created data set, which is D2. This includes all of the variables input in data set D1 as well as the new variables which were just created.

In lines 22 through 26, a new data set called D3 is created. A given subject's responses are retained in this data set only if that subject is a female. Notice that the SET statement sets D3 equal to D2 rather than D1. This way, the newly created variables (such as TRUST and SHOULD) appear in this all-female data set. In lines 30 through 32, a new data set called D4 is created, and it is also set equal to D2 (not D3). This new data set contains data only from males.

After this program is submitted for analysis, the SAS output contains three tables of means. The first table is based on lines 1-21, and gives the means based on all subjects. The second table is based on lines 23-28 and gives the means based on the responses of females. The third table is based on lines 30-35, and is based on the responses of males.

Concatenating and Merging Data Sets

The techniques taught to this point in this chapter are designed to help you transform data within a single data set (e.g., to recode a variable within a single data set). However, very often you need to perform transformations that involve combining more than one data set to create a new data set. For example, **concatenating** involves creating a new data set by combining two or more previously existing data sets. With concatenation, the same variables typically appear in both of the previously existing data sets, but the two sets contain data from *different subjects*. By concatenating the two previously existing sets, you create a new set that contains data from all subjects.

In contrast, **merging** involves combining data sets in a different way. With merging, each of the previously existing data sets typically contain data from the same subjects. However, the different, previously existing sets usually contain different *variables*. By merging these sets, you can create a single new data set that contains all of the variables found in the previously existing data sets. For example, assume that you conduct a study with 100 subjects. Data set A contains each subject's age, while data set B contains questionnaire responses from the same 100 subjects. By merging data sets A and B, you can create a new data set called C which, again, contains just 100 observations. A given observation in data set C contains a given subject's age as well as the questionnaire responses made by that same subject. Now that the data sets are merged, it is possible to correlate subject age with responses to the questionnaire (for example). This could not be done when AGE was in one data set and the questionnaire responses were in another.

The following section discusses the basics of concatenating and merging SAS data sets. For additional details and related procedures, see Chapters 4 and 9 of *SAS language: reference,* as well as Chapters 14, 15, and 17 of *SAS language and procedures: usage.*

Concatenating data sets. Imagine that you are conducting research that involves the Scholastic Assessment Test (SAT). You obtain data from four subjects: John, Sally, Fred, and Emma. You key information about these four subjects into a SAS data set called A. This data set contains three variables:

- NAME, which contains the subject's first name

- SATV, which contains the subject's score on the SAT verbal test

- SATM, which contains the subject's score on the SAT math test.

The contents of data set A appear in Table A.3.1. You can see that John has a score of 520 for SATV and a score of 500 for SATM, Sally had a score of 610 for SATV and 640 for SATM, and so forth.

Table A.3.1

Contents of Data Set A

NAME	SATV	SATM
John	520	500
Sally	610	640
Fred	490	470
Emma	550	560

Imagine that later you create a second data set called B that contains data from four different subjects: Susan, James, Cheri, and Will. Values for these subjects for SATV and SATM appear in Table A.3.2.

Table A.3.2

Contents of Data Set B

NAME	SATV	SATM
Susan	710	650
James	450	400
Cheri	570	600
Will	680	700

Assume that you would like to perform some analyses on a single data set that contains scores from all eight of these subjects. But you encounter a problem. The values in data set A were keyed differently than the values of data set B, making it impossible to read data from both sets with a single INPUT statement. For example, perhaps you keyed SATV in columns 10-12 in data set A, but keyed it in columns 11-13 in data set B. Because the variable was keyed in different columns in the two data sets, it is not possible to write a single INPUT statement that will input this variable (assuming that you use a formatted input approach).

One way around this problem is to input A and B as separate data sets and then concatenate them to create a single data set that contains all eight observations. You can then perform your analyses on the new data set. The following is the general form for concatenating multiple data sets into a single data set:

```
DATA  new-data-set-name;
     SET  data-set-1  data-set-2 ... data-set-n;
```

In the present situation, you wish to concatenate just two data sets (A and B) to create a new data set named C. This could be done in the following statements:

```
    DATA C;
       SET A B;
```

The entire program follows that places these statements in context. This program

- inputs data set A

- inputs data set B

- concatenates A and B to create C

- uses PROC PRINT to print the contents of data set C (PROC PRINT will be discussed in greater detail in Appendix A.4 of this text).

```
1      DATA   A;
2         INPUT  #1    @1   (NAME)   ($7.)
3                            @10 (SATV)  (3.)
4                            @14 (SATM)  (3.)   ;
5      CARDS;
6      John       520 500
7      Sally      610 640
8      Fred       490 470
9      Emma       550 560
10     ;
11
12     DATA   B;
13        INPUT  #1    @1   (NAME)   ($7.)
14                           @11 (SATV)  (3.)
15                           @15 (SATM)  (3.)   ;
16     CARDS;
17     Susan      710 650
18     James      450 400
19     Cheri      570 600
20     Will       680 700
21     ;
22
23     DATA   C;
24        SET   A   B;
25
26     PROC PRINT   DATA=C
27        RUN;
```

In the preceding program, data set A is input in program lines 1-10, and data set B is input in lines 12-21. In lines 23–24, the two data sets are concatenated to create data set C. In lines 26-27, PROC PRINT is used to print the contents of data set C, and the results of this procedure are reproduced as Output A.3.1. The results of Output A.3.1 show that data set C contains eight observations: the four observations from data set A along with the four observations from data set B. To perform additional statistical analyses on this combined data set, you would simply specify DATA=C in the PROC statement of your SAS program.

```
OBS     NAME     SATV     SATM

 1      John      520      500
 2      Sally     610      640
 3      Fred      490      470
 4      Emma      550      560
 5      Susan     710      650
 6      James     450      400
 7      Cheri     570      600
 8      Will      680      700
```

Output A.3.1: Results of Performing PROC PRINT on Data Set C

Merging data sets. As was stated earlier, you would normally merge data sets when

- you are working with two data sets

- both data sets contain information for the same subjects, but one data set contains one set of variables, while the other data set contains a different set of variables.

Once these two data sets have been merged, you will have a single data set that contains all of these variables. Having all the variables in one set allows you to assess the relationships between the variables, should you wish to do so.

As an illustration, assume that your sample consists of just four subjects: John, Sally, Fred, and Emma. Assume that you have obtained the social security number for each of these subjects, and that these numbers are included in a SAS variable named SOCSEC in both previously existing data sets. In data set D, you have SAT verbal test scores and SAT math test scores for these subjects (represented as SAS variables SATV and SATM, respectively). In data set E, you have college cumulative grade point average for the same four subjects (represented as GPA). Table A.3.3 and Table A.3.4 show the content of these two data sets.

Table A.3.3

<u>Contents of Data Set D</u>

NAME	SOCSEC	SATV	SATM
John	232882121	520	500
Sally	222773454	610	640
Fred	211447653	490	470
Emma	222671234	550	560

Table A.3.4

<u>Contents of Data Set E</u>

NAME	SOCSEC	GPA
John	232882121	2.70
Sally	222773454	3.25
Fred	211447653	2.20
Emma	222671234	2.50

Assume that, in conducting your research, you would like to compute the correlation between SATV and GPA (let's forget for the moment that you really shouldn't perform a correlation using such a small sample!). On the positive side, computing this correlation should be possible because you do have values for these two variables for all four of your subjects. On the negative side however, you will not be able to compute this correlation until both variables appear in the same data set. Therefore, it will be necessary to merge the variables contained in data sets D and E.

There are actually two ways of merging data sets. Perhaps the simplest way is the one-to-one merging approach. With **one-to-one merging**, observations are simply merged according to their order of appearance in the data sets. For example, imagine that you were to merge data sets D and E using one-to-one merging. In doing this, the SAS System would take the first observation from data set D and pair it with the first observation from data set E, and the result would become the first observation in the new data set (data set F). If the observations in data sets D and E were in exactly the correct sequence, this method would work fine. Unfortunately, if any of the observations were out of sequence, or if one data set contained more observations than another, then this approach could result in the incorrect pairing of observations. For this reason, this text recommends a different strategy for merging: the match-merging approach, described next. For users who are interested in learning more about one-to-one merging, consult Chapters 4 and 9 of *SAS language: reference* or Chapter 17 of *SAS language and procedures: usage*.

Match-merging seems to be the method that is least likely to produce undesirable results. With **match-merging**, both data sets must contain a common variable, so that values for this common variable can be used to combine observations from the two previously existing data sets into observations for the new data set. For example, consider data sets D and E from Table A.3.3 and Table A.3.4. The variable SOCSEC appears in both of these data sets, so it is a common variable. When the SAS System uses match-merging to merge these two data sets according to values on SOCSEC, it will

- read the social security number for the first subject in data set D

- look for a subject in data set E who has the same social security number

- merge the information from that subject's observation in data set D with his or her information from data set E (if it finds a subject in data set E with the same social security number)

- combine the information into a single observation in the new data set, F

- repeat this process for all subjects.

As the preceding suggests, the variable that you use as your common variable must be chosen very carefully. Ideally, each subject should be assigned a *unique value* for this common variable. This means that no two subjects should have the same value for the common variable. This objective should be achieved when social security numbers are used as the common variable, because no two people are given the same social security number.

The SAS program for match-merging data sets is somewhat more complex than the program for concatenating data sets. In part, this is because both previously existing data sets must be sorted according to values for the common variable prior to merging. This means that the observations must be rearranged in ascending order with respect to values for the common variable. Fortunately, this is easy to do with PROC SORT, a SAS System procedure that allows you to sort variables. This section shows how PROC SORT can be used to achieve this.

The general form for match-merging two previously existing data sets is presented as follows:

```
PROC SORT  DATA=data-set-1;
   BY  common-variable;
   RUN;
```

```
PROC SORT  DATA=data-set-2;
   BY  common-variable;
   RUN;


DATA  new-data-set-name;
   MERGE  data-set-1  data-set-2;
   BY  common-variable;
   RUN;
```

To illustrate, assume that you wish to match-merge data sets D and E from Table A.3.3 and Table A.3.4. To do this, use SOCSEC as the common variable. In the following program, these two data sets are input, sorted, and then merged using the match-merge approach:

```
1       DATA  D;
2          INPUT  #1   @1   (NAME)   ($7.)
3                       @9   (SOCSEC) (9.)
4                       @19  (SATV)   (3.)
5                       @23  (SATM)   (3.)   ;
6       CARDS;
7       John      232882121 520 500
8       Sally     222773454 610 640
9       Fred      211447653 490 470
10      Emma      222671234 550 560
11      ;
12
13
14      DATA  E;
15         INPUT  #1   @1   (NAME)   ($7.)
16                      @9   (SOCSEC) (9.)
17                      @19  (GPA)    (4.)   ;
18      CARDS;
19      John      232882121 2.70
20      Sally     222773454 3.25
21      Fred      211447653 2.20
22      Emma      222671234 2.50
23      ;
24
25      PROC SORT  DATA=D;
26          BY  SOCSEC;
27          RUN;
28
```

```
29      PROC SORT  DATA=E;
30          BY  SOCSEC;
31          RUN;
32
33      DATA  F;
34          MERGE  D  E;
35          BY  SOCSEC;
36          RUN;
37
38      PROC PRINT  DATA=F;
39          RUN;
```

In the preceding program, data set D was input in lines 1–11, and data set E was input in lines 14–23. In lines 25–31, both data sets were sorted according to values for SOSEC, and the two data sets were merged according to values of SOSEC in lines 33–36. Finally, the PROC PRINT on lines 38–39 requests a printout of the raw data contained in the new data set.

Output A.3.2 contains the results of PROC PRINT, which printed the raw data now contained in data set F. You can see that each observation in this new data set now contains the merged data from the two previously existing data sets D and E. For example, the line for the subject named Fred now contains his scores on the verbal and math sections of the SAT (which came from data set D), as well as his grade point average score (which came from data set E). The same is true for the remaining subjects. It would now be possible to correlate SATV with GPA, if that analysis were desired.

OBS	NAME	SOCSEC	SATV	SATM	GPA
1	Fred	211447653	490	470	2.20
2	Emma	222671234	550	560	2.50
3	Sally	222773454	610	640	3.25
4	John	232882121	520	500	2.70

Output A.3.2: Results of Performing PROC PRINT on Data Set F

Notice that the observations in Output A.3.2 are not in the same order in which they appeared in Tables A.3.3 and A.3.4. This is because they have now been sorted according to values for SOCSEC by the PROC SORT statements in the preceding SAS program.

Conclusion

After completing this chapter, you should be prepared to modify data sets, isolate subgroups of subjects for analysis, and perform other tasks that are often required when performing empirical research in the social sciences. At this point, you should be prepared to move on to the stage of analyzing data to determine what they *mean*. Some of the most simple statistics for this purpose (descriptive statistics and related procedures) are covered in the following chapter.

References

Hatcher, L. & Stepanski, E. (1994). *A step-by-step approach to using the SAS system for univariate and multivariate statistics.* Cary, NC: SAS Institute Inc.

SAS Institute Inc. (1990). *SAS language: reference, version 6, first edition.* Cary, NC: SAS Institute Inc.

SAS Institute Inc. (1989). *SAS language and procedures: usage, version 6, first edition.* Cary, NC: SAS Institute Inc.

EXPLORING DATA WITH PROC MEANS, PROC FREQ, PROC PRINT, AND PROC UNIVARIATE

> **Overview**. This appendix illustrates the use of four procedures: PROC MEANS, which may be used to calculate means, standard deviations, and other descriptive statistics for quantitative variables; PROC FREQ, which may be used to construct frequency distributions; PROC PRINT, which may be used to create a printout of the raw data set; and PROC UNIVARIATE, which may be used to test for normality and produce stem-and-leaf plots. Once data are keyed, these procedures can be used to screen for errors, test statistical assumptions, and obtain simple descriptive statistics to be reported in the research article.

Introduction: Why Perform Simple Descriptive Analyses?

The procedures discussed in this appendix are useful for (at least) three important purposes. The first purpose involves the concept of data screening. **Data screening** is the process of carefully reviewing the data to ensure that they were keyed correctly and are being read correctly by the computer. Before conducting any of the more sophisticated analyses to be described in this manual, you should carefully screen your data to make sure that you are not analyzing "garbage": numbers that were accidently punched while keying, impossible values on variables that no one could have obtained, numbers which were keyed in the wrong column, and so forth. The process of data screening does not guarantee that your data are correct, but it does increase the likelihood.

Second, these procedures are useful because they allow you to explore the shape of your data distribution. Among other things, understanding the shape of your data will help you choose the appropriate measure of central tendency (i.e., the mean versus the median). In addition, some statistical procedures require that sample data be drawn from a normally-distributed population, or at least that the sample data do not display a *marked* departure from normality. You can use the procedures discussed here to produce graphic plots of the data, as well as test the null hypothesis that the data are from a normal population.

Finally, the nature of an investigator's research question itself may require the use of a procedure such as PROC MEANS or PROC FREQ to obtain a desired statistic. For example, if your research question is "what is the average age at which women married in 1991?" you could obtain data from a representative sample of women who married in that year, analyze their ages with PROC MEANS, and review the results to determine the mean age.

Similarly, in almost any research article it is desirable to report demographic information about the sample. For example, if a study is performed on a sample that includes subjects from a variety of demographic groups, it is desirable to report the percent of subjects of each gender, the percent of subjects by race, the mean age, and so forth. You can also use PROC MEANS and PROC FREQ to obtain this information.

Example: A Revised Volunteerism Survey

To help illustrate these procedures, assume that you conduct a scaled-down version of your study on volunteerism. You construct a new questionnaire which asks just one question related to helping behavior. The questionnaire also contains an item that assesses subject sex, and another that determines the subject's class in college (e.g., freshman, sophomore). A reproduction of the questionnaire follows:

```
Please indicate the extent to which you agree or disagree with the
following statement:

   1  "I feel a personal responsibility to help needy people in my
      community." (please check your response below)

      (5) _____ Agree Strongly
      (4) _____ Agree Somewhat
      (3) _____ Neither Agree nor Disagree
      (2) _____ Disagree Somewhat
      (1) _____ Disagree Strongly


   2.  Your sex (please check one):

      (F) _____ Female
      (M) _____ Male


   3.  Your classification as a college student:

      (1) _____ Freshman
      (2) _____ Sophomore
      (3) _____ Junior
      (4) _____ Senior
      (5) _____ Other
```

Notice that this instrument has been printed so that keying the data will be relatively simple. With each variable, the value which will be keyed appears to the left of the corresponding subject response. For example, with question 1 the value "5" appears to the left of "Agree Strongly". This means that the number "5" will be keyed for any subject checking that response. For subjects checking "Disagree Strongly", a "1" will be keyed. Similarly, notice that, for question 2, the letter "F" appears to the left of "Female", so an "F" will be keyed for subjects checking this response.

The following format is used when keying the data:

Column	Variable Name	Explanation
1	RESNEEDY	Responses to question 1: Subject's perceived responsibility to help the needy
2	(blank)	(blank)
3	SEX	Responses to question 2: Subject's sex
4	(blank)	(blank)
5	CLASS	Responses to question 3: Subject's classification as a college student

You administer the questionnaire to 14 students. The following is the entire SAS program used to analyze the data, including the raw data:

```
1       DATA D1;
2          INPUT   #1   @1   (RESNEEDY)   (1.)
3                        @3   (SEX)        ($1.)
4                        @5   (CLASS)      (1.)    ;
5       CARDS;
6       5 F 1
7       4 M 1
8       5 F 1
9         F 1
10      4 F 1
11      4 F 2
12      1 F 2
13      4 F 2
14      1 F 3
15      5 M
16      4 F 4
17      4 M 4
18      3 F
19      4 F 5
20      ;
```

```
21          PROC MEANS    DATA=D1;
22             VAR RESNEEDY CLASS;
23             RUN;
24          PROC FREQ    DATA=D1;
25             TABLES SEX CLASS RESNEEDY;
26             RUN;
27          PROC PRINT    DATA=D1;
28             VAR RESNEEDY SEX CLASS;
29             RUN;
```

The data obtained from the first subject appear on line 6 of the preceding program. This subject has a value of "5" on the RESNEEDY variable (indicating that she checked "Agree Strongly"), has a value of "F" on the SEX variable (indicating that she is a female), and has a value on "1" on the CLASS variable (indicating that she is a freshman).

Notice that there are some missing data in this data set. On line 9 in the program, you can see that this subject indicated that she was a female freshman, but did not answer question 1. That is why the corresponding space in column 1 is left blank. In addition, there appears to be missing data for the CLASS variable on lines 15 and 18. Missing data are common in research of this nature (questionnaire research).

Computing Descriptive Statistics with PROC MEANS

You can use PROC MEANS to analyze quantitative (numeric) variables. For each variable analyzed, it provides the following information:

- The number of useful cases on which calculations were performed (abbreviated "N" in the output)

- The mean

- The standard deviation

- The minimum (smallest) value observed

- The maximum (largest) value observed

These statistics are produced by default, and some additional statistics (to be described later) may also be requested as an option.

Here is the general form for PROC MEANS:

```
PROC MEANS   DATA=data-set-name
             option-list
             statistic-keyword-list ;
   VAR  variable-list  ;
   RUN;
```

The PROC MEANS statement. The PROC MEANS statement begins with "PROC MEANS" and ends with a semicolon. It is recommended (and on some platforms it is required) that the statement should also specify the name of the data set to be analyzed with the DATA= option.

The "option-list" appearing in the preceding program indicates that you can request a number of options with PROC MEANS, and a complete list of these appears in the *SAS procedures guide* (1990). Some options especially useful for social science research are:

MAXDEC=n

> Specifies the maximum number of decimal places (digits to the right of the decimal point) to be used when printing results; possible range is 0 to 8.

VARDEF=divisor

> Specifies the devisor to be used when calculating variances and covariances. Two possible divisors are:

> > VARDEF=DF   Divisor is the degrees of freedom for the analysis: ($n–1$). This is the default.

> > VARDEF=N   Divisor is the number of observations, n.

The "statistic-keyword-list" appearing in the program indicates that you can request a number of statistics to replace the default output. Some statistics that may be of particular value in social science research include the following; see the *SAS procedures guide* (1990) for a more complete listing:

NMISS   The number of observations in the sample that displayed missing data for this variable.

RANGE   The range of values displayed in the sample.

SUM   The sum.

CSS   The corrected sum of squares.

USS   The uncorrected sum of squares.

VAR   The variance.

STDERR   The standard error of the mean.

SKEWNESS   The skewness displayed by the sample. **Skewness** refers to the extent to which the sample distribution departs from the normal curve because of a long "tail" on one side of the distribution. If the long tail appears on the right side of the sample distribution (where the higher values appear), it is described as being **positively skewed**. If the long tail appears on the left side of the distribution (where the lower values appear), it is described as being **negatively skewed**.

KURTOSIS   The kurtosis displayed by the sample. **Kurtosis** refers to the extent to which the sample distribution departs from the normal curve because it is either peaked or flat. If the sample distribution is relatively peaked (tall and skinny), it is described as being **leptokurtic**. If the distribution is relatively flat, it is described as being **platykurtic**.

T   The obtained value of Student's *t* test for testing the null hypothesis that the population mean is zero.

PRT   The *p* value for the preceding *t* test; that is, the probability of obtaining a *t* value this large or larger if the population mean were zero.

To illustrate the use of these options and statistic keywords, assume that you wish to use the MAXDEC= option to limit the printing of results to two decimal places, use the VAR keyword to request that the variances of all quantitative variables be printed, and use the KURTOSIS keyword to request that the kurtosis of all quantitative variables be printed. You could do this with the following PROC MEANS statement:

```
PROC MEANS   DATA=D1   MAXDEC=2   VAR   KURTOSIS ;
```

The VAR statement. Here again is the general form of the statements requesting the MEANS procedure, including the VAR statement:

```
PROC MEANS   DATA=data-set-name
             option-list
             statistic-keyword-list ;
  VAR  variable-list  ;
  RUN;
```

In the place of "variable-list" in the preceding VAR statement, you may list the quantitative variables to be analyzed. Each variable name should be separated by at least one blank space. If no VAR statement is used, SAS will perform PROC MEANS on all of the quantitative variables in the data set. This is true for many other SAS procedures as well, as explained in the following note:

What happens if I do not include a VAR statement? For many SAS System procedures, failure to include a VAR statement causes the system to perform the requested analyses on *all* variables in the data set. For data sets with a large number of variables, leaving off the VAR statement may therefore unintentionally result in a very long output file.

The program used to analyze your data set included the following statements; RESNEEDY and CLASS were specified in the VAR statement so that descriptive statistics would be calculated for both variables:

```
PROC MEANS    DATA=D1;
   VAR RESNEEDY CLASS;
   RUN;
```

Variable	N	Mean	Std Dev	Minimum	Maximum
RESNEEDY	13	3.6923077	1.3155870	1.0000000	5.0000000
CLASS	12	2.2500000	1.4222262	1.0000000	5.0000000

Output A.4.1: Results of the MEANS Procedure

Reviewing the output. Output A.4.1 contains the results created by the preceding program. Before doing any more sophisticated analyses, you should always perform PROC MEANS on each quantitative variable and carefully review the output to ensure that everything looks right. Under the heading "Variable" is the name of each variable being analyzed. The statistics for that variable appear to the right of the variable name. Below the heading "N" is the number of valid cases, or useful cases, on which calculations were performed. Notice that, in this instance, calculations were performed on only 13 cases for RESNEEDY. This may come as a surprise, because the data set actually contains 14 cases. However, recall that one subject did not respond to this question (question 1 on the survey). It is for this reason that N is equal to 13 rather than 14 for RESNEEDY in these analyses.

You should next review the mean for the variable to verify that it is a reasonable number. Remember that, with question 1, responses could range from 1 (for "Disagree Strongly") to 5 (for "Agree Strongly"). Therefore, the mean response should be somewhere between 1.00 and 5.00 for the RESNEEDY variable. If it is outside of this range, you will know that some type of error has been made. In the present case the mean for RESNEEDY is 3.69, which is within the predetermined range, so everything looks correct so far.

Using the same reasoning, it is wise to next check the column headed "Minimum". Here you will find the lowest value on RESNEEDY which appeared in the data set. If this is less than 1.00, you will again know that some type of error was made, because 1 was the lowest value that could have been assigned to a subject. On the printout, the minimum value is 1.00, which indicates no problems. Under "Maximum", the largest value observed for that variable is reported. This should not exceed 5.00, because 5 was the largest score a subject could obtain on item 1. The reported maximum value is 5.00, so again it looks like there were no obvious errors in keying the data or writing the program.

Once you have reviewed the results for RESNEEDY, you should also inspect the results for CLASS. If any of the observed values are out of bounds, you should carefully review the program for programming errors, and the data set for miskeyed data. In some cases, you may choose to use PROC PRINT to print out the raw data set because this makes the review easier. PROC PRINT is described later in this appendix.

Creating Frequency Tables with PROC FREQ

The FREQ procedure produces frequency distributions for quantitative variables as well as classification variables. For example, you may use PROC FREQ to determine the percent of subjects who "agreed strongly" with a statement on a questionnaire, the percent who "agreed somewhat," and so forth.

The PROC FREQ and TABLES statements. The general form for the procedure is as follows:

```
PROC FREQ   DATA=data-set-name;
   TABLES   variable-list  /   options;
   RUN;
```

In the TABLES statement, you list the names of the variables to be analyzed, with each name separated by at least one blank space. Below are the PROC FREQ and TABLES statements from the program presented earlier in this chapter (analyzing data from the volunteerism survey):

```
    PROC FREQ   DATA=D1;
      TABLES SEX CLASS RESNEEDY;
      RUN;
```

Reviewing the output. These statements will cause the SAS System to create three frequency distributions: one for the SEX variable, one for CLASS, and one for RESNEEDY. A reproduction of this output appears in Output A.4.2.

```
                                   Cumulative  Cumulative
        SEX    Frequency   Percent  Frequency    Percent
        ---------------------------------------------------
        F          11       78.6       11         78.6
        M           3       21.4       14        100.0



                                   Cumulative  Cumulative
       CLASS   Frequency   Percent  Frequency    Percent
       ----------------------------------------------------
         1          5       41.7        5         41.7
         2          3       25.0        8         66.7
         3          1        8.3        9         75.0
         4          2       16.7       11         91.7
         5          1        8.3       12        100.0

                    Frequency Missing = 2



                                   Cumulative  Cumulative
      RESNEEDY  Frequency  Percent  Frequency    Percent
      ----------------------------------------------------
         1          2       15.4        2         15.4
         3          1        7.7        3         23.1
         4          7       53.8       10         76.9
         5          3       23.1       13        100.0

                    Frequency Missing = 1
```

Output A.4.2: Results of the FREQ Procedure

Output A.4.2 shows that the variable name for the variable being analyzed appears on the far left side of the frequency distribution, just above the dotted line. The values assumed by the variable appear below this variable name. The first distribution provides information about the SEX variable, and below the word "SEX" appear the values "F" and "M." Information about female subjects appears to the right of "F", and information about males appears to the right of "M". When reviewing a frequency distribution, it is useful to think of these different values as representing categories to which a subject may belong.

Under the heading "Frequency", the output indicates the number of individuals who belong in a given category. Here, you can see that 11 subjects were female, while 3 were male. Below

"Percent", the percent of subjects in each category appears. The table shows that 78.6% of the subjects were female, while 21.4% were male.

Under "Cumulative Frequency" is the number of observations that appear in the current category plus all of the preceding categories. For example, the first (top) category for SEX was "female". There were 11 subjects in that category, so the cumulative frequency was 11. The next category was "male", and there were 3 subjects in that category. The cumulative frequency for the "male" category was therefore 14 (because 11 + 3 = 14). In the same way, the "Cumulative Percent" category provides the percent of observations that appear in the current category plus all of the preceding categories.

The next table presents results for the CLASS variable. Notice that just below this table appears "Frequency Missing = 2". This indicates that there were missing data for two subjects on the CLASS variable, a fact that may be verified by reviewing the data as they appear in the preceding program: CLASS values are blank for two subjects. The existence of two missing values means that there are only 12 valid cases for the CLASS variable. In support of this, the "Cumulative Frequency" column shows an ultimate value (at the bottom of the table) of only 12.

If no subject appears in a given category, the value representing that category will not appear in the frequency distribution at all. This is demonstrated with the third table, which presents the frequency distribution for the RESNEEDY variable. Notice that, under the "RESNEEDY" heading, you may find only the values "1", "3", "4", and "5". The value "2" does not appear because none of the subjects checked "Disagree Somewhat" for question 1.

How do I include missing data in my frequency table? By default, missing values are not printed as values in the frequency table and do not go into the calculation of the percentages, cumulative frequencies, or cumulative percentages. However, it is possible to include missing values in the table by including the MISSPRINT option in the TABLES statement. In the resulting table, missing values will be identified with a period ("."). In addition, it is also possible to request that missing values be included in the computation of statistics by including the MISSING option in the TABLES statement. An example of statements requesting these options is:

```
PROC FREQ   DATA=D1;
   TABLES SEX CLASS RESNEEDY  /  MISSPRINT   MISSING;
   RUN;
```

Printing Raw Data with PROC PRINT

PROC PRINT can be used to create a printout of your raw data as it exists in the computer's internal memory. The output of PROC PRINT shows each subject's value on each of the variables requested. You may use this procedure with both quantitative variables and classification variables. The general form is:

```
PROC PRINT   DATA=data-set-name;
   VAR  variable-list  ;
   RUN;
```

In the variable list, you may request any variable that has been specified in the INPUT statement, as well as any new variable which has been created from existing variables. If you do not include the VAR statement, then all existing variables will be printed. The program presented earlier in this appendix included the following PROC PRINT statements:

```
PROC PRINT   DATA=D1;
   VAR RESNEEDY SEX CLASS;
   RUN;
```

These statements produce Output A.4.3.

OBS	RESNEEDY	SEX	CLASS
1	5	F	1
2	4	M	1
3	5	F	1
4	.	F	1
5	4	F	1
6	4	F	2
7	1	F	2
8	4	F	2
9	1	F	3
10	5	M	.
11	4	F	4
12	4	M	4
13	3	F	.
14	4	F	5

Output A.4.3: Results of the PRINT Procedure

The first column of output is headed "OBS" for "observation." This variable is created by SAS to give an observation number to each subject. The second column provides the raw data for the RESNEEDY variable, the third column displays the SEX variable, and the last displays the CLASS variable. The output shows that observation 1 (subject 1) displayed a value of 5 on RESNEEDY, was a female, and displayed a value of 1 on the CLASS variable. Notice that SAS prints periods where missing values appeared.

PROC PRINT is helpful for verifying that your data are keyed correctly and that SAS is reading the data correctly. It is particularly useful for studies with a large number of variables such as when you use a questionnaire with a large number of questions. In these situations, it is often difficult to visually inspect the data as they exist in the SAS program file. In the SAS program file, subject responses are often keyed immediately adjacent to each other, making it difficult, for example, to determine just which number represents question 24 as opposed to question 25. When the data are printed using PROC PRINT, however, the variables are separated from each other and are clearly labeled with their variable names.

After keying questionnaire data, you should compare the results of PROC PRINT with several of the questionnaires as they were actually filled out by subjects. Verify that a subject's responses on the PROC PRINT output correspond to the original responses on the questionnaire. If not, it is likely that mistakes were made in either keying the data or in writing the SAS program.

Testing for Normality with PROC UNIVARIATE

The normal distribution is a symmetrical, bell-shaped theoretical distribution of values. The shape of the normal distribution is portrayed in Figure A.4.1.

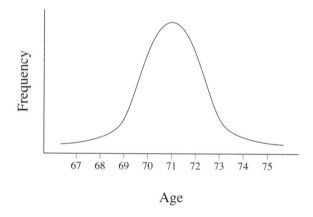

Figure A.4.1: The Normal Distribution

To understand the distribution in Figure A.4.1, assume that you are interested in conducting research on people who live in retirement communities. Imagine for a moment that it is possible to assess the age of every person in this population. To summarize this distribution, you prepare a figure similar to Figure A.4.1: the variable AGE is plotted on the horizontal axis, and the frequency of persons at each age is plotted on the vertical axis. Figure A.4.1 suggests that many of your subjects are around 71 years of age, since the distributions of ages "peaks" near the age of 71. This suggests that the mean of this distribution will likely be somewhere around 71. Notice also that most of your subjects' ages are between 67 (near the lower end of the distribution) and 75 (near the upper end of the distribution). This is the approximate range of ages that we would expect for persons living in a retirement community.

Why test for normality? Normality is an important concept in data analysis because there are at least two problems that may result when data are not normally distributed. The first problem is that markedly non-normal data may lead to incorrect conclusions in inferential statistical analyses. Many inferential procedures are based on the assumption that the sample of observations was drawn from a normally distributed population. If this assumption is violated, the statistic may give misleading findings. For example, the independent groups *t* test assumes that both samples in the study were drawn from normally distributed populations. If this assumption is violated, then performing the analysis may cause you to incorrectly reject the null hypothesis (or incorrectly fail to reject the null hypothesis). Under these circumstances, you should instead analyze the data using some procedure that does not assume normality (perhaps some nonparametric procedure).

The second problem is that markedly non-normal data may have a biasing effect on correlation coefficients, as well as more sophisticated procedures that are based on correlation coefficients. For example, assume that you compute the Pearson correlation between two variables. If one or both of these variables are markedly non-normal, this may cause your obtained correlation coefficient to be much larger (or much smaller) than the actual correlation between these variables in the population. Your obtained correlation is, essentially, just garbage. To make matters worse, many sophisticated data analysis procedures (such as principal component analysis) are actually performed on a matrix of correlation coefficients. If some or all of these correlations are distorted due to departures from normality, then the results of the analyses may again be misleading. For this reason, many experts recommend that researchers routinely check their data for major departures from normality prior to performing sophisticated analyses such as principal component analysis (e.g., Rummel, 1970).

Departures from normality. Assume that you draw a random sample of 18 subjects from your population of persons living in retirement communities. There are a wide variety of ways that your data may display a departure from normality.

Figure A.4.2 shows the distribution of ages in two samples of subjects drawn from the population of retirees. This figure is somewhat different from Figure A.4.1 because the distributions have been "turned on their sides" so that age is now plotted on the vertical axis rather than on the horizontal axis (this is so that these figures will be more similar to the stem-and-leaf plots produced by PROC UNIVARIATE, discussed in a later section) . Each small circle in a given distribution of Figure A.4.2 represents one subject. For example, in the distribution for Sample A, you can see that there is one subject at age 75, one subject at age 74, two subjects at age 73, three subjects at age 72, and so forth. The ages of the 18 subjects in Sample A range from a low of 67 to a high of 75.

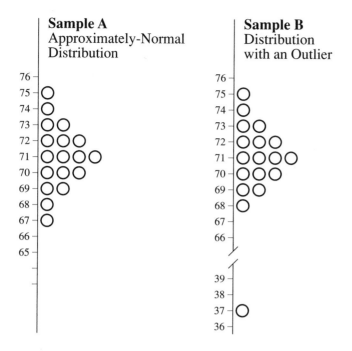

Figure A.4.2: Sample with an Approximately Normal Distribution and a Sample with an Outlier

The data in Sample A form an approximately normal distribution (called *approximately normal* because it is difficult to form a perfectly normal distribution using a small sample of just 18 cases). An inferential test (discussed later) will show that Sample A does not demonstrate a significant departure from normality. Therefore, it would probably be appropriate to include the data in Sample A in an independent samples *t* test, for example.

In contrast, there are problems with the data in Sample B. Notice that its distribution is very similar to that of Sample A, except that there is an outlier at the lower end of the distribution. An **outlier** is an extreme value that differs substantially in size from the other values in the distribution. In this case, the outlier represents a subject whose age is only 37. Obviously, this person's age is markedly different from that of the other subjects in your study. Later, you will see that this outlier causes the data set to demonstrate a significant departure from normality, making the data inappropriate for some statistical procedures. When you observe an outlier such as this, it is important to determine whether it should be either corrected or simply deleted from the data set. Obviously, if the outlier exists because an error was made in keying the data, it should be corrected.

A sample may also depart from normality because it displays kurtosis. **Kurtosis** refers to the peakedness of the distribution. The two samples displayed in Figure A.4.3 demonstrate different types of kurtosis:

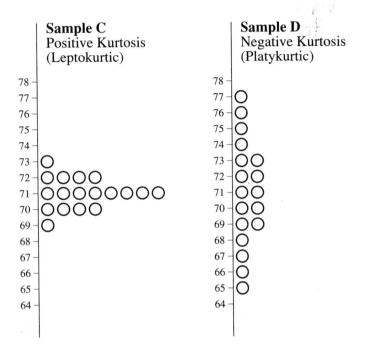

Figure A.4.3: Samples Displaying Positive versus Negative Kurtosis

Sample C in Figure A.4.3 displays **positive kurtosis**, which means that the distribution is relatively peaked (tall and skinny) rather than flat. Notice that, with Sample C, there are a relatively large number of subjects who cluster around the central part of the distribution (around age 71). This is what makes the distribution peaked (relative to Sample A, for example). Distributions with positive kurtosis are also called **leptokurtic**.

In contrast, Sample D in the same figure displays **negative kurtosis**, which means that the distribution is relatively flat. Flat distributions are sometimes also described as being **platykurtic**.

In addition to kurtosis, distributions may also demonstrate varying degrees of **skewness**, or sidedness. A distribution is skewed if the tail on one side of the distribution is longer than the tail on the other side. The distributions in Figure A.4.4 show two different types of skewness:

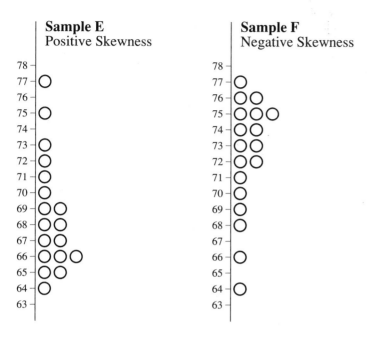

Figure A.4.4: Samples Displaying Positive versus Negative Skewness

Consider Sample E in Figure A.4.4. Notice that the biggest number of subjects in this distribution tend to cluster around the age of 66. The tail of the distribution that stretches above 66 (from 67 to 77) is relatively long, while the tail of the distribution that stretches below 66 (from 65 to 64) is relatively short. Clearly, this distribution is skewed. A distribution is said to display **positive skewness** if the longer tail of a distribution points in the direction of *higher* values. You can see that Sample E displays positive skewness, because its longer tail points toward larger numbers such as 75, 77, and so forth.

On the other hand, if the longer tail of a distribution points in the direction of lower values, the distribution is said to display **negative skewness**. You can see that Sample F of Figure A.4.4 displays negative skewness because in that sample the longer tail points downward, in the direction of lower values (such as 66 and 64).

General form for PROC UNIVARIATE. Like the MEANS procedure, PROC UNIVARIATE provides a number of descriptive statistics for quantitative variables, including the mean, standard deviation, kurtosis, and skewness. However, PROC UNIVARIATE has the added advantage of also printing a significance test for the null hypothesis that the data come from a normally distributed population. The procedure also provides plots that will help you understand the shape of your sample's distribution, along with additional information that will help you understand *why* your data depart from normality (if, indeed, they do). This text describes just a few of the features of PROC UNIVARIATE; for a complete listing, see the *SAS procedures guide* (1990).

Here is the general form for the PROC UNIVARIATE statements that produce the output discussed in this chapter:

```
PROC UNIVARIATE   DATA=data-set-name   NORMAL   PLOT;
   VAR variable-list;
   ID identification-variable;
   RUN;
```

In the preceding program, the NORMAL option requests a significance test for the null hypothesis that the sample data are from a normally distributed population. The Shapiro-Wilk statistic is printed for samples of 2000 or less; for larger samples the Kolmogorov statistic is printed. See Chapter 42 of the *SAS procedures guide* (1990) for details.

The PLOT option of the preceding program produces a stem-and-leaf plot, a box plot, and a normal probability plot, each of which is useful for understanding the shape of the sample's distribution. This text shows how to interpret the stem-and-leaf plot; see Schlotzhauer and Littell (1991) for guidance in interpreting box plots and normal probability plots.

The names of the variables to be analyzed should be listed in the VAR statement. The ID statement is optional, and is useful for identifying outliers. PROC UNIVARIATE prints an "extremes" table that lists the five largest and five smallest values in the data set. These values are identified by the identification variable listed in the ID statement. For example, assume that AGE (subject age) is listed in the VAR statement, and SOCSEC (for subject social security number) is listed in the ID statement. PROC UNIVARIATE will print the social security numbers for the subjects with the five largest and five smallest values on AGE. This should make it easier to identify the specific subject who contributes an outlier to your data set (this use of the extremes table is illustrated here).

Results for an approximately normal distribution. For purposes of illustration, assume that you wish to analyze the data that are illustrated as Sample A of Figure A.4.2 (the approximately normal distribution). You prepare a SAS program in which subject age is keyed in a variable called AGE, and subject identification numbers are keyed in a variable called SUBJECT. Here is the entire program that will input these data and analyze them using PROC UNIVARIATE:

```
 1             DATA D1;
 2                INPUT   #1   @1   (SUBJECT)   (2.)
 3                             @4   (AGE)       (2.)    ;
 4             CARDS;
 5              1 72
 6              2 69
 7              3 75
 8              4 71
 9              5 71
10              6 73
11              7 70
12              8 67
13              9 71
14             10 72
15             11 73
16             12 68
17             13 69
18             14 70
19             15 70
20             16 71
21             17 74
22             18 72
23             ;
24             PROC UNIVARIATE   DATA=D1   NORMAL   PLOT;
25                VAR AGE;
26                ID SUBJECT;
27                RUN;
```

The preceding program requests that PROC UNIVARIATE be performed on the variable AGE. Values of the variable SUBJECT will be used to identify outlying values of AGE in the extremes table.

With LINESIZE=80, the preceding program would produce two pages of output. This output would contain:

- a **moments table** that includes the mean, standard deviation, variance, skewness, kurtosis, and normality test, along with other statistics

- a **quantiles table** that provides the mode, median, 25th percentile, 75th percentile, and related information

- an **extremes table** that provides the five highest values and five lowest values on the variable being analyzed

- a stem-and-leaf plot, box plot, and normal probability plot.

Output A.4.4 includes the moments table, quantiles table, and extremes table for the analysis of Sample A.

```
                         Univariate Procedure

Variable=AGE


                              Moments

        N                 18  Sum Wgts           18
        Mean              71  Sum              1278
        Std Dev     2.057983  Variance     4.235294
        Skewness           0  Kurtosis     -0.13576
        USS            90810  CSS                72
        CV          2.898568  Std Mean     0.485071
        T:Mean=0    146.3702  Pr>|T|         0.0001
        Num ^= 0          18  Num > 0            18
        M(Sign)            9  Pr>=|M|        0.0001
        Sgn Rank        85.5  Pr>=|S|        0.0001
        W:Normal     0.98356  Pr<W           0.9666


                         Quantiles(Def=5)

        100% Max          75        99%          75
         75% Q3           72        95%          75
         50% Med          71        90%          74
         25% Q1           70        10%          68
          0% Min          67         5%          67
                                     1%          67

        Range              8
        Q3-Q1              2
        Mode              71


                              Extremes

        Lowest      ID        Highest      ID
             67(      8)           72(      18)
             68(     12)           73(       6)
             69(     13)           73(      11)
             69(      2)           74(      17)
             70(     15)           75(       3)
```

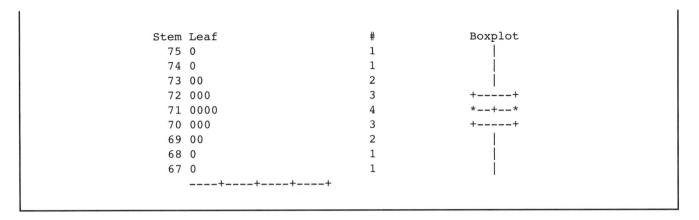

```
       Stem Leaf                        #              Boxplot
         75 0                           1                 |
         74 0                           1                 |
         73 00                          2                 |
         72 000                         3              +-----+
         71 0000                        4              *--+--*
         70 000                         3              +-----+
         69 00                          2                 |
         68 0                           1                 |
         67 0                           1                 |
            ----+----+----+----+
```

Output A.4.4: Tables from PROC UNIVARIATE for Sample A

On the far-left side of Output A.4.4, the note "Variable=AGE" indicates that AGE is the name of the variable being analyzed by PROC UNIVARIATE. The moments table is the first table reproduced in Output A.4.4. On the upper-left side of the moments table is the heading "N", and to the right of this you can see that the analysis was based on 18 observations. Below "N" are the headings "Mean" and "Std Dev"; to the right of these you can see that the mean and standard deviation for AGE were 71 and 2.06, respectively.

To the right of "Skewness" you can see that the skewness statistic for AGE is zero. In interpreting the skewness statistic, keep in mind the following:

- A skewness value of zero means that the distribution is not skewed; in other words, this means that the distribution is symmetrical, that neither tail is longer than the other.

- A positive skewness value means that the distribution is positively skewed, that the longer tail points toward higher values in the distribution (as with Sample E of Figure A.4.4).

- A negative skewness value means that the distribution is negatively skewed, that the longer tail points toward lower values in the distribution (as with Sample F of Figure A.4.4).

Since the AGE variable of Sample A displays a skewness value of zero, we know that neither tail is longer than the other in this sample.

A closer look at the moments table of Output A.4.4 shows that it actually consists of two columns of statistics. The column on the left provides statistics such as the sample size, the mean, the standard deviation, and so forth. The column on the right contains headings such as "Sum Wgts", "Sum", and "Variance". Notice that in this right-hand column, the fourth entry down has the heading "Kurtosis" (this appears just below "Variance"). To the right of "Kurtosis", you can see that the kurtosis statistic for AGE is –0.13576, which rounds to –.14. When interpreting this kurtosis statistic, keep in mind the following:

- A kurtosis value of zero means that the distribution displays no kurtosis; in other words, the distribution is neither relatively peaked nor is it relatively flat, compared to the normal distribution.

- A positive kurtosis value means that the distribution is relatively peaked, or leptokurtic.

- A negative kurtosis value means that the distribution is relatively flat, or platykurtic.

The small negative kurtosis value of –.14 in Output A.4.4 indicates that Sample A is slightly flat, or platykurtic.

The last entry in the left column of the moments table (headed "W:Normal") provides the Shapiro-Wilk statistic that tests the null hypothesis that the sample data were drawn from a normally distributed population. To the right of "W:Normal", you can see that the value for this statistic is 0.98356. To the immediate right of this statistic is its corresponding p value. This p value appears as the last entry in the right column of the moments table, to the right of the heading "Pr<W". In the present case, the p value is 0.9666. Remember that this statistic tests the null hypothesis that the sample data are from a normal distribution. This p value is very large at .9666, meaning that there are 9,666 chances in 10,000 that you would obtain the present results if the data were drawn from a normal population. In other words, it is very likely that you would obtain the present results if the sample were from a normal population. Because this statistic gives so little evidence to reject the null hypothesis, you can tentatively accept it, and assume that the sample was drawn from a normally-distributed population (this makes sense when you review the shape of the distribution of Sample A in Figure A.4.2: the sample data clearly display an approximately normal distribution). In general, you should reject the null hypothesis of normality only when the p value is less than .05.

Results for a distribution with an outlier. The data of Sample A in Figure A.4.2 displayed an approximately normal distribution. For purposes of contrast, assume that you now use PROC UNIVARIATE to analyze the data of Sample B from Figure A.4.2. You will remember that Sample B was similar in shape to sample A, except that Sample B contained an outlier: the lowest value in Sample B was 37, which was an extremely low score compared to the other values in the sample (if necessary, turn back to Figure A.4.2 at this time to verify this).

The raw data from Sample B follow. Columns 1-2 contain values of SUBJECT, the subject identification number, and columns 4-5 contain AGE values. Notice that these data are identical to those of Sample A, except for subject 8. In Sample A, subject 8's age was listed as 67; in Sample B, it is listed as 37.

```
 1 72
 2 69
 3 75
 4 71
 5 71
 6 73
 7 70
 8 37
 9 71
10 72
11 73
12 68
13 69
14 70
15 70
16 71
17 74
18 72
```

When analyzed with PROC UNIVARIATE, the preceding data would again produce two pages of output. Some of the results of this analysis are presented in Output A.4.5.

```
                          Univariate Procedure

Variable=AGE

                              Moments

            N              18    Sum Wgts         18
            Mean     69.33333    Sum            1248
            Std Dev  8.267584    Variance   68.35294
            Skewness -3.90499    Kurtosis   16.03325
            USS         87690    CSS            1162
            CV        11.9244    Std Mean   1.948688
            T:Mean=0 35.57949    Pr>|T|       0.0001
            Num ^= 0       18    Num > 0          18
            M(Sign)         9    Pr>=|M|      0.0001
            Sgn Rank     85.5    Pr>=|S|      0.0001
            W:Normal 0.458479    Pr<W         0.0001
```

```
                          Quantiles(Def=5)

            100% Max         75        99%        75
             75% Q3          72        95%        75
             50% Med         71        90%        74
             25% Q1          70        10%        68
              0% Min         37         5%        37
                                        1%        37
            Range            38
            Q3-Q1             2
            Mode             71


                             Extremes

          Lowest      ID        Highest      ID
              37(      8)           72(      18)
              68(     12)           73(       6)
              69(     13)           73(      11)
              69(      2)           74(      17)
              70(     15)           75(       3)


     Stem Leaf                       #           Boxplot
        7 5                          1              |
        7 0001111222334             13           +-----+
        6 899                        3              +
        6
        5
        5
        4
        4
        3 7                          1              *
          ----+----+----+----+
     Multiply Stem.Leaf by 10**+1
```

Output A.4.5: Tables from PROC UNIVARIATE for Sample B

By comparing the moments table of Output A.4.5 (for Sample B) to that of Output A.4.4 (for Sample A), you can see that the inclusion of the outlier has had a dramatic effect on some of the descriptive statistics for AGE. The mean of Sample B is now 69.33, down from the mean of 71 found for Sample A. More dramatic is the effect that the outlier has had on the standard deviation. With the approximately normal distribution, the standard deviation was only 2.05; with the outlier included, the standard deviation was much larger at 8.27.

Output A.4.5 shows that the skewness index for Sample B is −3.90. A negative skewness index such as this is just what you would expect: the outlier has, in essence, created a long tail that points toward the lower values in the AGE distribution, and you will remember that this normally results in a negative skewness index.

Output A.4.5 shows that the test for normality for Sample B results in a Shapiro-Wilk statistic of .458479 (to the right of "W:Normal") and a corresponding p value of .0001 (to the right of "Pr<W"). Because this p value is below .05, you reject the null hypothesis and tentatively conclude that Sample B was not drawn from a normally distributed population. In other words, you can conclude that Sample B displays a statistically significant departure from normality.

The extremes table for Sample B appears just below the quantiles table in Output A.4.5. On the left side of the extremes table, below the heading "Lowest", PROC UNIVARIATE prints the lowest values observed for the variable specified in the VAR statement (AGE, in this case). Here, you can see that the lowest values were 37, 68, 69, 69, and 70. To the immediate right of each value (appearing within parentheses) is the subject identification number for the subject who contributed that value to the data set. The subject identification variable is the variable specified in the ID statement (SUBJECT, in this case). Reviewing these values in parentheses shows you that subject 8 contributed the AGE value of 37, subject 12 contributed the AGE value of 68, and so forth. Compare the results in this extremes table with the actual raw data (reproduced earlier) to verify that these are in fact the specific subjects who provided these values on AGE.

On the right side of the extremes table, similar information is provided, although in this case it is provided for the five *highest* values observed in the data set. Under the heading "Highest" (and reading from the bottom up), you can see that the highest value on age was 75, and it was provided by subject 3; the next highest value was 74, provided by subject 17; and so forth.

This extremes table is useful for quickly identifying the specific subjects who have contributed outliers to a data set. For example, in the present case you were able to determine immediately that it was subject 8 who contributed the low outlier on AGE. Using the extremes table may be unnecessary when working with a very small data set (as in the present situation), but it can be invaluable when dealing with a large data set. For example, if you know that you have an outlier in a data set with 1,000 observations, the extremes table can immediately identify the subject who contributed the outlier. This will save you the tedious chore of examining data lines for each of the 1,000 observations individually.

Understanding the stem-and-leaf plot. A stem-and-leaf plot provides a visual representation of your data, using conventions somewhat similar to those used with Figures A.4.2, A.4.3, and A.4.4. Output A.4.6 provides the stem-and-leaf plot for Sample A (the approximately-normal distribution):

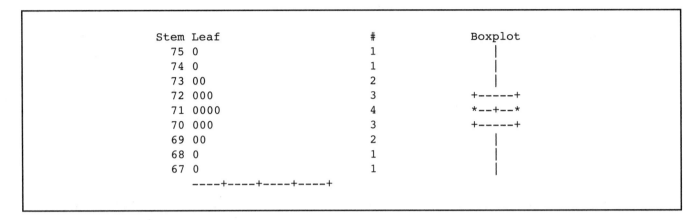

```
          Stem Leaf                   #            Boxplot
            75 0                       1               |
            74 0                       1               |
            73 00                      2               |
            72 000                     3            +-----+
            71 0000                    4            *--+--*
            70 000                     3            +-----+
            69 00                      2               |
            68 0                       1               |
            67 0                       1               |
               ----+----+----+----+
```

Output A.4.6: Stem-and-Leaf Plot from PROC UNIVARIATE for Sample A (Approximately Normal Distribution)

To understand a stem-and-leaf plot, it is necessary to think of a given subject's score on AGE as consisting of a "stem" and a "leaf." The **stem** is that part of the value that appears to the left of the decimal point, and the **leaf** consists of that part that appears to the right of the decimal point. For example, subject 8 in Sample A had a value on AGE of 67.0 For this subject, the stem is 67 (because it appears to the left of the decimal point), and the leaf is 0 (because it appears to the right). Subject 12 had a value on age of 68.0, so the stem for this value is 68, and the leaf is again 0.

In the stem-and-leaf plot of Output A.4.6, the vertical axis (running up and down) plots the various stems that could be encountered in the data set (these appear under the heading "Stem"). Reading from the top down, these stems are 75, 74, 73, and so forth. Notice that at the very bottom of the plot is the stem 67. To the right of this stem appears a single leaf (a single 0). This means that there was only one subject in Sample A with a stem-and-leaf of 67.0 (that is, a value on AGE of 67.0). Move up one line, and you see the stem 68. To the right of this, again one leaf appears (that is, one zero appears), meaning that only one subject had a score on AGE of 68.0. Move up an additional line, and you see the stem 69. To the right of this, two leaves appear (that is, two zeros appear). This means that there were two subjects with a stem-and-leaf of 69.0 (two subjects with values on AGE of 69.0). Continuing up the plot in this fashion, you can see that there were three subjects with at age 70, four subjects at age 71, three at age 72, two at age 73, one at 74, and one at 75.

On the right side of the stem-and-leaf plot appears a column headed "#". This column prints the number of observations that appear at each stem. Reading from the bottom up, this column again confirms that there was one subject with a score on age of 67, one with a score of 68, two with a score of 69, and so forth.

Reviewing the stem-and-leaf plot of Output A.4.6 shows that its shape is very similar to the shape portrayed for Sample A in Figure A.4.2. This is to be expected since both figures use similar conventions and both describe the data of Sample A. In Output A.4.6, notice that the shape of the distribution is symmetrical: neither tail is longer than the other. This, too, is to be expected since Sample A demonstrated zero skewness.

In some cases, the stem-and-leaf plot produced by UNIVARIATE will be somewhat more complex than the one reproduced in Output A.4.6. For example, Output A.4.7 includes the stem-and-leaf plot produced by Sample B from Figure A.4.2 (the distribution with an outlier). Consider the stem-and-leaf at the very bottom of this plot. The stem for this entry is 3, and the leaf is 7, meaning that the stem-and-leaf is 3.7. Does this mean that some subject had a score on AGE of 3.7? Not at all.

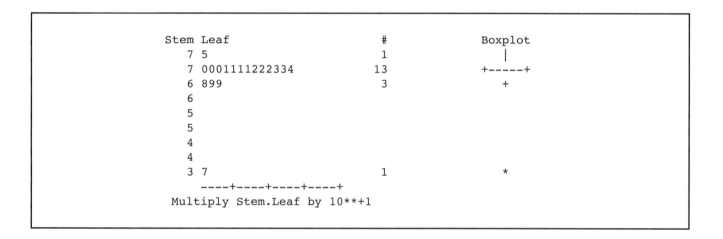

```
        Stem Leaf                      #            Boxplot
           7 5                         1               |
           7 0001111222334            13            +-----+
           6 899                       3               +
           6
           5
           5
           4
           4
           3 7                         1               *
             ----+----+----+----+
         Multiply Stem.Leaf by 10**+1
```

Output A.4.7: Stem-and-Leaf Plot from PROC UNIVARIATE for Sample B (Distribution with Outlier)

Notice the note at the bottom of this plot, which says "Multiply Stem.Leaf by 10**+1". This means "Multiply the stem-and-leaf by 10 raised to the first power." 10 raised to the first power, of course, is merely 10. This means that to find a subject's *actual* value on AGE, you must multiply a stem-and-leaf for that subject by 10.

For example, consider what this means for the stem-and-leaf at the bottom of this plot. This stem-and-leaf was 3.7. To find the actual score that corresponds to this stem-and-leaf, you would perform the following multiplication:

```
3.7 X 10 = 37
```

This means that for the subject who had a stem-and-leaf of 3.7, the actual value of AGE was 37.

Move up one line in the plot, and you come to the stem "4". However, note that there are no leaves for this stem, which means that there were no subjects with a stem of 4.0. Reading up the plot, note that no leaves appear until you reach the stem "6". The leaves on this line suggest that there was one subject with a stem-and-leaf of 6.8, and two subjects with a stem-and-leaf of 6.9. Multiply these values by 10 to determine their actual values on AGE:

```
6.8 X 10 = 68
6.9 X 10 = 69
```

Move up an additional line, and note that there are actually two stems for the value 7. The first stem (moving up the plot) includes stem-and-leaf values from 7.0 through 7.4, while the next stem includes stem-and-leaf values from 7.5 through 7.9. Reviewing values in these rows, you can see that there are three subjects with a stem-and-leaf of 7.0, four with a stem-and-leaf of 7.1, and so forth.

The note at the bottom of the plot told you to multiply each stem-and-leaf by 10 raised to the first power. However, sometimes this note will tell you to multiply by 10 raised to a different power. For example, consider the following note:

```
Multiply Stem.Leaf by 10**+2
```

This note tells you to multiply by 10 raised to the second power, or 100. Notice what some of the actual values on AGE would have been if this note had appeared (needless to say, such large values would not have made sense for the AGE variable):

```
6.8 X 100 = 680
6.9 X 100 = 690
```

All of this multiplication probably seems somewhat tedious at this point, but there is a simple rule that you can use to ease the interpretation of the note that sometimes appears at the bottom of a stem-and-leaf plot. With respect to this note, remember that the power to which 10 is raised indicates the *number of decimal places* you should move the decimal point in the stem-and-leaf. Once you have moved the decimal point this number of spaces, your stem-and-leaf will represent the actual value that you are interested in. For example, consider the following note:

```
Multiply Stem.Leaf by 10**+1
```

This note tells you to multiply the stem-and-leaf by 10 raised to the power of one; in other words, move the decimal point *one space to the right*. Imagine that you start with a stem-and-leaf of 3.7. Moving the decimal point one space to the right results in an actual value on AGE of 37. If you begin with a stem-and-leaf of 6.8, this becomes 68.

On the other hand, consider if the plot had included this note:

```
Multiply Stem.Leaf by 10**+2
```

It would have been necessary to move the decimal point *two* decimal spaces to the right. In this case, a stem-and-leaf of 3.7 would become 370, and 6.8 would become 680 (again, these values would not make sense for the AGE variable; they are used only for purposes of demonstration). Finally, remember that, if no note appears at the bottom of the plot, it is not necessary to move the decimal points in the stem-and-leaf values at all.

Results for distributions demonstrating skewness. Output A.4.8 provides some results from the PROC UNIVARIATE analysis of Sample E from Figure A.4.4. You will recall that this sample demonstrated a positive skew.

```
                          Univariate Procedure

Variable=AGE


                               Moments

              N                  18   Sum Wgts          18
              Mean        68.77778   Sum             1238
              Std Dev     3.622731   Variance    13.12418
              Skewness    0.869826   Kurtosis    0.110096
              USS            85370   CSS          223.1111
              CV          5.267299   Std Mean    0.853886
              T:Mean=0    80.54679   Pr>|T|        0.0001
              Num ^= 0         18   Num > 0           18
              M(Sign)            9   Pr>=|M|       0.0001
              Sgn Rank       85.5   Pr>=|S|       0.0001
              W:Normal    0.929487   Pr<W          0.1944


                           Quantiles(Def=5)

            100% Max          77        99%          77
             75% Q3           71        95%          77
             50% Med          68        90%          75
             25% Q1           66        10%          65
              0% Min          64         5%          64
                                        1%          64

            Range             13
            Q3-Q1              5
            Mode              66


                               Extremes

            Lowest    ID        Highest    ID
                64(     7)          71(    11)
                65(    15)          72(     6)
                65(    10)          73(    16)
                66(    14)          75(    13)
                66(     8)          77(     1)
```

```
              Stem Leaf                          #            Boxplot
               76 0                              1               |
               74 0                              1               |
               72 00                             2               |
               70 00                             2            +-----+
               68 0000                           4            *--+--*
               66 00000                          5            +-----+
               64 000                            3               |
                  ----+----+----+----+
```

Output A.4.8: Tables and Stem-and-Leaf Plot from PROC UNIVARIATE for Sample E
(Positive Skewness)

Remember that when the approximately normal distribution was analyzed, it displayed a
skewness index of zero. In contrast, note that the skewness index for Sample E in Output A.4.8
is 0.869826, which rounds to .87. This positive skewness index is what you would expect, given
the positive skew of the data. The skew is also reflected in the stem-and-leaf plot that appears in
Output A.4.8: notice the relatively long tail that points in the direction of higher values for age
(such as 74 and 76).

Although this sample displays a positive skew, it does not display a significant departure from
normality. In the moments table of Output A.4.8, you can see that the Shapiro-Wilk statistic (to
the right of "W:Normal") is .93; its corresponding p value (to the right of "Pr<W") is .1944.
Because this p value is greater than .05, you may not reject the null hypothesis that the data were
drawn from a normal population. With small samples such as the one investigated here, this test
is not very powerful (that is, is not very sensitive). This is why the sample was not found to
display a significant departure from normality, even though it was clearly skewed.

For purposes of contrast, Output A.4.9 presents the results of an analysis of Sample F from
Figure A.4.4. Sample F displayed a negative skew, and this is reflected in the skewness index of
−.87 that appears in Output A.4.9. Once again, the Shapiro-Wilk test at the bottom of the
moments table shows that the sample does not demonstrate a significant departure from
normality.

```
                           Univariate Procedure

Variable=AGE

                               Moments

                N              18   Sum Wgts          18
                Mean     72.22222   Sum            1300
                Std Dev  3.622731   Variance   13.12418
                Skewness -0.86983   Kurtosis   0.110096
                USS         94112   CSS        223.1111
                CV        5.01609   Std Mean   0.853886
```

```
          T:Mean=0    84.58064   Pr>|T|        0.0001

          Num ^= 0          18   Num > 0           18
          M(Sign)            9   Pr>=|M|       0.0001
          Sgn Rank        85.5   Pr>=|S|       0.0001
          W:Normal    0.929487   Pr<W          0.1944


                      Quantiles(Def=5)

          100% Max          77       99%          77
           75% Q3           75       95%          77
           50% Med          73       90%          76
           25% Q1           70       10%          66
            0% Min          64        5%          64
                                      1%          64

          Range             13
          Q3-Q1              5
          Mode              75


                          Extremes

          Lowest    ID      Highest    ID
              64(      11)       75(      13)
              66(       2)       75(      18)
              68(      10)       76(      12)
              69(       8)       76(      14)
              70(       5)       77(       7)


      Stem Leaf                    #      Boxplot
        76 000                     3         |
        74 00000                   5      +-----+
        72 0000                    4      *--+--*
        70 00                      2      +-----+
        68 00                      2         |
        66 0                       1         |
        64 0                       1         |
           ----+----+----+----+
```

Output A.4.9: Tables and Stem-and-Leaf Plot from PROC UNIVARIATE for Sample F (Negative Skewness)

The stem-and-leaf plot of Output A.4.9 reveals a long tail that points in the direction of lower values for AGE (such as 64 and 66). This, of course, is the type of plot that you would expect for a negatively skewed distribution.

Conclusion

Regardless of what other statistical procedures you use in an investigation, you should always *begin* the data analysis process by performing the simple analyses described here. This will help ensure that the data set and program do not contain any errors that, if left unidentified, could lead to incorrect conclusions and even to ultimate retractions of published findings. Once the data have survived this initial screening, you may move forward to the more sophisticated procedures described in this text.

References

Rummel, R. J. (1970). *Applied factor analysis.* Evanston, IL: Northwestern University Press.

SAS Institute Inc. (1990). *SAS procedures guide, version 6, third edition.* Cary, NC: SAS Institute Inc.

Schlotzhauer, S. D. & Littell, R. C. (1991). *SAS system for elementary statistical analysis.* Cary, NC: SAS Institute Inc.

Appendix A.5

PREPARING SCATTERGRAMS AND COMPUTING CORRELATIONS

> **Overview**: This appendix shows you how to use PROC PLOT to prepare bivariate scattergrams. These scattergrams can be used to verify that the relationships between the variables in your data set are linear (a necessary condition for most of the statistical procedures described in this text). This appendix also shows how to use PROC CORR to compute Pearson correlations between variables. Many of the statistical procedures described in this text allow you to input data in the form of a correlation matrix, and this appendix shows how to create such a matrix.

Introduction: When Are Pearson Correlations Appropriate?

The Pearson product-moment correlation coefficient (symbolized as r) can be used to assess the nature of the relationship between two variables when both variables are assessed on either an interval- or ratio-level of measurement. It is further assumed that both variables should include a relatively large number of values; for example, you would not use this statistic if one of the variables could assume only three values.

It would be appropriate to compute a Pearson correlation to investigate the nature of the relationship between SAT verbal test scores and grade point average (GPA). SAT verbal is assessed on an interval level of measurement, and may assume a wide variety of values (possible scores range from 200 to 800). Grade point ratio is also assessed on an interval level, and may also assume a wide variety of values from 0.00 through 4.00.

There are a number of additional assumptions that should be met before computing Pearson correlations between sets of variables (e.g., bivariate normal distribution). These assumptions are listed at the end of this appendix.

Interpreting the Coefficient

To more fully understand the nature of the relationship between the two variables studied, it is necessary to interpret two characteristics of a Pearson correlation coefficient. First, the **sign of the coefficient** tells you whether there is a positive relationship or a negative relationship between the two variables. A **positive correlation** indicates that as values on one variable increase values on the second variable also increase. A positive correlation is illustrated in Figure A.5.1, which illustrates the relationship between SAT verbal test scores and GPA in a fictitious sample of data.

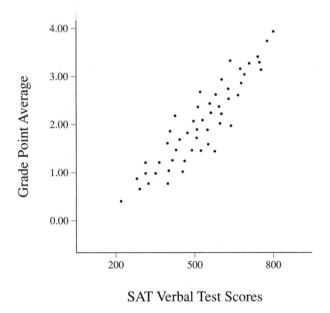

Figure A.5.1: A Positive Correlation

It can be seen that subjects who received low scores on the predictor variable (SAT verbal) also received low scores on the criterion variable (GPA); at the same time, subjects who received high scores on SAT verbal also received high scores on GPA. The two variables may therefore be said to be positively correlated.

With a **negative correlation**, on the other hand, as values on one variable increase, values on the second variable decrease. For example, you might expect to see a negative correlation between SAT verbal test scores and the number of errors that subjects make on a vocabulary test. The students with high SAT verbal scores will tend to make few mistakes, and the students with low SAT scores will tend to make many mistakes. This relationship is illustrated with fictitious data in Figure A.5.2.

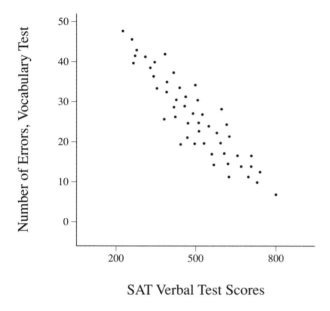

Figure A.5.2: A Negative Correlation

The second characteristic of a correlation coefficient is its **size**: the greater the absolute value of a correlation coefficient, the stronger the relationship between the two variables. Pearson correlation coefficients may range in size from −1.00 through 0.00 through +1.00. Coefficients of 0.00 indicate no relationship between the two variables. For example, if there were zero correlation between SAT scores and GPA, then knowing a person's SAT score would tell you nothing about what his or her GPA is likely to be. In contrast, correlations of −1.00 or +1.00 indicate perfect relationships. For example, if the correlation between SAT scores and GPA were 1.00, it would mean that knowing someone's SAT score would allow you to predict his or her GPA with pin-point accuracy. In the real world, however, SAT scores are not that strongly related to GPA, so you would expect the correlation between them to be less than 1.00.

Here is an approximate guide for interpreting the strength of the relationship between two variables, based on the absolute value of the coefficient:

```
±1.00 = Perfect correlation
 ±.80 = Strong correlation
 ±.50 = Moderate correlation
 ±.20 = Weak correlation
 ±.00 = No correlation
```

Remember that one considers the *absolute value* of the coefficient in interpreting its size. This is to say that a correlation of –.50 is just as strong as a correlation of +.50; a correlation of –.75 is just as strong as a correlation of +.75, and so forth.

Linear versus Nonlinear Relationships

The Pearson correlation is appropriate only if there is a linear relationship between the two variables. There is a **linear relationship** between two variables when their scattergram follows the form of a straight line. For example, it is possible to draw a straight line through the center of the scattergram presented in Figure A.5.3, and this straight line fits the pattern of the data fairly well. This means that there is a linear relationship between SAT verbal test scores and GPA.

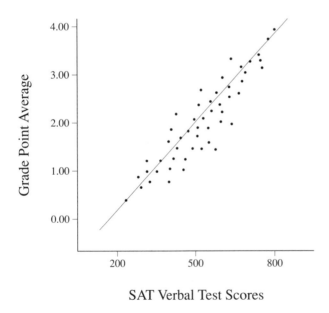

SAT Verbal Test Scores

Figure A.5.3: A Linear Relationship

In contrast, there is a **nonlinear relationship** between two variables if their scattergram does not follow the form of a straight line. For example, imagine that you have constructed a test of creativity, and have administered it to a large sample of college students. With this test, higher scores reflect higher levels of creativity. Imagine further that you obtain the SAT verbal test scores for these students, plot their SAT scores against their creativity scores, and obtain the scattergram presented in Figure A.5.4.

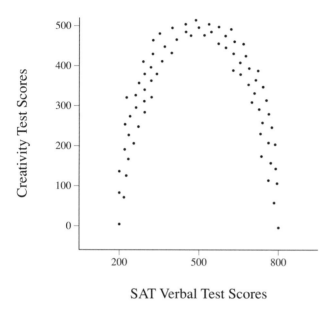

SAT Verbal Test Scores

Figure A.5.4: A Nonlinear Relationship

The scattergram in Figure A.5.4 reveals a nonlinear relationship between SAT scores and creativity. It shows that students with low SAT scores tend to have low creativity scores, students with moderate SAT scores tend to have high creativity scores, and students with high SAT scores tend to have low creativity scores. It is not possible to draw a good-fitting straight line through the data points of Figure A.5.4; this is why you can say that there is a *nonlinear* (or perhaps a *curvilinear*) relationship between SAT scores and creativity scores.

When you use the Pearson correlation to assess the relationship between two variables involved in a nonlinear relationship, the resulting correlation coefficient usually underestimates the actual strength of the relationship between the two variables. For example, computing the Pearson correlation between the SAT scores and creativity scores presented in Figure A.5.4 might result in a coefficient of .10, which would indicate only a very weak relationship between the two variables. And yet there is clearly a fairly strong relationship between SAT scores and creativity: The figure shows that, if you know what someone's SAT score is, you can predict with good accuracy what their creativity score will be.

The implication of all this is that you should always verify that there is a linear relationship between two variables before computing a Pearson correlation for those variables. One of the easiest ways of verifying that the relationship is linear is to prepare a scattergram similar to those presented in the preceding figures. Fortunately, this is very easy to do using the SAS System's PLOT procedure.

Producing Scattergrams with PROC PLOT

Here is the general form for requesting a scattergram with the PLOT procedure:

```
PROC PLOT    DATA=data-set-name;
   PLOT    criterion-variable*predictor-variable ;
   RUN;
```

The variable listed as the "criterion-variable" in the preceding program will be plotted on the vertical axis, and the "predictor-variable" will be plotted on the horizontal axis.

To illustrate this procedure, imagine that you have conducted a study dealing with the **investment model**, a theory of commitment in romantic associations (Rusbult, 1980). The investment model identifies a number of variables that are believed to influence a person's commitment to a romantic association. **Commitment** refers to the subject's intention to remain in the relationship. Here are some of the variables that are predicted to influence subject commitment:

Satisfaction: The subject's affective response to the relationship

Investment size: The amount of time and personal resources that the subject has put into the relationship

Alternative value: The attractiveness of the subject's alternatives to the relationship (e.g., the attractiveness of alternative romantic partners)

Assume that you have developed a 16-item questionnaire to measure these four variables. The questionnaire is administered to 20 subjects who are currently involved in a romantic relationship, and the subjects are asked to complete the instrument while thinking about their relationship. When they have completed the questionnaire, it is possible to use their responses to compute four scores for each subject. First, each subject receives a score on the **commitment scale**. Higher values on the commitment scale reflect greater commitment to the relationship. Each subject also receives a score on the **satisfaction scale**, where higher scores reflect greater satisfaction with the relationship. Higher scores on the **investment scale** mean that the subject believes that he or she has invested a great deal of time and effort in the relationship. Finally, with the **alternative value scale** higher scores mean that it would be attractive to the respondent to find a different romantic partner.

When the data have been keyed, it will be possible to use the PLOT procedure to prepare scattergrams for various combinations of variables. The following SAS program inputs some fictitious data and requests that a scattergram be prepared in which commitment scores are plotted against satisfaction scores:

```
1       DATA D1;
2          INPUT    #1    @1    (COMMIT)    (2.)
3                         @4    (SATIS)     (2.)
4                         @7    (INVEST)    (2.)
5                         @10   (ALTERN)    (2.)    ;
6       CARDS;
7        20 20 28 21
8        10 12  5 31
9        30 33 24 11
10        8 10 15 36
11       22 18 33 16
12       31 29 33 12
13        6 10 12 29
14       11 12  6 30
15       25 23 34 12
16       10  7 14 32
17       31 36 25  5
18        5  4 18 30
19       31 28 23  6
20        4  6 14 29
21       36 33 29  6
22       22 21 14 17
23       15 17 10 25
24       19 16 16 22
25       12 14 18 27
26       24 21 33 16
27       ;
28       PROC PLOT    DATA=D1;
29          PLOT COMMIT*SATIS;
30          RUN;
```

In the preceding program, scores on the Commitment scale are keyed in columns 1-2, and are given the SAS variable name COMMIT. Similarly, scores on the Satisfaction scale are keyed in columns 4-5, and are given the name SATIS; scores on the Investment scale appear in columns 7-8 and are given the name INVEST, and scores on the Alternative value scale appear as the last column of data, and are given the name ALTERN.

The data for the 20 subjects appear on lines 7-26 in the program. There is one line of data for each subject.

Line 28 of the program requests the PLOT procedure, specifying that the data set to be analyzed is data set D1. The PLOT command on line 29 specifies COMMIT as the criterion variable and SATIS as the predictor variable for this analysis. The results of this analysis appear in Output A.5.1.

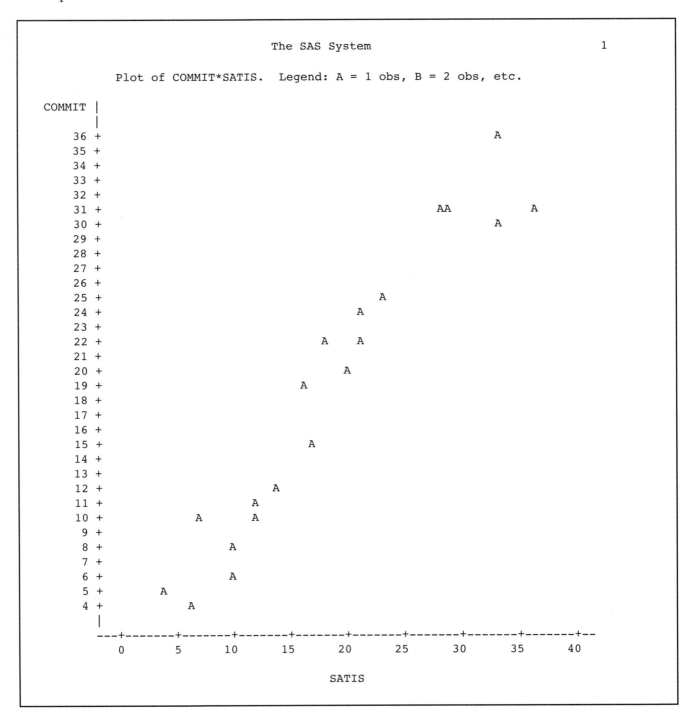

Output A.5.1: Scattergram of Commitment Scores Plotted against Satisfaction Scores

Notice that, in this output, the criterion variable (COMMIT) is plotted on the vertical axis, while the predictor variable (SATIS) is plotted on the horizontal axis. The shape of the scattergram shows that there is a linear relationship between SATIS and COMMIT. This can be seen from the fact that it would be possible to draw a good-fitting straight line through the center of the scattergram. Given that the relationship is linear, it seems safe to proceed with the computation of a Pearson correlation for this pair of variables.

The general shape of the scattergram also suggests that there is a fairly strong relationship between the two variables. Knowing where a subject stands on the SATIS variable allows you to predict, with some accuracy, where that subject will stand on the COMMIT variable. Later, you will compute the correlation coefficient for these two variables to see just how strong the relationship is.

Output A.5.1 shows that the relationship between SATIS and COMMIT is positive: large values on SATIS are associated with large values on COMMIT, and small values on SATIS are associated with small values on COMMIT. This makes intuitive sense; you would expect that the subjects who are highly satisfied with their relationships would also be highly committed to those relationships. To illustrate a negative relationship, you will next plot COMMIT against ALTERN. This is achieved by including the following statements in the preceding program:

```
PROC PLOT    DATA=D1;
   PLOT COMMIT*ALTERN;
   RUN;
```

These statements are identical to the earlier statements, except that ALTERN has now been specified as the predictor variable. These statements produced the scattergram presented in Output A.5.2.

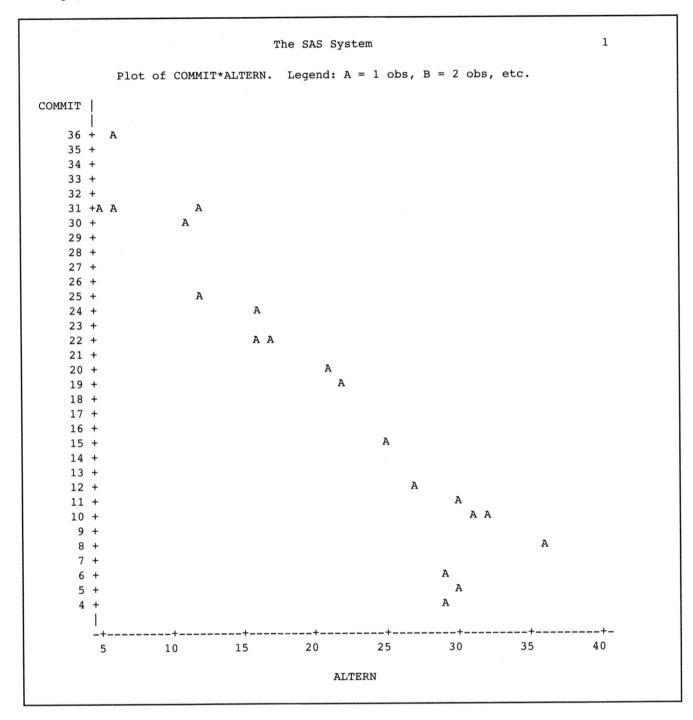

Output A.5.2: Scattergram of Commitment Scores Plotted against Alternative Value Scores.

Notice that the relationship between these two variables is negative. This is as you would expect; it makes intuitive sense that subjects who indicate that their alternatives to their current romantic partner are attractive would not be terribly committed to their current partner. The relationship between ALTERN and COMMIT also appears to be linear; it is therefore appropriate to assess the strength of the relationship between these variables with the Pearson correlation coefficient.

Computing Pearson Correlations with PROC CORR

The CORR procedure offers a number of options regarding what *type* of coefficient will be computed, as well as a number of options regarding the *way* they will appear on the printed page. Some of these options are discussed next.

Computing a single correlation coefficient. In some instances, you may want to compute the correlation between just two variables. Here is the general form for the statements that will accomplish this:

```
PROC CORR    DATA=data-set-name    options;
   VAR    variable1    variable2;
   RUN;
```

The choice of which variable is "variable1" and which is "variable2" is arbitrary. For a concrete example, assume that you want to compute the correlation between Commitment and Satisfaction. Here are the required statements:

```
   PROC CORR    DATA=D1;
      VAR COMMIT SATIS;
      RUN;
```

The preceding would result in a single page of output, reproduced here as Output A.5.3:

```
                        The SAS System                      1

                    Correlation Analysis

            2 'VAR' Variables:  COMMIT    SATIS


                    Simple Statistics

Variable      N       Mean      Std Dev        Sum      Minimum    Maximum

COMMIT        20    18.60000    10.05459    372.00000    4.00000    36.00000
SATIS         20    18.50000     9.51177    370.00000    4.00000    36.00000
```

```
     Pearson Correlation Coefficients / Prob > |R| under Ho: Rho=0 / N = 20

                                    COMMIT              SATIS

              COMMIT              1.00000             0.96252
                                     0.0                0.0001

              SATIS               0.96252             1.00000
                                     0.0001             0.0
```

Output A.5.3: Computing the Pearson Correlation between Commitment and Satisfaction

The first part of Output A.5.3 presents simple descriptive statistics for the variables being analyzed. This allows you to verify that everything looks right: that the correct number of cases were analyzed, that no variables were out of bounds, and so forth. The names of the variables appear below the "Variable" heading, and the statistics for the variables appear to the right of the variable names. These descriptive statistics show that 20 subjects provided usable data for the COMMIT variable, that the mean for COMMIT was 18.6 and the standard deviation was 10.05. It is always important to review the "Minimum" and "Maximum" columns to verify that no impossible scores appear in the data. With COMMIT, the lowest possible score was 4 and the highest possible score was 36. The "Minimum" and "Maximum" columns of Output A.5.3 show that no observed values were beyond these bounds, thus providing no evidence of miskeyed data (again, these proofing procedures do not guarantee that no errors were made in keying data, but they are useful for identifying some types of errors). Since the descriptive statistics provide no obvious evidence of keying or programming mistakes, you are now free to review the correlations themselves.

The bottom half of Output A.5.3 provides the correlations requested in the VAR statement. There are actually four correlation coefficients in the output because your statement requested that the system compute every possible correlation between the variables COMMIT and SATIS. This caused the computer to compute the correlation between COMMIT and SATIS, between SATIS and COMMIT, between COMMIT and COMMIT, and between SATIS and SATIS.

The correlation between COMMIT and COMMIT appears in the upper-left corner of the matrix of correlation coefficients in Output A.5.3. It can be seen that the correlation between these variables is 1.00, and this makes sense because the correlation of any variable with itself is always equal to 1.00. Similarly, in the lower-right corner, you see that the correlation between SATIS and SATIS is also 1.00.

The correlation you are actually interested in appears where the column headed COMMIT intersects with the row headed SATIS. The top number in the "cell" where this column and row intersect is .96252, which is the Pearson correlation between COMMIT and SATIS. You may round this to .96.

Just below the correlation is the *p* value associated with the correlation. This is the *p* value obtained from a test of the null hypothesis that the correlation between COMMIT and SATIS is zero in the population. More technically, the *p* value gives you the probability that you would obtain a sample correlation this large (or larger) in absolute magnitude if the correlation between COMMIT and SATIS were really zero in the population. For the present correlation of .96, the corresponding *p* value is .0001. This means that, given your sample size, there is only 1 chance in 10,000 of obtaining a correlation of .96 or larger if the population correlation were really zero. You may therefore reject the null hypothesis, and tentatively conclude that COMMIT is related to SATIS in the population. (The alternative hypothesis for this statistical test is that the correlation is not equal to zero in the population; this alternative hypothesis is two-sided, which means that it does not predict whether the population correlation is positive or negative, only that it is not equal to zero).

Determining sample size. The size of the sample used in computing the correlation coefficient may appear in one of two places on the output page. First, if all correlations in the analysis were based on the same number of subjects, the sample size would appear only once on the page, in the line above the matrix of correlations. This line appears just below the descriptive statistics. In Output A.5.3 the line indicates:

```
Pearson Correlation Coefficients / Prob > |R| under Ho: Rho=0 / N = 20
```

The "N =" portion of this output indicates the sample size. In Output A.5.3, the sample size was 20.

However, if one is requesting correlations between several different pairs of variables, it is possible that some correlations will be based on more subjects than others (due to missing data). In this case, the sample size will be printed for each individual correlation coefficient. Specifically, the sample size will appear immediately below the correlation coefficient and its associated *p* value, following this format:

```
correlation
p value
N
```

Computing all possible correlations for a set of variables. Here is the general form for computing all possible Pearson correlations for a set of variables:

```
PROC CORR    DATA=data-set-name    options;
   VAR    variable-list;
   RUN;
```

Each variable name in the preceding "variable-list" should be separated by at least one space. For example, assume that you now want to compute all possible correlations for the variables COMMIT, SATIS, INVEST, and ALTERN. The statements that will request these correlations are as follows:

```
PROC CORR    DATA=D1;
   VAR COMMIT SATIS INVEST ALTERN;
   RUN;
```

The preceding program produced the output reproduced here as Output A.5.4:

```
                            The SAS System                              1

                         Correlation Analysis

          4 'VAR' Variables:  COMMIT   SATIS    INVEST   ALTERN


                            Simple Statistics

Variable        N        Mean      Std Dev        Sum      Minimum     Maximum

COMMIT         20     18.60000    10.05459    372.00000    4.00000    36.00000
SATIS          20     18.50000     9.51177    370.00000    4.00000    36.00000
INVEST         20     20.20000     9.28836    404.00000    5.00000    34.00000
ALTERN         20     20.65000     9.78869    413.00000    5.00000    36.00000


      Pearson Correlation Coefficients / Prob > |R| under Ho: Rho=0 / N = 20

                  COMMIT           SATIS            INVEST          ALTERN

COMMIT           1.00000         0.96252          0.71043         -0.95604
                 0.0             0.0001           0.0004           0.0001


SATIS            0.96252         1.00000          0.61538         -0.93355
                 0.0001          0.0              0.0039           0.0001


INVEST           0.71043         0.61538          1.00000         -0.72394
                 0.0004          0.0039           0.0              0.0003


ALTERN          -0.95604        -0.93355         -0.72394          1.00000
                 0.0001          0.0001           0.0003           0.0
```

Output A.5.4: Computing All Possible Pearson Correlations

The correlations and p values in this output are interpreted in exactly the same way as with the preceding output. For example, to find the correlation between INVEST and COMMIT, you find the cell where the row for INVEST intersects with the column for COMMIT. The top number in this cell is .71043, which is the Pearson correlation between these two variables. Just below this correlation coefficient is the p value of .0004, meaning that there are only 4 chances in 10,000 of observing a sample correlation this large if the population correlation were really zero. In other words, the observed correlation is statistically significant.

Notice that the pattern of the correlations supports some of the predictions of the investment model: Commitment is positively related to Satisfaction and Investment size, and is negatively related to Alternative value. With respect to strength, the correlations range from being moderately strong to very strong (remember, however, that these data are fictitious).

What happens if I omit the VAR statement? It is possible to run PROC CORR without the VAR statement; this will cause every possible correlation to be computed between all quantitative variables in the data set. Use caution in doing this; with very large data sets, leaving off the VAR statement may result in a very long printout.

Computing correlations between subsets of variables. By using the WITH statement in the SAS program, it is possible to compute correlations between one subset of variables and a second subset of variables. The general form is as follows:

```
PROC CORR   DATA=data-set-name   options;
   VAR   variables-that-will-appear-as-columns;
   WITH   variables-that-will-appear-as-rows;
   RUN;
```

Any number of variables may appear in the VAR statement, and any number of variables may also appear in the WITH statement. To illustrate, assume that you want to prepare a matrix of correlation coefficients in which there is one column of coefficients, representing the COMMIT variable, and three rows of coefficients, representing the SATIS, INVEST, and ALTERN variables. The following statements will create this matrix:

```
PROC CORR   DATA=D1;
   VAR  COMMIT;
   WITH SATIS INVEST ALTERN;
   RUN;
```

Output A.5.5 presents the results generated by this program. Obviously, the correlations in this output are identical to those obtained in Output A.5.4, although Output A.5.5 is more compact. This is why it is often wise to use the WITH statement in conjunction with the VAR statement, as this can produce smaller and more manageable printouts than may be obtained if only the VAR statement is used.

```
                          The SAS System                              1

                        Correlation Analysis

              3 'WITH' Variables:   SATIS     INVEST     ALTERN
              1 'VAR'  Variables:   COMMIT


                          Simple Statistics

Variable        N        Mean      Std Dev         Sum     Minimum     Maximum

SATIS          20     18.50000     9.51177    370.00000    4.00000    36.00000
INVEST         20     20.20000     9.28836    404.00000    5.00000    34.00000
ALTERN         20     20.65000     9.78869    413.00000    5.00000    36.00000
COMMIT         20     18.60000    10.05459    372.00000    4.00000    36.00000


         Pearson Correlation Coefficients / Prob > |R| under Ho: Rho=0 / N = 20

                                      COMMIT

                   SATIS            0.96252
                                    0.0001

                   INVEST           0.71043
                                    0.0004

                   ALTERN          -0.95604
                                    0.0001
```

Output A.5.5: Computing Pearson Correlations for Subsets of Variables

Options used with PROC CORR. Here are some of the PROC CORR options that may be especially useful when conducting social science research. Remember that the option names should appear before the semicolon that ends the PROC CORR statement.

ALPHA

prints coefficient alpha (a measure of scale reliability) for the variables listed in the VAR statement. Chapter 3 of this text deals with coefficient alpha in greater detail.

COV

prints covariances between the variables. This is useful when you need a variance-covariance table, rather than a table of correlations.

KENDALL

prints Kendall's tau-b coefficient, a measure of bivariate association for variables assessed at the ordinal level.

NOMISS

drops from the analysis any observations (subjects) with missing data on any of the variables listed in the VAR statement. Using this option ensures that all correlations will be based on exactly the same observations (and, therefore, on the same *number* of observations).

NOPROB

prevents the printing of *p* values associated with the correlations.

RANK

for each variable, reorders the correlations from highest to lowest (in absolute value) and prints them in this order.

SPEARMAN

prints Spearman correlations, which are appropriate for variables measured on an ordinal level.

Appendix: Assumptions Underlying the Pearson Correlation Coefficient

• **Interval-level measurement**. Both the predictor and criterion variables should be assessed on an interval- or ratio-level of measurement.

• **Random sampling**. Each subject in the sample will contribute one score on the predictor variable, and one score on the criterion variable. These pairs of scores should represent a random sample drawn from the population of interest.

• **Linearity**. The relationship between the criterion variable and the predictor variable should be linear. This means that, in the population, the mean criterion scores at each value of the predictor variable should fall on a straight line. The Pearson correlation coefficient is not appropriate for assessing the strength of the relationship between two variables involved in a curvilinear relationship.

- **Bivariate normal distribution**. The pairs of scores should follow a bivariate normal distribution; that is, scores on the criterion variable should form a normal distribution at each value of the predictor variable. Similarly, scores of the predictor variable should form a normal distribution at each value of the criterion variable. When scores represent a bivariate normal distribution, they form an *elliptical* scattergram when plotted (i.e., their scattergram is shaped like a football: fat in the middle and tapered on the ends). However, the Pearson correlation coefficient is robust against violations of this assumption when the sample size is greater than 25.

Reference

Rusbult, C.E. (1980). Commitment and satisfaction in romantic associations: A test of the investment model. *Journal of Experimental Social Psychology, 16*, 172-186.

Appendix B

DATA SETS

Overview. This appendix provides data sets that were used in analyses reported in Chapter 1, "Principal Component Analysis", Chapter 2, "Exploratory Factor Analysis", and Chapter 3, "Assessing Scale Reliability with Coefficient Alpha". The data sets are presented here rather than in the chapters in which they were discussed because they are longer than most data sets used in the text.

Data Set from Chapter 1: Principal Component Analysis

Fictitious data from the Prosocial Orientation Inventory:

```
DATA D1;
   INPUT   #1  @1  (V1-V6)   (1.) ;
CARDS;
556754
567343
777222
665243
666665
353324
767153
666656
334333
567232
445332
555232
546264
436663
265454
757774
635171
667777
657375
```

```
545554
557231
666222
656111
464555
465771
142441
675334
665131
666443
244342
464452
654665
775221
657333
666664
545333
353434
666676
667461
544444
666443
676556
676444
676222
545111
777443
566443
767151
455323
455544
;
```

Data Sets from Chapter 2: Exploratory Factor Analysis

Fictitious data from the investment model study:

```
DATA D1;
   INPUT   #1  @1  (V1-V6)   (1.)
               @8  (COMMIT)  (2.) ;
```

```
CARDS;
776122 24
776111 28
111425  4
222633 24
551664  4
666524  4
633112 24
766212 23
454444 17
111332  8
444343 21
556212 20
543332 11
677222 27
666234 15
557322  6
555221 18
544111 17
424232 28
445435 13
767232 13
444422  8
653211 15
555323 16
655123 17
221121 11
666421 20
454332  9
655321 25
444332 12
433222 27
777314 16
555212  7
443221  4
243334 11
666111 15
423412 11
555222 18
223332  8
333335 13
445433 22
444323 12
455556 22
444112 17
```

```
334445 12
444321 16
655222 15
433344 15
557332 20
655222 13
;
```

Intercorrelations between 25 scales from the Job Search Skills Questionnaire, decimals omitted, $N = 220$.

```
VALUES
ABILIT   50
ASSESS   37 40
STRAT    32 45 62
EXPER    39 41 46 49
ORGCHAR  58 51 31 40 48
RESOCCU  37 40 53 57 62 47
RESEMPL  39 38 54 58 57 46 69
GOALS    53 45 32 34 44 56 40 41
BARRIER  40 40 41 48 50 50 49 54 53
MOTIVAT  49 46 39 38 41 53 44 43 58 51
RESUMES  32 40 38 53 50 43 50 53 40 44 46
LETREC   42 39 48 52 52 45 51 62 42 53 45 64
DIRECT   30 34 43 55 48 35 50 60 33 45 37 70 65
APPLIC   37 35 41 43 46 34 44 52 33 47 42 49 59 49
IDENEMP  35 36 49 62 59 42 75 72 40 53 42 58 60 67 48
CARDEV   27 24 51 51 59 31 50 49 30 44 33 49 51 55 42 59
AGENCY   28 32 45 53 42 34 46 61 28 36 32 53 49 59 44 56 55
FAIRS    20 27 42 53 47 26 51 48 27 43 30 50 53 55 44 59 63 52
ADVERT   37 42 46 62 53 43 58 67 39 54 41 60 61 65 54 70 52 62 58
COUNS    24 30 53 56 49 33 56 59 33 48 34 51 55 61 41 64 70 62 65 67
UNADV    27 32 45 54 45 33 52 62 37 49 44 45 54 52 52 61 51 58 49 68 65
NETWORK  34 36 46 57 53 42 53 55 42 48 47 49 50 53 39 59 52 48 46 66 56 62
INTERV   31 38 33 52 43 42 50 59 38 48 39 51 63 54 59 55 42 48 46 62 52 62 55
SALARY   28 35 43 52 45 37 64 67 36 50 39 52 56 52 44 68 49 56 52 64 62 63 54 64
```

Data Set from Chapter 3:
Assessing Scale Reliability with Coefficient Alpha

The data set described in Chapter 3 is identical to the data set from Chapter 1, "Principal Component Analysis", which appears earlier.

Appendix C

CRITICAL VALUES OF THE
CHI-SQUARE DISTRIBUTION

Q ν	0·250	0·100	0·050	0·025	0·010	0·005	0·001
1	1·32330	2·70554	3·84146	5·02389	6·63490	7·87944	10·828
2	2·77259	4·60517	5·99146	7·37776	9·21034	10·5966	13·816
3	4·10834	6·25139	7·81473	9·34840	11·3449	12·8382	16·266
4	5·38527	7·77944	9·48773	11·1433	13·2767	14·8603	18·467
5	6·62568	9·23636	11·0705	12·8325	15·0863	16·7496	20·515
6	7·84080	10·6446	12·5916	14·4494	16·8119	18·5476	22·458
7	9·03715	12·0170	14·0671	16·0128	18·4753	20·2777	24·322
8	10·2189	13·3616	15·5073	17·5345	20·0902	21·9550	26·125
9	11·3888	14·6837	16·9190	19·0228	21·6660	23·5894	27·877
10	12·5489	15·9872	18·3070	20·4832	23·2093	25·1882	29·588
11	13·7007	17·2750	19·6751	21·9200	24·7250	26·7568	31·264
12	14·8454	18·5493	21·0261	23·3367	26·2170	28·2995	32·909
13	15·9839	19·8119	22·3620	24·7356	27·6882	29·8195	34·528
14	17·1169	21·0641	23·6848	26·1189	29·1412	31·3194	36·123
15	18·2451	22·3071	24·9958	27·4884	30·5779	32·8013	37·697
16	19·3689	23·5418	26·2962	28·8454	31·9999	34·2672	39·252
17	20·4887	24·7690	27·5871	30·1910	33·4087	35·7185	40·790
18	21·6049	25·9894	28·8693	31·5264	34·8053	37·1565	42·312
19	22·7178	27·2036	30·1435	32·8523	36·1909	38·5823	43·820
20	23·8277	28·4120	31·4104	34·1696	37·5662	39·9968	45·315
21	24·9348	29·6151	32·6706	35·4789	38·9322	41·4011	46·797
22	26·0393	30·8133	33·9244	36·7807	40·2894	42·7957	48·268
23	27·1413	32·0069	35·1725	38·0756	41·6384	44·1813	49·728
24	28·2412	33·1962	36·4150	39·3641	42·9798	45·5585	51·179
25	29·3389	34·3816	37·6525	40·6465	44·3141	46·9279	52·618
26	30·4346	35·5632	38·8851	41·9232	45·6417	48·2899	54·052
27	31·5284	36·7412	40·1133	43·1945	46·9629	49·6449	55·476
28	32·6205	37·9159	41·3371	44·4608	48·2782	50·9934	56·892
29	33·7109	39·0875	42·5570	45·7223	49·5879	52·3356	58·301
30	34·7997	40·2560	43·7730	46·9792	50·8922	53·6720	59·703
40	45·6160	51·8051	55·7585	59·3417	63·6907	66·7660	73·402
50	56·3336	63·1671	67·5048	71·4202	76·1539	79·4900	86·661
60	66·9815	74·3970	79·0819	83·2977	88·3794	91·9517	99·607
70	77·5767	85·5270	90·5312	95·0232	100·425	104·215	112·317
80	88·1303	96·5782	101·879	106·629	112·329	116·321	124·839
90	98·6499	107·565	113·145	118·136	124·116	128·299	137·208
100	109·141	118·498	124·342	129·561	135·807	140·169	149·449
X	+0·6745	+1·2816	+1·6449	+1·9600	+2·3263	+2·5758	+3·0902

Table C.1: Critical Values of the Chi-Square Distribution

This table is abridged from Table 8 in E. S. Pearson and H. O. Hartley (Eds), *Biometrika tables for statisticians* (3rd ed., Vol. 1), Cambridge University Press, New York, 1970, by permission of the Biometrika Trustees.

Index

BOOKS by USERS

SAS Institute's Author Service

Call your local SAS® office to order these other books and tapes available through the Books by Users℠ program:

An Array of Challenges — Test Your SAS® Skills
by **Robert Virgile**...................................Order No. A55625

Applied Multivariate Statistics with SAS® Software
by **Ravindra Khattree**
and **Dayanand N. Naik**........................Order No. A55234

Applied Statistics and the SAS® Programming Language, Fourth Edition
by **Ronald P. Cody**
and **Jeffrey K. Smith**...........................Order No. A55984

Beyond the Obvious with SAS® Screen Control Language
by **Don Stanley** ....Order No. A55073

The Cartoon Guide to Statistics
by **Larry Gonick**
and **Woollcott Smith**............................Order No. A55153

Categorical Data Analysis Using the SAS® System
by **Maura E. Stokes, Charles E. Davis,**
and **Gary G. Koch** ....Order No. A55320

Common Statistical Methods for Clinical Research with SAS® Examples
by **Glenn A. Walker**..............................Order No. A55991

Concepts and Case Studies in Data Management
by **William S. Calvert**
and **J. Meimei Ma**.................................Order No. A55220

Essential Client/Server Survival Guide, Second Edition
by **Robert Orfali, Dan Harkey,**
and **Jeri Edwards**.................................Order No. A56285

Extending SAS® Survival Analysis Techniques for Medical Research
by **Alan Cantor**.....................................Order No. A55504

A Handbook of Statistical Analysis using SAS
by **B.S. Everitt**
and **G. Der** ....Order No. A56378

The How-To Book for SAS/GRAPH® Software
by **Thomas Miron** ....Order No. A55203

In the Know ... SAS® Tips and Techniques From Around the Globe
by **Phil Mason** ....Order No. A55513

Learning SAS® in the Computer Lab
by **Rebecca J. Elliott** ....Order No. A55273

The Little SAS® Book: A Primer
by **Lora D. Delwiche**
and **Susan J. Slaughter**......................Order No. A55200

Mastering the SAS® System, Second Edition
by **Jay A. Jaffe** ....Order No. A55123

The Next Step: Integrating the Software Life Cycle with SAS® Programming
by **Paul Gill** ....Order No. A55697

Painless Windows 3.1: A Beginner's Handbook for SAS® Users
by **Jodie Gilmore** ....Order No. A55505

Painless Windows: A Handbook for SAS® Users
by **Jodie Gilmore** ....Order No. A55769

Professional SAS® Programming Secrets, Second Edition
by **Rick Aster**
and **Rhena Seidman** ....Order No. A56279

Professional SAS® User Interfaces
by **Rick Aster** ....Order No. A56197

Quick Results with SAS/GRAPH® Software
by **Arthur L. Carpenter**
and **Charles E. Shipp** ....Order No. A55127

Quick Start to Data Analysis with SAS®
by **Frank C. Dilorio**
and **Kenneth A. Hardy**.........................Order No. A55550

Reporting from the Field: SAS® Software Experts Present Real-World Report-Writing Applications ....Order No. A55135

SAS® Applications Programming: A Gentle Introduction
by **Frank C. Dilorio** ....Order No. A55193

SAS® Foundations: From Installation to Operation
by **Rick Aster** ....Order No. A55093

SAS® Programming by Example
by **Ron Cody**
and **Ray Pass** ....Order No. A55126

SAS® Programming for Researchers and Social Scientists
by **Paul E. Spector** ....Order No. A56199

Heterick Memorial Library
Ohio Northern University

DUE	RETURNED	DUE	RETURNED
1. 6-16-10		13.	
2.][AOR]-1 2011		14.	
3. Oct 15 '13 JAN 15		15.	
4. 4-21-14 JUN 3 - 2014		16.	
5.		17.	
6.		18.	
7.		19.	
8.		20.	
9.		21.	
10.		22.	
11.		23.	
12.		24.	